CAMBRIDGE LIBRARY COLLECTION

Books of enduring scholarly value

Botany and Horticulture

Until the nineteenth century, the investigation of natural phenomena, plants and animals was considered either the preserve of elite scholars or a pastime for the leisured upper classes. As increasing academic rigour and systematisation was brought to the study of 'natural history', its subdisciplines were adopted into university curricula, and learned societies (such as the Royal Horticultural Society, founded in 1804) were established to support research in these areas. A related development was strong enthusiasm for exotic garden plants, which resulted in plant collecting expeditions to every corner of the globe, sometimes with tragic consequences. This series includes accounts of some of those expeditions, detailed reference works on the flora of different regions, and practical advice for amateur and professional gardeners.

An Introduction to the Natural System of Botany

Employed early in his career by Sir Joseph Banks, the botanist John Lindley (1799–1865) is best known for his recommendation that Kew Gardens should become a national botanical institution, and for saving the Royal Horticultural Society from financial disaster. As an author, he is best remembered for his works on taxonomy and classification. A partisan of the 'natural' system of Jussieu rather than the Linnaean, Lindley writes, in his preface to this 1830 work, that it was originally created for his own use, to avoid having recourse to 'rare, costly and expensive publications' available only in the libraries of the wealthy. His intention is to give a 'systematic view of the organisation, natural affinities, and geographical distribution of the whole vegetable kingdom', as well as of the uses of plants 'in medicine, the arts, and rural or domestic economy'. The work is important in the history of taxonomy.

Cambridge University Press has long been a pioneer in the reissuing of out-of-print titles from its own backlist, producing digital reprints of books that are still sought after by scholars and students but could not be reprinted economically using traditional technology. The Cambridge Library Collection extends this activity to a wider range of books which are still of importance to researchers and professionals, either for the source material they contain, or as landmarks in the history of their academic discipline.

Drawing from the world-renowned collections in the Cambridge University Library and other partner libraries, and guided by the advice of experts in each subject area, Cambridge University Press is using state-of-the-art scanning machines in its own Printing House to capture the content of each book selected for inclusion. The files are processed to give a consistently clear, crisp image, and the books finished to the high quality standard for which the Press is recognised around the world. The latest print-on-demand technology ensures that the books will remain available indefinitely, and that orders for single or multiple copies can quickly be supplied.

The Cambridge Library Collection brings back to life books of enduring scholarly value (including out-of-copyright works originally issued by other publishers) across a wide range of disciplines in the humanities and social sciences and in science and technology.

An Introduction to the Natural System of Botany

Or, A Systematic View of the Organisation, Natural Affinities, and Geographical Distribution, of the Whole Vegetable Kingdom

JOHN LINDLEY

CAMBRIDGE
UNIVERSITY PRESS

CAMBRIDGE
UNIVERSITY PRESS

University Printing House, Cambridge, CB2 8BS, United Kingdom

Cambridge University Press is part of the University of Cambridge.

It furthers the University's mission by disseminating knowledge in the pursuit of education, learning and research at the highest international levels of excellence.

www.cambridge.org
Information on this title: www.cambridge.org/9781108076654

This edition first published 1830
This digitally printed version 2015

ISBN 978-1-108-07665-4 Paperback

This book reproduces the text of the original edition. The content and language reflect the beliefs, practices and terminology of their time, and have not been updated.

Selected botanical reference works available in the
CAMBRIDGE LIBRARY COLLECTION

al-Shirazi, Noureddeen Mohammed Abdullah (compiler), translated by
Francis Gladwin: *Ulfáz Udwiyeh, or the Materia Medica* (1793)
[ISBN 9781108056090]

Arber, Agnes: *Herbals: Their Origin and Evolution* (1938)
[ISBN 9781108016711]

Arber, Agnes: *Monocotyledons* (1925) [ISBN 9781108013208]

Arber, Agnes: *The Gramineae* (1934) [ISBN 9781108017312]

Arber, Agnes: *Water Plants* (1920) [ISBN 9781108017329]

Bower, F.O.: *The Ferns (Filicales)* (3 vols., 1923–8) [ISBN 9781108013192]

Candolle, Augustin Pyramus de, and Sprengel, Kurt: *Elements of the Philosophy
of Plants* (1821) [ISBN 9781108037464]

Cheeseman, Thomas Frederick: *Manual of the New Zealand Flora*
(2 vols., 1906) [ISBN 9781108037525]

Cockayne, Leonard: *The Vegetation of New Zealand* (1928)
[ISBN 9781108032384]

Cunningham, Robert O.: *Notes on the Natural History of the Strait of Magellan
and West Coast of Patagonia* (1871) [ISBN 9781108041850]

Gwynne-Vaughan, Helen: *Fungi* (1922) [ISBN 9781108013215]

Henslow, John Stevens: *A Catalogue of British Plants Arranged According to
the Natural System* (1829) [ISBN 9781108061728]

Henslow, John Stevens: *A Dictionary of Botanical Terms* (1856)
[ISBN 9781108001311]

Henslow, John Stevens: *Flora of Suffolk* (1860) [ISBN 9781108055673]

Henslow, John Stevens: *The Principles of Descriptive and Physiological Botany*
(1835) [ISBN 9781108001861]

Hogg, Robert: *The British Pomology* (1851) [ISBN 9781108039444]

Hooker, Joseph Dalton, and Thomson, Thomas: *Flora Indica* (1855)
[ISBN 9781108037495]

Hooker, Joseph Dalton: *Handbook of the New Zealand Flora* (2 vols., 1864–7)
[ISBN 9781108030410]

Hooker, William Jackson: *Icones Plantarum* (10 vols., 1837–54)
[ISBN 9781108039314]

Hooker, William Jackson: *Kew Gardens* (1858) [ISBN 9781108065450]

Jussieu, Adrien de, edited by J.H. Wilson: *The Elements of Botany* (1849)
[ISBN 9781108037310]

Lindley, John: *Flora Medica* (1838) [ISBN 9781108038454]

Müller, Ferdinand von, edited by William Woolls: *Plants of New South Wales*
(1885) [ISBN 9781108021050]

Oliver, Daniel: *First Book of Indian Botany* (1869) [ISBN 9781108055628]

Pearson, H.H.W., edited by A.C. Seward: *Gnetales* (1929)
[ISBN 9781108013987]

Perring, Franklyn Hugh et al.: *A Flora of Cambridgeshire* (1964)
[ISBN 9781108002400]

Sachs, Julius, edited and translated by Alfred Bennett, assisted by W.T. Thiselton
Dyer: *A Text-Book of Botany* (1875) [ISBN 9781108038324]

Seward, A.C.: *Fossil Plants* (4 vols., 1898–1919) [ISBN 9781108015998]

Tansley, A.G.: *Types of British Vegetation* (1911) [ISBN 9781108045063]

Traill, Catherine Parr Strickland, illustrated by Agnes FitzGibbon Chamberlin:
Studies of Plant Life in Canada (1885) [ISBN 9781108033756]

Tristram, Henry Baker: *The Fauna and Flora of Palestine* (1884)
[ISBN 9781108042048]

Vogel, Theodore, edited by William Jackson Hooker: *Niger Flora* (1849)
[ISBN 9781108030380]

West, G.S.: *Algae* (1916) [ISBN 9781108013222]

Woods, Joseph: *The Tourist's Flora* (1850) [ISBN 9781108062466]

For a complete list of titles in the Cambridge Library Collection please visit:
www.cambridge.org/features/CambridgeLibraryCollection/books.htm

AN

INTRODUCTION

TO THE

NATURAL SYSTEM OF BOTANY:

OR,

A SYSTEMATIC VIEW

OF

THE ORGANISATION, NATURAL AFFINITIES, AND
GEOGRAPHICAL DISTRIBUTION,

OF THE WHOLE

VEGETABLE KINGDOM;

TOGETHER WITH THE USES OF THE MOST IMPORTANT SPECIES IN MEDICINE,
THE ARTS, AND RURAL OR DOMESTIC ECONOMY.

By JOHN LINDLEY, F.R.S. L.S. G.S.

MEMBER OF THE IMPERIAL ACADEMY NATURÆ CURIOSORUM ;
OF THE BOTANICAL SOCIETY OF RATISBON ; OF THE PHYSIOGRAPHICAL SOCIETY OF LUND ;
OF THE HORTICULTURAL SOCIETY OF BERLIN ;
HONORARY MEMBER OF THE LYCEUM OF NATURAL HISTORY OF NEW YORK, &c. &c.
AND PROFESSOR OF BOTANY IN THE UNIVERSITY OF LONDON.

"C'est ainsi que sont formées les familles très naturelles et généralement avouées. On extrait de tous les genres qui composent chacune d'elles les caractères communs à tous, sans excepter ceux qui n'appartiennent pas à la fructification, et la réunion de ces caractères communs constitue celui de la famille. Plus les ressemblances sont nombreuses, plus les familles sont naturelles, et par suite le caractère général est plus chargé. En procédant ainsi, on parvient plus sûrement au but principal de la Science, qui est, non de nommer une plante, mais de connoître sa nature et son organisation entière."—JUSSIEU.

LONDON:

LONGMAN, REES, ORME, BROWN, AND GREEN,

PATERNOSTER ROW.

M.DCCC.XXX.

LONDON:
J. MOYES, TOOK'S COURT, CHANCERY LANE.

TO

THE COURT OF EXAMINERS

OF

THE SOCIETY OF APOTHECARIES,

LONDON.

GENTLEMEN,

As Guardians of the education of a very considerable part of the Medical Profession, the subject of the following pages cannot be otherwise than interesting to you. If a knowledge of the Plants from which medicinal substances are obtained, is in itself an object of importance, as it most undoubtedly is, the Science which teaches the art of judging of the hidden qualities of unknown vegetables by their external characters is of still greater moment. To what extent this can safely be carried, it is not, in the actual state of human knowledge, possible to foresee; but it is at least certain, that it depends entirely upon a careful study of the natural relations of the Vegetable Kingdom.

Measures have lately been taken by the SOCIETY OF APOTHECARIES, which cannot fail to exercise a most beneficial influence upon Botany, and which must have been viewed with feelings of deep interest by all friends of the Science. As a humble individual, whose life is devoted to its investigation, I am anxious to take the present opportunity of expressing my sentiments upon the subject, by very respectfully offering for your acceptance a Work, which it is hoped will be found useful to the Student of Medical Botany.

I have the honour to be,

GENTLEMEN,

Your most obedient Servant,

JOHN LINDLEY.

University of London,
August, 1830.

PREFACE.

THE materials from which the following pages have been prepared were originally collected for the private use of the Author, to remove the inconvenience he constantly experienced from a necessity of referring daily to rare, costly, and extensive publications, often to be found only in the libraries of the wealthy. A belief that what was indispensable to himself might also prove useful to the public, afterwards led to the commencement of the present Work, the appearance of which has been accelerated by the growing want of some Introduction to that method of investigating the productions of the Vegetable Kingdom which, under the name of the Natural System, has gradually displaced more popular classifications, well adapted indeed to captivate the superficial inquirer, but exercising so baneful an influence upon Botany, as to have rendered it doubtful whether it even deserved a place among the sciences.

When the printing was commenced, we had no English Introduction whatever to the subject of which it treats; but, soon afterwards, a translation was published by Dr. Clinton, of the fourth edition of Richard's *Nouveaux Élémens de la Botanique*, in which much information is to be found. Had this work appeared calculated to answer the purpose of even a temporary Introduction, the matter now made public would have still remained in the cabinet of the Author; but the plan of M. Richard, indepen-

b

dently of other considerations, did not admit of so much detail as seemed desirable, and was scarcely adapted to render the Natural System of Botany popular in a country like Great Britain, where it has to contend with a great deal of deeply-rooted prejudice.

Two principal objects require to be kept in view, in a scientific work intended for common use : in the first place, there must be no sacrifice of science to popularity; but secondly, it is desirable that as much facility be afforded the student as the nature of the subject will admit. In reconciling these two apparently contradictory conditions lies the difficulty of rendering an arrangement in Natural History which is not merely superficial, generally intelligible. To be understood by the mass of mankind, it must be freed from all unnecessary technicalities, and must be essentially founded upon such peculiarities as it requires no unusual powers of vision, or of discrimination, to seize and apply : on the other hand, it is found by experience, that unless it depends upon a consideration of every point of structure, however numerous or various, however obscure or difficult of access, it will not answer the end for which all classifications ought to be designed, that of enabling the observer to judge of an unknown fact by a known one, and to determine the mutual relations which one body or being bears to another.

In attempting to steer a middle course, the Author is by no means satisfied that he shall be found to have attained the end he has proposed to himself. Botany is a most extensive science, involving a hundred thousand gradations of structure, with myriads of minor modifications, and extending over half the organic world ; the anatomical structure of the beings it comprehends is so minute, and their laws of life are so obscure, as to elude the keenest sight and to baffle the subtlest reasoning : so that to render it as easy of attainment as the world, misled by specious fallacies, is apt to believe it to be, is hopeless.

There are, however, no difficulties so great but they may be diminished; and even a determination of the relation which one part of the animated world bears to another, may be simplified by analysis, and an exposition of the principles upon which such relations are to be judged of.

With this view, in the first place, the value of the characters of which botanists make use are here carefully investigated, for the sake of pointing out the relative importance of the principal modifications of structure in the vegetable kingdom. In the second place, the characters of the orders are analysed by means of tables, in which the distinctive characters of each are reduced to their simplest denomination. It is true that this kind of analysis is attended by the evil of distracting attention from that general and universal study of organisation which the science demands, thus having a manifest tendency to render the Natural System artificial; and that it is also apt to mislead the inexperienced or incautious observer, in consequence of the many exceptions to which distinctive characters are frequently liable. But such evils are nothing compared with the confusion and perplexity an unaided inquirer must experience in disentangling the distinctions of orders for himself. It should also be borne in mind, that analytical tables are mere artificial aids in investigation, to be abandoned as soon as they cease to be indispensable. Many variations in the form of such tables may be easily made; and, in fact, the student cannot exercise himself better than in contriving them for himself, as he may readily do by beginning from some other point than that commenced with here.

The mode in which the tables of this book are to be employed will be best explained by an example, the *reader being supposed to be in possession of the preliminary knowledge which is afforded by the Introduction.* Let a Cistus be the subject of inquiry. Upon examining the tables, the first question which the student must ask himself is, Whether it belongs to Vascular or Cellular plants, to Dicotyledons

or Monocotyledons: the structure of the leaves tells him
this, and he decides for Dicotyledons. He next inquires
if it has the seeds naked or in a capsule; and ascertaining
that the latter is the case, he knows it belongs to Angiospermæ. He then finds it to be polypetalous, and that
the stamens are hypogynous, or those of the division called
Thalamifloræ. Having proceeded thus far, he is led to
inquire whether the carpella are in a state of combination, or distinct; and finding the former to be the case,
he sees that his plant is referable to what are called
Syncarpæ, among Polypetalous Dicotyledons with hypogynous stamens. Now the artificial divisions of this section are seen to depend, in the first instance, upon the
structure of the ovarium: that organ is examined, and
is found to be 1-celled, with the ovules parietal. Among
plants of this nature the placentæ are either linear and
contracted, or branched all over the surface of the valves;
there is no difficulty in ascertaining this point, and it is
found that the plant in question has the former character.
Then comes an inquiry whether the sepals are 2, or invariably 4, or 5 (occasionally varying to 4, 6, or 7); they are
found to be 5; and here the analysis is reduced to the
decision between whether the ovules have a foramen at
the extremity opposite the hilum, or next the hilum; the
former being ascertained to be the case, no doubt can
remain of the plant belonging to the natural order Cistineæ. This operation may appear rather tedious, but after
a little practice it is gone through quickly; and when
the conclusion sought for is attained, the station of the
plant is not only ascertained, but also that all vegetables
having the same characters are herbaceous or shrubby
plants, with gay ephemeral flowers, usually growing in
rocky places, and possessing no known qualities except
that of secreting, in some instances, a sort of resinous
substance used as a stomachic and tonic.

Examples need not be multiplied, one instance shewing what the method of analysis is, as well as more.

The plan adopted, independently of the part now adverted to, is this: To every collection of orders, whether called class, division, subdivision, tribe, section, or otherwise, such remarks upon the value of the characters assigned to it are prefixed as the personal experience of the Author, or that of others, shews them to deserve. To every order the NAME is given which is most generally adopted, or which appears most unexceptionable, with its SYNONYMES, a citation of a few authorities connected with each, and their date: so that, from these quotations, the reader will learn at what period the order was first noticed, and also in what works he is to look for further information upon it. To this succeeds the DIAGNOSIS, which comprehends the distinctive characters of the order reduced to their briefest form, and its most remarkable features, without reference to exceptions. The latter are adverted to in what are called ANOMALIES. Then follows the ESSENTIAL CHARACTER; a brief description of the order, in all its most important particulars. This is succeeded by a paragraph styled AFFINITIES, in which are discussed the relations which the order bears to others, and the most remarkable circumstances connected with its structure in case it exhibits any particular instance of anomalous organisation. GEOGRAPHY points out the distribution of the genera and species over the surface of the globe; and the head PROPERTIES comprehends all that is certainly known of the use of the species in medicine, the arts, domestic or rural economy, and so forth. A few genera are finally named as EXAMPLES of each order.

The arrangement of the orders is not precisely that of any previous work, nor indeed do any two Botanists adopt exactly the same plan; a circumstance which arises out of the very nature of the subject, the impossibility of expressing affinities by any lineal arrangement (the only one which can be practically employed), and the different value that different observers attach to the same characters. This is, however, of no practical importance, so long as the limits of the orders themselves are unchanged; for the latter are the basis of the system, to which all other considerations are subordinate. Such a collection of orders as that here given cannot certainly be called " the Natural System" of the Vegetable Kingdom, in the proper sense of those words; but it is what Botanists take as a substitute for it, until some fixed principle shall be discovered upon which combinations can be formed subordinate to the first great classes of Vasculares and Cellulares, of Exogenæ and Endogenæ. It is also certain, that in the actual state of Botany we are more usefully employed in determining the characters of natural groups by exact observation, than in speculating upon points which we have not yet the means of discussing properly.

In conclusion, the Author has only to add, that this Work must not be viewed as an Introduction to Botany. Those who would understand it must previously possess such an elementary acquaintance with the science as they may collect from his *Outline of the First Principles of Botany,* or some other work in which the modern views of vegetable organisation are explained. This, and the following introductory sketch of the principal modifications of structure, will be found to convey as much information as is absolutely required with reference to the immediate subject of the Work.

INTRODUCTION.

THE notion of classing species according to the likeness they bear to each other, which is the foundation of the Natural System, must have originated with the first attempts of man to reduce natural history to a science. When our forefathers spoke of "grass, and herbs yielding seed, and fruit trees yielding fruit, of moving creatures that have life in the water, of fowl that fly above the earth, and cattle and creeping thing," they employed the very same principles of arrangement which are now in use,—rudely sketched, indeed, but not more so than the imperfection of knowledge rendered unavoidable. At that time no means existed of appreciating the value of minute or hidden organs, the functions or even existence of which were unknown; but objects were collected into groups, characterised by common, external, and obvious signs. From such principles no naturalists except botanists have deviated; no one has thought of first combining under the name of animal kingdom quadrupeds and birds, insects and fishes, reptiles and mollusca, and then of subdividing them by the aid of a few arbitrary signs, in such a way that a portion of each should be found in every group—quadrupeds among birds and fishes, reptiles amongst insects and mammalia; but each great natural group has been confined within its own proper limits. Botany alone, of all the branches of natural history, has been treated otherwise; and this in modern times.

The first writers who acknowledged any system departed in no degree from what they considered a classification of plants according to their general resemblances. Theophrastus has his water-plants and parasites, pot-herbs and forest trees, and corn-plants; Dioscorides, aromatics and gum-bearing plants, eatable vegetables and corn-herbs; and the successors, imitators, and copiers of those writers, retained the same kind of arrangement for many ages.

At last, in 1570, a Fleming, of the name of Lobel, improved the vulgar modes of distinction, by taking into account characters of a more definite nature than those which had been employed by his predecessors; and thus was laid the foundation of the modern accurate mode of studying vegetation. To this author succeeded many others, who, while they disagreed upon the value to be ascribed to the small number of modifications of structure with which they were acquainted, adhered to the ancient plan of making their classification coincide with natural affinities. Among them the most distinguished were Cæsalpinus, an Italian, who published in 1583, our countryman John Ray, and the more celebrated Tournefort, who wrote in the end of the seventeenth century. At this time the materials of Botany had increased so much, that the introduction of more precision into arrangement became daily an object of greater importance; and this led to the contrivance of a plan which should be to Botany what the alphabet is to language, a key by which what is really known of the science might be readily ascertained. With this in view, Rivinus invented, in 1690, a system depending upon the conformation of the corolla; Kamel, in 1693, upon the fruit alone; Magnol, in 1720, on the calyx and corolla; and finally, Linnæus, in 1731, on variations in the sexual organs. The method of the last author has enjoyed a degree of celebrity which has rarely fallen to the lot of human contrivances, chiefly on account of its clearness and simplicity; and in its day it undoubtedly effected its full proportion of good. Its author, however, probably intended it as a mere substitute for the Natural System, for which he found the world in his day unprepared, to be relinquished as soon as the principles of the latter could be settled, as seems obvious from his writings, in which he calls the Natural System *primum et ultimum in botanicis desideratum.* He could scarcely have expected that his artificial method should exist when the science had made sufficient progress to enable botanists to revert to the principles of natural arrangement, the temporary abandonment of which had been solely caused by the difficulty of defining its groups. This difficulty no longer exists; means of defining natural assemblages, as certain as those employed for limiting artificial divisions, have been discovered by modern botanists; and the time has arrived when the ingenious expedients of Linnæus, which could only be justified by the state of Botany when he first entered upon his career, must be finally relinquished. We now know something of the phenomena of vegetable life; by modern improvements in optics, our microscopes are capable of revealing to us the structure of the minutest organs, and the nature of their combination; repeated observations have explained the laws under which the external forms of plants are modified; and it is upon these considerations that the Natural System depends. What, then, should now hinder us from using the powers we possess, and bringing the science to that state in which only it can really be useful or interesting to mankind?

Its uncertainty and difficulty deter us, say those who, acknowledging the manifest advantages of the Natural System, nevertheless continue to make use of the artificial method of Linnæus. I do not know of any other objections than these, which I hope to set aside by the following remarks.

First, as to its uncertainty. That it is not open to this charge, no one will, I think, assert; on the contrary, it is admitted on all hands that it fully participates in those imperfections to which human contrivances are subject, particularly such as, like natural history, are from their nature not susceptible of mathematical accuracy. But while no claim is advanced on its behalf to superiority in this respect over artificial methods, it may be safely stated, that it is not more uncertain than the celebrated sexual system of Linnæus, the only one with which it is worth comparing it. By uncertain, I mean that the characters of the classes and orders of the Natural system are not more subject to exceptions than those of the Linnean, as perhaps may be proved from documents in the hands of every English reader. We are so accustomed to believe that the certainty of the sexual system is equal to its simplicity, that this opinion has acquired the nature of a fixed prejudice, and we are perhaps not prepared to assent to the truth of a contrary proposition. Without, however, travelling out of the way, or seeking for proofs of it among books or plants with which the reader is unacquainted, the following table of exceptions to the sexual system, taken from Smith's Compendium of the Flora Britannica, may possibly carry some weight with it :—

Linnean Class or Order.	Total number of Genera in Smith's Compendium.	Number of Genera which contain Species at variance with the Characters of the Classes and Orders.
Monandria	5	3
Triandria Monogynia	9	2
Tetrandria	21	5
Pentandria Monogynia	41	5
Pentandria Digynia, excluding Umbellatæ	8	3
Pentandria Trigynia	5	1
Pentandria Hexagynia	1	1
Hexandria Trigynia	5	1
Hexandria Polygynia	1	1
Octandria	12	5
Decandria	21	8
Dodecandria	6	2
Monœcia	24	4
Diœcia	14	2
	173	43

From this it appears, that out of 173 genera belonging to fourteen Linnean sections, no fewer than forty-three genera, or nearly one quarter, contain species at variance with the characters of the classes and orders in which they are placed. Were general works on Botany examined in the same manner, it would be found that the proportion of exceptions is at least as great as that indicated by the foregoing table, which comprehends only those species, the variations of which are constant and uniform, and does not include mere accidental deviations, such as the tendency of Tetrandrous flowers to become Pentandrous, of Pentandrous to become Tetrandrous, or of both to become Polygamous.

Although this is not stated for the purpose of extolling the Natural System at the expense of the Linnean, but rather, as has just been remarked, for the sake of doing away with a vulgar prejudice, yet I cannot forbear expressing my doubt whether any fourteen natural orders can be named in which the proportion of exceptions is so considerable as this, namely, more than one in five.

Upon the supposed peculiar difficulties of the Natural System I have elsewhere made some general remarks (*Synopsis*, p. x.), which need not be repeated here. It will be better now to inquire more particularly in what the difficulty consists.

It is said that the primary characters of the classes are not to be ascertained without much laborious research; and that not one step can be advanced until this preliminary difficulty is overcome. Those who hold a language of this kind must be so unacquainted with the subject, that their arguments, if they can be called by such a name, scarcely deserve a reply. The objection has, however, been made, and must be answered.

In natural history many facts have been originally discovered by minute and painful research, which, when once ascertained, are readily to be detected by some more simple process, of which Botany is perhaps the most striking proof that can be adduced. The first question to be determined by a student of Botany, who wishes to inform himself of the name, affinities, and uses of a plant, appears to be, whether his subject contains spiral vessels or not, because the two great divisions of the vegetable kingdom, called Vasculares and Cellulares, are characterised by the presence or absence of these minute organs. It is true, we have learned by careful observation, and multiplied microscopical analyses, that vascular plants have spiral vessels, and cellular plants have none; but it is not true, that in practice so minute and difficult an inquiry needs to be instituted, because it has also been ascertained that all plants that bear flowers have spiral vessels, and are therefore Vascular; and that vegetables which have no flowers are destitute of spiral vessels, and are therefore Cellular ; so that the inquiry of the student, instead of being directed in the first instance to an obscure but highly curious microscopical fact, is at once arrested by the two most obvious peculiarities of the vegetable kingdom.

Among vascular plants two great divisions have been formed; the names of which, Monocotyledons and Dicotyledons, are derived from the former having usually but one lobe to the seed, and the latter two,—a structure much more difficult to ascertain than the presence or absence of spiral vessels, and more subject to exceptions. But no botanist would proceed to dissect the seed of a plant for the purpose of determining to which of these divisions it belonged, except in some special cases. We know that the minute organisation of the seed corresponds with a peculiar structure of the stem, leaves, and flowers, the most highly developed, and most easily examined parts of vegetation ; a botanist, therefore, prefers to examine the stem, or the leaf of a plant, to see whether it is a Monocotyledon or a Dicotyledon, and does not find it necessary to anatomise the seed.

The presence or absence of albumen, the structure of the embryo, the position of the seeds or ovula, the nature of the fruit, the modifications of the flower, will, I presume, be hardly brought forward as other difficult points for the student of the Natural System, because, whether the one system or the other be employed, he must make himself acquainted with such facts, for the purpose of determining genera. The common Toad-flax cannot be discovered by its characters in any book of botany, without the greater part of this kind of inquiry being gone through.

In the determination of genera, however, facility is entirely on the side of the Natural System. Jussieu has well remarked, " that whatever trouble is experienced in remembering, or applying the characters of natural orders, is more than compensated for by the facility of determining genera, the characters of which are simple in proportion as those of orders are complicated. The reverse takes place in arbitrary arrangements, where the distinctions of classes and sections are extremely simple and easy to remember, while those of genera are in proportion numerous and complicated."

Let me not, however, be misunderstood in what I have been saying of the *supposed* difficulties of the Natural System. Far be it from me to state that there are no difficulties for the botanical student to overcome ; on the contrary, there is no science which demands more minute accuracy of observation, more patient research, or a more constant exercise of the reasoning faculties, than that of Botany. But no subject of human inquiry can be pursued loosely and usefully at the same time; for we may rest assured, that that which can be studied superficially is little deserving of being studied at all.

It may perhaps be urged, that the Natural System is still in so
unsettled a state, that botanists disagree among themselves about the
limits and relative position of the orders; an argument to which
some weight undoubtedly attaches. But, at the same time, it must
be remarked, that all sciences of observation proceed towards a
settled state by slow degrees; that Botany is one upon which
there is at least as much to learn as is at present known; and
that the differences of opinion, just alluded to, affect the orders
themselves but little, and the principles of the science not at all,
but apply rather to the particular series in which the orders should
stand with relation to each other—a point which is not likely to
be settled at present, and which is of very little importance for
any useful purpose.

The last kind of difficulty, and the only one of which I admit
the force, is the want of an introductory work upon the subject; and
this, I presume to hope, will be diminished by the appearance of the
present publication.

The principle upon which I understand the Natural System of
Botany to be founded is, that the affinities of plants may be determined by a consideration of all the points of resemblance between
their various parts, properties, and qualities; and that thence an
arrangement may be deduced in which those species will be placed
next each other which have the greatest degree of relationship; and
that consequently the quality or structure of an imperfectly known
plant may be determined by those of another which is well known.
Hence arises its superiority over arbitrary or artificial systems, such
as that of Linnæus, in which there is no combination of ideas, but
which are mere collections of isolated facts, not having any distinct
relation to each other.

This is the only intelligible meaning that can be attached to the
term Natural System, of which Nature herself, who creates species
only, knows nothing. It is absurd to suppose that our genera, orders,
classes, and the like, are more than mere contrivances to facilitate
the arrangement of our ideas with regard to species. A genus,
order, or class, is therefore called natural, not because it exists in
Nature, but because it comprehends species naturally resembling
each other more than they resemble any thing else.

The advantages of such a system, in applying Botany to useful
purposes, are immense, especially to medical men, with whose profession the science has always been identified. A knowledge of the
properties of one plant is a guide to the practitioner, which enables
him to substitute some other with confidence, which is naturally
allied to it; and physicians, on foreign stations, may direct their

inquiries, not empirically, but upon fixed principles, into the qualities of the medicinal plants which nature has provided in every region for the alleviation of the maladies peculiar to it. To horti- culturists it is not less important : the propagation or cultivation of one plant is usually applicable to all its kindred ; the habits of one species in an order will often be those of the rest ; many a gardener might have escaped the pain of a poisoned limb, had he been ac- quainted with the laws of affinity ; and, finally, the phenomena of grafting, that curious operation, which is one of the grand features of distinction between the animal and vegetable kingdoms, and the success of which is wholly controlled by ties of blood, can only be understood by the student of the Natural System.

In every kind of arrangement, which has the natural relationship of objects for its basis, there are two principal inconveniences to overcome. The first is, that as objects resemble each other more or less in a multitude of different respects, it is impossible to indicate all their affinities in a lineal arrangement ; and yet no other arrange- ment than a lineal one can be practically employed. The conse- quence of this is, that while the orders themselves are really natural, the same title often cannot be applied to the arrangement of them in masses. For example, Cupuliferæ and Betulineæ are obviously connected by the most intimate relationship, and, as collections of species, each of them is perfectly natural ; yet one of them stands among Apetalous plants, the other among Achlamydeous ones ; hence the two latter groups are artificial. In fact, it appears from what we at present know, that no large combinations of orders are natural which are not founded upon anatomical differences ; thus, Cellulares and Vasculares, Exogenæ and Endogenæ, Gymnospermous and Angiospermous Dicotyledons, are natural divisions ; but Apetalæ, Polypetalæ, Monopetalæ, Achlamydeæ, and all their subordinate sec- tions, are entirely artificial.

The second inconvenience is, that the characters which vegetables exhibit are of such uncertain and variable degrees of importance, that it is often difficult to say what value should be attached to any given modification of structure. As this is a practical question, which requires to be well understood, I shall endeavour to explain in some detail the nature and relative value of those peculiarities of which botanists make use in determining vegetable affinities ; repeating, as a general rule which is not open to exception, that characters which are purely physiological,—that is to say, which depend upon differences of internal anatomical structure,—are of much more value than varieties of form, position, number, and the like, which are mere modifications of external organs.

It is a maxim of the Linnean school, that the parts of fructification should be employed in characterising classes, orders, and genera, to the exclusion of all modifications of the leaves or stem. This, although theoretically insisted upon, was practically abandoned by Linnæus himself, and is to be received with great caution. The organs of fructification are only entitled to a superior degree of consideration, when found by experience to be less liable to variation than those of vegetation.

All plants are composed of what are called elementary organs, that is to say, of a vegetable membrane appearing under the form of parenchyma or cellular tissue in different states, of spiral vessels, and of ducts, or tubes: these organs enter into the composition of plants in various ways, and are not all even necessary to their existence : sometimes spiral vessels disappear, and again both these and the ducts cease to be developed,—cellular tissue, which is the basis of vegetation, alone remaining. Upon the peculiar arrangement of these minute organs, external form necessarily depends ; and as it is found by experience, that while the anatomical structure of plants is subject to little or no variation, it is difficult to define their external modifications with accuracy, the reason of the superior importance of physiological characters will be apparent.

Some, and by far the greater part of, plants are propagated by productions called seeds, which are the result of an action believed to be analogous to the sexual intercourse of animals ; others are multiplied by bodies called sporules, of the real nature of which little is yet known, further than that they do not appear to result from the communication of sexes. Hence plants are naturally and primarily divided into two great divisions, called SEXUAL and ASEXUAL.

Physiologists have discovered that these peculiarities are connected with others in anatomical structure of no less importance. For instance, plants propagated by seeds, and possessing distinct sexes, have spiral vessels ; while those which are increased by bodies not depending upon the presence of sexual apparatus, are universally destitute of spiral vessels. To the latter statement there is no known exception,—species to which spiral vessels have been ascribed being found to possess nothing more nearly related to those organs than ducts, or false tracheæ. The former character is not absolutely without exception ; the singular genus Rafflesia being described both by Brown and Blume as without spiral vessels, Caulinia fragilis not having them according to Amici, and Lemna being destitute of them according to the evidence of others. But these exceptions are not regarded of much importance.

It therefore appears that two great divisions, established upon different principles, agree in the kind of plants they comprehend ;

VASCULARES, or those which have spiral vessels, being the same as SEXUAL plants, and CELLULARES, or those which have no spiral vessels, answering to ASEXUAL plants.

Sexual organs being considered essential to a flower (no apparatus whatever from which they are absent being understood to constitute one), two other unexceptionable characters belong to these same divisions; all Vasculares, or Sexual plants, bear flowers, and all Cellulares, or Asexual plants, are flowerless; the former are also called Phænogamous, the latter Cryptogamous.

Two great but unequal divisions being thus established, upon both anatomical and external characters, botanists have inquired whether similar differences of a secondary character could not be discovered among each of them. Observations upon Cellulares have led to the establishment of three groups of unequal importance, which are not, however, universally received. Vasculares have been found to comprehend two great but unequal tribes, differing essentially in the laws which govern their growth. It has been ascertained that a large number of them grows by the addition of successive layers of new matter to the outside, and that another, but smaller number, increases by additions to the inside; the youngest or most newly formed parts being in the one case on the outside, and in the other case in the inside. For this reason, one of these divisions has been called Exogenous, and the other Endogenous. It is difficult to conceive how the external increase of Exogenæ could take place without some adequate protection to the young newly formed tissue from the atmosphere and accidental injury, and, accordingly, the substance called bark is created by nature for that purpose, within which the new deposit takes place : as this last is formed annually, the age of an Exogenous plant is indicated in the trunk by imaginary lines called concentric circles, which are in fact caused by the cessation of growth in one year, and the renewal of it in another. The centre of this system is a cellular substance called pith. Therefore, a section of the trunk of an Exogenous plant exhibits bark on the outside, pith in the centre, and concentric deposits of woody matter between these two, all connected in a solid mass by plates of cellular tissue, radiating from the centre to the circumference, and called medullary rays. Endogenæ, the addition to which is internal, have no need of an external coating to protect their newly formed matter from injury, and are therefore destitute of bark ; moreover, as the layers of new matter are not concentric, but irregular, and do not either correspond with particular seasons of growth, nor commence round any distinct centre of vegetation, there is no distinction of bark, woody concentric deposits, and pith ; the connecting tissue by which the parts are all tied together is mixed up with the substance of the whole, and does not radiate regularly in plates from the centre to the circumference, and consequently there are no medullary rays. Nothing can be more clearly made out than the existence of these two modes of growth in vascular plants; and the nature of them will be at once understood by an inspection of a section of an Oak branch, and of a Cane.

Upon Exogenæ I do not know that any remarks need be made, they being exceedingly uniform in the great features of their structure; except in Coniferæ and Cycadeæ, which, without deviating from the mode of growth of Exogenæ, exhibit a peculiar modification of the woody tissue. But Endogenæ are perhaps divisible into two subordinate forms, which have been pointed out by Agardh. First, Grasses, which, as this distinguished writer well remarks, are the least monocotyledonous of all; they have a distinct pith, hollow branched stems like Umbelliferæ, and buds at the axillæ of the leaves; but they have no bark, no medullary rays, and their direction of increase is inwards: and, secondly, Palms, which are endogenous in the strictest sense of the word.

From this it appears, that Vasculares, or Flowering plants, are distinguished into such as are Exogenous and such as are Endogenous; and that while the former are incapable of any further anatomical division, the latter contain perhaps two different forms. It must, however, be borne in mind, that a great deal is yet to be learned upon this subject. Vegetable anatomy has not yet been studied sufficiently with a view to generalization, and is, besides, a subject yet in its infancy. Nothing can be more probable than that differences in the tissue, or in the relative position or structure of vessels, will one day be found to accompany external differences far beyond what has yet been observed.

Anatomical differences in plants having been apparently exhausted, inquiry has been turned to the degree in which modifications of the compound or external organs are capable of being employed to determine natural affinities; and, it has been found that these, although of secondary importance only, nevertheless deserve the utmost attention, as they frequently afford the only characters of which it is practicable to make use.

The Root, properly so called, offers no characters that have been found uniform in particular families; in fact, the modifications of which it is susceptible are so few, that it is difficult to conceive in what way they can be applied. Certain forms of root-like stems and buds have, however, been observed, to which some attention should be paid. In the first place, neither bulb nor rhizoma is known in Exogenous plants, while in Endogenæ they are sometimes characteristic of particular orders. Thus, all Marantaceæ and Scitamineæ, and most Irideæ, have a rhizoma in one form or other, and bulbs are a usual character of Asphodeleæ and Amaryllideæ; in the former, however, the bulb is often represented by a rhizoma, or cormus, as in Brodiæa, Leucocoryne, and their allies, or by those succulent fibres called fasciculate roots, as in Asphodelus itself; and in the latter the bulb is sometimes entirely absorbed by succulent perennial leaves, as in Clivia.

External variations in the figure of the STEM are sometimes available as distinctions of orders. Thus, a twining stem is almost without exception in Menispermeæ, a square stem is universal in Labiatæ, and an angular one in Stellatæ; but more frequently its figure affords no indication whatever of affinities.—Texture of the stem is of scarcely more value. Cacteæ, it is true, have always the cellular tissue in excess, and derive by that circumstance one of their great distinctions from Grossulaceæ; but even in Cacteæ the Pereskias are scarcely more succulent than other plants; and Euphorbiaceæ and Asclepiadeæ exhibit instances both of the most decided state of anamorphosis, and of the normal condition of stems in general.—In the internal arrangement of the layers of Exogenous stems, I am not aware of any character which distinguishes orders besides those to which I have already adverted; except in Calycantheæ, which are distinctly known by the presence of four incomplete centres of vegetation surrounding the principal one, and so forming four angles which are visible externally. (See Mirbel's figure, in the Annales des Sciences, vol. xiv. p. 367.) But as I have before observed, very little is really known upon this subject.

The LEAVES are subject to modifications not less important in determining the mutual relations of plants, than the functions which they perform in the vegetable economy. Their characters depend upon their relative position, their degree of division, their venation, and the presence or absence of pellucid dots within their substance.—All Cinchonaceæ (Rubiaceæ) have opposite entire leaves; in Labiatæ, Apocyneæ, Gentianeæ, Monimieæ, and many others, they are also uniformly opposite; but in the genus Fuchsia, in which they are usually opposite, species exist in which they are not only alternate, but both the one and the other on the same plant; and alternate-leaved species exist in Compositæ, Scrophularineæ, and Malpighiaceæ, orders the leaves of which are generally opposite. In Cupuliferæ, Umbelliferæ, Ternströmiaceæ, Hamamelideæ, and Urticeæ, they are uniformly alternate; but in Combretaceæ and Leguminosæ, orders usually having alternate leaves, they are occasionally opposite; and Halorageæ, Ericineæ, and Ficoideæ, are orders in which the genera have their leaves arranged in no certain manner.

c

I do not know how far this irregularity is connected with the following observations of Schlechtendahl, which, however, deserve attention. " Those leaves," he says, " which are connected either by
their base, or by the intervention of a stipula, I call opposite, and
those which are not so connected, spuriously opposite (pseudo-
opposita). Opposite leaves are never disjoined, as in Rubiaceæ
and Caryophylleæ; spuriously opposite ones, which are much more
common, being easily disjoined, readily become alternate. Branches
obey the same laws as leaves." *Linnæa*, 1. 207.—All Spondiaceæ,
Rhizoboleæ, &c. have compound leaves; in many others they are
always simple; and in such orders as Acerineæ, Aurantiaceæ,
Geraniaceæ, Rutaceæ, and Sapindaceæ, both simple and compound
leaves are found. This character, therefore, is not considered of so
much value as many others.—Neither is the degree of division of
the margin usually important, toothed and entire leaves being
often found in the same order. Nevertheless, there is no instance
of toothed leaves in Cinchonaceæ, Gentianeæ, Guttiferæ, or Malpighiaceæ; and they are very rare in Endogenous plants.—
Characters derived from the arrangement of veins are known to
be in many cases of the utmost importance; and it is probable,
that when this subject shall have been more accurately studied,
they will be found of even more value than has been yet supposed.
The great obstacle to employing characters derived from venation,
exists in the want of words to express clearly and accurately the
different modes in which veins are arranged. I have endeavoured
to remove this by some observations in the Botanical Register;
and I am persuaded the subject deserves the particular attention
of botanists. It is already known that the internal structure and
peculiar growth of Exogenæ and Endogenæ are externally indicated by the arrangement of the veins of their leaves,—those of
Exogenæ diverging abruptly from the midrib, and then branching
and anastomosing in various ways, so as to form a reticulated plexus
of veins of unequal size; while those of Endogenæ run straight from
the base to the apex, or diverge gradually from the midrib, not
ramifying in their course, but being simply connected with each
other by transverse bars, examples of which are afforded on the one
hand by the Rose, and on the other by the Iris and Arrow-root.
Although a few exceptions exist to both these laws, yet the grand
characters of the leaves of those classes are such as I describe. But,
independently of this, many other orders are distinguished without
exception by modifications of venation. Thus, all Melastomaceæ
have three or more collateral ribs connected by branched transverse
bars, something in the way of Endogenæ; all Myrtaceæ have one or
two fine veins running parallel with the margin, and just within it; all

Cupuliferæ have the principal lateral veins running straight out from the midrib to the margin ; Betulineæ are distinguished by this among other characters from Salicineæ ; and the same peculiarity separates the genuine genera of Dilleniaceæ, called Delimaceæ by Decandolle, from those of which Hibbertia is the representative.—Leaves which contain reservoirs of oily secretions, indicated by the presence of pellucid glands within their substance, are almost always universal in a given order. Thus, Myrtaceæ, properly so called, (with the exception of the paradoxical pomegranate,) are distinguished by these glands from Melastomaceæ ; in one genus of which, however, (Diplogenea,) slight traces of them are to be found : they are present in all Aurantiaceæ ; by this character Wintereæ are distinguished from Magnoliaceæ, Amyrideæ from Connaraceæ, &c. &c. In the orders Phytolacceæ, Petiveraceæ, Labiatæ, and Zygophylleæ, there are, however, genera with and without pellucid dots.

At the base of some leaves are frequently found little membranous or foliaceous appendages called STIPULÆ, which are in fact leaves in an imperfect state of developement. Their presence may therefore be understood to indicate a peculiar degree of composition in the leaves to which they belong, and they really indicate affinities in a very remarkable manner. In studying them, however, care must be taken not to confound genuine foliaceous appendages, to which alone the name of stipulæ properly appertains, with dilatations, or membranous or glandular processes of the petiole, such as are found in Ranunculaceæ, Grossulaceæ, Apocyneæ, Umbelliferæ, and others. The presence of stipulæ is universal in Cinchonaceæ, which are thus distinguished from Stellatæ, in Betulineæ, Salicineæ, Magnoliaceæ, Artocarpeæ, and many others : a particular modification of them, called the ochrea, is the peculiar distinction of Polygoneæ ; and they are universally absent in Myrtaceæ properly so called, Guttiferæ, Gentianeæ, Malpighiaceæ, and many others. The orders Cistineæ, Saxifrageæ, and Loganieæ, are among the very few cases in which genera exist both with and without stipulæ. (See Von Martius *Nov. Gen. et Sp.* 2. 135.)

The little starved leaves found at the base of many flowers, and technically called BRACTEÆ, are rarely employed as distinctions of orders, offering scarcely any modifications of importance. In Cruciferæ they are never present, and in Marcgraaviaceæ they are usually hollow, being folded together by their two edges, like the leaves of which carpella are formed.

Forms of INFLORESCENCE are occasionally, but not often, found characteristic of particular tribes. Thus all Compositæ, Calycereæ, and Dipsaceæ, have their flowers in heads ; all Umbelliferæ bear umbels ; all Labiatæ have axillary cymes called verticillastri ; all

Plantagineæ, Cyperaceæ, and Gramineæ, have dense simple imbricated
spikes; all Betulineæ, Cupuliferæ, and Salicineæ, bear amenta or
catkins; and most Coniferæ have a strobilus or cone; in the latter,
however, the flowers are sometimes solitary, as in Taxus, and then
the usual form of inflorescence is departed from.

The outer envelope of the flower, called the Calyx, is used in a
variety of ways to distinguish orders; but the characters it affords are
far from being of equal or uniform importance. Its absence implies
the absence of the corolla also, which cannot possibly be present
when the calyx is away, unless, as in Compositæ, it is obliterated by
the pressure of surrounding bodies. By its absence all the orders
called Achlamydeous are characterised, such as Salicineæ, Piperaceæ,
Saurureæ, &c.; but in Betulineæ it is present in the male flowers,
and in Euphorbia itself, among Monochlamydeæ, it is wholly wanting.
These exceptions do not, however, affect the general importance of
characters derived from its presence or absence. If it is unaccompanied by the corolla, plants are said to be Monochlamydeous; and
this is a point of very uniform value. I know of no true Monochlamydeous orders in which the presence of a corolla forms an exception,
unless the faucial scales of Thymelææ are considered the rudiments
of a corolla.—The sepals or leaves of which it is composed are either
distinct or combined; and from this circumstance characters are
sometimes advantageously derived. Thus, in Scleranthcæ the calyx
is always monosepalous, and in Chenopodeæ it is as regularly
polysepalous; but in Caryophylleæ both forms are observable.—The
number of sepals is sometimes a character of importance, as in
Cruciferæ, in which they are always 4, in Papaveraceæ, which have
never more than 2, and in the greater part of Endogenous plants,
which have usually 3. This character, however, requires to be used
with circumspection, as there are many more instances of the number
of sepals being variable than regular. Thus in Lineæ and Malvaceæ
they are 3-4-5; in Guttiferæ they vary from 2 to 6; in Homalineæ
from 5 to 15; and in Samydeæ from 3 to 7.—The æstivation of the
calyx is always to be well considered, as certain forms are often
among the best known indications of affinity. Malvaceæ, Tiliaceæ,
Elæocarpeæ, Tremandreæ, Sterculiaceæ, and Bombaceæ, have it exclusively valvate among polypetalous dicotyledons with hypogynous
stamens; Ternströmiaceæ have the sepals constantly imbricated in
a particular way; Vites have the lobes of the calyx distinct and
wide apart from a very early period of their existence: but in
Penæaceæ both valvate and imbricate æstivation exists.—In some
plants the sepals are all of equal size; in others they are very
unequal either in form, direction, or texture; in the former case they
are said to be regular, in the latter irregular, and by this difference

certain orders are characterised. Thus Sapindaceæ and Polygaleæ have a calyx constantly irregular; many orders are constantly regular; but it frequently happens that both regular and irregular calyces co-exist in the same order, as in Rosaceæ, Labiatæ, Leguminosæ, and a great many others.—In most orders the sepals occupy one series or verticillus only; others have them in two series, and this has not been found to be connected with any material differences otherwise; but when the number of series is increased much beyond two, they cease to be separately distinguishable, and form an imbricated calyx, which is frequently confounded with the corolla, as in Calycantheæ and Wintereæ. I know of no order in which genera with an imbricated calyx of this kind and a calyx of the common kind co-exist. It is one of the principal points which separate Calycantheæ from Rosaceæ.—The most important character connected with the calyx is, however, its cohesion or non-cohesion with the ovarium; or, as botanists incorrectly call it, its being superior or inferior. Many orders are positively characterised by this, as Compositæ, Umbelliferæ, Caprifoliaceæ, Orchideæ, and very many more; and, as it usually happens that it exists without exception, it becomes one of the most useful means of distinction of which we are in possession. Pomaceæ are, for instance, by this means at once known from Rosaceæ, Scævoleæ from Brunoniaceæ, and Cinchonaceæ from Apocyneæ. No instance of a superior calyx has been found in Ranunculaceæ, Cruciferæ, Papaveraceæ, Rutaceæ, and a number of others. But there are some singular exceptions to this law. Thus, among Anonaceæ, an order with indefinite superior ovaria, we find Eupomatia, in which they are inferior. In Anacardiaceæ, which have almost universally a superior ovarium, a genus is said by Mr. Brown to exist in which it is inferior; in Melastomaceæ all degrees of cohesion take place between the calyx and the ovarium; and in Saxifrageæ this uncertainty of structure is still more remarkable. It should, however, be observed, that in the two latter orders the tendency to cohesion between the calyx and ovarium may be almost always ascertained by careful dissection; and even in Parnassia, an anomalous genus which is referred to Saxifrageæ, usually having an ovarium completely superior, there exists a species in which it is partially inferior. I have said that the difference between a superior and inferior calyx consists only in the cohesion of that organ with the ovarium in the one case, and its separation from it in another; and this is the view which is always taken of it, all that part which intervenes between the segments and the pedicel being considered the tube of the calyx. But I strongly suspect that we have yet to learn that theory has in this case carried botanists too far, and that there are cases in which the apparent origin of the

calyx is the real origin. Upon this supposition, what is now called
the tube of the calyx may be sometimes a peculiar extension or
hollowing out of the apex of the pedicel, of which we see an example
in Eschscholtzia, and of which Rosa and Calycanthus, and perhaps
all supposed tubes without apparent veins, may also be instances.
In this case the whole of our ideas about superior and inferior
calyxes will require modification. But upon this subject I cannot
enter here : I have in the following Work spoken of these points of
structure according to the received opinions of botanists.

The second floral envelope we call the COROLLA. It consists of
a number of leaves equal to those of the calyx, and alternating with
them ; in addition to which they are usually coloured.—If the corolla
is present, a plant is said to be dichlamydeous, and much importance
is attached to this peculiarity ; far more, I think, than it deserves.
It constantly separates plants having much natural affinity, as
Euphorbiaceæ far from Rhamneæ, Amarantaceæ widely from Illecebreæ ; and it is also one to which there are numberless exceptions.
This is, however, not the case with monopetalous dicotyledons,
Primulaceæ and Oleaceæ being almost the only instances of orders
among those which are truly monopetalous, containing apetalous
genera.—The difference between a monopetalous and a polypetalous
corolla is this, that in the one the leaves out of which the corolla is
formed are distinct, and in the other united. Great value is attached to this, and it is in fact a difference of first-rate importance :
thus, all Ranunculaceæ, Rosaceæ, Cruciferæ, Papaveraceæ, Terebintaceæ, and a multitude of others, are, without exception, polypetalous ; and all Boragineæ, Labiatæ, Scrophularineæ, and Bignoniaceæ, are equally, without exception, monopetalous : but in the
polypetalous orders of Crassulaceæ, Diosmeæ, Polygaleæ, Ternströmiaceæ, &c., there are many monopetalous genera ; and monopetalous
Caprifoliaceæ are usually associated with Hedera and Cornus, which
are as much polypetalous as any other plants.—The æstivation of
the corolla rarely furnishes characters connected with the natural
properties of plants ; nevertheless, Compositæ are essentially distinguished by their valvate, and Asclepiadeæ and Apocyneæ by their
contorted æstivation, an exception to the one existing only in the
genus Leptadenia, and in the other in Gardneria. The æstivation
of both calyx and corolla has as yet received too little attention
for its value to be judged of generally.—The regularity or irregularity of the corolla is most commonly important : thus, Orchideæ,
Polygaleæ, Bignoniaceæ, Fumariaceæ, are irregular without exception ; the regular flowers of Boragineæ will almost distinguish them from Labiatæ, which have as frequently irregular ones ;
yet Echium in Boragineæ is irregular, and Caprifoliaceæ exhibit

all the gradations from a corolla of the most irregular form to one of the most perfect symmetry. In Compositæ both are found continually in the same head; and Lobeliaceæ, which may be almost always distinguished from Campanulaceæ by their irregularity, become nearly regular in Isotoma.—The venation of the petals is scarcely ever employed for distinction, little being at present known of it. Compositæ are distinguished by the peculiar arrangement of the veins of their corolla; and they are always oblique in Hypericineæ.

From within the corolla arise certain metamorphosed leaves, which are called the SEXES of plants. From the manner in which they are combined, good characters may sometimes be derived, but frequently no characters at all. Thus, Xanthoxyleæ are known from Diosmeæ and Terebintaceæ by their unisexual flowers; all Euphorbiaceæ, Begoniaceæ, Amentaceæ, Coniferæ, Myriceæ, are unisexual. But Vites, Gramineæ, Cyperaceæ, Chenopodeæ, Umbelliferæ, and even Ranunculaceæ, contain hermaphrodite and diclinous genera; and it is familiar to every one, that flowers of all these kinds (that is, male, female, and hermaphrodite), stand side by side in Compositæ.

Of these sexes the STAMENS are what are called the male organs, and are undoubtedly the apparatus by means of which vivification is communicated to the ovula or eggs. They either arise immediately from below the ovarium, having no adhesion to the calyx, when they are said to be *hypogynous,* or they contract an adhesion of greater or smaller extent with either the calyx or corolla, when they become *perigynous,* or, finally, they appear to proceed from the apex of an inferior ovarium, in which case they are named *epigynous;* but it is usually now understood that all stamens take their origin from below the ovarium; and if this opinion be well founded, there will be no material difference between those which are perigynous and those which are epigynous; and these two modifications are accordingly confounded together by most modern botanists. M. Ad. Brongniart, however, conceives epigynous stamens to be essentially distinct from perigynous, founding his opinion upon the genus Raspailia, which has a superior ovarium, from the top of which arise the stamens; but it is possible perhaps to explain this apparent anomaly. To the difference between perigynous and hypogynous stamens the French school attaches the greatest value, not being willing to admit any genus with hypogynous stamens into an order with perigynous ones, and *vice versâ;* and there is somewhere an observation, that of such primary importance is this distinction, that while poisonous orders are to be known by their stamens being hypogynous, all in which they are perigynous are wholesome. Setting aside, however, this hypothesis, which has not the general applica-

tion that has been ascribed to it, there is no doubt that insertion of
stamens does very often go along with essential differences of other
kinds; for example, it distinguishes with precision Rosaceæ from
Ranunculaceæ, Violaceæ from Passifloreæ, Reaumurieæ from Nitra-
riáceæ, Aurantiaceæ from Burseraceæ. But, on the other hand,
there is not only frequently, as may be well supposed, so slight a
degree of adhesion between the stamens and calyx as to render it
difficult to say whether the former are perigynous or hypogynous, as
in Galacineæ, Tamariscineæ, and many others; but there are orders
which do really exhibit instances of both modes. Thus Eschscholtzia
has decidedly perigynous stamens, and yet it is undoubtedly a genus
of Papaveraceæ, the character of which is to have them hypogynous;
and all kinds of gradations, from the one form to the other, are
observable in Saxifrageæ. The stamens of Macrostylis, among the
hypogynous order Diosmeæ, are manifestly perigynous. In Geraniaceæ the genus Geranium has the stamens hypogynous, and
Pelargonium perigynous. Caryophylleæ are arranged among genera
with hypogynous stamens, yet some of them (Larbrea and Adenarium)
are perigynous; in Illecebreæ part of the genera are perigynous, and
part hypogynous. The perigynous stamens of Turneraceæ divide
them from Cistineæ, to which they are closely allied. — The manner
in which the stamens cohere is sometimes an indication of affinity;
for instance, they are monadelphous in Malvaceæ and Meliaceæ,
diadelphous in great numbers of Leguminosæ, polyadelphous in
Hypericineæ; but more commonly this character is unimportant, as
in Malvaceæ themselves, which have sometimes distinct stamens;
Leguminosæ, which have very often such; in Ternströmiaceæ, which
have both united and disunited ones. — It not unfrequently occurs
that the conversion of the petals into stamens takes place imperfectly, in which case a part of the stamens are said to be sterile, and
this is sometimes a useful character for detecting affinities. Thus, in
many Büttneriaceæ one-fifth are sterile and petaloid, in Galacineæ
every other one, in Aquilarineæ two-thirds, in Bignoniaceæ the
uppermost of 5 is rudimentary. — A peculiarity of a similar nature is
the want of symmetry which sometimes exists between the petals or
sepals, and stamens. Supposing the flower to be formed without
abortion of any kind, and by a regular alternation of metamorphoses,
as is usually the case, the petals will be always some multiple of
the sepals, and the stamens of the petals; and of course any irregularity in this respect will destroy the supposed symmetry. This is
often a point of much importance to observe; for example, in
Boragineæ the stamens are always equal to the segments of the
corolla, and the flowers of that order are consequently symmetrical;
in Labiatæ, on the contrary, one at least of the stamens is constantly

missing, and the flowers are therefore regularly unsymmetrical, a character by which these orders may be constantly known, when the form of their corolla will not distinguish them. In Phytolacceæ there is a constant tendency to a want of symmetry; and this is one of the characters by which that order is known from Chenopodeæ. That part of the stamen which contains the fertilizing matter or pollen is known by the name of the ANTHER, and is a case usually consisting of two parallel or slightly diverging cells, containing pollen, and opening by a longitudinal fissure; but from this plan many deviations take place, which are of great value in determining affinities. Thus, all Malvaceæ, properly so called, and Epacrideæ, have but one cell; in Laurineæ and Berberideæ the valves are hinged by their upper margin; in Ericeæ the pollen is emitted by pores; in Melastomaceæ the same takes place, along with a peculiar conformation of the lower part of the anther; in Hamamelideæ dehiscence is effected by the falling off of the face of the anthers: but in Solaneæ, the genera of which have usually their anthers bursting longitudinally, the genus Solanum itself opens by pores.—The mode in which the anther is united with the filament is sometimes taken into account, as in Anonaceæ, Nymphæaceæ, Humiriaceæ, and Aroideæ, or Typhaceæ, in which they are always adnate; and Gramineæ, in which they are as regularly versatile. But this modification appears of no great moment, nor indeed does any peculiarity of the connectivum, all kinds of forms of which are found in, Labiatæ; and even in the small order of Penæaceæ we have anthers with the connectivum excessively fleshy, and in the ordinary state.

POLLEN rarely affords any marks by which affinities are to be traced. The most remarkable deviations from it exist in Asclepiadeæ and Orchideæ; the former having it always in a state of concretion, resembling wax, by which they are known from Apocyneæ, and the latter having it frequently so, but also containing numerous genera, the pollen of which is scarcely distinguishable from its ordinary powdery state.

Immediately between the stamens and the ovarium is sometimes found a fleshy ring or fleshy glands, called a DISK, and supposed for very good reasons to represent an inner row of imperfectly developed stamens. The presence of this disk is constant in Umbelliferæ, Compositæ, Labiatæ, Boragineæ, Rosaceæ, and many others, while its absence is equally universal in others. It is not, however, much used as a principal mark of distinction, its real value not having been yet ascertained. There are some highly curious modifications of it in Rhamneæ and Meliaceæ. It is a very remarkable fact, that in Gentianeæ and their allies, which have the pericarpial leaves right

and left with respect to the common axis of inflorescence, it is never
truly present; while in Scrophularineæ and their allies, the pericarpial leaves of which are anterior and posterior, it is as uniformly
present in one shape or other.

The last modification of leaves in the fructification consists in
their conversion into what is called the female organ, or OVARIUM;
that is to say, into the case which contains the young seeds or ovules.
Now that the structure of this part is well understood, we know that
an ovarium either consists of one or several connected pericarpial
leaves, called carpella, arranged around a common axis, or of several
combined into a single body. Upon this difference the distinction
depends of what I call apocarpous ovaria, or those of which the
carpella are distinct; and syncarpous are those of which the carpella
are compactly combined. These differences appear to me of much
importance, and subject to as few exceptions as any modifications
that botanists make use of. Thus Berberideæ are distinguished
from Papaveraceæ, Nelumboneæ from Nymphæaceæ, Amyrideæ from
Burseraceæ, Boragineæ from Ehretiaceæ, and the like. But, at the
same time, it will be seen that cases exist of both forms being found
in the same natural order, as Xanthoxyleæ. This, however, is rare.—
The cohesion of the ovarium with the calyx, or its separation from it,
has been already treated of in speaking of the calyx. — An ovarium
may be either one-celled, in consequence of its consisting of a single
carpellum, in which case it will belong to the apocarpous division;
or it may consist of several carpella strictly cohering, and therefore
syncarpous, but nevertheless one-celled, in consequence of the obliteration of the dissepiments. Peculiarities of this latter nature are
almost always of ordinal importance, *at least if the placentæ are
parietal;* for instance, the latter is the structure of Papaveraceæ,
Homalineæ, Flacourtiaceæ, Cucurbitaceæ, Papayaceæ, and Violaceæ,
to which there is no exception; but Caryophylleæ and Bruniaceæ,
the usual structure of which is to be one-celled, have the placentæ
in the centre; and in both these orders there are genera, the ovarium
of which contains several cells.—Another point that deserves particular attention is the relation borne to the axis of inflorescence by
the pericarpial leaves, of which an ovarium is formed. What the
exact value of this character may be, is not yet known; but it is
certain that Gentianeæ and their allies have their principal leaves
right and left of the axis, while Scrophularineæ and their allies,
which are sometimes to be distinguished with difficulty, have the
pericarpial leaves anterior and posterior with respect to the axis.
Rosaceæ and Leguminosæ differ in a nearly similar way.—Connected
with the apocarpous or syncarpous state of the ovarium is the union
or separation of the styles, which, therefore, scarcely require distinct

mention. It is as well, however, to remark, that the separation of styles is commonly a sign of the apocarpous state of the ovarium, provided the latter is not very apparent otherwise; and the cohesion of the styles is constantly an evidence of the contrary; and in this view the Elder and Hydrangea tribes may be justifiably separated from Caprifoliaceæ.

The STIGMA seldom offers any good characters. In some cases, however, advantage is taken of it, as in Lineæ, the capitate stigmas of which distinguish them from Caryophylleæ, in which they occupy the whole inner face of the styles; and in Goodenoviæ, Scævoleæ, and Brunoniaceæ, there is a peculiar membranous appendage enveloping the stigma, and called an indusium, which distinguishes those orders from all others.

The number of the OVULA (that is to say, whether they are definite or indefinite,) is frequently an important difference, as, for example, between Campanulaceæ and Compositæ, Goodenoviæ and Scævoleæ; but while I think considerable value usually attaches to this, it must not be forgotten that there are exceptions to it in several instances, especially in Caprifoliaceæ, if Hydrangea really belongs to that order, and Fumariaceæ and Cruciferæ.—The position of the ovula is much more essential than their number, and may be considered as one of the most valuable forms of structure that can be taken into account. It is uniform in Compositæ, Valerianeæ, Umbelliferæ, and others, and it constitutes an absolute distinction between Artocarpeæ and Urticeæ; but in Sanguisorbeæ, Pedalineæ, and Styraceæ, both erect and suspended ovules co-exist; this union of the two positions occurs in a most remarkable degree in Penæaceæ; and among Violaceæ, the genus Conohoria offers, according to M. A. St. Hilaire, (*Pl. Usuelles*, No. 10,) an instance of three kinds of direction in as many species; in C. Lobolobo, the ovula are ascending; in C. Castaneæfolia, they are suspended; and in C. Rinorea, one is suspended, one ascending, and the intermediate peritropal, or at right angles with the placentæ.—The situation of the foramen of the ovulum is a circumstance which should always be taken into account, because it indicates with certainty the future position of the radicle, which it is of first-rate importance to ascertain, but which will be more properly spoken of in considering the value of distinctions drawn from that source.

The ripened ovarium is the FRUIT. The differences in its structure are of the same nature as those of the ovarium, and need not be repeated. Its texture and mode of dehiscence are the principal sources of distinctions, but they perhaps deserve as little attention as any of which botanists make use. It is true that the fruit of all Grossulaceæ is baccate, of all Labiatæ indehiscent, and of all Primulaceæ capsular; but Marcgraaviaceæ, Melastomaceæ, Myrtaceæ, Ranunculaceæ, and Rosaceæ, and a crowd of other orders, contain both baccate and capsular, dehiscent and indehiscent genera.

The characters obtained from the position of the SEED are of the same value as those from the position of the ovula; in addition to which, the peculiarities of the testa are made use of. In some Monocotyledonous orders, as Asphodeleæ and Smilaceæ, the texture is employed as a mark of distinction; its being winged or otherwise distinguishes Meliaceæ from Cedreleæ, and the presence of a fungous swelling about the hilum is a good characteristic of Polygaleæ.— Linnean botanists make a distinction between naked and covered seeds, attributing the former character to Labiatæ, Boragineæ, &c.; but the sense in which they use the term is so manifestly erroneous, that botanists were at one time led to believe that no such things as naked seeds existed. It is now, however, known, from the accurate observations of Mr. Brown, that certain tribes of plants do exist in which the seeds are really naked, that is to say, susceptible of impregnation and maturation without the intervention of any peri-carpial covering. These are Coniferæ and Cycadeæ, orders exceed-ingly remarkable in other respects, especially in the structure of their vascular tissue. In consequence of these peculiarities, they have been distinguished by A. Brongniart as a class of the same dignity as Dicotyledons and Monocotyledons. Without assenting to this proposition, to which I think there are great objections, it is impos-sible to doubt that the naked seeds of these orders constitute a secondary character of as much importance as any of which botanists have knowledge.

The substance which surrounds the embryo is called the ALBUMEN, and its absence or presence constitutes a valuable mark of distinction. There can be no doubt that when it exceeds the bulk of the embryo very considerably, as in Ranunculaceæ, Papaveraceæ, Umbelliferæ, Grasses, and the like, it is of such importance, that no plant destitute of albumen is likely to be found appertaining to such orders; but, on the other hand, I doubt very much whether its presence or absence deserves much attention in orders which are called by German botanists subalbuminous,—that is to say, where the embryo and albumen are of nearly equal bulk; for it should be remembered, that it always exists in seeds at some period of their existence, and that its remains may very well be expected to be found in almost any seeds; thus, in fact, both albuminous and exalbuminous seeds are found in Proteaceæ (*Brown in Linn. Trans.* 10. 36); and even in Rosaceæ, which are as free from remains of albumen as any order, it is said to be distinctly present in Neillia, and in others traces are to be seen adhering to the inner membrane of the testa.—The texture of the albumen is frequently consulted with advantage: in all Rubiaceæ it is horny or fleshy ; Euphorbiaceæ, oily ; Grasses, Polygoneæ, Chenopodeæ, mealy ; in Annonaceæ, it is ruminated, &c. ; but among Apocyneæ, which have solid albumen, it is ruminated in Alyxia.

The direction of the EMBRYO within the testa, which is indicated in the ovulum by the foramen, is one of the very few characters to which we know of no exceptions; and if it were a less obscure point of structure, it would consequently be one of the most useful. For example, in all Cistineæ, Urticeæ, and Polygoneæ, the radicle is not turned towards the hilum, as in other tribes, but takes an opposite direction ; and these orders are distinguished from their allies by this, better than by any other known character.

The number of COTYLEDONS is generally believed to be one of the most important means of distinguishing the great natural divisions called Monocotyledons, Dicotyledons, and Acotyledons ; and it is a most curious fact, that this goes along with anatomical structure. There are, however, plants among Monocotyledons with two cotyledons, as the common Wheat ; and among Dicotyledons with only one, as Penæa and some Myrtaceæ ; or even none, as Cuscuta and Utricularia ; or several, as Schizopetalon in Cruciferæ, Benthamia in Boragineæ, Ceratophylleæ, and most Coniferæ.—To the relative position of the cotyledons there are not the same objections, whence the character of Dicotyledons has been found to consist in the cotyledons being opposite to each other ; of Monocotyledons, in their being alternate with each other, if there is more than one ; and of Acotyledons, in germination taking place from no particular point, rather than in their number.

The only remaining character of vegetation which I find it necessary to notice is a singular and very uncommon one, which distinguishes a few small families of plants. This consists in the presence of the remains of the AMNIOS around the embryo in its perfect state : the amnios always surrounds the embryo in an early state, but is most commonly absorbed before the formation of the embryo is completed ; but in Saurureæ, Piperaceæ, and Nymphæaceæ, its remains surround the embryo in the form of a sac, which was mistaken by Richard, who did not understand its nature, for a peculiar appendage of the embryo, or rather for a particular form of the radicle,—an hypothesis which that distinguished botanist supported with great skill, but which is now generally abandoned.

I have now gone through the whole of the characters of which botanists make use in distinguishing and determining the affinities of plants, and I think it must be apparent that the difficulties

XXXIV   INTRODUCTION.

connected with the subject are neither slight nor easily to be
overcome. If these observations are properly attended to, no one
can be at a loss to understand, that to define any group of plants,
of what rank soever, is impracticable; that differences of structure
are of an uncertain and unequal value; and that the affinities of
plants are never to be absolutely made out by solitary characters,
but depend upon more or less intricate combinations, the power
of judging of which is the same test of a skilful botanist, as an
appreciation of symptoms is that of a physician.

General View of the following Analytical Table.

NATURAL DIVISIONS.

Class I.
VASCULARES,
or
FLOWERING
PLANTS.

Sub-class I.—EXOGENÆ, or DICOTYLEDONOUS PLANTS.
  1.—Angiospermæ.........
  2.—Gymnospermæ.

Sub-class II.—ENDOGENÆ, or MONOCOTYLEDONOUS PLANTS.
  1.—Glumaceæ.
  2.—Petaloideæ.

Class II.
CELLULARES,
or
FLOWERLESS
PLANTS.
  1.—Filicoideæ.
  2.—Muscoideæ.
  3.—Aphyllæ.

ARTIFICIAL DIVISIONS.

*POLYPETALÆ.

†Thalamifloræ. { † Apocarpæ.
                { ‡‡Syncarpæ.

††Calycifloræ. { † Apocarpæ.
                { ‡‡Syncarpæ.

**APETALÆ.
***ACHLAMYDEÆ.
****MONOPETALÆ.

ARTIFICIAL ANALYSIS

OF

THE ORDERS.

CLASS I.—VASCULARES, OR FLOWERING PLANTS.

Plants having distinct flowers and sexes.

SUB-CLASS I.—EXOGENÆ, OR DICOTYLEDONOUS PLANTS.

Leaves reticulated. Stem with wood, pith, bark, and medullary rays. Flowers with a quinary division. Cotyledons 2 or more, opposite.

TRIBE I.—ANGIOSPERMÆ.

Seeds enclosed in a pericarpium.

* POLYPETALÆ.
Petals distinct.

† THALAMIFLORÆ.
Stamens hypogynous, or adhering to the sides of the ovarium. (Some Diosmeæ perigynous.)

‡ APOCARPÆ.
Carpella more or less distinct, sometimes solitary.

Flowers unisexual - - - - - -	23. MENISPERMEÆ.
Flowers hermaphrodite.	
Fruits immersed in a fleshy disk - - - -	6. NELUMBONEÆ.
Fruits not immersed in a fleshy disk.	
Anthers bursting by valves curling backwards -	22. BERBERIDEÆ.
Anthers bursting by longitudinal slits.	
Stipulæ present.	
Leaves with transparent dots - -	17. WINTEREÆ.
Leaves without dots - - - -	15. MAGNOLIACEÆ.
Stipulæ absent.	
Albumen ruminated - - - -	13. ANONACEÆ.
Albumen solid.	
Seeds with an arillus - - -	16. DILLENIACEÆ.
Seeds without an arillus.	
Ovarium solitary - - -	8. PODOPHYLLEÆ.
Ovaria more than one.	
Leaves sheathing at the base -	3. RANUNCULACEÆ.
Leaves with a taper petiole -	7. HYDROPELTIDEÆ.
Albumen none.	
Leaves with pellucid dots -	111. AMYRIDEÆ.
Leaves without pellucid dots.	
Stigmas capitate or terminal -	110. CONNARACEÆ.
Stigmas linear. Petals sepaloid -	117. CORIARIEÆ.

d

‡‡ SYNCARPÆ.

Carpella cohering in a solid (multilocular) pericarpium.

¶ Ovarium many-celled, with the ovula attached to the face
 of the dissepiments - - - - } 5. NYMPHÆACEÆ.

¶¶ Ovarium 1-celled, with the ovula parietal.
  Placentæ linear, contracted.
    Sepals 2.
      Corolla regular - - - - 4. PAPAVERACEÆ.
      Corolla irregular - - - - 10. FUMARIACEÆ.
    Sepals invariably 4.
      Stamens tetradynamous. Disk glandular, or 0.
      Ovarium sessile - - - - } 9. CRUCIFERÆ.
      Stamens indefinite. Disk continuous, enlarged.
      Ovarium stalked - - - - } 11. CAPPARIDEÆ.
    Sepals 5 (occasionally varying to 4, 6, or 7).
      Ovula with the foramen at the extremity opposite
        the hilum - - - - - } 134. CISTINEÆ.
      Ovula with the foramen at the extremity next the hilum.
        Stamens indefinite - - - 135. BIXINEÆ.
        Stamens definite.
          Vernation circinate - - - 137. DROSERACEÆ.
          Vernation straight.
            Capsule with loculicidal dehiscence.
              Stipulæ present. Sepals distinct.
              Seeds naked - - } 130. VIOLACEÆ.
              Stipulæ absent. Sepals combined.
              Seeds comose - - } 142. TAMARISCINEÆ.
            Capsule with septicidal dehiscence.
              Stipulæ 0. Sepals concrete - 141. FRANKENIACEÆ.
    Placentæ branched over the surface of the valves - 12. FLACOURTIACEÆ.

¶¶¶ Ovarium 2- or more-celled, with the ovula attached to
the axis ; or only 1-celled, with the ovula adhering to a
placenta in the centre.
  Æstivation of the calyx valvate.
    Anthers bursting by pores.
      Petals lacerated, imbricated in æstivation - 30. ELÆOCARPEÆ.
      Petals entire, involute in æstivation - - 128. TREMANDREÆ.
    Anthers bursting longitudinally.
      Filaments distinct. Disk glandular - - 29. TILIACEÆ.
      Filaments connate. Disk 0.
        Anthers bilocular - - - - 27. STERCULIACEÆ.
        Anthers unilocular.
          Stamens monadelphous - - 24. MALVACEÆ.
          Stamens penta- or polyadelphous - 26. BOMBACEÆ.
  Æstivation of the calyx imbricate or open.
    Stamens indefinite.
      Styles several.
        Seeds smooth - - - - 36. HYPERICINEÆ.
        Seeds villous - - - - 37. REAUMURIEÆ.
      Style single.
        Stigma peltate, petaloid, persistent - - 136. SARRACENIEÆ.
        Stigma not dilated, withering.
          Anthers subulate, opening by a linear pore
            at the apex - - - } 31. DIPTEROCARPEÆ.
          Anthers opening longitudinally.
            Leaves with stipulæ - - - 25. CHLENACEÆ.
            Leaves without stipulæ.
              Leaves compound - - - 99. RHIZOBOLEÆ.
              Leaves simple.
                Leaves opposite - - 34. GUTTIFERÆ.
                Leaves alternate.
                  Seeds indefinite - - 35. MARCGRAAVIACEÆ.
                  Seeds definite - - - 32. TERNSTRÖMIACEÆ.
    Stamens definite.
      Flowers unsymmetrical. (That is, the segments
      of the calyx, the petals, and the stamens, not
      regular multiples of each other.) *Anisomeria.*

Sepals very unequal. Stamens irregularly arranged upon a hypogynous disk. (Petals usually with some interior appendage.) - - - - } 100. SAPINDACEÆ.
Ovules definite, erect.
  Fruit dehiscent - - 98. HIPPOCASTANEÆ.
  Fruit indehiscent.
    Stamens distinct - - 101. ACERINEÆ.
    Stamens cohering at the base in a fleshy cup - - } 94. HIPPOCRATEACEÆ.
Ovules definite, pendulous - - 68. OLACINEÆ.
  Ovarium 1-celled, with a central columnar placenta.
    Stamens monadelphous. Fruit dehiscent - - } 129. POLYGALEÆ.
    Stamens distinct. One of the sepals spurred - - } 124. TROPÆOLEÆ.
Ovules indefinite - - - 126. BALSAMINEÆ.
Flowers symmetrical. (That is, the segments of the calyx, the petals, and the stamens, regular multiples of each other.) *Isomeria.*
Embryo coiled round mealy albumen - 140. CARYOPHYLLEÆ.
Embryo straight, or a little curved; albumen, if present, not mealy.
  Stamens combined in a long tube; anthers subsessile..
    Seeds definite, not winged; anthers all fertile - } 105. MELIACEÆ.
    Seeds indefinite, winged; anthers partly sterile - } 106. CEDRELEÆ.
  Stamens distinct, except at the base; anthers with long filaments.
    Seeds indefinite.
      Embryo minute, in fleshy albumen - } 121. PITTOSPOREÆ.
      Embryo in the axis of fleshy albumen - - } 123. OXALIDEÆ.
      Embryo destitute of albumen.
        Fruit drupaceous. Trees 95. BREXIACEÆ.
        Fruit capsular. Herbs - 143. ELATINEÆ.
    Seeds definite.
Ovarium deeply lobed, with the style arising from the base of the carpella, which are seated on a succulent receptacle - } 118. OCHNACEÆ.
Ovarium not seated on a succulent receptacle.
Ovula erect - - - - - 104. VITES.
Ovula pendulous.
  One of the sepals spurred - - - 125. HYDROCEREÆ.
  None of the sepals spurred.
    Leaves with pellucid dots.
      Fruit succulent - - - 108. AURANTIACEÆ.
      Fruit capsular or drupaceous.
        Flowers unisexual - - - 114. XANTHOXYLEÆ.
        Flowers hermaphrodite.
          Endocarp not separable from the sarcocarp 116. RUTACEÆ.
          Endocarp separating from the sarcocarp as a 2-valved coccus - - } 115. DIOSMEÆ.
    Leaves without pellucid dots.
      Fruit 1-celled - - - 102. ERYTHROXYLEÆ.
      Fruit many-celled.
        Stamens arising from hypogynous scales.
          Leaves opposite, with stipulæ - 119. ZYGOPHYLLEÆ.
          Leaves exstipulate - - 120. SIMARUBACEÆ.
        Stamens immediately hypogynous.
          Cotyledons shrivelled - - 122. GERANIACEÆ.
          Cotyledons flat.
            Styles distinct. Stigmas capitate - 139. LINEÆ.
            Styles concrete, or nearly so.

Seeds without albumen. Connec- } 103. MALPIGHIACEÆ.
tivum small - -
Seeds with albumen. Connectivum } 107. HUMIRIACEÆ.
dilated - -

†† CALYCIFLORÆ.
Stamens perigynous ; distinct from the corolla when it is monopetalous.

‡ APOCARPÆ.
Carpella distinct. In Pomaceæ they cohere more or less ; but the styles are distinct.

Calyx adhering more or less to the ovaria.
  Stamens definite.
    Herbaceous plants (without stipulæ) - - 38. SAXIFRAGEÆ.
    Shrubs with opposite leaves (and interpetiolar stipulæ) 39. CUNONIACEÆ.
  Stamens indefinite.
    Fruit capsular. Seeds indefinite - - 40. BAUERACEÆ.
    Fruit pomaceous. Seeds definite - - 74. POMACEÆ.
Calyx distinct from the ovarium.
  Leaves with stipulæ.
    Ovaria several - - - - 73. ROSACEÆ.
    Ovaria solitary.
      Ovula peritropal. Fruit a legume - - 77. LEGUMINOSÆ.
      Ovula erect - - - - 76. CHRYSOBALANEÆ.
      Ovula suspended - - - - 75. AMYGDALEÆ.
  Leaves without stipulæ.
    Sepals numerous, imbricated - - 18. CALYCANTHEÆ.
    Sepals in a single whorl.
      Seeds definite, without albumen - - 113. ANACARDIACEÆ.
      Seeds indefinite, with albumen.
        Ovarium with hypogynous scales. Vegetation } 147. CRASSULACEÆ.
        succulent - - -
        Ovarium without hypogynous scales. Vegeta- } 38. SAXIFRAGEÆ (*bis*).
        tion normal - - - -

‡‡ SYNCARPÆ.
Carpella combined into a multilocular pericarpium.

¶ Ovarium superior.
  Ovarium 1-celled, with parietal placentæ.
    Embryo in the midst of fleshy albumen.
      Æstivation of the corolla twisted.
        Throat of the calyx with a membranous corona - 132. MALESHERBIACEÆ.
        Throat of the calyx without a membranous corona. 133. TURNERACEÆ.
      Æstivation of the corolla imbricated - 131. PASSIFLOREÆ.
    Embryo without albumen. Flowers rather irregular - 28. MORINGEÆ.
  Ovarium 1-celled, with the ovula not parietal, but either
  pendulous, or attached to a free central placenta.
    Sepals 2. Stamens opposite the petals - - 144. PORTULACEÆ.
    Sepals 5. Stamens opposite the sepals - - 150. ILLECEBREÆ.
  Ovarium with several cells.
    Calyx tubular, covering the fruit - - 52. SALICARIÆ.
    Calyx deeply divided or polysepalous.
      Flowers regular.
        Ovarium deeply lobed. Style lateral .. - 92. STACKHOUSEÆ.
        Ovarium undivided. Style terminal.
          Disk not developed.
            Ovula indefinite.
              Stamens all fertile. Petals concrete. } 145. FOUQUIERACEÆ.
              (Succulent). - - -
              Stamens alternately barren - - 146. GALACINEÆ.
            Ovula definite - - - 149. NITRARIACEÆ.
          Disk developed.
            Disk glandular - - - 69. CHAILLETIACEÆ.
          Disk annular.
            Stamens equal in number to the petals.
              Stamens opposite the petals - 96. RHAMNEÆ.
              Stamens alternate with the petals.

Leaves simple, without stipulæ   93. CELASTRINEÆ.
Leaves compound, with stipulæ   97. STAPHYLEACEÆ.
Stamens some multiple of the num-
   ber of the petals.
    Ovula in pairs - - 112. BURSERACEÆ.
    Ovula solitary - - 109. SPONDIACEÆ.
  Flowers irregular - - - - 127. VOCHYACEÆ (*bis*).
¶¶ Ovarium inferior.
Ovarium 1-celled, with parietal placentæ.
  Stamens partly sterile. Petals and sepals dissimilar - 51. LOASEÆ.
  Stamens all fertile.
    Petals and sepals alike.
      Vegetation normal - - - - 70. HOMALINEÆ.
      Vegetation succulent - - - 46. CACTI.
    Petals and sepals different - - - 45. GROSSULACEÆ.
Ovarium with several cells, and the placentæ in the axis ; or,
  if with only one cell, then with the ovula not parietal, but
  erect or pendulous.
    Sepals with a spur - - - - 127. VOCHYACEÆ.
    Sepals without a spur.
      Leaves with pellucid dots (opposite and entire) - 56. MYRTACEÆ.
      Leaves without pellucid dots.
        Embryo lying on the outside of (mealy) albumen 148. FICOIDEÆ.
        Embryo in the axis of the seed.
          Anthers inflexed in æstivation (long).
            Leaves 1-ribbed. Cotyledons convolute. ⎱ 55. MEMECYLEÆ.
            Seeds few - - - ⎰
            Leaves 3- or more ribbed. Cotyledons flat. ⎱ 54. MELASTOMACEÆ.
            Seeds numerous - - - ⎰
          Anthers not inflexed in æstivation (roundish).
            Ovula indefinite.
              Stamens indefinite.
                Seeds without albumen - - 33. LECYTHIDEÆ.
                Seeds with albumen - - 43. PHILADELPHEÆ.
              Stamens definite.
                Divisions of the calyx 5 (rarely 4) 44. ESCALLONIEÆ.
                Divisions of the calyx 4 - 47. ONAGRARIÆ.
            Ovula definite.
              Ovula erect - - - 49. CIRCÆACEÆ.
              Ovula pendulous.
                Stamens equal to the sepals, or
                  fewer.
                  Albumen wanting. (Cotyle- ⎱ 50. HYDROCARYES.
                  dons unequal) - - ⎰
                  Embryo in the axis of albumen.
                    Sepals depauperated, with ⎱ 48. HALORAGEÆ.
                    an open æstivation - ⎰
                    Sepals imbricated. Ovarium ⎱ 41. BRUNIACEÆ.
                    half superior - .. ⎰
                  Embryo minute in the base of
                  albumen.
                    Cells of ovarium 2 - 2. UMBELLIFERÆ.
                    Cells of ovarium more ⎱ 1. ARALIACEÆ.
                    than 2 - - ⎰
                Stamens some multiple of the sepals.
                Stipulæ present.
                  Leaves alternate. (Stipulæ ⎱ 42. HAMAMELIDEÆ.
                    deciduous) - ⎰
                    Leaves opposite. (Stipulæ in- ⎱ 53. RHIZOPHOREÆ.
                    terpetiolar - - ⎰
                Stipulæ absent.
                  Cotyledons convolute. (Petals ⎱ 57. COMBRETACEÆ.
                    oblong) - - ⎰
                  Cotyledons flat. (Petals linear) 58. ALANGIEÆ.

** APETALÆ.
  Petals usually absent.

¶ Ovula indefinite.
  Ovarium with several cells.

Æstivation of the calyx valvate - - - 62. ARISTOLOCHIÆ.
Æstivation of calyx imbricate.
  Flowers regular. Leaves exstipulate. Ovarium superior - - - - - } 138. NEPENTHEÆ.
  Flowers irregular. Leaves with large membranous stipulæ - - - - - } 157. BEGONIACEÆ.
Ovarium with 1 cell, and parietal placentæ.
  Fruit indehiscent - - - - 63. CYTINEÆ.
  Fruit dehiscent.
    Flowers unisexual or deformed.
      Embryo straight - - - 90. DATISCEÆ.
      Embryo reniform - - - 89. RESEDACEÆ.
    Flowers hermaphrodite.
      Stamens perigynous. Leaves dotted - - 71. SAMYDEÆ.
      Stamens hypogynous, unilateral - - 161. LACISTEMEÆ.
¶ ¶ Ovula definite.
Their point of attachment at or near the apex of the cell.
  Valves of the anthers curling upwards - 21. LAURINEÆ.
  Valves of the anthers bursting longitudinally.
    Ovaria several, distinct in each calyx - - 19. MONIMIEÆ.
    Ovaria single, sometimes lobed or spiked, unisexual. Oyula two or more in each cell.
      Flowers amentaceous.
        Ovarium inferior. Albumen 0 - - 82. CUPULIFERÆ.
        Ovarium superior. Albumen fleshy - 81. STILAGINEÆ.
      Flowers collected upon a fleshy receptacle. Ovula always single in each cell - } 80. ARTOCARPEÆ.
      Flowers (solitary), with loose inflorescence.
        Ovarium 4-celled - - - 61. PENÆACEÆ (bis).
        Ovarium 2-celled, indehiscent - - 79. ULMACEÆ.
        Ovarium 3- or many-celled - - 88. EUPHORBIACEÆ.
        Ovarium 1-celled.
          Calyx many-parted - - 165. CERATOPHYLLEÆ.
          Calyx tubular.
            Calyx superior - - - 64. SANTALACEÆ.
            Calyx inferior.
              Fruit 2-valved - - 67. AQUILARINEÆ.
              Fruit indehiscent.
                Leaves with stipulæ - 72. SANGUISORBEÆ.
                Leaves without stipulæ.
                  Flowers naked - - 65. THYMELÆÆ.
                  Flowers in an involucellum - } 66. HERNANDIEÆ.
Their point of attachment at or near the base of the cell.
  Valves of the anthers curling upwards - - 20. ATHEROSPERMEÆ.
  Valves of the anthers bursting longitudinally.
    Calyx superior - - - - 87. JUGLANDEÆ.
    Calyx inferior.
      Stamens combined in a cylinder - - 14. MYRISTICEÆ.
      Stamens distinct.
        Embryo a homogeneous solid mass - 61. PENÆACEÆ (bis).
        Embryo with distinct radicle and cotyledons.
          Radicle at the end remote from the hilum.
             Stipulæ distinct - - - 78. URTICEÆ.
             Stipulæ ochreate - - - 156. POLYGONEÆ.
          Radicle next the hilum.
             Stamens hypogynous - - 158. NYCTAGINEÆ.
             Stamens perigynous.
               Calyx tubular.
                 Embryo curved round albumen 152. SCLERANTHEÆ.
                 Embryo straight.
                    Stamens opposite the sepals 60. PROTEACEÆ.
                    Stamens alternate with the sepals - - } 59. ELÆAGNEÆ.
             Calyx of several leaves, or deeply divided.
               Embryo without albumen - 155. PETIVERACEÆ.
               Embryo curved round albumen.

Stamens opposite the sepals.
 Albumen mealy - 154. PHYTOLACCEÆ.
Stamens alternate with the
 sepals.
  Calyx scarious, bracteo- } 151. AMARANTACEÆ.
   late - -
  Calyx herbaceous, ebrac- } 153. CHENOPODEÆ.
   teate -
  Embryo in the axis of fleshy albumen 91. EMPETREÆ.

*** ACHLAMYDEÆ.
Calyx and corolla both absent, at least in the female flowers.

Ovarium 2- or more-celled ; or if 1-celled, with 2 placentæ.
 Seeds indefinite.
  Flowers solitary - - - - 163. PODOSTEMEÆ.
  Flowers amentaceous - - - 84. SALICINEÆ.
 Seeds definite.
  Seeds pendulous - - - - 83. BETULINEÆ.
  Seeds peltate - - - 164. CALLITRICHINEÆ.
  Seeds ascending - - - 159. SAURUREÆ.
Ovarium 1-celled, with but 1 placenta.
 Ovules pendulous.
  Leaves opposite. Flowers spiked - - 160. CHLORANTHEÆ.
  Leaves alternate. Flowers amentaceous - 85. PLATANEÆ.
 Ovules erect.
  Embryo naked. Flowers amentaceous - - 86. MYRICEÆ.
  Embryo enclosed in a sac - - - 162. PIPERACEÆ.

**** MONOPETALÆ.
Petals cohering in a tube.

¶ Ovarium more or less inferior.
Ovarium with parietal placentæ.
 Placentæ 2. Corolla irregular. Albumen - - 209. GESNEREÆ.
 Placentæ 3. Corolla regular. Albumen 0 - - 181. CUCURBITACEÆ.
Ovarium with the placentæ either in the axis, or at the apex,
 or the base.
 Flowers gynandrous - - - - 177. STYLIDEÆ.
 Flowers not gynandrous.
  Stigma with an indusium.
   Seeds indefinite - - - - 176. GOODENOVIÆ.
   Seeds definite - - - - 178. SCÆVOLEÆ.
  Stigma naked.
   Ovarium 1-celled, with a definite number of ovules.
    Ovules erect. Anthers connate - - 186. COMPOSITÆ.
    Ovules pendulous.
     Stamens alternate with the lobes of the corolla.
      Anthers partly connate. Filaments mona- } 187. CALYCEREÆ.
       delphous - - -
      Anthers distinct.
       Seeds with albumen - - 184. DIPSACEÆ.
       Seeds without albumen - - 185. VALERIANEÆ.
     Stamens opposite the lobes of the corolla - 192. LORANTHEÆ.
Ovarium 2- or more-celled ; or 1-celled, with in-
 definite ovules.
 Leaves opposite.
  With stipulæ - - - 190. CINCHONACEÆ.
  Without stipulæ.
   Seeds definite.
    Radicle inferior - - - 189. STELLATÆ.
    Radicle superior - - 191. CAPRIFOLIACEÆ.
   Seeds indefinite - - - 203. COLUMELLIACEÆ.
 Leaves alternate.
  Ovules definite - - - - 167. STYRACEÆ.
  Ovules indefinite.
   Corolla plaited, many-lobed - - 168. BELVISIACEÆ.
   Corolla with not more than 5 lobes.
    Flowers irregular - - - 175. LOBELIACEÆ.
    Flowers regular.

Fruit capsular - - - - 174. CAMPANULACEÆ.
Fruit succulent - - 172. VACCINIEÆ.
¶ ¶ Ovarium superior.
☞ Flowers regular.
Ovarium deeply 4-lobed - - - - 222. BORAGINEÆ.
Ovaria 2, cohering by their stigma - - 196. APOCYNEÆ (bis).
Ovarium entire.
Ovarium 1-celled, without incomplete dissepiments.
Placentæ 5, parietal - - - - 180. PAPAYACEÆ.
Placenta free, central, single.
Fruit indehiscent - - - - 206. MYRSINEÆ.
Fruit dehiscent - - - - 207. PRIMULACEÆ.
Placentæ 2, parietal, or at the bottom of the cavity of
the ovarium.
Stigma with an indusium - - - 179. BRUNONIACEÆ.
Stigma naked.
Ovulum solitary, pendulous from the tip of an } 183. PLUMBAGINEÆ.
umbilical cord - - -
Ovula several, attached to two placentæ - 226. HYDROPHYLLEÆ.
Ovarium 2- or more-celled ; or, if 1-celled, with incomplete
dissepiments.
Ovula definite.
Anthers 1-celled - - - - 171. EPACRIDEÆ (bis).
Anthers 2-celled.
Stamens 2.
Seeds pendulous - - - 205. OLEACEÆ.
Seeds erect - - - - 204. JASMINEÆ.
Stamens 4 ; corolla scarious - - - 182. PLANTAGINEÆ
Stamens 3, or 5, or more. (bis).
Seeds peltate - - - - 194. LOGANIACEÆ
Seeds pendulous. (bis).
Seeds without albumen.
Cotyledons plano-convex - 223. HELIOTROPICEÆ.
Cotyledons plaited longitudinally - 225. CORDIACEÆ.
Seeds with albumen.
Calyx and corolla, 5-lobed - 224. EHRETIACEÆ.
Calyx and corolla, 3-6-lobed.
Stamens some multiple of the lobes } 202. EBENACEÆ.
of the corolla - -
Stamens equal in number to the } 166. ILICINEÆ.
lobes of the corolla -
Seeds erect or ascending.
Corolla imbricated in æstivation. Cotyle-
dons plano-convex.
Seed-coat bony, with a long scar on } 169. SAPOTEÆ.
one side - -
Seed-coat membranous - (bis) 200. POLEMONIACEÆ.
Corolla plaited in æstivation. Cotyle- } 199. CONVOLVULACEÆ.
dons shrivelled - - -
Ovula indefinite.
Æstivation contorted.
Corolla not agreeing in the number of its divisions } 193. POTALIACEÆ.
with the calyx. Seeds peltate, sessile -
Corolla agreeing with the calyx in the number
of its divisions. Seeds attached to the placenta
by a little cord.
Pollen waxy. Stigma greatly dilated - 195. ASCLEPIADEÆ.
Pollen powdery. Stigma simple - 196. APOCYNEÆ.
Æstivation imbricated, plaited, or valvate.
Styles several - - - - 201. HYDROLEACEÆ.
Style 1.
Anthers 1-celled - - - - 171. EPACRIDEÆ.
Anthers 2-celled.
Cells of the anther hard and dry, with
appendages.
Seeds apterous. Embryo in the axis } 170. ERICEÆ.
of albumen. (Shrubs.)
Seeds winged. Embryo minute, at } 173. PYROLACEÆ.
the base of albumen. (Herbs.)

Cells of the anther succulent, without
appendages.
　　Ovarium 3-celled　　　-　　　- 200. POLEMONIACEÆ.
　　Ovarium 2- or 4-celled.
　　　　Filaments flaccid.　Pericarp mem- } 182. PLANTAGINEÆ.
　　　　　branous, dehiscing transversely　 }
　　　　Filaments rigid.　Pericarp hard
　　　　　or fleshy.
　　　　　Leaves alternate　　-　　　- 213. SOLANEÆ.
　　　　　Leaves opposite.
　　　　　　Æstivation valvate　-　　- 198. SPIGELIACEÆ.
　　　　　　Æstivation　imbricate　or
　　　　　　　convolute.
　　　　　　　Stipulæ　between　the } 194. LOGANIACEÆ.
　　　　　　　　petioles　　-　　　 }
　　　　　　　Stipulæ absent　-　　- 197. GENTIANEÆ.
☞ Flowers irregular.
Ovarium deeply lobed　　-　　-　　-　　- 221. LABIATÆ.
Ovarium entire.
　　Fruit indehiscent, or not opening by valves.
　　　Fruit 1-celled　　-　　-　　-　　- 188. GLOBULARINEÆ.
　　　Fruit 2- or 4-celled ; the cells all normal.
　　　　Radicle inferior　-　　-　　-　　- 220. VERBENACEÆ.
　　　　Radicle superior.
　　　　　Ovules erect　　-　　-　　- 219. SELAGINEÆ.
　　　　　Ovules pendulous　　-　　-　　- 218. MYOPORINEÆ.
　　Fruit with several cells, all of which beyond 2 are } 215. PEDALINEÆ (bis).
　　　spurious　　-　　-　　-　　-　　- }
　　Fruit dehiscent.
　　　Ovarium 1-celled, with a central placenta　-　　- 208. LENTIBULARIÆ.
　　　Ovarium 2-celled, or 1-celled, with two opposite pa-
　　　　rietal placentæ.
　　　　Albumen none.
　　　　　Seeds attached to rigid hooked processes　- 214. ACANTHACEÆ.
　　　　　Seeds adhering immediately to the placentæ.
　　　　　　Seeds winged　　-　　-　　- 217. BIGNONIACEÆ.
　　　　　　Seeds apterous.
　　　　　　　Fruit siliquose,　1-celled,　or spuriously } 216. CYRTANDRACEÆ.
　　　　　　　　2-celled　　-　　-　　-　　- }
　　　　　　　Fruit woody, short, spuriously 4- or 6- } 215. PEDALINEÆ.
　　　　　　　　celled　　-　　-　　-　　- }
　　　　Albumen present.
　　　　　Radicle pointing to the hilum.
　　　　　　Ovarium 2-celled　-　　-　　- 211. SCROPHULARINEÆ.
　　　　　　Ovarium with more cells than 2　-　- 170. ERICEÆ (bis).
　　　　　Radicle pointing to the extremity of the seed
　　　　　　which is most remote from the hilum.
　　　　　　Embryo in the axis.　Ovarium 2-celled 212. RHINANTHACEÆ.
　　　　　　Embryo minute in the apex.　Ovarium } 210. OROBANCHEÆ.
　　　　　　　1-celled　　-　　-　　- }

TRIBE II.—GYMNOSPERMIÆ.

Seeds destitute of a pericarpium.

Resinous.　Leaves simple, Trunk branched　　-　　- 228. CONIFERÆ.
Mucilaginous.　Leaves pinnated, Trunk unbranched　-　　- 227. CYCADEÆ.

SUB-CLASS II.—ENDOGENÆ, or MONOCOTYLEDONOUS PLANTS.

Leaves with parallel veins. Stem with no distinction of wood, bark, and pith. Flowers with a ternary division. Cotyledon 1 *; or, if* 2, *alternate.*

TRIBE I.—PETALOIDEÆ.

Calyx and corolla both developed, in 3 or 6 divisions ; or, if absent, then the sexual apparatus naked.

* TRIPETALOIDEÆ.
Calyx herbaceous.   Corolla petaloid.

Ovarium superior.
  Placentæ covering the whole lining of the carpella   - 230. BUTOMEÆ.
  Placentæ occupying the inner suture of the carpella.
    Carpella several, distinct   - - - 229. ALISMACEÆ.
    Carpella concrete.
      Capsule 3-celled, 3-valved   - - - 232. COMMELINEÆ.
      Capsule 1-celled, with parietal placentæ. (Flowers } 233. XYRIDEÆ.
        capitate)   - - -
Ovarium inferior.
  Embryo exalbuminous. (Water plants.)   - - 231. HYDROCHARIDEÆ.
  Embryo albuminous.
    Stamens 6   - - - - 234. BROMELIACEÆ.
    Stamen 1.
      Anther 2-celled, terminal   - - - 241. SCITAMINEÆ.
      Anther 1-celled, lateral   - - - 242. MARANTACEÆ.

** HEXAPETALOIDEÆ.
Calyx and corolla nearly equal in size, and uniform in colour ; both fully developed and petaloid ; (the number of divisions usually 3 or 6).

Ovarium inferior.
  Stamens and style concrete - - - - 240. ORCHIDEÆ.
  Stamens and style distinct.
    Stamens 3, opposite the sepals.
      Anthers turned outwards, bursting lengthwise - 239. IRIDEÆ.
      Anthers turned inwards, bursting transversely - 236. BURMANNIEÆ.
    Stamens 5-6, or more ; or if 3, opposite the petals.
      Flowers hermaphrodite.
        Veins of the leaves diverging from the midrib } 243. MUSACEÆ.
          towards the margin - -
        Veins of the leaves parallel with the midrib.
          Perianthium deeply parted, the sepals equitant
            with respect to the petals.
            Seeds rostellate, with a hard black coat. } 235. HYPOXIDEÆ.
              Flowers regular - -
            Seeds with a membranous, or soft spongy } 238. AMARYLLIDEÆ.
              coat. Flowers more or less irregular
          Perianthium tubular, the sepals not equitant   237. HÆMODORACEÆ.
        Flowers unisexual. Perianthium short, spreading   250. DIOSCOREÆ.
  Ovarium superior.
    Anthers turned outwards - - - - 245. MELANTHACEÆ.
    Anthers turned inwards.
      Perianthium irregular, involute after flowering - 246. PONTEDEREÆ.
      Perianthium regular.
        Fruit drupaceous, or fibrous. Albumen cartila- }
          ginous, or fleshy. Embryo included, remote from } 252. PALMÆ.
          the hilum. Leaves divided - -
        Fruit capsular, or succulent. Embryo next the
          hilum. Leaves undivided.

Perianthium subglumaceous. Testa pale and soft. Style 1. - - - } 244. JUNCEÆ.

Perianthium coloured. Testa black and brittle. Style 1.

 Flowers from the axillæ of solitary bracteæ 247. ASPHODELEÆ.
 Flowers surrounded by petaloid bracteæ - 248. GILLIESIEÆ.

Perianthium dilated and coloured. Testa soft or spongy. Style 1. - - } 251. LILIACEÆ.

Styles 3 or 1, trifid. Testa membranous. Leaves broad. Stem often twining or branching } 249. SMILACEÆ.

Fruit capsular. Embryo external, remote from the hilum. Flowers glumaceous, capitate - } 253. RESTIACEÆ.

*** SPADICEÆ.

Calyx and corolla absent, or imperfectly developed in the form of herbaceous scales, which are equal in size, and uniform in colour: (the number of scales usually 2 or 4).

Ovarium inferior - - - • - 257. BALANOPHOREÆ.
Ovarium superior.
 Flowers on a spadix.
  Fruit consisting of fibrous drupes, collected in parcels into many-celled pericarpia - - } 254. PANDANEÆ.
  Fruit simple, succulent or dry.
   Spadix in a spatha. Anthers subsessile, cordate. Segments of the perianthium sessile - } 256. AROIDEÆ.
   Spadix naked, or nearly so. Anthers cuneate. Filaments long, lax. Segments of perianthium in the male flowers unguiculate - - - } 255. TYPHACEÆ.
 Flowers on a rachis, or solitary.
  Leafy and caulescent.
   Ovules pendulous - - - - 258. FLUVIALES.
   Ovules erect - - - • - 259. JUNCAGINEÆ.
  Leafless and stemless - - - - 260. PISTIACEÆ.

TRIBE II.—GLUMACEÆ.

Flowers destitute of true calyx and corolla, but enveloped in imbricated bracteæ.

Leafsheaths entire. Embryo undivided, included within the albumen. Stem angular - - - } 262. CYPERACEÆ.
Leafsheaths slit. Embryo lenticular, on the outside of the albumen, with a naked plumula. Stem cylindrical } 261. GRAMINEÆ.

CLASS II.—CELLULARES.

Neither sexes, flowers, nor spiral vessels.

* FILICOIDEÆ.
 A distinct axis and vascular system.

Reproductive organs in terminal cones - - - 263. EQUISETACEÆ.
Reproductive organs dorsal, in thecæ or naked - - 264. FILICES.
Reproductive organs in axillary thecæ - - - 265. LYCOPODIACEÆ.
Reproductive organs in thecæ enclosed within indehiscent involucra - - - - - } 266. MARSILEACEÆ.

** MUSCOIDEÆ.
A distinct axis, but no vascular system.

Theca closed by an operculum -   -   -   - 267. MUSCI.
Theca dehiscing without an operculum   -   -   - 268. HEPATICÆ.
Theca indehiscent, deciduous. Branches leafless and verticillate 269. CHARACEÆ.

*** APHYLLÆ.
Neither distinct axis nor vascular system.

Aerial ; always growing exposed to the air.
  Sporules lying in superficial receptacles   -   - 270. LICHENES.
  Sporules internal   -   -   -   -   - 271. FUNGI.
Aquatic : always growing under water   -   - 272. ALGÆ.

THE

NATURAL ORDERS OF PLANTS.

CLASS I. VASCULARES, OR FLOWERING PLANTS.

COTYLEDONEÆ, *Juss. Gen.* p. 70. (1789.)— EMBRYONATÆ, *Richard. Anal.* p. 50. (1808.)
— VASCULARES, *Dec. Fl. Fr.* 1. 68. (1815); *Lindl. Synops.* p. 3. (1829.)— PHA-
NEROGAMOUS *or* PHÆNOGAMOUS PLANTS *of authors.*

ESSENTIAL CHARACTER. — Substance of the plant composed of cellular tissue, woody
fibre, ducts, and spiral vessels. *Leaves* composed of parenchyma, and of veins consisting
of woody fibre and spiral vessels. *Cuticle* with stomata. *Flowers* consisting of floral envelopes, stamens, and pistilla. *Seeds* distinctly attached to a placenta, covered with a testa,
and containing an embryo with one or more cotyledons ; germinating at two fixed points,
the plumula and radicle.

The presence of flowers, of spiral vessels, and of cuticular stomata, will at
all times distinguish these from Cellulares, or flowerless plants, in which ducts
sometimes exist, but which never have spiral vessels. Vasculares approach
Cellulares by Podostemeæ, some of which resemble Azolla in habit, by Fluviales, which are near Algæ, especially by Coniferæ and Cycadeæ, which are
closely akin to Lycopodiaceæ and Filices, and also by Casuarina, which must,
in any natural ordination, stand near Equisetaceæ. Besides the more obvious
points of difference just adverted to, Vasculares differ from Cellulares in their
embryo ; not, however, in the number of the cotyledons, as is generally supposed in consequence of the common names of Dicotyledones, Monocotyledones, and Acotyledones, but in the germination of the seeds of the two former
always taking place from two fixed points, and in the latter from no fixed point.

Vasculares are divided into the sub-classes *Exogenæ* or *Dicotyledonous,*
and *Endogenæ* or *Monocotyledonous* plants.

SUB-CLASS I. EXOGENÆ, OR DICOTYLEDONS.

DICOTYLEDONES, *Juss. Gen.* 70. (1789); *Desf. Mem. Inst.* 1. 478. (1796.)— EXOR-
HIZEÆ *and* SYNORHIZEÆ, *Rich. Anal.* (1808.)— DICOTYLEDONEÆ *or* EXO-
GENÆ, *Dec. Theor.* p. 209. (1813.)—PHANEROCOTYLEDONEÆ *or* SEMINIFERÆ,
Agardh. Aph. 74. (1821.)

ESSENTIAL CHARACTER. — *Trunk* more or less conical, consisting of three parts, one
within the other ; viz. bark, wood, and pith, of which the wood is enclosed within the two
others ; increasing by an annual deposit of new wood and cortical matter between the
wood and bark. *Leaves* always articulated with the stem, often opposite, their veins
branching and reticulated. *Flowers,* if with a distinct calyx, often having a quinary
division. *Embryo* with two or more opposite cotyledons, which often become green and
leaf-like after germination ; radicle naked, *i.e.* elongating into a root without penetrating
any external case.

Their reticulated leaves, distinctly articulated with the stem, usually distinguish these plants from Endogenæ, from which they are also known by
the following points : Exogenæ have a distinct deposition of pith, wood, and
bark ; Endogenæ have all these confounded : Exogenæ, if trees, are conical
and branched (example, an Oak); Endogenæ are cylindrical and simple-

B

stemmed (example, a Palm). Besides which, the following characters, although far less absolute, deserve attention : Exogenæ in germination protrude their radicle at once; while in Endogenæ it is contained within the substance of the embryo, through which it ultimately bursts : Exogenæ have two or more cotyledons; Endogenæ have but one. Exogenæ approach Endogenæ by Grasses and Asphodeleæ, which branch like themselves, and by Smilaceæ and Aroideæ, which have foliage resembling that of many Exogenæ. The number of divisions of their flower is hardly ever ternary, but usually some multiple of two, or four, or five. In this country the trees and shrubs, and larger herbaceous plants, are nearly all Exogenous; while our native Endogenæ are chiefly confined to grasses, sedges, orchises, bulbs, and submerged water-plants.

Exogenous plants have their seeds either enclosed in a pericarpium (*Angiospermæ*), or naked (*Gymnospermæ*).

TRIBE I. ANGIOSPERMÆ.

These comprehend all Exogenous plants the seeds of which are enclosed within a pod, or shell, or coat proceeding from the ovarium; in short, the whole of that sub-class, with the exception of Cycadeæ and Coniferæ. They are all fecundated through the medium of a stigma and style; while Gymnospermæ, having no stigma or style, have the vivifying influence of the pollen communicated directly to the seed through its foramen. The latter must not be confounded with the naked-seeded plants of Linnæus, which all belong to Angiospermæ, and which are either minute fruits, or divisions of a compound pistillum : they are always known by the presence of a style and stigma.

This tribe is divided into *Polypetalous, Apetalous, Achlamydeous,* and *Monopetalous* plants ; of which the first three may be considered extremely artificial divisions if taken separately, but forming together a tolerably natural whole ; while the Monopetalous division is also, in a great measure, natural. I shall therefore treat of Exogenæ under two heads only.

1. POLYPETALOUS, APETALOUS, AND ACHLAMYDEOUS PLANTS.

Polypetalous plants have both a calyx and corolla ; Apetalous plants have only a calyx, without a corolla ; and Achlamydeous ones have neither : but these distinctions are merely artificial, and even in that point of view very imperfect,—Polypetalous orders constantly containing Apetalous genera, and orders with the strictest natural affinity differing in the absence or presence of floral envelopes. Even Decandolle himself suggests (*Mémoire sur les Combretacées*, p. 2), that it is doubtful whether the division of Monochlamydeæ (which are the same as Apetalæ) is not entirely artificial.

While, therefore, I have availed myself of these differences in framing the diagnoses, and forming the artificial table, I have, in the following detailed account of the orders, thrown the three divisions together, so that the mutual relations of the orders may be obscured as little as possible. In using the artificial tables, if an Apetalous plant cannot be referred to any order of Apetalæ, its place should be sought for among Polypetalæ, to some order of which it will probably be found to be an exception : it is very little likely to belong to Monopetalæ, the Apetalous genera of which are extremely rare. There are no plants of Achlamydeæ with a calyx except some Betulineæ, the flowers of which have a membranous veinless covering, of the nature of a calyx.

These orders pass into Monopetalæ through Caprifoliaceæ, among which Hedera is nearly allied to Araliaceæ, and through Salicariæ which are very near Labiatæ, Meliaceæ which touch upon Styraceæ, and Passifloreæ which stand next to Cucurbitaceæ.

LIST OF THE ORDERS.

1. Araliaceæ.
2. Umbelliferæ.
3. Ranunculaceæ.
4. Papaveraceæ.
5. Nymphæaceæ.
6. Nelumboneæ.
7. Hydropeltideæ.
8. Podophylleæ.
9. Cruciferæ.
10. Fumariaceæ.
11. Capparideæ.
12. Flacourtianeæ.
13. Anonaceæ.
14. Myristiceæ.
15. Magnoliaceæ.
16. Dilleniaceæ.
17. Wintereæ.
18. Calycantheæ.
19. Monimieæ.
20. Atherospermeæ.
21. Laurineæ.
22. Berberideæ.
23. Menispermeæ.
24. Malvaceæ.
25. Chlenaceæ.
26. Bombaceæ.
27. Sterculiaceæ.
28. Moringeæ.
29. Tiliaceæ.
30. Elæocarpeæ.
31. Dipterocarpeæ.
32. Ternströmiaceæ.
33. Lecythideæ.
34. Guttiferæ.
35. Marcgraaviaceæ.
36. Hypericineæ.
37. Reaumurieæ.
38. Saxifrageæ.
39. Cunoniaceæ.
40. Baueraceæ.
41. Bruniaceæ.
42. Hamamelideæ.
43. Philadelpheæ.
44. Escallonieæ.
45. Grossulaceæ.
46. Cacti.
47. Onagrariæ.
48. Halorageæ.
49. Circæaceæ.
50. Hydrocaryes.
51. Loaseæ.
52. Salicariæ.
53. Rhizophoreæ.
54. Melastomaceæ.
55. Memecyleæ.
56. Myrtaceæ.
57. Combretaceæ.
58. Alangieæ.
59. Elæagneæ.
60. Proteaceæ.
61. Penæaceæ.
62. Aristolochiæ.
63. Cytineæ.
64. Santalaceæ.
65. Thymelææ.
66. Hernandieæ.
67. Aquilarineæ.
68. Olacineæ.
69. Chailletiaceæ.
70. Homalineæ.
71. Samydeæ.
72. Sanguisorbeæ.
73. Rosaceæ.
74. Pomaceæ.
75. Amygdaleæ.
76. Chrysobalaneæ.
77. Leguminosæ.
78. Urticeæ.
79. Ulmaceæ.
80. Artocarpeæ.
81. Stilagineæ.
82. Cupuliferæ.
83. Betulineæ.
84. Salicineæ.
85. Plataneæ.
86. Myriceæ.
87. Juglandeæ.
88. Euphorbiaceæ.
89. Resedaceæ.
90. Datisceæ.
91. Empetreæ.
92. Stackhouseæ.
93. Celastrineæ.
94. Hippocrateaceæ.
95. Brexiaceæ.
96. Rhamneæ.
97. Staphyleaceæ.
98. Hippocastaneæ.
99. Rhizoboleæ.
100. Sapindaceæ.
101. Acerineæ.
102. Erythroxyleæ.
103. Malpighiaceæ.
104. Vites.
105. Meliaceæ.
106. Cedreleæ.
107. Humiriaceæ.
108. Aurantiaceæ.
109. Spondiaceæ.
110. Connaraceæ.
111. Amyrideæ.
112. Burseraceæ.
113. Anacardiaceæ.
114. Xanthoxyleæ.
115. Diosmeæ.
116. Rutaceæ.
117. Coriarieæ.
118. Ochnaceæ.
119. Zygophylleæ.
120. Simarubaceæ.
121. Pittosporeæ.
122. Geraniaceæ.
123. Oxalideæ.
124. Tropæoleæ.
125. Hydrocereæ.
126. Balsamineæ.
127. Vochyaceæ.
128. Tremandreæ.
129. Polygaleæ.
130. Violaceæ.
131. Passifloreæ.
132. Malesherbiaceæ.
133. Turneraceæ.
134. Cistineæ.
135. Bixineæ.
136. Sarracennieæ.
137. Droseraceæ.
138. Nepentheæ.
139. Lineæ.
140. Caryophylleæ.
141. Frankeniaceæ.
142. Tamariscineæ.
143. Elatineæ.
144. Portulaceæ.
145. Fouquieraceæ.
146. Galacineæ.
147. Crassulaceæ.
148. Ficoideæ.
149. Nitrariaceæ.
150. Illecebreæ.
151. Amarantaceæ.
152. Scleranthæ.
153. Chenopodeæ.
154. Phytolacceæ.
155. Petiveraceæ.
156. Polygoneæ.
157. Begoniaceæ.
158. Nyctagineæ.
159. Saurureæ.
160. Chlorantheæ.
161. Lacistemeæ.
162. Piperaceæ.
163. Podostemeæ.
164. Callitrichineæ.
165. Ceratophylleæ.

I. ARALIACEÆ. The Aralia Tribe.

Araliæ, *Juss. Gen.* 217. (1789.) — Araliaceæ, *A. Richard in Dictionnaire Classique d'Histoire Naturelle,* 1. 506. (1822.)

Diagnosis. Polypetalous dicotyledons, with definite perigynous stamens, concrete carpella, an inferior ovarium of several cells, pendulous solitary ovula, leaves sheathing at the base, umbellate flowers, and embryo in the base of fleshy albumen.

Anomalies. None.

Essential Character. — *Calyx* superior, entire or toothed. *Petals* definite, 5 or 6, deciduous, valvate in æstivation. *Stamens* definite, 5 or 6, or 10 or 12, arising from within the border of the calyx, and from without an epigynous disk. *Ovarium* inferior, with more cells than 2; *ovula* solitary, pendulous; *styles* equal in number to the cells; *stigmas* simple. *Fruit* succulent, or dry, consisting of several 1-seeded cells. *Seeds* solitary, pendulous; *albumen* fleshy, having a minute *embryo* at the base, with its radicle pointing to the hilum.—*Trees, shrubs,* or *herbaceous* plants, with, in all respects, the habit of Umbelliferæ.

Affinities. Distinguished from Umbelliferæ solely by their many-celled fruit and more shrubby habit. Connected with Caprifoliaceæ through Hedera.

Geography. China, India, North America, and the Tropics of the New World, are the chief abodes of the species of this small order.

Properties. The Ginseng, which is the root of Panax quinquefolium, is much valued by the Chinese for its beneficial influence upon the nerves, and for other supposed properties. It is, however, discarded from European practice. *Ainslie,* 1. 154. There appears to be no reasonable doubt that the Ginseng has really an invigorating and stimulant power when fresh. The virtues that are ascribed to it by the Chinese, although perhaps imaginary to a great extent, are nevertheless founded upon a knowledge of its good effects; which, after the statements made by Father Jartoux, cannot reasonably be called in question. An aromatic gum resin is exuded by the bark of Aralia umbellifera, and others.

Examples. Aralia, Gastonia, Panax.

II. UMBELLIFERÆ. The Umbelliferous Tribe.

Umbelliferæ, *Juss. Gen.* 218. (1789); *Koch in N. Act. Bonn.* 12. 73. (1824); *Dec. and Duby,* p. 213. (1828); *Lindl. Synops.* 111. (1829); *Dec. Mémoire* (1829.)

Diagnosis. Polypetalous dicotyledons, with five perigynous stamens, concrete carpella, an inferior didymous ovarium with two styles and solitary pendulous ovula, leaves sheathing at the base, umbellate flowers, and a minute embryo in the base of fleshy albumen.

Anomalies. Sometimes there are three carpella.

Essential Character. — *Calyx* superior, either entire or 5-toothed. *Petals* 5, inserted on the outside of a fleshy disk; usually inflexed at the point; æstivation imbricate, rarely valvate. *Stamens* 5, alternate with the petals, incurved in æstivation. *Ovarium* inferior, 2-celled, with solitary pendulous ovula; crowned by a double fleshy disk; *styles* 2, distinct; *stigmata* simple. *Fruit* consisting of 2 carpella, separable from a common axis, to which they adhere by their face (*the commissure*); each carpellum traversed by elevated *ridges,* of which 5 are primary, and 4, alternating with them, secondary; the ridges are separated by *channels,* below which are often placed, in the substance of the pericarp, certain linear receptacles of coloured oily matter, called *vittæ*. *Seed* pendulous, usually adhering inseparably to the pericarpium, rarely loose; *embryo* minute, at the base of abundant horny *albumen; radicle* pointing to the hilum.—*Herbaceous* plants, with fistular furrowed stems. *Leaves* usually divided, sometimes simple, sheathing at the base. *Flowers* in umbels, white, pink, yellow, or blue, generally surrounded by an involucrum.

AFFINITIES. It is unnecessary to insist upon the relation of this order and Araliaceæ, which scarcely differ. With Saxifrageæ it agrees in habit, if Hydrocotyle is compared with Chrysosplenium, and if the sheathing and divided leaves of the two orders are considered. To Geraniaceæ, Decandolle remarks that they are allied, in consequence of the cohesion of the carpella around a woody axis, and of the umbellate flowers which grow opposite the leaves, and also because the affinity of Geraniaceæ to Vites, and of the latter to Araliaceæ, is not to be doubted. To me it appears, that the most certain affinity of Umbelliferæ is with Ranunculaceæ, with which they agree in habit, in properties, in the presence of a large quantity of albumen, of solitary seeds in the carpella, a minute embryo, and distinct styles; and from which they differ in their inferior fruit and definite perigynous stamens, rather than in any thing else of real importance. The arrangement of this order has only within a few years arrived at any very definite state; the characters upon which genera and tribes could be formed were for a long while unsettled: it is, however, now generally admitted, that the number and development of the ribs of the fruit, the presence or absence of reservoirs of oil called vittæ, and the form of the albumen, are the leading peculiarities which require to be attended to. Upon this subject see Koch's *Dissertation*, Lagasca in the *Otiosas Españolas*, and Decandolle's *Mémoire*, — especially the last. I do not give the characters of the sub-orders or tribes, because they are rather to be considered artificial divisions than natural groups.

GEOGRAPHY. Natives chiefly of the northern parts of the northern hemisphere, inhabiting groves, thickets, plains, marshes, and waste places. According to the investigation of M. Decandolle, the following is the proportion of the order found in different parts of the world:—

In the Old World......... 663
In America.................. 159 or In the northern hemisphere 679
In Australia................ 54 In the southern ditto.............. 205
In scattered islands 14

PROPERTIES. The properties of this order require to be considered under two points of view: firstly, those of the vegetation; and, secondly, those of the fructification. The character of the former is, generally speaking, suspicious, and often poisonous in a high degree; as in the case of Hemlock, Fool's Parsley, and others, which are deadly poisons. Nevertheless, the stems of the Celery, the leaves of Parsley and Samphire, the roots of the Skirret, the Carrot, the Parsnep, and the tubers of Œnanthe pimpinelloides and Bunium bulbocastanum, are wholesome articles of food. The fruit, vulgarly called the seeds, is in no case dangerous, and is usually a warm and agreeable aromatic, as Caraway, Coriander, Dill, Anise, &c. From the stem, when wounded, sometimes flows a stimulant, tonic, aromatic, gum-resinous concretion, of much use in medicine; as Opoponax, which is procured from Pastinaca opoponax in the Levant, and Assafœtida from the Ferula of that name in Persia. Gum ammoniac is supposed to be obtained from Heracleum gummiferum. It is a gum resin of a pale yellow colour, having a faint but not unpleasant odour, with a bitter, nauseous taste. Internally applied, it is a valuable deobstruent and expectorant. It is said by Dr. Paris to be, in combination with rhubarb, a useful medicine in mesenteric affections, by correcting viscid secretions. *Ainslie*, 1. 160. The substance called Galbanum is produced by some plant of this order, which is supposed to be what botanists call Bubon Galbanum. It is a stimulant of the intestinal canal and uterus, and is found to allay that nervous irritability which often accompanies hysteria. *Ainslie*, 1. 143. Æthusa

Cynapium has been found by Professor Ficinus, of Dresden, to contain a peculiar alkali, which he calls Cynopia. *Turner*, 654. The fruit of Ligusticum ajawain of *Roxb.* is prescribed in India in diseases of horses and cows. *Ainslie*, 1. 38.

EXAMPLES. Chærophyllum, Pastinaca, Eryngium, Hydrocotyle, &c.

III. RANUNCULACEÆ. THE CROW-FOOT TRIBE.

RANUNCULI, *Juss. Gen.* (1789.)—RANUNCULACEÆ, *Dec. Syst.* 1. 127. (1818); *Prodr.* 1.2. (1824); *Lindl. Synops.* p. 7. (1829.)

DIAGNOSIS. Polypetalous dicotyledons, with hypogynous stamens, anthers bursting by longitudinal slits, several distinct simple carpella, exstipulate leaves sheathing at their base, solid albumen, and seeds without arillus.

ANOMALIES. In Garidella and Nigella the carpella cohere more or less. In Thalictrum, some species of Clematis, and some other genera, there are no petals. Pæonia has a persistent calyx.

ESSENTIAL CHARACTER.—*Sepals* 3-6, hypogynous, deciduous, generally imbricate in æstivation, occasionally valvate or duplicate. *Petals* 5-15, hypogynous, in one or more rows, distinct, sometimes deformed in correspondence with metamorphosis in the stamens. *Stamens* indefinite in number, hypogynous; *anthers* adnate, in the true genera turned outwards. *Pistilla* numerous, seated on a torus, 1-celled or united into a single many-celled pistillum; *ovarium* one or more seeded, the *ovula* adhering to the inner edge; *style* one to each ovarium, short, simple. *Fruit* either consisting of dry nuts or caryopsides, or baccate with one or more seeds, or follicular with one or two valves. *Seeds* albuminous; when solitary, either erect or pendulous. *Embryo* minute. *Albumen* corneous.—*Herbs*, or very rarely *shrubs*. *Leaves* alternate or opposite, generally divided, with the petiole dilated and forming a sheath half clasping the stem. *Hairs*, if any, simple. *Inflorescence* variable.

AFFINITIES. This is an order which has a strong affinity with many others, some of which are widely apart from each other. Its most immediate resemblance is with Dilleniaceæ, Magnoliaceæ, and their allies, to which it approaches in the position, number, and structure of its parts of fructification generally, differing however in an abundance of particulars; as from Dilleniaceæ, in the want of arillus, deciduous calyx, and whole habit; from Magnoliaceæ, in the want of stipulæ, and sensible qualities; from Papaveraceæ and Nymphæaceæ, in the distinct, not concrete, carpella, watery, not milky, fluids, acrid, not narcotic, properties. More distant analogy may be traced with Rosaceæ, with which they agree in their numerous carpella, the number of their floral divisions and indefinite stamens; but differ in those stamens being hypogynous instead of perigynous, in the presence of large albumen surrounding a minute embryo, want of stipulæ and acrid properties. With Umbelliferæ they accord in the last particular, and also in their sheathing leaves, habit, and abundant albumen, with a minute embryo; but those plants differ in their calyx being concrete with the ovarium, and in their stamens being invariably definite; no doubt, however, can be entertained, that in any really natural arrangement Ranunculaceæ and Umbelliferæ should be placed near each other. Another analogy has been indicated by botanists between this order and Alismaceæ, with which it agrees in its numerous ovaria, and in habit; but that order is monocotyledonous. A great peculiarity of Ranunculaceæ consists in the strong tendency exhibited by many of the genera to produce their sepals, petals, and stamens, in a state different from that of other plants; as, for example, in Delphinium, Aquilegia, and Aconitum, in which they are furnished with

7

a spur, and in Ranunculus itself, which has a nectariferous gland at the base of the petals. An instance is described of the polypetalous regular corolla of Clematis viticella being changed into a monopetalous irregular one, like that of Labiatæ. *Nov. Act. Acad. N. C.* 14. p. 642. t. 37.

GEOGRAPHY. The largest proportion of this order is found in Europe, which contains more than 1-5th of the whole; North America possesses about 1-7th, India 1-25th, South America 1-17th; very few are found in Africa, except upon the shores of the Mediterranean: eighteen species have, according to Decandolle, been discovered in New Holland. They characterise a cold damp climate, and are, when met with in the Tropics, found inhabiting the sides and summits of lofty mountains: in the lowland of hot countries they are almost unknown.

PROPERTIES. Acridity, causticity, and poison, are the general characters of this suspicious order, which, however, contains species in which those qualities are so little developed as to be innoxious. The caustic principle is, according to Krapfen, as cited by Decandolle, of a very singular nature; it is so volatile that, in most cases, simple drying, infusion in water, or boiling, are sufficient to dissipate it: it is neither acid nor alkaline: it is increased by acids, sugar, honey, wine, spirit, &c. and is only effectually destroyed by water. The leaves of Knowltonia vesicatoria are used as vesicatories in Southern Africa. Ranunculus glacialis is a powerful sudorific; Aconitum Napellus and Cammarum are diuretic. The Hepatica, Actæa racemosa, and Delphinium consolida, are regarded as simple astringents. *Dec.* The roots of several Hellebores are drastic purgatives; those of the perennial Adonises are, according to Pallas, emmenagogues; and those of several Aconitums, especially Napellus and Cammarum, are acrid in a high degree. *Ibid.* The root of the Aconitum of India, one of the substances called Bikh, or Bish, is a most virulent poison. *Trans. Med. and Phil. Soc. Calc.* 2. 407. Authors are, however, not well agreed what the precise plant is which produces this Bikh, although all agree in referring it to Ranunculaceæ. In India, it seems there are three principal kinds of Bish, varying from each other in their properties, but all belonging to a genus which Dr. Hamilton refers to Caltha. According to this author, the Bishma, or Bikhma, is a strong bitter, very powerful in the cure of fevers: the Bish, Bikh, or Kodoya Bikh, has a root possessing poisonous properties of the most dreadful kind, whether taken into the stomach, or applied to wounds: the Nir Bishi, or Nirbikhi, has no deleterious properties, but is used in medicine. *Brewster,* 1. 250. For some important information on this Bikh, Vish, Visha, or Ativisha, which Dr. Wallich considers his Aconitum ferox, see *Plant. As. Rar.* vol. 1. p. 33. tab. 41. The root of Pæony is acrid and bitter, but is said to possess antispasmodic properties. Ranunculus flammula and sceleratus are powerful epispastics, and are used as such in the Hebrides, producing a blister in about an hour and a half. Their action is, however, too violent, and the blisters are difficult to heal, being apt to pass into irritable ulcers. *Ed. Ph. J.* 6. 156. Beggars use them for the purpose of forming artificial ulcers, and also the leaves of Clematis recta and flammula. From the seeds of Delphinium staphysagria, the chemical principle called Delphine was procured by MM. Lassaigne and Fenuelle; it exists in union with oxalic acid. *Ibid.* 3. 305. The root of Hydrastis canadensis has a strong and somewhat narcotic smell, and is exceedingly bitter; it is used in North America as a tonic, under the name of *Yellow root. Barton,* 2. 21. The root of Coptis trifolia, or Gold-thread, is a pure and powerful bitter, devoid of any thing like astringency; it is a popular remedy in the United States for aphthous affections of the mouth in children. *Ibid.* 2. 100. The wood and bark of Xanthorhiza apiifolia are a very pure tonic bitter. The shrub contains both a gum and resin, each of which is intensely bitter. *Ibid.* 2. 205. The seeds of Nigella sativa were formerly employed instead of pepper; those of Delphinium Staphisagria are vermifugal and caustic; those of Aquilegia are simply tonic. *Dec.*

M. Decandolle makes the following divisions in this order:—

I. TRUE RANUNCULACEÆ.

Anthers bursting outwardly.

§ 1. CLEMATIDEÆ.

Dec. Syst. 1. 131. (1818); *Prodr.* 1. 2. (1824.)
Æstivation of the calyx valvate, or induplicate. *Petals* none, or plane. *Carpella* indehiscent, 1-seeded, terminated by a bearded tail (which is the indurated style). *Seed* pendulous. *Leaves* opposite.
EXAMPLES. Clematis, Naravelia.

§ 2. ANEMONEÆ.

Dec. Syst. 1. 168. (1818); *Prodr.* 1. 10. (1824.)
Æstivation of calyx and corolla imbricated. *Petals* none, or plane. *Carpella* 1-seeded, indehiscent, usually terminated by a tail or point. *Seed* pendulous. *Leaves* radical, or alternate.
EXAMPLES. Anemone, Thalictrum.

§ 3. RANUNCULEÆ.

Dec. Syst. 1. 228. (1818); *Prodr.* 1. 25. (1824.)
Æstivation of calyx and corolla imbricated. *Petals* 2-lipped, or furnished with an interior scale at the base. *Carpella* 1-seeded, dry, indehiscent. *Seed* erect. *Leaves* radical, or alternate.
EXAMPLES. Ranunculus, Myosurus.

§ 4. HELLEBOREÆ.

Dec. Syst. 1. 306. (1818); *Prodr.* 1. 44. (1824.)
Æstivation of calyx and corolla imbricated. *Petals* either none, or irregular, 2-lipped, and nectariferous. *Calyx* petaloid. *Carpella* capsular, dehiscent, many-seeded.
EXAMPLES. Eranthis, Trollius, Aconitum.

II. SPURIOUS RANUNCULACEÆ.

Anthers bursting inwardly.
EXAMPLES. Actæa, Xanthorhiza, Pæonia.

IV. PAPAVERACEÆ. THE POPPY TRIBE.

PAPAVERACEÆ, *Juss. Gen.* 236. (1789) *in part; Dec. Syst.* 2. 67. (1818); *Prodr.* 1. 117. (1824); *Lindl. Synops.* 16. (1829.)

DIAGNOSIS. Polypetalous dicotyledons, with hypogynous stamens, concrete carpella, a 1-celled ovarium, narrow parietal placentæ, 2 sepals, and a regular corolla.

ANOMALIES. Bocconia has no petals, and a monospermous capsule. Hypecoum has the inner petals 3-lobed. Eschscholtzia has perigynous stamens.

ESSENTIAL CHARACTER.— *Sepals* 2, deciduous. *Petals* hypogynous, either 4, or some multiple of that number, placed in a cruciate manner. *Stamens* hypogynous, either 8, or some multiple of 4, generally very numerous, inserted in 4 parcels, one of which adheres to the base of each petal ; *anthers* 2-celled, innate. *Ovarium* solitary ; *style* short, or none ; *stigmas* alternate with the placentæ, 2 or many ; in the latter case stellate upon the flat apex of the ovarium. *Fruit* 1-celled, either pod-shaped, with 2 parietal placentæ, or capsular, with several placentæ. *Seeds* numerous ; *albumen* between fleshy and oily ; *embryo* minute, straight, at the base of the albumen, with plano-convex *cotyledons.*—*Herbaceous plants or shrubs*, with a milky juice. *Leaves* alternate, more or less divided. *Peduncles* long, 1-flowered ; *flowers* never blue.

AFFINITIES. The siliquose-fruited genera, such as Glaucium and Eschscholtzia, indicate the near affinity of this order to Cruciferæ, from which they differ in the want of a dissepiment to the fruit, in the stamens being indefinite, and in the presence of copious albumen. Through Papaver they approach Nymphæaceæ, and through Sanguinaria Podophylleæ, from all which they are distinguished with facility. Their relationship to Fumariaceæ is more obscure, and is only to be understood by considering Cruciferæ to be their connecting link. The anomalies in the order are of little importance, with the exception of Eschscholtzia, which has its stamens arising from the throat of a flatly campanulate calyx, instead of being hypogynous : this plant, however, may, instead of being an exception to the character, be considered as affording a proof that all is not calyx which intervenes between the base of the sepals and the base of the ovarium. I conceive that it would be more natural to understand the apparent base of the calyx of Eschscholtzia as a hollow apex of the peduncle ; but if this be admitted, it will become doubtful whether many supposed tubes of the calyx are not hollowed peduncles also ; as, for example, Calycanthus, Rosa, Scleranthus, Margyricarpus, &c. I have already made some remarks upon this subject in the Introduction, which see. A comparison of the structure of Papaveraceæ and Cruciferæ, by Mirbel, is to be found in the *Ann. des Sc.* 6. 266.

GEOGRAPHY. Europe, in all directions, is the principal seat of Papaveraceæ, almost two-thirds of the whole order being found in it. Two species only are, according to Decandolle, peculiar to Siberia, three to China and Japan, one to the Cape of Good Hope, one to New Holland, and six to Tropical America. Several are found in North America, beyond the tropic ; and it is probable that the order will yet receive many additions from that region. Most of them are annuals. The perennials are chiefly natives of mountainous tracts.

PROPERTIES. Every one knows what narcotic properties are possessed by the poppy, and this character prevails generally in the order. Their seed is universally oily, and in no degree narcotic. The oil obtained from the seeds of Papaver somniferum is found to be perfectly wholesome, and is, in fact, consumed on the continent in considerable quantity. It is also employed extensively for adulterating olive oil. Its use was at one time prohibited in France by decrees issued in compliance with popular clamour ; but it is now openly sold, the government and people having both grown wiser. *See Ed. P. J.* 2. 17. Meconopsis napalensis, a Nipal plant, is described as being extremely poisonous, especially its roots. *Don. Prodr.* 98. The Sanguinaria canadensis, or Puccoon, is emetic and purgative in large doses, and in smaller quantities is stimulant, diaphoretic, and expectorant. *Barton*, 1. 37. The seeds of Argemone mexicana are used in the West Indies as a substitute for ipecacuanha ; and the juice is considered by the native doctors of India as a valuable remedy in ophthalmia, dropt into the eye and over the tarsus ; also as a good application to chancres. It is purgative and deobstruent. *Ainslie*, 2. 43. The Brazilians administer the juice of their Cardo santo, Argemone mexicana, to persons or animals bitten by serpents, but, it would appear, without much success. *Prince Max. Trav.* 214. The narcotic principle of opium is an alkaline substance, called Morphia. The same drug contains a peculiar acid, called the Meconic; and a vegetable alkali, named Narcotine, to which the unpleasant stimulating properties are attributed by Magendie. *Turner*, 6. 47.

Examples. Papaver, Chelidonium, Eschscholtzia.

V. NYMPHÆACEÆ. The Water Lily Tribe.

Nymphæaceæ, *Salisbury, Ann. Bot.* 2. p. 69. (1805); *Dec. Propr. Med. ed.* 2. p. 119. (1816); *Syst.* 2. 39. (1821); *Prodr.* 1. 113. (1824); *Lindl. Synops.* 15. (1829.)

Diagnosis. Polypetalous dicotyledons, with hypogynous stamens, concrete carpella, a many-celled ovarium, and ovula attached to the face of the dissepiments.

Anomalies. None.

Essential Character. — *Sepals* and *petals* numerous, imbricated, passing gradually into each other, the former persistent, the latter inserted upon the disk which surrounds the pistillum. *Stamens* numerous, inserted above the petals into the disk, sometimes forming, with the combined petals, a superior monopetalous corolla; *filaments* petaloid; *anthers* adnate, bursting inwards by a double longitudinal cleft. *Disk* large, fleshy, surrounding the ovarium more or less. *Ovarium* polyspermous, many-celled, with the stigmata radiating from a common centre upon a sort of flat urceolate cap. *Fruit* many-celled, indehiscent. *Seeds* very numerous, attached to spongy dissepiments, and enveloped in a gelatinous arillus. *Albumen* farinaceous. *Embryo* small, on the outside of the base of the albumen, enclosed in a membranous bag; *cotyledons* foliaceous.—*Herbs*, with peltate or cordate fleshy leaves, arising from a prostrate trunk, growing in quiet waters.

Affinities. There exists a great diversity of opinion among botanists as to the real structure of this order, and, consequently, as to its affinities. This has arisen chiefly from the anomalous nature of the embryo, which is not naked, as in most plants, but enclosed in a membranous sac or bag. By some, among whom was the late M. Richard, this sac or bag was considered a cotyledon, analogous to that of grasses, and enveloping the plumula; and hence the order was referred to Endogenæ, or Monocotyledons, and placed in the vicinity of Hydrocharideæ. By others, at the head of whom are Messrs. Mirbel and Decandolle, the sac is considered a membrane of a peculiar kind; and what Richard and his followers denominate plumula, is for them a 2-lobed embryo, wherefore they place the order in Exogenæ, or Dicotyledons. I do not think it worth citing all the arguments that have been adduced on each side the question, as botanists seem now to be generally agreed upon referring Nymphæaceæ to Dicotyledons. I observe, however, that Dr. Von Martius adheres to the opinion that Nymphæaceæ are monocotyledonous, and nearly related to Hydrocharideæ. See *Hortus Regius Monacensis*, p. 25. (1829.) Those who are curious to investigate the subject are referred to M. Decandolle's Memoir, in the first volume of the Transactions of the Physical and Natural History Society of Geneva. In this place it will be sufficient to advert briefly to the proof that is supposed to exist of their being Dicotyledons. In the first place, the structure of the stem is essentially that of Exogenæ. See Mirbel's examination of the anatomy of Nuphar luteum, in the *Annales du Muséum*, vol. 16. p. 20; and of Nelumbium, the close affinity of which with Nymphæaceæ no one can possibly doubt, in the same work, vol. 13. t. 34. In both these plants the bundles of fibres are placed in concentric circles, the youngest of which are outermost; but they all lie among a great quantity of cellular tissue: between each of these circles is interposed a number of air-cells, just as is found in Myriophyllum and Hippuris, both undoubted Dicotyledons in the opinion of every body except Link, who refers the latter to Endogenæ (see *Gewächsk.* 6. p. 288). Secondly, the leaves are those of Dicotyledons, and so is their convolute vernation, which is not known in Monocotyledons, and their insertion and distinct articulation with the stem. Thirdly, the flowers of Nymphæaceæ have so great an analogy generally with Dicotyledons, and particularly with that of Magnoliaceæ, and their fruit with Papaveraceæ, that it is difficult to doubt their belonging to the same class. Fourthly, the reasons which have been offered for considering the embryo monocotyledonous, however plausible they may have appeared while we were unacquainted with the true structure of the ovulum of plants, have no longer the importance that they were formerly supposed to possess. The sac, to which I have already alluded, to which so much unnecessary value has been attached, and which was mistaken for a cotyledon by Richard, is no doubt analogous to the sac of Saururus and Piper, and is nothing more than the remains of the innermost of the membranous coats of the ovulum, usually indeed absorbed, but in this and similar cases remaining and covering over the embryo. Mr. Brown (*Appendix to King's Voyage*) considers it the remains of the membrane of the amnios. M. Decandolle assigns a further reason for considering Nymphæaceæ Dicotyledons, that they are lactescent, a property not known in Monocotyledons. But in this he is mistaken; Limnocharis, a genus belonging to Butomeæ, is lactescent. Independently of the peculiarities to which I have now alluded, this order is remarkable in some other respects. It offers one of the best examples which can be adduced of the gradual passage of petals into stamens, and of sepals into petals : if attentively examined, the transition will be found so gradual that many intermediate bodies will be seen to be neither precisely petals nor stamens, but both in part. The development of the disk, which is so remarkable in Nelumboneæ, takes place here in various degrees. In some, as in Nuphar, it is merely an hypogynous expansion, out of which grow the stamens and petals; in others, as Nymphæa, it elevates itself as high as the top of the ovarium, to the surface of which it is adnate, and as the stamens are carried up along with it, we have these organs apparently proceeding from the surface of the ovarium : in another genus, the *Barclaya* of Dr. Wallich, the petals are also carried up with the stamens, on the outside of which they even cohere into a tube, so that in this genus we have a singular instance of an inferior calyx and a superior corolla in the same plant. Supposing this order to be exogenous and dicotyledonous, a fact about which there appears to me to be no doubt, its immediate affinity will be with Papaveraceæ, with some genera of which it agrees in the very compound nature of the fruit, from the apex of which the sessile stigmas radiate, in the presence of narcotic principles and a milky secretion, and in the great breadth of the placentæ. They are also closely akin to Magnoliaceæ, with which they agree in the imbricated nature of the petals, sepals, and stamens; to Nelumboneæ their close resemblance is evident ; with Ranunculaceæ they are connected through the tribe of Pæonies, with which they agree in the dilated state of the discus; which, in Pæonia papaveracea and Moutan, frequently rises as high as the top of the ovaria, and in the indefinite number of their hypogynous stamens; but in Ranunculaceæ the placentæ only occupy the edge of each of the carpella of which the fruit is made up ; so that in Nigella, in which the carpella cohere in the centre, the seeds are attached to the axis, while in Nymphæaceæ the placentæ occupy the whole surface of each side of the individual carpella of which the fruit is composed. But if such are the undoubted immediate affinities of Nymphæaceæ, it is certain that some strong analogies exist between them and Hydrocharideæ, to the vicinity of which they are referred by those who believe them to be Monocotyledonous. Taking Nelumboneæ for a transition order, they have some relation to Alismaceæ, the only monocotyledonous order in which there is an indefinite number of carpella in each flower, and to Hydrocharideæ, with which they agree in the structure, though not the vernation, of their leaves, and their habit. An analogy of a similar nature with this last may be also traced between them and Menyantheæ.

GEOGRAPHY. Floating plants, inhabiting the whole of the northern hemisphere, occasionally met with at the southern point of Africa, but generally rare in the southern hemisphere, and entirely unknown on the continent of South America.

PROPERTIES. The whole of this order has the reputation of being anti-aphrodisiac, sedative, and narcotic,—properties not very clearly made out, but generally credited. Their stems are certainly bitter and astringent, for which reason they have been prescribed in dysentery. After repeated washings, they are capable of being used for food. *Dec.—A. R.*

EXAMPLES. Nymphæa, Nuphar.

VI. NELUMBONEÆ.

NYMPHÆACEÆ, § Nelumboneæ, *Dec. Syst.* 2. 43. (1821); *Prodr.* 1. 113. (1824.)

DIAGNOSIS. Polypetalous dicotyledons, with hypogynous stamens, distinct simple carpella immersed in a fleshy dilated torus, and floating leaves.
ANOMALIES. None.

ESSENTIAL CHARACTER.—*Sepals* 4 or 5. *Petals* numerous, oblong, in many rows, arising from without the base of the disk. *Stamens* numerous, arising from within the petals, in several rows; *filaments* petaloid; *anthers* adnate, bursting inwards by a double longitudinal cleft. *Disk* fleshy, elevated, excessively enlarged, enclosing in hollows of its substance the *ovaria*, which are numerous, separate, monospermous, with a simple style and stigma. *Nuts* numerous, half buried in the hollows of the disk, in which they are, however, loose. *Seeds* solitary, or rarely 2; *albumen* none; *embryo* large, with two fleshy cotyledons and a highly developed *plumula*, enclosed in its proper membrane.—*Herbs*, with peltate fleshy *leaves* arising from a prostrate trunk, growing in quiet waters.

AFFINITIES. Closely related to Nymphæaceæ, with which they are usually united. They differ entirely in the structure of their fruit, but agree in their foliage and flowers. The order consists of a single genus. See Nymphæaceæ.

GEOGRAPHY. Natives of stagnant or quiet waters in the temperate and tropical regions of the northern hemisphere, both in the Old and the New World; most abundant in the East Indies. They were formerly common in Egypt, but are now extinct in that country, according to Delile.

PROPERTIES. Chiefly remarkable for the beauty of the flowers. The fruit of Nelumbium speciosum is believed to have been the Egyptian bean of Pythagoras. The nuts of all the species are eatable and wholesome. The root, or, more properly, the creeping stem, is used as food in China.

EXAMPLE. Nelumbium.

VII. HYDROPELTIDEÆ.

CABOMBEÆ, *Rich. Anal. Fr.* (1808.) — PODOPHYLLACEÆ, § Hydropeltideæ, *Dec.*
Syst. 2. 36. (1821); *Prodr.* 1. 112. (1824.)

DIAGNOSIS. Polypetalous dicotyledons, with hypogynous stamens, anthers bursting by longitudinal slits, several distinct simple carpella, exstipulate floating leaves not sheathing at the base, solid albumen, and seeds without arillus.

ANOMALIES. None.

ESSENTIAL CHARACTER. — *Sepals* 3 or 4, coloured inside. *Petals* 3 or 4, alternate with the sepals. *Stamens* definite or indefinite, hypogynous, arising from an obscure torus; *anthers* linear, turned inwards, continuous with the filament. *Ovaria* 2 or more, terminated by a short style. *Fruit* indehiscent, tipped by the indurated style. *Seeds* definite, pendulous ; *embryo* fungilliform, seated at the base of firm, somewhat fleshy *albumen.* — *Aquatic* plants, with floating leaves. *Flowers* axillary, solitary, yellow or purple.

AFFINITIES. Their nearest relation is to Nymphæaceæ, from which they are known by their definite seeds and distinct carpella. From Podophylleæ, to which they are united by Decandolle, they differ in their floating habit, definite seeds, and numerous ovaries. In the affinities of both these orders they otherwise partake. According to Richard, Cabomba is a monocotyledon : Hydropeltis is clearly related closely to Caltha.

GEOGRAPHY. American water-plants, found from Cayenne to New Jersey. The whole order consists of but two species.

PROPERTIES. Unknown.

EXAMPLES. Hydropeltis, Cabomba.

VIII. PODOPHYLLEÆ.

PODOPHYLLACEÆ, § Podophylleæ, *Dec. Syst.* 2. 32. (1821); *Prodr.* 1. 111. (1824);
Von Martius H. Reg. Monac. (1829) ; *a sect. of Papaveraceæ.*

DIAGNOSIS. Polypetalous dicotyledons, with hypogynous stamens, anthers bursting by longitudinal slits, a solitary simple carpellum, exstipulate leaves, solid albumen, and seeds without arillus.

ANOMALIES. None.

ESSENTIAL CHARACTER. — *Sepals* 3 or 4, deciduous or persistent. *Petals* in two or three rows, each of which is equal in number to the sepals. *Stamens* hypogynous, 12-18, arranged in two, three, or more rows ; *filaments* filiform ; *anthers* linear or oval, terminal, turned inwards, bursting by a double longitudinal line. *Torus* not enlarged. *Ovarium* solitary ; *stigma* thick, nearly sessile, somewhat peltate. *Fruit* succulent or capsular, 1-celled. *Seeds* indefinite, attached to a lateral placenta, sometimes having an arillus ; *embryo* small, at the base of fleshy albumen.—*Herbaceous* plants. *Leaves* broad, lobed. *Flowers* radical, solitary, white.

AFFINITIES. Very nearly allied to the herbaceous genera of Berberideæ, from which they scarcely differ, except in the dehiscence of their anthers. From Papaveraceæ, to which they have been recently referred by Von Martius, they are known by their watery, not milky, juice, by their solitary unilateral placentæ, and by their fleshy, not oily, albumen. From Ranunculaceæ they are divided, among other characters, by their anthers bursting inwardly ; in which, however, they agree with Decandolle's spurious genera, which that author suspects might be better even referred to Podophylleæ. Hydropeltideæ, which are joined to them by that learned botanist, are here considered a distinct order.

GEOGRAPHY. All inhabitants of the marshes of North America.

PROPERTIES. The root of the May Apple, Podophyllum peltatum, is one of the most safe and active cathartics that is known. *Barton, 2. 14.* Jeffersonia is also purgative. *Decand.*

EXAMPLES. Podophyllum, Jeffersonia.

IX. CRUCIFERÆ. THE CRUCIFEROUS TRIBE.

CRUCIFERÆ, *Juss. Gen.* 237. (1789); *Dec. Mémoire sur les Crucifères (no date)*; *Syst.* 2. 139. (1821); *Prodr.* 1. 131, (1824); *Lindl. Synops.* 20. (1829.)

DIAGNOSIS. Polypetalous dicotyledons, with hypogynous tetradynamous stamens.

ANOMALIES. Schizopetalum has 4 cotyledons ; sometimes the petals are abortive.

ESSENTIAL CHARACTER. — *Sepals* 4, deciduous, cruciate. *Petals* 4, cruciate, alternate with the sepals. *Stamens* 6, of which two are shorter, solitary, and opposite the lateral sepals, occasionally toothed ; and four longer, in pairs, opposite the anterior and posterior sepals ; generally distinct, sometimes connate, or furnished with a tooth on the inside. *Disk* with various green glands between the petals and the stamens and ovarium. *Ovarium* superior, unilocular, with parietal placentæ usually meeting in the middle, and forming a spurious dissepiment. *Stigmata* two, opposite the placentæ. *Fruit* a siliqua or silicula, 1-celled, or spuriously 2-celled ; 1- or many-seeded ; dehiscing by two valves separating from the replum ; or indehiscent. *Seeds* attached in a single row by a funiculus to each side of the placentæ, generally pendulous. *Albumen* none. *Embryo* with the radicle folded upon the cotyledons.—*Herbaceous* plants, annual, biennial, or perennial, very seldom suffruticose. *Leaves* alternate. *Flowers* usually yellow or white, seldom purple.

AFFINITIES. This order is among the most natural that are known, and its character of having what Linnæan botanists call tetradynamous stamens is scarcely subject to exception. It has a near relation to Capparideæ, Papaveraceæ, and Fumariaceæ. With Capparideæ it agrees in the number of the stamens of some species of that order, in the fruit having two placentæ and a similar mode of dehiscence, and in the quaternary number of the divisions of the flower. To Papaveraceæ it approaches in the number of the petals, an unusual number to prevail in dicotyledonous plants, and again in the structure of the fruit of some genera of that order, such as Glaucium and Chelidonium. With the siliquose-fruited Fumariaceæ it has much analogy, and even with the whole of that order in the number of its petals, supposing the common opinion of the nature of the floral envelopes of Fumariaceæ to be correct, or in the binary division of its flower, from which the quaternary is only a slight deviation, upon the hypothesis I have suggested in speaking of that order.

Cruciferæ may be said to be characterised essentially by their deviation from the ordinary symmetry observable in the relative arrangement of the parts of fructification of other plants, — deviations which are of a very interesting nature. Their stamens are arranged thus: two stand opposite each of the anterior and posterior sepals, and one opposite each of the lateral sepals; there being 6 stamens to 4 sepals, instead of either 4 or 8, as would be normal. Now in what way does this arise? is the whorl of stamens to be considered double, one of the series belonging to the sepals, and one to the petals, and, of these, one imperfect? I am not aware of any such explanation having been offered, nor do I know of any better one. It appears to me that the outer series is incomplete, by the constant abortion of the stamens belonging to the anterior and posterior sepals. But it is in their fruit that their great peculiarity consists. I transcribe the following observations upon this subject from the *Botanical Register*, fol. 1168, in which I have entered in some detail into the inquiry.

" It is well known, that in regularly-formed fruits the style or stigma universally and necessarily *alternates* with the placenta, for reasons which it would be superfluous to insist upon in this place. But in Cruciferæ the stigmata are *opposite* to the placentæ, terminating a sort of frame or replum, the two sides of which are often connected by a membranous septum, on the outside of which latter the ovula are arranged in a single row on each side; so that in many of the more highly developed plants of the order there are four placentæ opposed to each other by pairs, and forming the inner edge of each side of the replum, which itself terminates in the stigmas. To this replum is attached on each side a deciduous plate, or valve as it is called, which has no vascular connexion with either the replum, stigmata, or pedicel. In consequence of this singular arrangement of parts, it has been found extremely difficult to understand the exact nature of the Cruciferous pistillum, or to reduce it to the rules which are known to govern the formation of other compound pistilla.

" According to Mr. Brown, and, after him, to M. Decandolle, the pistillum of Cruciferæ is to be understood to consist of two confluent ovaria, united by their placentæ, two lamellæ from each of which project into the cavity of the ovarium, and, meeting in the centre, coalesce and form the septum. This, however, does not remove the difficulty of the stigmata being opposite the placentæ, instead of alternate with them. I am not aware that any explanation of this point has been published by Mr. Brown; but M. Decandolle (*Théorie Elémentaire*, ed. 1. p. 133) accounts for it thus. He assumes that there are several kinds of simple pistilla, some of which are not to be found in an isolated state, but the possible existence of which he conceives to be demonstrated by certain compound pistilla, that cannot be reduced to their simplest state without the admission of such a position. Among these supposititious simple pistilla is one called the *Siliquelle*, ' which is formed originally of three pieces, the two lateral producing ovula on their inner surface, and the outer (intermediate) one bearing no ovula; pistilla of this description make up the fruit of Nymphæaceæ, Papaveraceæ, and Cruciferæ. When two pistilla of this kind are united by the external edge of their lateral pieces, they form those fruits which are said to have intervalvular placentæ; each of these double placentæ is elongated into a style or stigma, simple in appearance, but in reality formed by two half styles grown together.'

" To maintain this theory, it is necessary to assume, in the first place, the existence of a simple pistillum, of a structure not only entirely hypothetical, but opposed to all we know of vegetable organisation; and, in the next place, that the stigmata of the order, although so simple in appearance that no trace whatever of composition can be found in them, are, nevertheless, each composed of two half stigmata in a state of cohesion.

" To us this explanation has always been unsatisfactory. It was difficult to believe that rules of structure, well ascertained to be uniform in other plants, should be deviated from in Cruciferæ, especially when the irregularity observable in the arrangement of other parts of their flower was taken into account. It always appeared more probable, that the anomalous nature of the pistillum depended upon some irregularity corresponding to that of the stamens, than upon peculiar laws appertaining to Cruciferæ alone.

" This seems to be at length proved by Eschscholtzia, the fruit of which is so similar to that of Cruciferæ, that the uniformity of the laws under which they are both formed is not likely to be disputed. In this plant the pistillum is unilocular, with four stigmata, of which the two opposite ones are smaller than the two others. Upon opening this pistillum we find that there are two parietal placentæ corresponding with the smaller stigmata, and that there are no placentæ opposite the larger stigmata; in other words, that it is formed of four simple pistilla, two of which are opposite and ovuliferous, with their placentæ in the usual place, alternating with themselves; and two nearly abortive, destitute of placentæ, consequently not ovuliferous, and so nearly suppressed by the superior energy of their two neighbours, that their existence would have been unknown but for the stigmata which indicate their presence. This is one way of understanding Eschscholtzia; but as the ovula are not inserted in the placentæ in a double row, but rather confusedly arranged in several rows, it may also be assumed that the lateral, imperfect, half-obliterated stigmata have a line of placentæ, with ovula appertaining to themselves, but so confounded with the placentæ of their lateral and more powerful neighbours, that, in consequence of their close approximation, they cannot be distinguished. We, however, incline to the former of these two opinions. Let this be as it may, upon either supposition, the structure of Cruciferous pistilla is, we think, susceptible of explanation. We shall, for convenience, reason upon the former of the two hypotheses.

" If we compare the fruit of Eschscholtzia and Cruciferæ, we shall at first, perhaps, be led to believe that while they have a certain degree of resemblance in some points, they nevertheless differ widely in others of more importance : we find both of them with two opposite parietal placentæ, connected with a quaternary arrangement of the other parts of the flower, and that in both instances their placentæ are opposite to stigmata. But we also see that in Cruciferæ dehiscence takes place by the separation of two valves from the sides of the siliqua, leaving the placentæ undivided ; while in Eschscholtzia it takes place through each placenta, half of which, therefore, adheres to each edge of the two valves into which the fruit finally separates. But if we look into their structure a little more narrowly, we shall perhaps find that these differences are not only capable of reconciliation, but that they explain each other.

" The fruit of Cruciferæ is separable into four parts; that is to say, into two valves without stigmata, and two double placentæ without valves : in Eschscholtzia there are two valves with placentæ and stigmata, and two stigmata without valves or placentæ. But suppose that the two valves of Cruciferæ had stigmata, as they should have (and a tendency to produce which actually exists in Iberis umbellata), and that the two stigmata of Eschscholtzia had valves, as would be regular, what would then be the difference between the two? It would be reduced to nearly this : that in Eschscholtzia the two placentiferous pieces would occupy the greater part of the pericarpium, the two sterile valves being very small ; while in Cruciferæ the two placentiferous pieces would be very small, the chief part of the pericarpium being occupied by the sterile valves."

Such was the idea I was led, by the curious structure of Eschscholtzia, to entertain in 1828, upon the fruit of Cruciferæ. I am aware that it is possible to explain the peculiar economy of the replum of Cruciferæ by that of Carmichaelia, and that the line of dehiscence in fruit is no evidence of the plan upon which it has been constructed. I also know that a less paradoxical way of understanding the structure of the Siliqua, is to take two confluent carpella, each of which has a 2-lobed or 2-horned stigma, for the type of such a fruit; upon which supposition each apparent stigma of the siliqua will be made up of two halves : and moreover I have been shewn by Mr. Brown some instances of monstrous formation, which seem to confirm such an opinion. Nevertheless, I wish to record, in this book, my view of the subject, whether it shall be ultimately found to be accurate or inaccurate, for the following reasons. In the first place, it will shew young botanists how narrowly it is necessary for them to observe the structure of plants, and how indispensable it is to bear constantly in mind the analogies that exist between the formation of one plant and another; in the second place, by pursuing the discussion, I hope to induce some one to set the question at rest, by means of such demonstration as it is capable of receiving; and thirdly, I still retain my opinion, notwithstanding what I have seen and heard since it was formed; relying chiefly upon the peculiarities of Eschscholtzia, which seems to me to be so intimately connected with the question at issue, and so obviously formed upon the same plan as Cruciferæ, whatever that plan may be, that what can be shewn to be true of one must be true of the other.

Almost all Cruciferæ have the calyx imbricated in æstivation; but Mr. Brown has noticed (*Denham*, p. 7) that in Savignya and Ricotia it is valvate. It is a very common character of Cruciferæ to be destitute of bracteæ.

GEOGRAPHY. An order eminently European; 166 species are found in northern and middle Europe, and 178 on the northern shore or islands of the Mediterranean; 45 are peculiar to the coast of Africa, between Mogador and Alexandria; 184 to Syria, Asia Minor, Tauria, and Persia; 99 to Siberia; 35 to China, Japan, or India; 16 to New Holland and the South Sea Islands; 6 to the Isle of France and the neighbouring islands; 70 to the Cape of Good Hope; 9 to the Canaries or Madeira; 2 to St. Helena; 2 to the West Indies; 41 to South America; 48 to North America; 5 to the islands between North America and Kamtchatka; and 35 are common to various parts of the world. This being their general geographical distribution, it appears that, exclusive of species that are uncertain, or common to several different countries, about 100 are found in the southern hemisphere, and about 800 in the northern, or 91 in the new, and the rest in the old world. Finally, if we consider them with regard to temperature, we shall find that there are, —

In the frigid zone of the northern hemisphere.....................................		205
In all the tropics (and chiefly in mountainous regions)...............		30
In the temperate zone { of the northern hemisphere... 548 } of the southern ditto .... 86 }		634

Such were the calculations of Decandolle in 1821 (*Syst.* 2. 142). Although requiring considerable modification, especially in the Siberian and North American numbers, which are much too low, they serve to give a general idea of the manner in which the order is dispersed over the globe.

PROPERTIES. The universal character of Cruciferæ is to possess antiscorbutic and stimulant qualities, combined with an acrid flavour. These are so uniform, that I shall only offer some very general remarks upon them; for which I am chiefly indebted to Decandolle's *Essai sur les Propriétés Médicales des Plantes*, to which I refer those who wish for more information. Cruciferæ contain a great deal of azote, to which it is supposed is due their animal odour when rotting. Mustard, Cress, Horseradish, and many others, are extremely stimulating and acrid. The seeds of Sinapis chinensis are considered by Hindoo and Mahometan practitioners as stimulant, stomachic, and laxative. *Ainslie*, 1. 230. The seeds of one species of Arabis (chinensis *Rottler*) are prescribed by the Indian doctors as stomachic and gently stimulant; but they apprehend its bringing on abortion if imprudently given. *Ibid.* 2. 12. When the acrid flavour is dispersed among an abundance of mucilage, various parts of these plants become a wholesome food; such as the root of the Radish and the Turnip, the herbage of the Water-cress, the Cabbage, the Sea-kale, and the stems of various plants of the cabbage tribe. Prince Maximilian, of Wied Neuwied, relates that the Brazilian Indians use a kind of cress, which in taste resembles that of Europe, as a good remedy

c

for asthma. *Travels*, 1. 35. Their seeds universally abound in a fixed oil, which is expressed from some species, as the Rape, for various economical purposes.

Linnæus divided this order, which is the same as his Tetradynamia, by the form of the fruit, under two heads, bearing the names of Siliquosa and Siliculosa. More recently, divisions have been founded upon the nature of the plicature of the cotyledons, and the position of the radicle with respect to them. It is difficult to say what degree of importance really deserves to be attached to these characters, which are however in general use, and which will probably continue to be employed for the purpose of distinction.

The following are the modifications used by Decandolle : —

1. The cotyledons are flat, with the radicle lying upon their edges. (*Pleurorhizeæ.*)

EXAMPLES. Cheiranthus, Arabis, Alyssum.

2. The cotyledons are flat, with the radicle lying upon their back. (*Notorhizeæ.*)

EXAMPLES. Sisymbrium, Erysimum, Lepidium.

3. The cotyledons are folded lengthwise. (*Orthoploceæ.*)

EXAMPLES. Brassica, Sinapis, Vella.

4. The cotyledons are coiled up spirally. (*Spirolobeæ.*)

EXAMPLES. Bunias, Erucaria.

5. The cotyledons, instead of being coiled up spirally, or folded lengthwise, are bent double. (*Diplecolobeæ.*)

EXAMPLES. Heliophila, Subularia.

X. FUMARIACEÆ. THE FUMITORY TRIBE.

FUMARIACEÆ, *Dec. Syst.* 2. 105. (1821) ; *Prodr.* 1. 125. (1824) ; *Lindl. Synops.* 18. (1829.)

DIAGNOSIS. Polypetalous dicotyledons, with a definite number of hypogynous diadelphous stamens, concrete carpella, a 1-celled ovarium, narrow parietal placentæ, 2 sepals, and an irregular corolla.

ANOMALIES.

ESSENTIAL CHARACTER.—*Sepals* 2, deciduous. *Petals* 4, cruciate, parallel ; the 2 outer, either one or both, saccate at the base ; the 2 inner callous and coloured at the apex, where they cohere and enclose the anthers and stigma. *Stamens* 6, in two parcels, opposite the outer petals, very seldom all separate ; *anthers* membranous, the outer of each parcel 1-celled, the middle one 2-celled. *Ovarium* superior, 1-celled ; *ovula* horizontal ; *style* filiform ; *stigma* with two or more points. *Fruit* various ; either an indehiscent 1- or 2-seeded nut, or a 2-valved polyspermous pod. *Seeds* horizontal, shining, with an arillus. *Albumen* fleshy. *Embryo* minute, out of the axis ; in the indehiscent fruit straight ; in those which dehisce somewhat arcuate.—*Herbaceous plants*, with brittle stems and a watery juice. *Leaves* usually alternate, multifid, often with tendrils. *Flowers* purple, white, or yellow.

AFFINITIES. The following are M. Decandolle's remarks upon this subject (*Syst.* 2. 106.) :—" Fumariaceæ are very near Papaveraceæ, on account of their 2-leaved deciduous calyx, of the structure of the fruit of such species as dehisce, and of their fleshy albumen ; but they differ, firstly, in their juice being watery, instead of milky ; secondly, in their petals being usually irregular and in cohesion with each other ; thirdly, in their diadelphous stamens, which bear indifferently 1- and 2-celled anthers." The same learned writer also points out the affinity that exists between them and Cruciferæ, which differ chiefly in the arrangement of their stamens, in the number of the leaves of the calyx, in their regular petals and exalbuminous seeds. I am, however, inclined to suspect, that the floral envelopes of Fumariaceæ are not rightly described.

I am by no means sure that it would not be more consonant to analogy to consider the parts of their flower divided upon a binary plan; thus understanding the outer series of the supposed petals as calyx, and the inner only as petals; while the parts now called sepals are perhaps more analogous to bracteæ; an idea which their arrangement, and the constant tendency of the outer series to become saccate at the base, which is not uncommon in the calyx of Cruciferæ, but never happens, as far as I know, in their petals, would seem to confirm. Of this, some further evidence may be found in the stamens. These are combined in two parcels, one of which is opposite each of the divisions of the outer series, and consists of one perfect 2-celled anther in the middle and two lateral 1-celled ones : now, supposing the lateral 1-celled anthers of each parcel to belong to a common stamen, the filament of which is split by the separation of the two parcels, an hypothesis to which I do not think any objection can be entertained, we shall find that the number of stamens of Fumariaceæ is 4, one of which is before each of the divisions of the flower; an arrangement which is precisely what we should expect to find in a normal flower consisting of 2 sepals and 2 petals, and the reverse of what ought to occur if the divisions of the flower were really all petals, as has been hitherto believed.

The economy of the sexual organs of Fumariaceæ is remarkable. The stamens are in two parcels, the anthers of which are a little higher than the stigma; the two middle ones of these anthers are turned outwards, and do not appear to be capable of communicating their pollen to the stigma; the four lateral ones are also naturally turned outwards, but by a twist of their filament their face is presented to the stigma. They are all held firmly together by the cohesion of the tops of the flower, which, never unclosing, offer no apparent means of the pollen being disturbed, so as to be shed upon the stigmatic surface. To remedy this inconvenience, the stigma is furnished with two blunt horns, one of which is inserted between and under the cells of the anthers of each parcel, so that without any alteration of position on the part of either organ, the mere contraction of the valves of the anthers is sufficient to shed the pollen upon that spot where it is required to perform the office of fecundation.

This order offers every gradation, from monospermous to polyspermous fruit, and between indehiscence, as in Fumaria itself, and dehiscence, as in Corydalis.

GEOGRAPHY. Their principal range is in the temperate latitudes of the northern hemisphere, where they inhabit thickets and waste places. Two are found at the Cape of Good Hope.

PROPERTIES. The character of Fumariaceæ is, to be scentless, a little bitter, in no degree milky, and to act as diaphoretics and aperients. *Dec.* The root of Fumaria cava and Corydalis tuberosa has been found to contain a peculiar alkali called Corydalin. *Turner*, 653.

EXAMPLES. Fumaria, Diclytra, Corydalis.

XI. CAPPARIDEÆ. THE CAPER TRIBE.

CAPPARIDEÆ, *Juss. Gen.* 242. (1789) ; *Ann. Mus.* 18. 474. (1811) ; *Dec. Prodr.* 1. 237. (1824).

DIAGNOSIS. Polypetalous dicotyledons, with hypogynous stamens, concrete carpella, a 1-celled pedicellate ovarium, narrow simple parietal placentæ, a continuous enlarged disk, and reniform seeds.

ANOMALIES. Some species of Niebuhria, Mærua, Boscia, Cadaba, and Thylacium, have no petals. The stamens are occasionally tetradynamous, according to Decandolle.

ESSENTIAL CHARACTER.—*Sepals* 4, either nearly distinct, equal, or unequal, or cohering in a tube, the limb of which is variable in form. *Petals* 4, cruciate, usually unguiculate and unequal. *Stamens* almost perigynous, very seldom tetradynamous, most frequently arranged in some high multiple of a quaternary number, definite or indefinite. *Disk* hemispherical, or elongated, often bearing glands. *Ovarium* stalked; *style* none, or filiform. *Fruit* either podshaped and dehiscent, or baccate, 1-celled, very rarely 1-seeded, most frequently with 2 polyspermous placentæ. *Seeds* generally reniform, without albumen, but with the lining of the testa tumid, attached to the margin of the valves; *embryo* incurved; *cotyledons* foliaceous, flattish.—*Herbaceous plants, shrubs*, or even *trees*, without true stipulæ, but sometimes with spines in their place. *Leaves* alternate, stalked, undivided, or palmate. *Flowers* in no particular arrangement.

AFFINITIES. Distinguished from Cruciferæ by their stamens being often indefinite, if definite never tetradynamous, or scarcely ever, and by their reniform seeds. They are related to Passifloreæ in their stipitate ovarium, and fleshy indehiscent fruit with parietal polyspermous placentæ; and to Flacourtiaceæ in the structure of their fruit, parietal placentæ, and indefinite stamens; from these last they are known by their narrow placentæ, exalbuminous seeds and peculiar habit; and from the former by a number of obvious characters. Mr. Brown remarks (*Denham*, 15,) that some species of Capparis, of which C. spinosa is an example, have as many as 8 placentæ.

GEOGRAPHY. These are chiefly found in the tropics and in the countries bordering upon them, where they abound in almost every direction. Of the capsular species, a single one, Cleome violacea, is found in Portugal; another, Polanisia graveolens, occurs as far to the north as Canada; and one or two others are met with in the southern provinces of the United States. Of the fleshy-fruited kinds, the common Caper, Capparis spinosa, a native of the most southern parts of Europe, is that which approaches the nearest to the north; Africa abounds in them.

PROPERTIES. M. Decandolle compares Capparideæ with Cruciferæ in regard to their sensible qualities; and they no doubt resemble each other in many respects; for instance, the Capers are stimulant, antiscorbutic, and aperient; the bark of the root of the Caper passes for a diuretic; and several species of Cleome have a pungent taste, like that of mustard. The root of Cleome dodecandra is used as a vermifuge in the United States. Cleome icosandra acts as a vesicatory, and is used in Cochin China as a sinapism. Dancer states that the bark of the root of Crateva gynandra blisters like Cantharides. *Ainslie*, 2. 88. But there is an exception to this in a plant called *Fruta de Burro*, which is found in the neighbourhood of Carthagena, the fruit of which is extremely poisonous. It is supposed to be a species of Capparis, nearly allied to the Capp. pulcherrima of Jacquin; and must not be confounded with the Fruta del Burro of Humboldt, found in Guiana, which is a valuable medicinal plant, belonging to Anonaceæ.

This order is divided into CLEOMEÆ, or the genera with herbaceous stems and capsular fruit, and CAPPAREÆ, or true Capers, which have shrubby stems and fleshy fruit.

EXAMPLES. Cleome, Capparis.

XII. FLACOURTIACEÆ.

FLACOURTIANEÆ, *Richard in Mem. Mus.* 1. 366. (1815); *Dec. Prodr.* 1. 255. (1829.)

DIAGNOSIS. Polypetalous dicotyledons, with hypogynous stamens, concrete carpella, and a 1-celled ovarium, with parietal placentæ branching all over the surface of the inside.

ANOMALIES. Ryania, Patrisia, Flacourtia, Roumea, and Stigmarota, that is to say, more than half the order, have no petals.

ESSENTIAL CHARACTER.—*Sepals* definite, from 4-7, cohering slightly at the base. *Petals* equal to the latter in number and alternate with them, seldom wanting, *Stamens* hypogynous, of the same number as the petals, or twice as many, or some multiple of them, occasionally changed into nectariferous scales. *Ovarium* roundish, distinct, sessile or slightly stalked ; *style* either none or filiform ; *stigmas* several, more or less distinct. *Fruit* 1-celled, either fleshy and indehiscent, or capsular, with 4 or 5 valves, the centre filled with a thin pulp. *Seeds* few, thick, usually enveloped in a pellicle formed by the withered pulp, attached to the surface of the valves in a branched manner, not in a line as in Violeæ and Passifloreæ ; *albumen* fleshy, somewhat oily ; *embryo* straight in the axis, with the radicle turned to the hilum, and therefore usually superior ; *cotyledons* flat, foliaceous.—*Shrubs* or small *trees*. *Leaves* alternate, simple, on short stalks, without stipulæ, usually entire, and coriaceous. *Peduncles* axillary, many-flowered. *Flowers* sometimes unisexual.

AFFINITIES. The unilocular fruit, over the whole of the inside of which the placentæ spread, is, according to Decandolle, sufficient to distinguish them from all other Dicotyledons. They resemble the Capparideæ with fleshy fruit in a number of particulars ; and M. Decandolle indicates an approach to Passifloreæ : this chiefly depends upon both orders having parietal placentæ, and the presence of a series of barren stamina, analogous to the corona of Passifloreæ. They have also some relation to Samydeæ.

GEOGRAPHY. Almost all natives of the hottest parts of the East and West Indies, and Africa. Two or three species are found at the Cape of Good Hope, and one or perhaps two in New Zealand.

PROPERTIES. Nothing is known of their sensible qualities. The fruit of some of the Flacourtias is eatable and wholesome ; that of Hydnocarpus venenata is used in Ceylon for poisoning fish, which afterwards become so unwholesome as to be unfit for food.

Decandolle has the following tribes (*Prodr.* 1. 255.) :—

1. PATRISIEÆ.

Flowers hermaphrodite, apetalous. Sepals 5, coloured inside, persistent. Stamens indefinite. Fruit capsular or berried. *Dec.* It is to be suspected that this tribe really belongs to Passifloreæ, on account of its affinity to Smeathmannia ; but their seeds are smooth, not pitted, and the placentæ do not occupy lines, but are spread over the whole surface. *Ibid.*

EXAMPLES. Ryanæa. Patrisia.

2. FLACOURTIEÆ.

Flowers diœcious, apetalous. Stamens indefinite. Fruit baccate, indehiscent. *Dec.*

EXAMPLES. Flacourtia, Roumea.

3. KIGGELARIEÆ.

Flowers diœcious. Petals? 5, alternate with the sepals. Stamens definite. Fruit somewhat baccate, finally dehiscing. *Dec.*

EXAMPLES. Kiggelaria, Melicytus.

4. ERYTHROSPERMEÆ.

Flowers hermaphrodite. Petals and stamens 5-7. Fruit indehiscent, somewhat baccate.

EXAMPLE. Erythrospermum.

XIII. ANONACEÆ. The Custard Apple Tribe.

Anonæ, *Juss. Gen.* 283. (1789.) — Anonaceæ, *Rich. Anal. Fr.* 17. (1808); *Dunal Monogr.* (1817); *Dec. Syst.* 1. 462. (1818); *Prodr.* 1. 83. (1824.) — Glypto-spermæ, *Vent. Tabl.* 3. 75. (1799.)

DIAGNOSIS. Polypetalous dicotyledons, with hypogynous stamens, anthers bursting by longitudinal slits, numerous distinct simple carpella, exstipulate leaves, and ruminated albumen.

ANOMALIES. Monodora has a solitary carpellum. In Anona palustris the ovaria are not distinct. Rollinia has the petals united. Stamens and carpella definite in Bocagea.

ESSENTIAL CHARACTER. — *Sepals* 3-4, persistent, usually partially cohering. *Petals* 6, hypogynous, in two rows, coriaceous, with a valvular æstivation. *Stamens* indefinite, covering a large hypogynous torus, packed closely together, very rarely definite. *Filaments* short, more or less angular. *Anthers* adnate, turned outwards, with an enlarged 4 cornered connectivum, which is sometimes nectariferous. *Ovaria* usually numerous, closely packed, separate or cohering, occasionally definite. *Styles* short; *stigmata* simple; *ovula* solitary, or a small number, erect or ascending. *Fruit* consisting of a number of carpella, which are either succulent or dry, sessile or stalked, 1- or many-seeded, distinct or concrete into a fleshy mass. *Seeds* attached to the suture in one or two rows; *testa* brittle; *embryo* minute, in the base of hard, fleshy, ruminate *albumen.*—*Trees* or *shrubs. Leaves* alternate, simple, almost always entire, without stipulæ. *Flowers* usually green or brown, axillary, solitary, or 2 or 3 together, shorter than the leaves; the *peduncles* of abortive flowers sometimes indurated, enlarged, and hooked.

AFFINITIES. No doubt can be entertained of the close affinity of this order to Magnoliaceæ, from which, however, it differs in the want of stipulæ, in the form of the anthers, and in the peculiar condition of the ovarium: agreeing in the ternary division of the parts of fructification, and their indefinite stamens and ovaria. An affinity has been pointed out between them and Menispermeæ; but it appears to me to be very weak. The great feature of the order is its ruminated albumen, to which there is no exception, and scarcely any parallel. The parietal insertion of ovula, ascribed to this order by Decandolle, is not universal. The ovula are erect in Anona, Guatteria, and Anaxagorea. *A. St. H. in Pl. Usu.* 33. A remarkable plant is described by Mr. Brown, in the Appendix to Flinder's Voyage, under the name of Eupomatia laurina, in which the stamens are manifestly perigynous, and the tube of the calyx coherent with the ovarium. This genus is referred by its learned discoverer to Anonaceæ, with which there can be no doubt but that it has a very striking analogy; but its structure is nevertheless so peculiar, that I hesitate, with M. Decandolle, in absolutely identifying it with Anonaceæ. I have remarked in Anona laurifolia that the pollen is arranged in two distinct rows in each cell of the anther, and that when that organ bursts, the grains of pollen fall out, cohering in a single row, so as to have the appearance of a necklace. Supposing Wintereæ not to be stipulate, as St. Hilaire asserts, this order will be more nearly related to them than to Magnoliaceæ. Connected with Berberideæ through Bocagea.

GEOGRAPHY. The tropics of the old and new world are the natural land of these plants: thence they spread, in a few instances, to the northward and the southward.

PROPERTIES. The general character is, to have a powerful aromatic taste and smell in all the parts. The bark of Uvaria tripetaloidea yields, being tapped, a viscid matter, which hardens in the form of a fragrant gum. *Dec.* The flowers of many species, especially of Artabotrys odoratissima and Cananga virgata, are exceedingly fragrant. The dry fruits of many species are very aromatic; those of Uvaria aromatica are the Piper æthiopicum of the shops. Xylopia sericea, a large tree found in forests near Rio Janeiro, where it is called Pindaïba, bears a highly aromatic fruit, with the flavour of pepper, for which it may be advantageously substituted. Its bark is tough, and readily separated into fibres, from which excellent cordage is manufactured. *Plantes Usuelles*, no. 33. Of other species the fruit is succulent and eatable, containing a sugary mucilage, which predominates over the slight aromatic flavour that they produce. Of this kind are the Custard Apples of the East and West Indies, the Cherimoyer of Peru, and others. In Asimina triloba an acid is present of a very active nature, according to Duhamel; but this is not certain. The Anona sylvatica, called *Araticu do mato*, in Brazil, has a light white wood, very fit for the use of turners, and for the same purposes as the lime-tree of Europe. Its fruit is described as good for the dessert. *Plantes Usuelles*, 29. The wood of the root of A. palustris is employed in Brazil for corks. *Ib.* 30. The Indians on the Orinoco, particularly in Atures and Maypura, have an excellent febrifuge, called *Frutta de Burro*, which is the fruit of Uvaria febrifuga. *Humboldt, Cinch. Forests*, p. 22. *Eng. ed.*

EXAMPLES. Anona, Unona, Guatteria.

XIV. MYRISTICEÆ. THE NUTMEG TRIBE.

MYRISTICEÆ, *R. Brown, Prodr.* 399. (1810.)

DIAGNOSIS. Apetalous dicotyledons, with diœcious flowers, a 3-lobed calyx, ruminated albumen, and columnar stamens.

ANOMALIES.

ESSENTIAL CHARACTER. — *Flowers* diœcious, with no trace of a second sex. *Calyx* trifid, with valvular æstivation. MALES. *Filaments* completely united in a cylinder. *Anthers* 3-12, definite, 2-celled, turned outwards, and bursting longitudinally; either connate or distinct. FEMALES. *Calyx* deciduous. *Ovary* superior, sessile, with a single erect ovulum; *style* very short; *stigma* somewhat lobed. *Fruit* baccate, dehiscent, 2-valved. *Seed* nut-like, enveloped in a many-parted *arillus; albumen* ruminate, between fatty and fleshy; *embryo* small; *cotyledons* foliaceous; *radicle* inferior; *plumula* conspicuous.—*Tropical trees*, often yielding a red juice. *Leaves* alternate, without stipulæ, not dotted, quite entire, stalked, coriaceous; usually, when full grown, covered beneath with a close down. *Inflorescence* axillary or terminal, in racemes, glomerules, or panicles; the *flowers* each with one short cucullate bractea. *Calyx* coriaceous, mostly downy outside, with the hairs sometimes stellate, smooth in the inside. — *R. Br.* chiefly.

AFFINITIES. Usually placed, on account of their apetalous flowers, in the vicinity of Laurineæ, from which they are distinguished by the structure of their calyx, anthers, and fruit; perhaps more nearly allied to Anonaceæ, on account of their 3-lobed calyx,—a remarkable peculiarity in Dicotyledons, —their ruminated albumen, minute embryo, and sensible properties. Mr. Brown places them between Proteaceæ and Laurineæ, remarking, that they are not closely akin to any other order.

GEOGRAPHY. Natives exclusively of the tropics of India and America.

PROPERTIES. The bark abounds in an acrid juice, which is viscid and stains red; the rind of their fruit is caustic; the arillus and albumen, the former known under the name of Mace, and the latter of Nutmeg, are important aromatics, abounding in a fixed oil of a consistence analogous to fat, which, in a species called Virola sebifera, is so copious as to be extracted easily by immersing the seeds in hot water. The common Nutmeg is the produce of Myristica moschata; but an aromatic fruit is also borne by other species. The Nutmeg of Santa Fé is the Myristica Otoba. *Humb. Cinch. For.* p. 29. *Eng. ed.*

EXAMPLES. Myristica, Knema.

XV. MAGNOLIACEÆ. THE MAGNOLIA TRIBE.

MAGNOLIÆ, *Juss. Gen.* 280. (1789); MAGNOLIACEÆ, *Dec. Syst.* 1. 439. (1818); *Prodr.* 1. 77. (1824.)

DIAGNOSIS. Polypetalous dicotyledons, with hypogynous stamens, anthers bursting by longitudinal slits, numerous distinct simple carpella, and stipulate leaves without transparent dots.

ANOMALIES. The flowers of Mayna are diœcious.

ESSENTIAL CHARACTER. — *Sepals* 3-6, deciduous. *Petals* 3-27, hypogynous, in several rows. *Stamens* indefinite, distinct, hypogynous. *Anthers* adnate, long. *Ovaria* numerous, simple, arranged upon the torus above the stamens, 1-celled; *ovules* either ascending or suspended; *'styles* short; *stigmas* simple. *Fruit* either dry or succulent, consisting of numerous carpella, which are either dehiscent or indehiscent, distinct or partially connate, always numerous, and arranged upon an elongated axis, sometimes terminated by a membranous wing. *Seeds* solitary, or several, attached to the inner edge of the carpella. *Embryo* minute, at the base of fleshy albumen.—Fine *trees* or *shrubs.* *Leaves* alternate, not dotted, coriaceous, articulated distinctly with the stem; with deciduous stipulæ, which, when young, are rolled together like those of Ficus. *Flowers* large, solitary, often strongly odoriferous.

AFFINITIES. Nearly related to Dilleniaceæ, from which they are chiefly distinguished by the ternary, not quinary, arrangement of the parts of the flower; from Anonaceæ, to which they also approach, their stipulæ and solid albumen separate them. Their stipulation points out their affinity with Urticeæ; their imbricate petals and sepals, and numerous ovaria, with Calycantheæ, and through them with Monimieæ.

GEOGRAPHY. The focus of this order is undoubtedly North America, where the woods, the swamps, and the sides of the hills, abound with them. Thence they straggle, on the one hand, into the West India Islands, and, on the other, into India, through China and Japan. Mr. Brown remarks (*Congo,* 465), that no species have been found on the continent of Africa, or in any of the adjoining islands. Twenty-eight species are all that M. Decandolle enumerates.

PROPERTIES. The general character of the order is, to have a bitter tonic taste, and fragrant flowers. The latter produce a decided action upon the nerves, which, according to Decandolle, induces sickness and headach from Magnolia tripetala, and, on the authority of Barton, is so stimulating on the part of Magnolia glauca as to produce paroxysms of fever, and even an attack of inflammatory gout. The bark has been found to be destitute of tannin and gallic acid, notwithstanding its intense bitterness. The bark of the root of Magnolia glauca is an important tonic. *Barton,* 1. 87. The same property is found in the Liriodendron tulipifera, which has even been said to be equal to Peruvian bark. Michelia Doltsopa is one of the finest trees in Nipal, yielding an excellent fragrant wood, much used in that country for house-building. *Don. Prodr.* 226. Magnolia excelsa has a valuable timber, called *Champ,* at first greenish, but soon changing into a pale yellow; the texture is fine. *Wallich. Tent.* 7. The cones of Magnolia acuminata yield, in Virginia, a spirituous tincture, which is employed with some success in rheumatic affections; and the seeds of most species are remarkable for their bitterness: those of M. Yulan are employed in China as febrifuges, under the name of *Tsin-y*. *Dec*. No Magnoliaceæ are aromatic. EXAMPLES. Magnolia, Liriodendron.

XVI. DILLENIACEÆ.

DILLENIACEÆ, *Dec. Syst.* 1. 395. (1818); *Prodr.* 1. 67. (1824.)

DIAGNOSIS. Polypetalous dicotyledons, with hypogynous stamens, anthers bursting by longitudinal slits, distinct simple carpella, exstipulate leaves, solid albumen, and arillate seeds.

ANOMALIES. In several genera of the section Delimaceæ there is but one carpellum; and in Dillenia and Colbertia the carpella partly cohere.

ESSENTIAL CHARACTER. — *Sepals* 5, persistent, 2 exterior, 3 interior. *Petals* 5, deciduous, hypogynous, in a single row. *Stamens* indefinite, hypogynous, arising from a torus, either distinct or polyadelphous, and either placed regularly around the pistillum or on one side of it. *Filaments* dilated either at the base or apex. *Anthers* adnate, 2-celled, usually bursting longitudinally, always turned inwards. *Ovaria* definite, more or less distinct, with a terminal *style* and simple *stigma ; ovules* ascending. *Fruit* consisting either of from 2 to 5 distinct unilocular carpella, or of a similar number cohering together ; the carpella either baccate or 2-valved, pointed by the style. *Seeds* fixed in a double row to the inner edge of the carpella, either several or only 2, occasionally solitary by abortion ; surrounded by a pulpy arillus. *Testa* hard. *Embryo* minute, lying in the base of fleshy albumen.—*Trees, shrubs,* or *under-shrubs. Leaves* usually alternate, almost always without stipulæ, very seldom opposite, most commonly coriaceous, with strong veins running straight from the midrib to the margin, entire or toothed, often separating from the base of the petiole, which remains adhering to the stem. *Flowers* solitary, in terminal racemes or panicles, often yellow.

AFFINITIES. These are nearly akin to Magnoliaceæ, from which they are distinguished by their want of stipulæ and quinary arrangement of the parts of fructification; and to Ranunculaceæ, from which their persistent calyx, stamens, and whole habit, divide them. They are universally characterised by the presence of arillus; a peculiarity which certainly exists in Hibbertia, notwithstanding M. Decandolle's definition of that genus. The most genuine form of the order is known by the veins of the leaves running straight from the midrib to the margin.

GEOGRAPHY. According to Decandolle, 50 of this order are found in Australasia, 21 in India and its neighbourhood, 3 in equinoctial Africa, and 21 in equinoctial America; but since the publication of the *Systema* several have been added, both to the Indian and South American species.

PROPERTIES. Dilleniaceæ are generally astringent. The Brazilians make use of a decoction of Davilla rugosa in swellings of the legs and testicles, very common maladies in hot and humid parts of South America. *Pl. Usuelles*, no. 22. Davilla elliptica is also astringent, and furnishes the vulnerary called *Cambaïbinha* in Brazil. *Ibid.* 23. In Curatella Cambaïba the same astringent principle recommends its decoction as an excellent wash for wounds. *Ibid.* 24. The young calyces of Dillenia scabrella and speciosa have a pleasantly acid taste, and are used in curries by the inhabitants of Chittagong and Bengal. *Wallich.* Almost all Delimaceæ have the leaves covered with asperities, which are sometimes so hard that the leaves are even used for polishing.

Two tribes are distinguished in this family : —

1. § DELIMACEÆ.

§ Delimaceæ. *Dec. Syst.* 1. 396. (1818); *Prodr.* 1. 67. (1824.)
Filaments filiform, dilated at the apex, and bearing on each side a round
distinct cell of the anther. *Ovaria* from 1 to 5. *Styles* filiform, acute.
Carpella capsular, bladdery, or baccate, usually 1 or 2-seeded.—*Trees* or
shrubs, which sometimes twine. *Dec.*
EXAMPLES. Tetracera, Delima.

2. § DILLENEÆ.

Dilleneæ. *Salisb. Parad. Lond. n.* 73. (1806); § *Dec. Syst.* 1. 411.
(1818); *Prodr.* 1. 70. (1824).
Filaments not dilated at the apex, anthers elongate, adnate. *Ovaria*
usually from 2 to 5, distinct, rarely solitary; or from 5 to 20, partially
connate.—*Trees* or *shrubs*, very seldom twining. *Dec.* *Flowers* often fragrant or fœtid.
EXAMPLES. Dillenia, Hibbertia.

XVII. WINTEREÆ. THE WINTER'S BARK TRIBE.

WINTEREÆ, *R. Brown in Decand. Syst.* 1. 548. (1818.)— ILLICIEÆ, *Dec. Prodr.* 1. 77.
(1824.) *a section of* Magnoliaceæ.

DIAGNOSIS. Polypetalous aromatic dicotyledons, with hypogynous
stamens, anthers bursting by longitudinal slits, distinct simple carpella, and
stipulate leaves, with transparent dots.

ANOMALIES. The flowers of Tasmannia are diœcious or polygamous,
and the carpella solitary.

ESSENTIAL CHARACTER.—*Flowers* hermaphrodite or unisexual. *Sepals* 2-6, sometimes not distinguishable from the petals, either deciduous or persistent. *Petals* 2-30, in
several rows when more than 5. *Stamens* short, indefinite, hypogynous, distinct. *Anthers*
adnate. *Ovaria* definite, arranged in a single whorl, 1-celled, with several suspended
ovules, which are attached to the suture. *Stigmata* simple, sessile. *Fruit* either dry or
succulent, consisting of a single row of carpella, which are either dehiscent or indehiscent,
and distinct. *Seeds* solitary or several, with or without arillus. *Embryo* very small, straight,
in the base of fleshy albumen.—*Shrubs* or small *trees.* *Leaves* alternate, dotted, coriaceous,
persistent, with convolute deciduous stipulæ. *Flowers* solitary, often brown or chocolate
colour, and sweet-scented.

AFFINITIES. Closely related to Magnoliaceæ, from which they differ
chiefly in their dotted leaves and aromatic qualities. They are also closely
allied to Calycantheæ, from which their hypogynous stamens, alternate
stipulate leaves, and albuminous seeds, sufficiently distinguish them. They
also partake of the affinities of Magnoliaceæ, with Anonaceæ, &c. According to St. Hilaire, the supposed stipulæ of Wintereæ are only imperfectly
developed leaves which enfold the buds. *Pl. Usuelles*, no. 26—28. But
what are stipules except starved leaves? The same author remarks, that
Bonpland considered the embryo as destitute of albumen, which was, however, a mistake, it being undoubtedly as it is here described. For several
good remarks upon Drimys, see the *Pl. Usuelles* as quoted.

GEOGRAPHY. A very small order, with an extensive range. Of the 10
species enumerated by Decandolle, 2 are found in New Holland, 2 in the
hotter parts of America, 2 in the southern and 2 in the northern territories of
the same continent, 1 in China and Japan, and 1 in New Zealand.

PROPERTIES. All that writers have stated about the aromatic stimulant
properties of Magnoliaceæ should be applied to this order, formerly con-

founded with them. The seeds of Illicium anisatum are considered in India to be powerfully stomachic and carminative. A very fragrant volatile oil is also obtained from them. *Ainslie*, 2. 20. The Chinese burn them in their temples, and Europeans employ them to aromatise certain liquors, such as the Anisette de Bourdeaux. Drymis Winteri yields the Winter's Bark, which is known for its resemblance to that of cinnamon. *A. R.* A bark called Melambo Bark, possessing similar properties, is described by M. Cadet in the *Journal de Pharmacie*, 1815, p. 20. The bark of Drimys granatensis, called *Casca d' Anta* in Brazil, is much used against colic. It is tonic, aromatic, and stimulant, and resembles, in nearly all respects, the Drimys Winteri, or Winter's Bark. *Plantes Usuelles*, 26—28.

EXAMPLES. Illicium, Wintera.

XVIII. CALYCANTHEÆ. THE CAROLINA ALLSPICE TRIBE.

CALYCANTHEÆ, *Lindl. in Bot. Reg. fol.* 404. (1819); *Dec. Prodr.* 3. 1. (1828.) — CALYCANTHINÆ, *Link. Enum.* 2. 66. (1822.)

DIAGNOSIS. Polypetalous dicotyledons, with definite perigynous stamens, numerous imbricated sepals, ovaria enclosed in a fleshy tube, convolute albumen, anthers turned outwards, opposite exstipulate leaves, and stems with 5 axes of growth.

ANOMALIES.

ESSENTIAL CHARACTER. — *Sepals* and *petals* confounded, indefinite, imbricated, combined in a fleshy tube. *Stamens* indefinite, inserted in a fleshy rim at the mouth of the tube, the inner sterile. *Anthers* adnate, turned outwards. *Ovaries* several, simple, 1-celled, with one terminal style, adhering to the inside of the tube of the calyx; *ovula* solitary, or sometimes 2, of which one is abortive, ascending. *Nuts* enclosed in the fleshy tube of the calyx, 1-seeded, indehiscent. *Seed* ascending; *albumen* none; *cotyledons* convolute, with their face next the axis; *radicle* inferior.—*Shrubs*, with square stems, having 4 woody imperfect axes surrounding the central ordinary one. *Leaves* opposite, simple, scabrous, without stipulæ. *Flowers* axillary, solitary.

AFFINITIES. It is not very clear to what order this is most nearly related. Jussieu originally placed it at the end of Rosaceæ (*Gen.*); he subsequently referred it to Monimieæ; and I afterwards formed it into a particular family. With Monimieæ it is less nearly related than it appears to be, the principal points of resemblance being the disposition of several nuts within a fleshy calyx in both orders; for Calycantheæ can scarcely be considered apetalous, as Monimieæ are, on account of the obvious petals of Chimonanthus. The imbricated sepals, in Calycanthus chocolate-coloured and becoming confounded with the petals, the fragrance of the flowers, and the plurality of ovaria, seem to indicate an affinity with Wintereæ, especially with Illicium; but the decidedly perigynous stamens and fleshy calyx enclosing the ovaria in its tube, the highly developed embryo, and want of albumen, are great objections to such an approximation. Combretaceæ agree in having an exalbuminous embryo with convolute cotyledons; but with this their resemblance ceases. Myrtaceæ also agree in this same particular, in the case of Punica; and their opposite leaves, without stipulæ, frequent fragrance, and perigynous stamens, strengthen the affinity indicated by the embryo. Rosaceæ, to which Jussieu originally referred Calycanthus, agree in the perigynous insertion of their stamens, in the peculiar structure of their calyx, the tube of which in Rosa is entirely analogous to that of Calycantheæ, in the superposition of their ovules when two are present, and in the high developement of their exalbuminous embryo: upon

the whole, therefore, no order appears to have so much affinity with Calycantheæ as Rosaceæ; and the sagacity of Jussieu, in originally referring
Calycanthus to that order, is completely confirmed by the discovery recently
made by the Rev. Mr. Lowe, that the cotyledons of Chamæmeles, a genus of
Pomaceæ, which Jussieu includes in Rosaceæ, are convolute. This, I think,
fixes the station of Calycantheæ in the neighbourhood of Rosaceæ, Pomaceæ,
and Myrtaceæ, to which it is nearly equally allied, and from which it is
distinguished by its imbricated sepals, and anthers, partly fertile and partly
sterile, being turned outwards. This order is also characterised by the
singular structure of the wood, a peculiarity originally remarked by Mirbel in
one species, and which I have since ascertained to exist in all. In the stems
of these plants there is the usual deposit of concentric circles of wood around
the pith, and, in addition, four very imperfect centres of deposition on the
outside next the bark; a most singular structure, which may be called, without much inaccuracy, an instance of exogenous and endogenous growth
combined in the same individual. A good figure of this interesting fact
has been given by Mirbel in the *Annales des Sciences Naturelles*, vol. 14.
p. 367.

GEOGRAPHY. Natives of North America and Japan.

PROPERTIES. The aromatic fragrance of the flowers is their only known
quality.

EXAMPLES. Calycanthus, Chimonanthus.

XIX. MONIMIEÆ.

MONIMIEÆ, *Juss. in Ann. Mus.* 14. 130. (1809); *Dec. Ess. Med.* 265. (1816.)

DIAGNOSIS. Apetalous dicotyledons, with definite pendulous ovula,
numerous distinct ovaria, and anthers bursting longitudinally.

ANOMALIES.

ESSENTIAL CHARACTER.—*Flowers* unisexual. *Calyx* tubular, toothed or lobed at
the apex, with valvular æstivation. *Stamens* indefinite, covering all the inside of the calyx;
anthers 2-celled, bursting longitudinally. *Ovaria* several, superior, distinct, enclosed within
the tube of the calyx, each with its own *style* and *stigma*; *ovule* pendulous. *Fruit* consisting of several 1-seeded nuts, enclosed within the enlarged calyx. *Seed* pendulous; *embryo*
in the midst of an abundant *albumen*; *radicle* superior.—*Trees* or *shrubs*, without aroma.
Leaves opposite, without stipulæ. *Hairs* stellate. *Flowers* axillary, in short racemes.

AFFINITIES. Allied to Urticeæ, from which they differ in the presence
of several ovaria within each calyx, in their pendulous ovula, in the radicle
being turned towards the hilum, and in the presence of abundant albumen;
also to Laurineæ, from which they particularly differ in the dehiscence of their
anthers, and in the number of their ovaria; and to Atherospermeæ, which
agree in sensible qualities, and in the number of their ovaria, but which differ
in the dehiscence of the anthers, and in the erect position of the ovules. With
Calycantheæ they have also a good deal of relation. Mr. Brown considers that what is here called a calyx is more properly an involucrum.
Flinders, 553.

GEOGRAPHY. All natives of South America.

PROPERTIES. All the parts of the bark and leaves exhale an aromatic
odour, which is compared by travellers to that of Laurels or Myrtles.
Decand.

EXAMPLES. Monimia, Ruizia.

XX. ATHEROSPERMEÆ.

ATHEROSPERMEÆ, *R. Brown in Flinders,* 553. (1814.)

DIAGNOSIS. Apetalous aromatic dicotyledons, with definite erect ovula, and anthers bursting by recurved valves.
ANOMALIES.

ESSENTIAL CHARACTER.—*Flowers* unisexual or hermaphrodite. *Calyx* tubular, divided at the top into several segments, usually placed in two rows, the inner of which is partly petaloid; to these are superadded some scales in the female and hermaphrodite flowers. *Stamens* in the males very numerous in the bottom of the calyx, with scales among them; in the hermaphrodites fewer, and arising from the orifice of the calyx; *anthers* adnate, 2-celled, bursting with a valve which separates from the base to the apex. *Ovaria* more than one, usually indefinite, each with a single erect ovulum; *styles* simple, arising either from the side or the base; *stigmas* simple. *Nuts* terminated by the persistent styles become feathery, enclosed in the enlarged tube of the calyx. *Seed* solitary, erect; *embryo* short, erect, at the base of soft, fleshy albumen; *radicle* inferior.—*Trees.* *Leaves* opposite, without stipulæ. *Flowers* axillary, solitary.

AFFINITIES. The anthers of this order are the same as those of Laurineæ and Berberideæ, from the latter of which they differ entirely, but with the former of which they agree in their aromatic odour. The order is nearly related to Monimieæ, with which it is even combined by Jussieu; but it differs in the position of the ovula, and in the structure of the anthers.
GEOGRAPHY. Natives of New Holland and South America. Only two genera are known.
PROPERTIES. Aromatic shrubs.
EXAMPLES. Pavonia, Atherosperma.

XXI. LAURINEÆ. THE CINNAMON TRIBE.

LAURI, *Juss. Gen.* 80. (1789); LAURINÆ, *Vent. Tabl.* (1799); *R. Brown Prodr.* 401. (1810).

DIAGNOSIS. Apetalous aromatic dicotyledons, with definite suspended ovules, and anthers bursting by recurved valves.
ANOMALIES. Cassytha is aphyllous and parasitical.

ESSENTIAL CHARACTER.—*Calyx* 4-6-cleft, with imbricated æstivation, the limb sometimes obsolete. *Stamens* definite, perigynous, opposite the segments of the calyx, and usually twice as numerous; the 3 innermost, which are opposite the 3 inner segments of the calyx, sterile or deficient; the 6 outermost scarcely ever abortive; *anthers* adnate, 2-4-celled; the cells bursting by a longitudinal persistent valve from the base to the apex; the outer anthers valved inwards, the inner valved outwards. *Glands* usually present at the base of the inner filaments. *Ovarium* single, superior, with a single pendulous ovulum; *style* simple; *stigma* obtuse. *Fruit* baccate or drupaceous, naked or covered. *Seed* without albumen; *embryo* inverted; *cotyledons* large, plano-convex, peltate near the base!; *radicle* very short, included, superior; *plumula* conspicuous, 2-leaved.—*Trees,* often of great size. *Leaves* without stipulæ, alternate, seldom opposite, entire or very rarely lobed. *Inflorescence* panicled or umbelled. Sometimes leafless twining *under-shrubs* or *parasitical herbs,* with spiked flowers, each having 3 bracteæ. *R. Br.*

AFFINITIES. Distinguished from all apetalous dicotyledons, except Atherospermeæ, by the peculiar dehiscence of their anthers, and divided from that order by the ovulum being pendulous, not erect. In sensible qualities they resemble Myristiceæ, which are at once known by their unisexual flowers and columnar stamens. The genus Cassytha, a parasitical leafless plant, is remarkable for differing from the order in nothing whatever, except its very peculiar habit.

GEOGRAPHY. Trees inhabiting the tropics of either hemisphere; in a very few instances only, straggling to the northward in North America and Europe. No genus is known to exist in any part of the continent of Africa, except the paradoxical Cassytha. This is the more remarkable, as several species of Laurus have been found both in Teneriffe and Madeira, and some other genera exist in Madagascar, and in the Isles of France and Bourbon. *Brown, Congo,* 464.

PROPERTIES. It would be difficult to name another order at once so important and uniform in its qualities as this, the species being universally aromatic, warm, and stomachic. Cinnamon and Cassia are the produce of various species; the most genuine are yielded by Laurus Cinnamomum and L. Cassia; but L. Culilaban and Malabathrum can both be substituted for these spices: the Cinnamon of the Isle of France is Laurus cupularis, that of Peru is L. Quixos. The Cinnamon of Santa Fé is produced by Laurus Cinnamomoides. *Humb. Cinch. For.* 27. *Eng. ed.* The Sassafras nuts of the London shops are the fruit of the Laurus Pucheri of the Flora Peruviana. *Ibid.* Camphor is yielded by Laurus Camphora and other species; even by the Cinnamon tree itself. The properties of all these are due to the presence of a volatile oil; but they also contain in many cases a fixed oil which is supposed to constitute the principal part of the fruit of Persea gratissima, so much esteemed in the West Indies under the name of the Avocado Pear; the same oil appears in the form of a greasy exudation in the fruit of Litsea sebifera. A species of Laurus in Sumatra, called by Dr. Jack, Parthenoxylon, yields an oil useful in rheumatic affections; and an infusion of the roots is drank as sassafras, the qualities of which it resembles. *Ed. P. J.* 6. 398. The bark of Laurus Benzoin is highly aromatic, stimulant, and tonic, and is extensively used in North America in intermittent fevers. The oil of the fruit is said to be stimulant. *Barton,* 2. 95. A plant of this family found in the forests of Spanish Guiana yields a volatile oil, with a warm and pungent taste and aromatic smell. It is employed externally as a discutient, and internally as a diaphoretic, diuretic, and resolvent. *Ed. P. J.* 12. 417. The volatile oil obtained from some species of Laurus found in vast forests between the Oronoko and the Parime, is produced in great abundance by merely making an incision into the bark with an axe, as deep as the liber. It gushes out in such quantity, that several quarts may be obtained by a single incision. It has the reputation of being a powerful discutient. For further information, see *Brewster's Journal,* 1. 134. In addition to these qualities, there is present in some species an acrid, red, or violet juice, like that found in Myristiceæ; it is particularly abundant in L. parvifolia, globosa, fœtens, and caustica.

EXAMPLES. Laurus, Cinnamomum, Tetranthus, Cassytha.

XXII. BERBERIDEÆ. THE BERBERRY TRIBE.

BERBERIDEÆ, *Vent. Tabl.* 3. 83. (1799); *Dec. Syst.* 2. 1. (1821); *Prodr.* 1. 105. (1824); *Lindl. Synops.* 14. (1829.)

DIAGNOSIS. Polypetalous dicotyledons, with hypogynous stamens equal in number to the petals and opposite them, anthers opening by recurved valves, and a single simple carpellum.

ANOMALIES.

ESSENTIAL CHARACTER.—*Sepals* 3-4-6, deciduous, in a double row, surrounded externally by petaloid scales. *Petals* hypogynous, either equal to the sepals in number, and opposite to them, or twice as many, generally with an appendage at the base in the inside. *Stamens* equal in number to the petals, and opposite to them; *anthers* generally with two separated cells, opening elastically with a valve from the bottom to the top. *Ovarium* solitary, 1-celled; *style* rather lateral; *stigma* orbicular. *Fruit* berried or capsular. *Seeds* attached to the bottom of the cell on one side, 1,2, or 3; *albumen* between fleshy and corneous; *embryo* straight in the axis; *cotyledons* flat.—*Shrubs* or *herbaceous perennial* plants, for the most part smooth. *Leaves* alternate, compound, without *stipulæ.*

AFFINITIES. Botanists appear of one opinion in considering Menispermeæ the nearest order to this, agreeing in having the stamens opposite the petals, the floral envelopes regularly imbricated, 3 or 4 in each row, never 5, the fruit usually baccate, and fleshy albumen. These, however, differ in their habit, the separation of the sexes in distinct flowers, and the presence of several distinct carpella, while in Berberideæ there is never more than one, which is perfectly simple, as is demonstrated by the position of the placentæ, the single style, &c. With Podophylleæ they are connected through Leontice and Diphylleia, which have a near relation to Jeffersonia and Podophyllum itself. In the singular structure of their anthers there is a striking analogy with Laurineæ, Atherospermeæ, and Hamamelideæ, orders not otherwise akin to Berberideæ. Leontice thalictroides offers one of the few instances of seeds being absolutely naked, that is to say, not covered by any integument originating in the pericarpium. In this plant the ovarium is ruptured in an early state by the expansion of the ovulum, which, having been impregnated, continues to grow, and ultimately arrives at maturity, although deprived of its pericarpial covering. The spines of the common Berberry are a curious state of leaf, in which the parenchyma is displaced, and the ribs have become indurated. They, as well as all the simple leaves of ordinary appearance, are articulated with the petiole, and are therefore compound leaves reduced to a single foliole; whence the supposed genus Mahonia does not differ essentially from Berberis in foliage any more than in fructification. Berberideæ are related to Anonaceæ through the genus Bocagea; their ovarium is generally like that of Anonaceæ. Aug. St. Hilaire remarks, that the opposition of the stamens to the petals, and the erect ovules, place them in alliance with Vites. *Fl. Braz.* 1. 47.

GEOGRAPHY. Natives chiefly of mountainous places in the temperate parts of the northern hemisphere. Some have, however, been found in South America as far as the Straits of Magellan; none in Africa, Australasia, or in the South Sea islands. *Dec.* There are several species of Berberry in Chile.

PROPERTIES. The berries of Berberis vulgaris and other species are acid and astringent, and form with sugar an agreeable refreshing preserve. Their acid is the oxalic. The stem and bark of the Berberry are excessively astringent, and are employed for that reason by dyers. *Dec.* The root yields a yellow dye. *A. Rich.*

EXAMPLES. Berberis, Leontice, Achlys.

XXIII. MENISPERMEÆ. THE COCCULUS TRIBE.

MENISPERMEÆ, *Juss. Gen.* 284. (1789); *Dec. Syst.*1. 508. (1818).—MENISPERMACEÆ, *Dec. Prodr.* 1. 95. (1824.)

DIAGNOSIS. Polypetalous dicotyledons, with hypogynous stamens opposite the petals, distinct simple carpella, minute unisexual flowers, and twining shrubby stems.

ANOMALIES. In Agdestis, a doubtful genus of the order, the flowers are hermaphrodite. Cissampelos, Stauntonia, Pselium, and Schizandra, have no petals in their male flowers. Schizandra is scarcely a twiner.

ESSENTIAL CHARACTER.—*Flowers* (by abortion?) unisexual, usually diœcious and very small. *Sepals* and *petals* confounded, in one or several rows, each of which is composed of either 3 or 4 parts, hypogynous, deciduous. *Stamens* monadelphous, or occasionally distinct, sometimes opposite the petals and equal to them in number, sometimes 3 or 4 times as many. *Anthers* adnate, turned outwards or proceeding immediately from the point of the filament. *Ovaries* sometimes numerous, each with one style, cohering slightly at the base, sometimes completely soldered together into a many-celled body, which is occasionally in consequence of abortion 1-celled. *Drupes* usually berried, 1 seeded, oblique or lunate, compressed. *Seed* of the same shape as the fruit ; *embryo* curved, or turned in the direction of the circumference ; *albumen* wanting, or in very small quantity ; *cotyledons* flat, sometimes lying face to face, sometimes distant from each other and lying in separate cells of the seed ; *radicle* superior, but its position is sometimes obscured by the curvature of the seed. —*Shrubs*, with a flexible tough tissue, and sarmentaceous habit. *Leaves* alternate, entire or occasionally divided, mucronate. *Flowers* small, usually racemose.

AFFINITIES. The relation that is borne by these plants to Berberideæ has been pointed out under that order : some Anonaceæ agree with them in having a twining habit, and the whole resemble them in the ternary division of their flowers ; they are, however, abundantly distinct : M. Decandolle points out a resemblance with Sterculiaceæ, consisting in the monadelphous stamens and peltate leaves ; but it is of little moment. The ternary and quaternary arrangement of the flowers is very remarkable among Dicotyledons. According to Aug. St. Hilaire, this order is related to Euphorbiaceæ through Phyllanthus, the male flowers of which are in certain species absolutely the same as those of Cissampelos. It also approaches Malvaceæ by those genera which, like Caperonia, have stipulate leaves, and distinct caducous petals separated from the calyx by the gynophore. *Fl. Braz.* 59. The position of the seed is altered materially from that of the ovulum in the progress of the growth of the fruit. According to Aug. St. Hilaire, the ovulum of Cissampelos is attached to the middle of the side of a straight ovarium, which after fecundation gradually incurves its apex until the style touches the base of the pericarp, when the two surfaces being thus brought into contact unite, and a drupe is formed, the seed of which is curved like a horse-shoe, and the cavity of which is divided by a spurious incomplete dissepiment, consisting of two plates : the attachment of the seed is at the top of the false dissepiment, on each side of which it extends equally. *Pl. Usuelles*, no. 35. The whole order requires careful revision by means of living plants, and is well worth the especial attention of some Indian botanist.

GEOGRAPHY. The whole of this order consists of fewer than a hundred species, which are common in the tropics of Asia and America, but uncommon out of those latitudes : all Africa contains but 5, North America 6, and Siberia 1. The species are universally found in woods, twining round other plants.

PROPERTIES. The root of several species is bitter and tonic, and the seeds of some of them narcotic. The root of Menispermum palmatum *Lam.* or the Columbo root, is esteemed highly on account of its powerful antiseptic, tonic, and astringent properties. See *Bot. Mag. fol.* 2970. Menispermum cordifolium of Willd., called *Gulancha* in Bengal, is used extensively in a variety of diseases by the native practitioners of India, especially in such as are attended by febrile symptoms not of a high inflammatory kind, and in fevers of debility : the parts used are the root, stems, and leaves, from which a decoction called *Páchana* is prepared. A sort of extract called *Pálo* is obtained from the stem, and is considered an excellent remedy in urinary affections and gonorrhœa. *Trans. M. & P. Soc. Calc.* 3. 298. Cocculus platyphylla is used by the Brazilians in intermittent fevers and liver complaints. Its properties, like those of Cocculus cinerescens, are highly esteemed, and appear to be due to the presence of a bitter and tonic principle. In the seed of Cocculus suberosus the bitter crystallisable poisonous principle has been detected, called picrotoxia. *Pl. Usuelles*, 42. The roots of the *Orelha de Onça* of Brazil, Cissampelos ovalifolia, are bitter, and their decoction is employed with success in intermittent fevers. *Ibid.* no. 34. Cissampelos ebracteata, also called *Orelha de Onça*, is reputed an antidote to the bite of serpents. *Ib.* no. 35. The root of Cissampelos pareira and Abuta amara is both diuretic and aperient, and known under the name of Pareira brava. *Dec.* The Abuta candicans of Cayenne, where it is known by the name of Liane amère, is extremely bitter. *Ibid.* The drug called in the shops Cocculus indicus is the seed of Menispermum Cocculus, and is well known for its narcotic properties, especially in poisoning fishes. Nevertheless, according to Decandolle, the berries of Menispermum edule *Lam.* are eaten with impunity in Egypt ; but they are acrid, and a very intoxicating liquor is obtained from them by distillation. The bitter poisonous principle of Cocculus indicus is the above-mentioned vegetable alkali, *picrotoxia.* It has been supposed that a peculiar acid, called the *menispermic*, also existed in the same plant ; but this is now known to have been merely a mixture of sulphuric and oxalic acids. *Turner*, 653.

EXAMPLES. Cocculus, Menispermum, Cissampelos.

XXIV. MALVACEÆ. The Mallow Tribe.

MALVACEÆ, *Juss. Gen.* 271. (1789) *in part.* ; *Brown in Voy. to Congo*, p. 8. (1818) ; *Kunth Diss.* p. 1. (1822) ; *Dec. Prodr.* 1. 429. (1824) ; *Lindl. Synops.* p. 40. (1829) ; MALVACEÆ, § Malveæ, *Aug. St. Hil. Fl. Bras. mer.* 1. 173. (1827.)

DIAGNOSIS. Polypetalous dicotyledons, with hypogynous monadelphous stamens, concrete carpella, an ovarium of several cells, and the placentæ in the axis, a calyx with valvate æstivation, 1-celled anthers bursting longitudinally, no disk, crumpled cotyledons, and alternate stipulate leaves with stellate pubescence.

ANOMALIES. In Malope the carpella are numerous, and distinct, not arranged in a single row, as in the rest of the order.

ESSENTIAL CHARACTER.—*Sepals* 5, very seldom 3 or 4, more or less united at the base, with a valvate æstivation, often bearing external bracteæ forming an involucrum. *Petals* of the same number as the sepals, hypogynous, with a twisted æstivation, either distinct or adhering to the tube of the stamens. *Stamens* usually indefinite, sometimes of the same number as the petals, hypogynous ; *filaments* monadelphous ; *anthers* 1-celled, reniform, bursting transversely. *Ovarium* formed by the union of several carpella round a common axis, either distinct or coherent ; *styles* the same number as the carpella, either united or distinct ; *stigmata* variable. *Fruit* either capsular or baccate, its carpella being either monospermous or polyspermous, sometimes united in one, sometimes separate or separable ; dehiscence either loculicidal or septicidal. *Seeds* sometimes hairy ; *albumen* none, or in small quantity ; *embryo* curved, with twisted and doubled *cotyledons.*—*Herbaceous* plants, *trees*, or *shrubs.* Leaves alternate, more or less divided, stipulate. *Hairs* stellate. *Peduncles* usually axillary.

AFFINITIES. The relation of Malvaceæ with Sterculiaceæ, Tiliaceæ, Bombaceæ, and Elæocarpeæ, is clearly indicated by their general accordance in structure, and especially by the valvate æstivation of their calyx. With other orders they also agree in numerous points ; as, with Ranunculaceæ in the indefinite stamens and distinct aggregated carpella of Malope ; with

Ternströmiaceæ in their monadelphous stamens; with Chlenaceæ in the presence of an involucrum below the flower, and monadelphous stamens; with Lineæ in their mucilaginous properties, definite seeds, many-celled fruit, and unguiculate petals; and through the medium of this last order with Caryophylleæ.

GEOGRAPHY. These plants are found in great abundance in the tropics, plentifully in the hotter parts of temperate regions, but gradually diminishing to the north. Thus in Sicily they form $\frac{1}{86}$ of the flowering plants (*Presl*), in France $\frac{1}{145}$ (*Humboldt*), in Sweden $\frac{1}{233}$ (*Wahl.*), in Lapland unknown, in the temperate parts of North America $\frac{1}{125}$, in the equinoctial parts of the same continent $\frac{1}{47}$; or, taking into account only the vegetation of the valleys, they, according to Humboldt, form $\frac{1}{50}$ of the flowering plants in the tropics, $\frac{1}{200}$ in the temperate zone, and are not found in the frigid zone. But these calculations no doubt include at least Bombaceæ and Sterculiaceæ.

PROPERTIES. The uniform character is to abound in mucilage, and to be totally destitute of all unwholesome qualities. The use to which Mallows and Marsh-mallows are applied in Europe is well known. Similar properties are possessed by extra-European species. Sida cordifolia mixed with rice is used to alleviate the bloody flux. Emollient fomentations are prepared from Sida mauritiana by the Hindoo doctors. *Ainslie*, 1. 205. The flowers of Bençao de Deos, Abutilon esculentum, are used in Brazil as a boiled vegetable. *Pl. Usuelles*, 51. A decoction of Sphæralcea Cisplatina is administered in the same country in inflammations of the bowels, and is generally employed for the same purposes as the Marsh-mallow in Europe. *Ib*. 52. Pavonia diuretica is prescribed in Brazil as a diuretic; it is supposed to act rather as an emollient. *Ibid*. 53. The wood is always very light, and of little value. Rocket-sticks are obtained from the light straight stems of Sida micrantha. *Ibid*. 49. The chewed leaves of another species, S. carpinifolia, are applied in Brazil to the punctures of wasps. *Ib.* 50. The bark is often so tenacious as to be manufactured into cordage. Malva crispa was found by Cavanilles to be fit for this purpose; and several species of Hibiscus are employed in like manner in tropical countries. From the fibres of the bark of Hibiscus arboreus the whips are manufactured with which the negro slaves are lashed in the West India Islands. The plant is called *Mohoe* or *Mohaut*. *Hamilt. Prodr*. 49. The petals of some are astringent; this property exists in. Malva Alcea (*Dec*.) and in Hibiscus Rosa sinensis, of which the Chinese make use to blacken their eyebrows and the leather of their shoes. *Ib*. The leaves of Althea rosea are said to yield a blue colouring matter not inferior to indigo. *Ed. P. J.* 14. 376. A decoction of the root and stem of Urena lobata is employed in Brazil as a remedy in windy cholic; the flowers are used as an expectorant in dry and inveterate coughs. The bark furnishes good cordage. *Pl. Us.* 56. A few species, such as Hibiscus Sabdariffa and surattensis, &c., are slightly acid. The unripe fruit of the Ochro, or Hibiscus esculentus, is a favourite ingredient in soups, which are thickened by the mucilaginous quality of this plant. The musky seeds of Hibiscus Abelmoschus are considered cordial and stomachic, and by the Arabians are mixed with coffee. *Ainslie*, 2. 73. The root of Sida lanceolata is intensely bitter, and is considered a valuable stomachic. *Ainslie*, 2. 179. It has been supposed that the root of Althæa officinalis contains a peculiar alkaline principle called *Althein;* but it has since been stated by M. Plisson that it does not exist; what was taken for it having been Asparagin. *Brewster*, 8. 369. The *Cotton* of commerce is the hairy covering of the seeds of several species of Gossypium.

EXAMPLES. Malva, Lavatera, Hibiscus.

XXV. CHLENACEÆ.

CHLENACEÆ, *Thouars Hist. Veg. Afr. Austr.* 46. (1806); *Dec. Prodr.* 1. 521. (1824.)

DIAGNOSIS. Polypetalous dicotyledons, with hypogynous indefinite monadelphous stamens, concrete carpella, an ovarium with several cells, and suspended ovules, an imbricated calyx enclosed in an involucrum, stipulate leaves, and round anthers bursting longitudinally.

ANOMALIES. Leptolæna has definite stamens.

ESSENTIAL CHARACTER. — *Involucrum* 1-2-flowered, persistent, of variable form and texture. *Sepals* 3, small; æstivation imbricated? *Petals* 5 or 6, hypogynous, broader at the base, sometimes cohering there. *Stamens* either very numerous, or sometimes only 10; *filaments* either cohering at the base into a tube, or adhering to the tube of petals; *anthers* roundish, adnate, or loose, 2-celled. *Ovarium* single, 3-celled; *style* 1, filiform; *stigma* triple. *Capsule* 3-celled, or 1-celled by abortion. *Seeds* solitary or numerous, attached to the centre, suspended; *embryo* green, central; *albumen* fleshy according to Jussieu, or horny according to Du Petit Thouars; *cotyledons* foliaceous, wavy.—*Trees* or *shrubs*. *Leaves* alternate, with stipulæ, entire. *Stipulæ* deciduous. *Flowers* in panicles or racemes. *Dec.*

AFFINITIES. The monadelphous stamens and involucrated flowers indicate an affinity with Malvaceæ. But Jussieu refers them rather to the vicinity of Ebenaceæ, considering the order monopetalous, and the seeds albuminous. Very little is, in fact, known of these plants.

GEOGRAPHY. They are only eight certain species, which are all natives of Madagascar.

PROPERTIES. Handsome shrubs, with fine flowers, often red; but nothing is known of their qualities.

EXAMPLES. Sarcolæna, Leptolæna, Rhodolæna.

XXVI. BOMBACEÆ. THE COTTON TREE TRIBE.

BOMBACEÆ, *Kunth. Diss. Malv.* p. 5. (1822); *Dec. Prodr.* 1. 475. (1824); *A. St. Hilaire Fl. Br. merid.* 1. 257. (1827); *a section of* Malvaceæ.

DIAGNOSIS. Polypetalous dicotyledons, with hypogynous polyadelphous stamens, concrete carpella, an ovarium of several cells with the placentæ in the axis, a calyx with valvate æstivation, 1-celled anthers bursting longitudinally, no disk, flat cotyledons, and alternate stipulate leaves with stellate pubescence.

ANOMALIES. In Cheirostemon there are no petals, and the stamens are united in a 1-sided 5-lobed body.

ESSENTIAL CHARACTER. — *Sepals* 5, cohering in a campanulate or cylindrical tube, which is either truncate, or with 5 divisions: at the base of this, on the outside, are sometimes a few minute bracteæ. *Petals* 5, regular; or sometimes none, but in that case the inside of the calyx is coloured. *Stamens* 5, 10, 15, or more; *filaments* cohering at the base into a tube, which is soldered to the tube of the petals, divided at the apex into 5 parcels, each of which bears one or more anthers, among which are sometimes some barren threads; *anthers* 1-celled, linear, reniform or anfractuose. *Ovarium* consisting of 5 carpella, rarely of 10, either partly distinct or cohering strictly, and dehiscing in various ways; *styles* as many as the carpella, either distinct or more or less coherent; *ovula* 2, or many more. *Fruit* variable, capsular, or indehiscent, usually with 5 valves, septiferous in the middle. *Seeds* often enveloped in wool or pulp; sometimes albuminous, with flat *cotyledons;* sometimes exalbuminous, with shrivelled or convolute *cotyledons.*—*Trees* or *shrubs*. *Leaves* alternate, with stipulæ. *Pubescence* of the herbaceous parts stellate.

AFFINITIES. So near Malvaceæ, that they may perhaps be considered rather a section than a distinct order. They are, however, often possessed of a peculiar habit, being chiefly large trees, with broad umbrageous leaves, and fine showy flowers. Their calyx is thick, and has not the regular valvate æstivation of true Malvaceæ; they are also known by their pentadelphous stamens. The *Hand plant* of Mexico (Cheirostemon) owes its name to this latter circumstance; its five bundles of stamens being thick, coloured, and all turned to one side, so as to resemble a paw with five claws.

GEOGRAPHY. The station seems to be the hottest parts of the world; for the Plagianthus of Forster, referred here by M. Decandolle, probably does not belong to the order. The principal part of the species are South American or West Indian; a few Helicteres, one Eriodendron, one Bombax, and the Durio, being all that are recorded from the East Indies, and Adansonia and Ophelus being the only African plants of the order.

PROPERTIES. These, like Malvaceæ, are mucilaginous plants, having no known deleterious properties. Bombax pentandrum, the Cotton Tree of India, yields a gum, which is given in conjunction with spices in certain stages of bowel complaints. *Ainslie*, 2. 97. The largest tree in the world is the Adansonia, or Baobab Tree, the trunk of which has been found with a diameter of 30 feet; but its height is not in proportion. " It is emollient and mucilaginous in all its parts. The leaves dried and reduced to powder constitute *Lalo*, a favourite article with the Africans, which they mix daily with their food, for the purpose of diminishing the excessive perspiration to which they are subject in those climates; and even Europeans find it serviceable in cases of diarrhœa, fevers, and other maladies. The fruit is, perhaps, the most useful part of the tree. Its pulp is slightly acid and agreeable, and frequently eaten; while the juice is expressed from it, mixed with sugar, and constitutes a drink, which is valued as a specific in putrid and pestilential fevers." *Hooker Bot. Mag.* 2792. The dried pulp is mixed with water, and administered, in Egypt, in dysentery. It is chiefly composed of a gum, like Gum Senegal, a sugary matter, starch, and an acid which appears to be the malic. *Delile, Cent.* 12. The fruit of the Durian is considered one of the most delicious productions of nature; it is remarkably fœtid, and therefore disagreeable to those who are unaccustomed to it, but it universally becomes in the end a favourite article of the dessert. It is found in the islands of the Indian Archipelago, where it is cultivated extensively; see *Hort. Trans.* 5. 106. The seeds of many of the species are enveloped in long hairs, like those of the true Cotton: it is found, however, that they cannot be manufactured, in consequence of no adhesion existing between the hairs. This is said to arise from the hairs being perfectly smooth, and destitute of certain asperities found upon the hairs of the true Cotton, to which that plant owes its valuable properties. The woolly coat of the seeds of the *Arvore de Paina* (Chorisia speciosa), and several species of Eriodendron and Bombax, is employed in different countries for stuffing cushions, and for similar domestic purposes. *Pl. Us.* 63. Helicteres Sacarolha, called by the latter name only in Brazil, is used against venereal disorders: a decoction of the root is administered. It is supposed that its effects depend upon its mucilaginous properties. *Ibid.* 64.

EXAMPLES. Bombax, Matisia, Montezuma, Eriodendron.

XXVII. STERCULIACEÆ.

STERCULIACEÆ, *Vent. Malm.* 2. 91. (1799.) — HERMANNIACEÆ, *Juss.* — BYTTNERIACEÆ, *Brown in Flinders,* 2. 540. (1814); *Kunth. Diss.* p. 6. (1822); *Dec. Prodr.* 1. 481. (1824); *Aug. St. Hil. Fl. Bras. mer.* 1. 139. (1827); *a section of* Malvaceæ.

DIAGNOSIS. Polypetalous dicotyledons, with hypogynous monadelphous stamens, concrete carpella, an ovarium of several cells, and the pla-

centæ in the axis, a calyx with valvate æstivation, 2-celled anthers bursting
longitudinally, no disk, and alternate stipulate leaves with stellate pubescence.

ANOMALIES. The carpella of Sterculia and Erythropsis are distinct,
and their flowers have no petals. True Büttneriaceæ have five abortive
stamens. Waltheria has but one carpellum, four being abortive.

ESSENTIAL CHARACTER. — *Calyx* either naked or surrounded with an involucrum,
consisting of 5 sepals, more or less united at the base, with a valvular æstivation. *Petals* 5,
or none, hypogynous, convolute in æstivation, often saccate at the base, and variously
lengthened at the apex. *Stamens* definite or indefinite, monadelphous in various ways,
some among them being often sterile; *anthers* 2-celled, turned outwards. *Pistillum* consisting of 5, or rarely 3, carpella, either distinct or cohering into a single ovarium; *styles*
equal in number to the carpella, distinct or united; *ovula* erect. *Fruit* capsular, with 3 or
5 cells. *Seeds* with a strophiolate apex, often winged; *albumen* oily or fleshy, rarely wanting; *embryo* straight, with an inferior radicle; *cotyledons* either foliaceous, flat, and plaited,
or rolled round the plumula, or else very thick, but this only in the seeds without albumen.
— *Trees* or *shrubs*. *Pubescence* often stellate. *Leaves* alternate, simple, often toothed,
with stipulæ. *Peduncles* cymose.

AFFINITIES. I take this order as it is understood by Kunth and
Decandolle, without being at all certain that Büttneriaceæ, as proposed
by Mr. Brown, are not really distinct. As it now stands, it comprehends
plants very variable in some of their characters, as will appear from the
distinctions of the sections enumerated further on. Differing as these
do from each other, they are all distinguished from their nearest allies,
Malvaceæ, by their 2-celled anthers, and from Tiliaceæ and Elæocarpeæ by
their monadelphous stamens. Their valvate calyx is the great mark of combination which unites them with these last-mentioned orders. The fruit
of Sterculia often exhibits beautiful illustrations of the real nature of that
form of fruit which botanists call the follicle, and helps to demonstrate that
it, and hence all simple carpella, are formed of leaves, the sides of which
are inflexed, and the margins dilated into placentæ, bearing ovula. In Sterculia platanifolia, in particular, the follicles burst and acquire the form of
coriaceous leaves, bearing the seeds upon their margin. But, notwithstanding this peculiarity of the distinct carpella, on account of which
Sterculia would, as the type of an order, be referable to another artificial
section, it is impossible to doubt that Reevesia, a remarkable Chinese
plant, having the habit and peculiar conformation of anthers found in Sterculia, along with the petals and fruit of Pterospermum, completely identifies
the genus with polypetalous syncarpous orders.

The following are the sections: —

§ 1. TRUE STERCULIACEÆ.

Büttneriaceæ, § Sterculiaceæ, *Kunth*, l. c. (1822). § Sterculieæ, *Dec.*
Prodr 1. 481. (1824.)

Flowers frequently unisexual. Flowers with or without petals. Stamens often connected in a long column, bearing the anthers at the apex.
Fruit either deeply lobed, or concrete.—Trees. Leaves simple, entire, or
lobed; petioles with a swelling at both their base and apex.

EXAMPLES. Sterculia, Heritiera, Reevesia.

§ 2. DOMBEYACEÆ.

Büttneriaceæ, § Dombeyaceæ, *Kunth*, l. c. (1822). *Dec.* l. c. (1824.)

Calyx 5-lobed. Petals 5, rather large, unequal-sided, convolute in
æstivation. Stamens some multiple of the number of the petals, in a
single row, monadelphous, rarely all fertile, usually some sterile, thread-
or strap-shaped; some (usually 2 or 3 between each sterile stamen) fertile,
and more or less combined. Styles from 3 to 5, combined or distinct.

Ovula 2 or more in each cell, in two rows. Embryo straight, in the axis of fleshy albumen. Cotyledons leafy, often bifid, crumpled or flat. *Dec.*
EXAMPLES. Pentapetes, Astrapæa, Dombeya.

§ 3. WALLICHIEÆ.

Büttneriaceæ, § Wallichieæ, *Dec. Mem. Mus.* 10. 102. (1823); *Prodr.* 1. 501. (1824.)
Calyx 5-lobed, surrounded by an involucrum, consisting of from 3 to 5 leaves, and distant from the flower. Petals 5, flat. Stamens numerous, with long monadelphous filaments, of which the outermost are the smallest, arranged in a column like those of Malvaceæ. Anthers erect, 2-celled. *Dec.*
EXAMPLES. Eriolæna, Wallichia.

§ 4. HERMANNIACEÆ.

Hermanniaceæ, *Juss. ex Kunth Diss.* p. 11. (1822); *Nov. Gen.* 5. 312. (1821); *Dec. Prodr.* 1. 490. (1824); *a section of* Büttneriaceæ.
Flowers hermaphrodite. Calyx 5-lobed, persistent, either with or without an involucrum. Petals 5, twisted spirally before expansion. Stamens 5, monadelphous in a slight degree, all fertile and opposite the petals, with ovate 2-celled anthers. Carpella concrete. Albumen between fleshy and mealy. Embryo included; radicle inferior, ovate. Cotyledons flat, leafy, entire. *Dec.*—Shrubs, or herbaceous plants. Leaves alternate, simple, entire, or variously cut. Stipules 2, adhering to the petioles. Peduncles axillary, or opposite the leaves, or terminal, with 1, 3, or many flowers, which are usually in umbels. *Kunth.* M. Decandolle assigns these plants a curved embryo; but all Hermanniaceæ have it not.
EXAMPLES. Melochia, Hermannia, Riedleia.

§ 5. TRUE BÜTTNERIACEÆ.

Büttneriaceæ, *R. Brown*, l. c.; *Kunth*, l. c. p. 6.—Büttnerieæ, *Dec. Prodr.* 1. 484.
Petals usually hollowed out at the base, and expanded at the point into a sort of strap. Filaments 5, sterile, ligulate, opposite the petals; others fertile, alternate, solitary, or pentadelphous in threes, or with but a single anther. Ovarium 5-celled, the cells usually 2-seeded. Seeds sometimes without albumen, with thick cotyledons; sometimes albuminous, with foliaceous, plane, or convolute cotyledons. *Dec.*—Trees, shrubs, or very rarely herbaceous plants. Leaves alternate, entire, sometimes cut. Stipules twin. Peduncles axillary, opposite the leaves, and terminal, with 1 or many flowers. *Kunth.*
EXAMPLES. Theobroma, Guazuma, Commersonia, Büttneria.

§ 6. LASIOPETALEÆ.

Lasiopetaleæ, *Gay. Mem. Mus.* 7. 431. (1821).—Büttneriaceæ, § Lasiopetaleæ, *Kunth*, l. c. (1822); *Dec.* l. c. (1824.)
Calyx 5-parted, petaloid, persistent, or withering. Petals minute, like scales, or wanting. Filaments subulate, connate at the base; sometimes 5, opposite the petals; sometimes 10, alternately barren and fertile. Anthers incumbent, with contiguous lobes. Ovarium with from 3 to 5 cells, each of which contains from 2 to 8 ovules. Carpella 5, 2-valved, usually closely concrete, or partially distinct. Seeds strophiolate at the base. Albumen fleshy. Embryo erect. Cotyledons flat, foliaceous. *Dec.*—Shrubs. Leaves alternate, usually in threes, simple, entire, or lobed. Stipules twin (or perhaps none). Inflorescence cymose, corymbose, or racemose, opposite the leaves, very rarely produced within the leaves. Pedicels with bracteæ, sometimes articulated above the middle. *Kunth.*
EXAMPLES. Lasiopetalum, Seringia.

GEOGRAPHY. India, New Holland, the Cape of Good Hope, and South America, with the West Indies, are the chief countries inhabited by this order, taken collectively; but its various sections are each characterised by peculiarities of geographical distribution. Thus :—

Sterculiaceæ are principally found in India and equinoctial Africa; 5 or 6 only have been discovered in Mexico and South America.

Dombeyaceæ are all African or East Indian, mostly the latter, with the exception of Pentapetes ovata, found in New Spain.

Wallichieæ are half Indian and half South American; but 4 species only are on record in the whole.

Of *Hermanniaceæ* two-thirds are found exclusively at the Cape of Good Hope; the remainder are chiefly West Indian and South American; about one-tenth are natives of the East Indies, and two or three are found in the South Seas.

The *Büttneriaceæ* are principally natives of South America and the West Indies; about one-seventh is found in the East Indies, a similar number in New Holland, and a single species, Glossostemon Bruguieri, in Persia.

Lasiopetaleæ are exclusively from New Holland.

PROPERTIES. These, like the orders most nearly related to them, are chiefly remarkable for the abundance of mucilage they contain. The seeds of Sterculia acuminata afford the Kola spoken of by African travellers, which, when chewed or sucked, renders the flavour of water, even if half putrid, agreeable. The seeds of the Chichà, Sterculia Chicha, are eaten as nuts by the Brazilians. *Pl. Usuelles*, 46. The Gum Tragacanth of Sierra Leone is produced by a species of Sterculia (*St. Tragacantha Mihi*). The pod of Sterculia fœtida is, according to Horsfield, employed in gonorrhœa in Java. The leaves are considered repellent and aperient. A decoction of the fruit is mucilaginous and astringent. *Ainslie*, 2. 119. The bark of a species of Sterculia is employed in the Moluccas as an emmenagogue; and the seeds of all that genus are filled with an oil, which may be expressed and used for lamps. There is a slight acridity in the seeds of Sterculia. The Waltheria Douradinha is used in Brazil as a remedy for venereal disorders, for which its very mucilaginous nature renders it proper. *Pl. Usuelles*, 36. The fruit of Guazuma ulmifolia is filled with a sweet and agreeable mucilage, which the Brazilians suck with much pleasure. In Martinique the young bark is used to clarify sugar, for which the copious mucilage it yields when macerated qualifies it. In the same island the infusion of the old bark is esteemed as a sudorific, and useful in cutaneous diseases. *Ibid*. 47. The buttery, slightly bitter substance, called *Cocoa*, is obtained from the seeds of Theobroma Cacao, and from this Chocolate is prepared.

XXVIII. MORINGEÆ.

MORINGEÆ, *R. Brown in Denham*, p. 33. (1826.)

DIAGNOSIS. Polypetalous dicotyledons, with perigynous stamens, concrete carpella, a superior 1-celled ovarium with parietal placentæ, a 3-valved capsule, somewhat irregular flowers, and embryo without albumen.

ANOMALIES.

ESSENTIAL CHARACTER. — *Calyx* consisting of 5 nearly equal divisions (deciduous, *Dec.*), the tube lined with a fleshy disk; *æstivation* slightly imbricated. *Corolla* of 5 nearly equal petals, the uppermost of which is ascending. *Stamens* 10, arising from the top of the tube of the calyx; 5 opposite the sepals, sometimes sterile; *filaments* slightly petaloid, callous and hairy at the base; *anthers* simple, 1-celled, with a thick convex connectivum. *Ovarium* stipitate, superior, 1-celled, with 3 parietal placentæ; *style* filiform, terminal, not obliquely inserted; *stigma* simple. *Fruit* a long pod-like capsule, with 3 valves, and only 1 cell; the valves bearing the seeds along their middle. *Seeds* numerous, half buried in the fungous substance of the valves, sometimes winged; *embryo* without albumen; *radicle* straight, very small; *cotyledons* fleshy, plano-convex.—*Trees.* *Leaves* pinnate, with an odd one. *Flowers* in panicles.

AFFINITIES. Confounded with Leguminosæ, until separated by the authority of Mr. Brown, who does not, however, point out the real affinities of the order. M. Decandolle, who did not overlook its anomalous structure as a Leguminous plant, accounted for the compound nature of its fruit upon the supposition, that although unity of carpellum is the normal structure of Leguminosæ, yet the presence of more ovaria than one, in a few instances in that order, explained the constantly trilocular state of that of Moringa. To this, however, there are numerous and grave objections, which cannot fail to strike every botanist. To me it appears very near Bignoniaceæ, notwithstanding its polypetalous corolla, agreeing with that order in its compound fruit, winged seeds, irregular flowers, and compound leaves. It may be also compared with Malvaceæ, on account of its nearly valvate sepals, or rather with Büttneriaceæ on the same account, and because of its sterile stamens alternating with the fertile ones; its habit is, however, against the approximation, and it is probable that these coincidences indicate analogy rather than affinity.

GEOGRAPHY. Natives of the East Indies and Arabia.

PROPERTIES. The root of the Hyperanthera Moringa has a pungent odour, with a warm, biting, and somewhat aromatic taste; it is used as a stimulant in paralytic affections and intermittent fever; it is also employed as a rubefacient. *Ainslie*, 1. 175. The nuts (seeds) of this plant, called by the French *pois quéniques* and *chicot*, have been used in venereal affections. *Ibid.*

EXAMPLE. Moringa.

XXIX. TILIACEÆ. THE LINDEN TRIBE.

TILIACEÆ, *Juss. Gen.* 290. (1789) *in part.; Kunth. Malv. Diss.* p. 14. (1822); *Dec. Prodr.* 1. 503. (1824); *Lindl. Coll.* p. 54. (1829.)

DIAGNOSIS. Polypetalous dicotyledons, with hypogynous distinct stamens, concrete carpella, an ovarium with several cells, and the placentæ in the axis, a calyx with valvate æstivation, anthers bursting longitudinally, and hypogynous glands between the petals and ovarium.

ANOMALIES. Petals sometimes absent. Diplophractum is remarkable for having an extremely anomalous fruit, with several spurious cells, and with the placentæ apparently in the circumference instead of the axis. Apeiba has sometimes as many as 24 cells in the fruit. Mr. Brown notices the existence of an African genus of this order (Christiana, *Dec.*), remarkable in having a calyx of 3 lobes, while its corolla consists of 5 petals; the fruit composed of 5 single-seeded capsules, connected only at the base. *Cong.* 428.

ESSENTIAL CHARACTER. — *Sepals* 4 or 5, with a valvular æstivation, usually with no involucrum. *Petals* 4 or 5, entire, usually with a little pit at their base; very seldom wanting; most commonly the size of the sepals. *Stamens* generally indefinite, hypogynous, distinct; *anthers* 2-celled, dehiscing longitudinally; in Sparmannia the outer stamens are barren. *Disk* formed of glands, equal in number to the petals, at the base of which they are placed, adhering to the stalk of the ovarium. *Ovarium* single, composed of from 4 to 10 carpella; *style* one; *stigmata* as many as the carpella. *Fruit* dry, of several cells. *Seeds* numerous; *embryo* erect in the axis of fleshy *albumen*, with flat foliaceous *cotyledons.* — *Trees* or *shrubs*, very seldom *herbaceous* plants. *Leaves* simple, stipulate, toothed, alternate. *Flowers* axillary.

AFFINITIES. These resemble Sterculiaceæ, Malvaceæ, and the orders allied to them, in most respects, and especially in the valvate æstivation of their calyx. They are known by their glandular disk and distinct stamens, with 2-celled anthers.

GEOGRAPHY. The principal part of the order is found within the tropics all over the world, forming mean weed-like plants, or shrubs, or trees, with handsome, usually white or pink, flowers. A small number is peculiar to the northern parts of either hemisphere, where they form timber-trees.

PROPERTIES. They have all a mucilaginous, wholesome juice. The leaves of Corchorus olitorius are used in Egypt as a pot-herb. The berries of some of them are succulent and eatable. The species are more remarkable for the toughness of the fibres of their inner bark, which are used for various economical purposes. Fishing lines and nets are made in India of Corchorus capsularis; and the Russian mats of commerce are manufactured from the Tilia. The bark of Luhea paniculata is used in Brazil for tanning leather. The wood of Luhea divaricata, which is white and light, but very close grained, makes good musket-stocks, and wooden soles for shoes. The Brazilians call all such *Açoita cavallos*, because the sticks they use for driving their cattle are generally obtained from them. *Pl. Us.* 66.

EXAMPLES. Tilia, Sparmannia, Corchorus.

XXX. ELÆOCARPEÆ.

ELÆOCARPEÆ, *Juss. Ann. Mus.* 11. 223. (1808); *Dec. Prodr.* 1. 519. (1824.)

DIAGNOSIS. Polypetalous dicotyledons, with numerous hypogynous distinct stamens, concrete carpella, a many-celled ovarium with the placentæ in the axis, a calyx with valvate æstivation, anthers bursting by pores, and lacerated imbricated petals.

ANOMALIES. None, if Decadia, a genus of which little is known, with round anthers and 10 slightly serrated petals, be excluded.

ESSENTIAL CHARACTER. — *Sepals* 4 or 5, with a valvular æstivation, and no involucrum. *Petals* 4 or 5, hypogynous, lobed or fringed at the point. *Disk* glandular, somewhat projecting. *Stamens* from 15 to 20; *filaments* short, distinct; *anthers* long, filiform, 4-cornered, 2-celled, the cells opening by an oblong pore at the apex. *Ovarium* many-celled; *style* one. *Fruit* variable, either indehiscent, dry, or drupaceous, or valvular. *Seeds* 2 or more in each cell; *albumen* fleshy; *embryo* erect, with flat, leafy *cotyledons.* — *Trees* or *shrubs.* *Leaves* alternate, entire or serrated, simple, with deciduous *stipulæ.* *Flowers* racemose.

AFFINITIES. These differ from Tiliaceæ only in their fringed petals, and anthers opening by two pores at the apex. *Dec.* M. Kunth combines them with that order. *Diss. Malv.* p. 16.

GEOGRAPHY. Of the described species, 10 are found in the East Indies, 4 in South America, 2 in New Holland, and 2 in New Zealand; several more, however, exist in India.

PROPERTIES. Nothing more is known than that the fruit of some is eatable. They are handsome trees or shrubs, with showy flowers; and the furrowed, sculptured, bony fruit of the Elæocarpi, being freed from its pulp, forms handsome necklaces, which are not uncommonly set in gold, and sold in the shops.

EXAMPLES. Elæocarpus, Vallea.

XXXI. DIPTEROCARPEÆ. THE CAMPHOR TREE TRIBE.

DIPTEROCARPEÆ, *Blume Bijdr.* p. 222. (1825); *Fl. Javæ* (1829).

DIAGNOSIS. Polypetalous dicotyledons, with hypogynous indefinite stamens, subulate anthers opening towards the apex, concrete carpella, an ovarium of several cells with pendulous ovules in pairs, a tubular calyx with imbricated æstivation, and a fruit surrounded by the dilated unequal foliaceous calyx.

ANOMALIES.

ESSENTIAL CHARACTER.— *Calyx* tubular, 5-lobed, unequal, naked at the base; æstivation imbricated. *Petals* hypogynous, sessile, combined at the base; æstivation contorted. *Stamens* indefinite, hypogynous, distinct, or slightly and irregularly polyadelphous; *anthers* innate, subulate, opening longitudinally towards the apex; *filaments* dilated at the base. *Ovarium* superior, without a disk, few-celled; *ovules* in pairs, pendulous; *style* single; *stigma* simple. *Fruit* coriaceous, 1-celled by abortion, 3-valved or indehiscent, surrounded by the enlarged calyx. *Seed* single, without albumen; *cotyledons* twisted and crumpled, or unequal and obliquely incumbent; *radicle* superior.—Elegant *trees*, abounding in resinous juice. *Leaves* alternate, involute in vernation, with veins running out from the midrib to the margin; *stipules* deciduous, oblong, convolute, terminating the branches with a taper point. *Peduncles* terminal, or almost so, in racemes or panicles; *flowers* usually large.

AFFINITIES. Very near Elæocarpeæ, but also allied to Malvaceæ in the contorted æstivation of the corolla, and the crumpled cotyledons: they differ from the latter in having the stamens either distinct or partially combined, long narrow 2-celled anthers, and pendulous ovules; and from the former in their petals not being fringed, and in want of albumen. Their resinous juice, solitary superior ovarium, drupaceous fruit, numerous long anthers, irregular coloured calyx, and single exalbuminous seed, allies them, as Blume remarks, to Guttiferæ, from which their stipulæ and the æstivation of the corolla abundantly distinguish them. The enlarged foliaceous unequal segments of the calyx, while investing the fruit, point out this family at once.

GEOGRAPHY. Only found in the eastern islands of the Indian Archipelago, where, according to Blume, they form the largest trees of the forest.

PROPERTIES. Here belongs the famous Camphor tree of Sumatra, Dryobalanops Camphora, which is no doubt a species of Dipterocarpus. The camphor is found in a concrete state in the cavities and fissures in the heart of the tree. It is less volatile than the common camphor of commerce. *Ed. P. J.* 6. 400. See remarks upon this tree in Blume's *Flora Javæ*. Shorea robusta yields a balsamic resin used in the temples of India. The fruit of Vateria indica (Piney Tree) is boiled for the sake of a tallow, which rises to the surface of the water, and forms a hard cake when cool. In this state it is whitish, greasy to the touch, with rather an agreeable odour. It is extremely tenacious and solid, but melts at a temperature of 97½° Fahr. *Brewster*, 4. 186.

EXAMPLES. Dipterocarpus, Dryobalanops.

XXXII. TERNSTRÖMIACEÆ.

TERNSTRÖMIEÆ, *Mirb. Bull. Philom.* 381. (1813.)—TERNSTRÖMIACEÆ, *Dec. Mem.
Soc. H. N. Genev.* vol. 1. (1823); *Prodr.* 1. 523. (1824); *Cambessédes Mémoire*
(1828.)—THEACEÆ, *Mirb. Bull. Phil.* (1813.)—CAMELLIEÆ, *Dec. Theor. Elem,
ed.* 1. (1813); *Prodr.* 1. 529. (1824.)

DIAGNOSIS. Polypetalous dicotyledons, with hypogynous, indefinite,
monadelphous, or polyadelphous stamens, concrete carpella, an ovarium of
several cells, with the placentæ in the axis, a persistent imbricated many-
leaved calyx, alternate simple leaves, and definite seeds.

ANOMALIES. Cochlospermum has the ovarium 1-celled, with imperfect
septa, to the margins of which the ovula are attached. Leaves very rarely
opposite. *Cambessédes.*

ESSENTIAL CHARACTER.—*Flowers* very rarely polygamous. *Sepals* 5 or 7, imbricated in æstivation, concave, coriaceous, deciduous, the innermost often the largest. *Petals*
5, 6, or 9, equal in number to the sepals, often combined at the base. *Stamens* very numerous; *filaments* filiform, monadelphous, or polyadelphous; *anthers* versatile, or adnate. *Ovarium* superior, with several cells; *styles* from 3 to 7, filiform, more or less combined; *ovules*
pendulous, or erect, or peltate. *Capsule* 2-7-celled and capsular, with the dehiscence taking
place in various ways; sometimes coriaceous and indehiscent; usually with a central
column. *Seeds* large, attached to the axis, very few; *albumen* none, or in very small
quantity; *embryo* straight, bowed or folded back, the radicle turned to the hilum; *cotyledons*
very large, often filled with oil, occasionally plaited lengthwise; an arillus sometimes
present.—*Trees* or *shrubs.* Leaves alternate, coriaceous, without *stipulæ*, usually undivided,
now and then with pellucid dots. *Peduncles* axillary or terminal, articulated at the base.
Flowers generally white, seldom pink or red, very rarely (in Cochlospermum) yellow.

AFFINITIES. This order originated in 1813, with M. Mirbel, who
separated some of its genera from Aurantiaceæ, where they had been
placed by Jussieu, and at the same time founded another closely allied
order, under the name of Theaceæ. These opinions were substantially
adopted by Messrs. Kunth and Decandolle, the latter of whom, moreover,
formed several sections among his Ternströmiaceæ. It is, however, certain,
that no solid difference exists between this last order and Theaceæ or
Camellieæ, as they were called by Decandolle; and Cambessédes, after a
careful revision of the whole, has come to the conclusion, that even the sections proposed by Decandolle among Ternströmiaceæ are untenable. I shall
profit by M. Cambessédes' observations in all I have to say upon the order.
Ternströmiaceæ may be compared, in the first place, with Guttiferæ,
with which they accord more closely than with any thing else, and in the
affinities of which they entirely participate. They differ thus: in Ternströmiaceæ the leaves are alternate, to which there are scarcely any exceptions; they are always opposite in Guttiferæ. In the former the normal
number of the parts of the flower appears to be 5 and its multiples; in
Guttiferæ it is evidently two. In the former the calyx is always perfectly
distinct from the corolla; these two organs are usually confounded in the
latter. Ternströmiaceæ have the petals generally united at the base, and a
twisted æstivation; in Guttiferæ they are distinct, with a convolute æstivation. The seeds of the former are almost always either destitute of albumen,
or furnished with a membranous wing; the latter have neither the one nor
the other. The first have the radicle always near the hilum; the second
have it either near the hilum or turned in an opposite direction. Finally,
in Guttiferæ the cotyledons are very thick, and firmly glued together; and
this character, which is not observed in Ternströmiaceæ, is the more important, as it is not liable to any exception. Ternströmiaceæ are allied
to Hypericineæ through the medium of Carpodontos, a genus which, with
the foliage of the latter order, has the fruit of the former; and also of
certain plants of Hypericineæ, which, according to Cambessédes, have

a definite number of seeds. With Marcgraaviaceæ they agree through Norantea, which has the stamens slightly adherent to the base of the petals, and fixed anthers; but that order is entirely different in habit, and is well marked by its singular cucullate bracteæ, its fruit, and its wingless exalbuminous seeds. Many genera of Ternströmiaceæ, such as Kielmeyera and others, have the habit of Tiliaceæ, while the fruit of Laplacea is strikingly like that of Luhea; but the æstivation of the calyx and many other characters distinguish them.

GEOGRAPHY. Although the plants of this order which are known in European gardens are chiefly from China or North America, these form but an inconsiderable part of the whole: 7 or 8 are all that are contained in the first of these countries, and 4 in the latter; while between 60 and 70, all beautiful trees or shrubs, are natives of the woods of South America: about a score are known in the East Indies, and one in Africa.

PROPERTIES. These are ill understood, but little being known of the greater part of the species. The tea which is so extensively consumed by Europeans is produced by different species of Thea and Camellia. An excellent table oil is expressed from the seeds of Camellia oleifera. The different species and varieties of Camellia japonica are the glory of gardeners. The fruit of a species of Saurauja is said to be acidulous, and to resemble Tomatoes in flavour. *Dec.* The leaves of Kielmeyera speciosa are employed in Brazil for fomentations, for which they are well adapted, on account of the mucilage in which they abound. *Pl. Us.* 58. It is believed in Brazil, that a decoction of the roots of a plant called *Butua do curvo* (Wittelsbachia insignis *Mart.*, Maximilianea regia *Ibid.*, Cochlospermum insigne *Aug. St. H.*) has the power of healing internal abscesses. The Brazilians take it for all kinds of internal bruises. *Pl. Us.* 57.

EXAMPLES. Thea, Gordonia, Saurauja, Ternströmia.

XXXIII. LECYTHIDEÆ.

LECYTHIDEÆ, *Richard MSS. Poiteau Mem. Mus.* 13. 141. (1825); *Dec. Prodr.* 3. 290, (1828); *a sect. of* Myrtaceæ. *Ach. Richard in Ann. des Sc.* 1. 321. (1824.)

DIAGNOSIS. Polypetalous dicotyledons, with indefinite perigynous stamens, concrete carpella, an inferior ovarium of several cells, round anthers, indefinite ovula, and exalbuminous seeds.

ANOMALIES. Ovula sometimes definite.

ESSENTIAL CHARACTER.—*Calyx* superior, 2- to 6-leaved, or urceolate, with a divided limb; æstivation valvate or imbricated. *Corolla* consisting of 6 petals, sometimes cohering at the base, with an imbricated æstivation. *Stamens* indefinite, epigynous, either connected into a single petaloid cucullate unilateral body, or monadelphous at the base. *Ovarium* inferior, 2- to 6-celled; *ovula* indefinite, or definite attached to the axis; *stigma* simple. *Fruit* a woody capsule, either opening with a lid or remaining closed. *Seeds* several, covered by a thick integument; *embryo* without *albumen*, either undivided, or with two large plaited leafy or fleshy *cotyledons*, sometimes folded upon the *radicle*, which is next the *hilum*.—Large *trees*, with alternate entire or toothed *leaves*, with minute deciduous stipulæ, and without pellucid dots. *Flowers* large, showy, terminal, solitary, or racemose.

AFFINITIES. Combined by Decandolle and others with Myrtaceæ, from which they differ most essentially in their alternate, often serrated, leaves, without pellucid dots. To me they appear, notwithstanding the perigynous station of their stamens, to be more nearly allied to Ternströmiaceæ. For an account of the germination of Lecythis, see *Du Petit Thouars, Ess.* 3. 32.

GEOGRAPHY.   Natives of the hottest parts of South America, especially of Guiana.

PROPERTIES.   The fruit of Couroupita guianensis, called *Abricot sauvage* in Cayenne, is vinous and pleasant.   The most gigantic tree in the ancient forests of Brazil is that called the *Sapucaya*.   It is the Lecythis ollaria, the seeds of which are large and eatable.   *Pr. Max. Trav.* 83.   The fleshy seeds of all the species of Lecythis are eatable, but they leave a bitter unpleasant after-taste in the mouth.   The bark of L. ollaria is easily separable, by beating the liber into a number of fine distinct layers, which divide so neatly from each other, that, when separated, they have the appearance of thin satiny paper.   Poiteau says he has counted as many as 110 of these coatings.   The Indians cut them in pieces, as wrappers for their cigars.   The well-known Brazil nuts of the shops of London are the seeds of Bertholletia excelsa.   The lacerated parts of the flowers of Couroupita guianensis become blue upon exposure to the air.   The Gustavia urceolata is called *bois puant*, because its wood becomes, after similar exposure, excessively foetid.   *Poiteau,* l. c.

EXAMPLES.   Bertholletia, Lecythis, Gustavia.

XXXIV. GUTTIFERÆ.   THE MANGOSTEEN TRIBE.

GUTTIFERÆ, *Juss. Gen.* 243. (1789); *Dec. Prodr.* 1. 557. (1824); *Cambessédes Mémoire* (1828).

DIAGNOSIS.   Polypetalous dicotyledons, with hypogynous indefinite unequal stamens, adnate anthers, concrete carpella, an ovarium of several cells with the placentæ in the axis, a persistent imbricated many-leaved calyx, opposite simple leaves without stipulæ, and resinous juice.

ANOMALIES.   Havetia has the anthers *immersed* in a fleshy receptacle. The ovarium of Calophyllum is 1-celled, and the petals opposite the sepals.

ESSENTIAL CHARACTER.—*Flowers* hermaphrodite, or unisexual. *Sepals* from 2 to 6, usually persistent, round, membranous, and imbricated, frequently unequal and coloured. *Petals* hypogynous, from 4 to 10, passing insensibly into sepals. *Stamens* numerous, either distinct, or combined in one or more parcels; hypogynous, rarely definite; *filaments* of various lengths; *anthers* adnate, bursting inwards, sometimes very small, occasionally bursting outwards, sometimes 1-celled, and sometimes opening by a pore. *Disk* fleshy, occasionally 5-lobed. *Ovarium* solitary, superior, 1- or many-celled; *ovules* solitary, erect, or ascending, or numerous and attached to central placentæ; *style* none, or very short; *stigma* peltate, or radiate. *Fruit* either dry or succulent, 1 or many-celled, 1- or many-seeded, dehiscent or indehiscent. *Seeds* frequently nestling in pulp; their coat thin and membranous; always apterous, very frequently with an arillus; *albumen* none; *embryo* straight; *cotyledons* thick, inseparable; *radicle* either turned to or from the hilum. — *Trees* or *shrubs*, occasionally parasitical, yielding resinous juice. *Leaves* without stipulæ, opposite, very rarely alternate, coriaceous, entire, with a strong midrib, and often with the lateral veins running through to the margin. *Flowers* usually numerous, axillary, or terminal, white, pink, or red, articulated with their peduncle.

AFFINITIES.   In treating of Ternströmiaceæ I have made use of the excellent memoir of Cambessédes for the purpose of explaining the affinities of that order with this; and I draw the following comparisons from the same source; premising only, that European botanists are much in want of good observations upon living plants of Guttiferæ, and that there is no order that is more in need of elucidation from some skilful Indian botanist than this.   M. Cambessedes remarks, that Guttiferæ differ from Hypericineæ in their branches, their leaves, and their articulated peduncles; in the normal number of the parts of their flowers, which appears to be two and its multiples, instead of three or four, which obtains in Hypericineæ ; in their anthers united the whole length with the filament, and not articulated at the summit; in their seeds, which often have an arillus, and are solitary in each cell of the ovarium, a character found in no Hypericineæ (the mono-spermous cells of the fruit of some Vismias is due to abortion); finally, in the structure of the embryo, which is different in the two orders. Marcgraavi-aceæ are disinguished by their alternate leaves, the singular form of their lower bracteæ, their petals frequently united, and by their seeds being very small, and exceedingly numerous.

GEOGRAPHY. All natives of the tropics, the greater part of South Ame-rica ; a few are from Madagascar, none from the continent of Africa. They generally require situations combining excessive heat and humidity.

PROPERTIES. The species all abound in a viscid, yellow, acrid, and purgative gum-resinous juice resembling Gamboge. According to some, the Stalagmitis Gambogioides yields the gum-resin called Gamboge, which is obtained by removing the bark or by breaking the leaves and young shoots. This susbtance, or something approaching it very nearly, is also obtained from Garcinia celebica, and a plant named Gambogia gutta. The powerful drastic cathartic properties of Gamboge are well known. If dissolved in water, and examined beneath a very powerful microscope, this substance will be found to consist entirely of active molecules. According to Dr. Hamilton, there is no ground for supposing the Gamboge to be produced by Garcinia Cambogia, as some have believed. *L. Tr.* 13. 485. In the West Indies the juice of Mammea is employed to destroy the chiggers, little insects which attack the naked feet, introducing themselves into the flesh below the toe-nails. The bark of many kinds is astringent and slightly vermifugal. The berry of Garcinia Mangostana is believed to be the most grateful to the palate of all the fruits that are known. The Butter and Tallow-tree of Sierra Leone, which owes its name (Pentadesma butyracea) to the yellow greasy juice its fruit yields when cut, belongs to this order.

EXAMPLES. Garcinia, Calophyllum, Clusia.

XXXV. MARCGRAAVIACEÆ.

MARCGRAAVIACEÆ, *Juss. Ann. Mus.* 14. 397. (1809) ; *Dec. Prodr.* 1. 565. (1824.)

DIAGNOSIS. Polypetalous dicotyledons, with hypogynous indefinite sta-mens, concrete carpella, an ovarium of several cells with the placentæ in the axis, a persistent imbricated many-leaved calyx, alternate simple leaves and indefinite seeds.

ANOMALIES. The corolla is calyptriform in Antholoma and Marc-graavia.

ESSENTIAL CHARACTER. — *Sepals* from 2 to 7, usually coriaceous and imbricated. *Corolla* hypogynous ; sometimes monopetalous, calyptriform, entire, or torn at the point, sometimes consisting of five petals. *Stamens* indefinite, inserted either on the receptacle or on a hypogynous membrane ; *filaments* dilated at the base ; *anthers* long, innate, bursting inwards. *Ovarium* single, superior, usually furrowed, many-celled, many-seeded ; *style* single ; *stigma* single or capitate ; *ovula* numerous, attached to a central placenta. *Capsule* coriaceous, consisting of several valves which separate slightly ; *dissepiments* pro-ceeding from the middle of the valves, but not meeting in the centre, so that the fruit is 1-celled. *Seeds* very minute and numerous, nestling in pulp.—*Shrubs*, having sometimes a scrambling habit. *Leaves* alternate. *Flowers* in umbels or spikes. *Peduncles* naked, or furnished with either simple or cucullate hollow *bracteæ.*

AFFINITIES. The station of this order is uncertain ; it approaches Eben-aceæ in its monopetalous corolla cut round at the base, in the anthers attached by their base, and the alternate leaves : Ericeæ in the anthers and disk of the genus Antholoma : Hypericineæ and Guttiferæ in the hypogynous stamens, the polypetalous corolla of some genera, placentation and numerous seeds ; wherefore Jussieu stationed the order near Clusia. *Dec. Prodr.* 1. 565. (1824.) M. Turpin has somewhere remarked, that the bracteæ of this order offer a clear explanation of the conversion of a degenerated leaf into an ovulum.

GEOGRAPHY. All found in equinoctial America, except Antholema, which is a native of New Caledonia.

PROPERTIES. Handsome and curious plants, remarkable for their singular cucullate bracteæ. Nothing is known of their qualities.

M. Decandolle distinguishes

Sub-order I. MARCGRAAVIEÆ.

Corolla calyptriform. Stamens inserted in the receptacle.

Sub-order II. NORANTEÆ.

Petals 5. Stamens pressed close to the corolla, and as if inserted into it.
EXAMPLES. Norantea, Marcgraavia.

XXXVI. HYPERICINEÆ. THE TUTSAN TRIBE.

HYPERICA, *Juss. Gen.* 254. (1789.)—HYPERICINEÆ; *Chois. Prodr. Hyp.* 32.(1821); *Dec. Prodr.* 1. 541. (1824); *Lindl. Synops.* p. 41. (1829.)

DIAGNOSIS. Polypetalous dicotyledons, with hypogynous indefinite symmetrical polyadelphous stamens, concrete carpella, an ovarium of several cells with the placentæ in the axis, an irregular calyx with imbricate æstivation, indefinite seeds, and resinous yellow juice.

ANOMALIES. Laneritia has 10 monadelphous stamens. Some species of Vismia have solitary seeds, according to Cambessédes.

ESSENTIAL CHARACTER.—*Sepals* 4-5, either more or less cohering, or wholly distinct, persistent, unequal, with glandular dots. *Petals* 4-5, hypogynous, with a twisted æstivation and oblique venation, often having black dots. *Stamens* indefinite, hypogynous, in three or more parcels ; *anthers* versatile. *Ovary* single, superior ; *styles* several, rarely connate ; *stigmas* simple, occasionally capitate. *Fruit* a capsule or berry, of many valves and many cells ; the edges of the former being curved inwards. *Seeds* minute, indefinite, usually tapering, attached to a placenta in the axis or on the inner edge of the dissepiments ; *embryo* straight, with an inferior *radicle* and no *albumen.*—*Herbaceous* plants, *shrubs*, or *trees*, with a resinous juice. *Leaves* opposite, entire, dotted, occasionally alternate and crenated. *Flowers* generally yellow. *Inflorescence* variable.

AFFINITIES. Nearly allied to Guttiferæ, from which they chiefly differ in their small round and versatile anthers, numerous styles, and polyspermous capsules. To Cistineæ they approximate in many points, differing principally in their fruit, polyadelphous stamens, and dotted leaves. With Saxifrageæ they appear to me to have a strong relation, through the medium of Parnassia, the fringed glands of which are analogous to the polyandrous fascicles of Hypericum. The leaves of Hypericineæ are very commonly marked with dots, which are either transparent, or black and opaque.

GEOGRAPHY. These are very generally spread over the surface of the earth, inhabiting mountains and valleys, marshes and dry plains, meadows and heaths. The following is the distribution of them, according to M. Choisy :— Europe, 19 ; North America, 41 ; South America, 21 ; West Indies, 1 ; Asia, 24 ; New Holland, 5 ; Africa and the neighbouring islands,

7; Azores and Canaries, 5; common to Europe and Asia, 4; common to Europe, Asia, and Africa, 1. (*Choisy Prodr.* 1821.)

PROPERTIES. The juice of many species is slightly purgative and febrifugal; it is most copious in the Vismias, and is analogous to Gamboge, has a resinous smell, and gives out to spirit of wine, or oil, a red colour, which may be employed in dyeing. Hypericum hircinum is fœtid. A gargle for sore throats is prepared in Brazil from the Hypericum connatum, commonly called *Orelha de Gato. Pl. Us.* 61. A decoction of the leaves of another species, Hypericum laxiusculum, or *Allecrim brabo*, is reputed in the same country as a specific against the bites of serpents. *Ib.* 62.

EXAMPLES. Hypericum, Vismea, Elodea.

The following sections are employed by M. Choisy:—

Sub-order I. TRUE HYPERICINEÆ.

Seeds taper. Styles usually from 3 to 5.

Tribe 1. VISMIEÆ. Fruit a berry. Flowers in distinct, leafless, racemose, or corymbose panicles. Trees or shrubs. Leaves stalked.

Tribe 2. HYPERICEÆ. Fruit a capsule. Flowers terminal or axillary. Herbaceous plants or under-shrubs. Leaves usually sessile.

Sub-order II. ANOMALOUS HYPERICINEÆ.

Seeds flat, winged. Styles more than 5.

XXXVII. REAUMURIEÆ.

REAUMURIEÆ, *Ehrenberg in Ann. des Sc.* 12. 78. (1827.)

DIAGNOSIS. Polypetalous dicotyledons, with indefinite hypogynous stamens, concrete carpella, an imbricated calyx, an ovarium of several cells, several styles, and villous seeds definite in number.

ANOMALIES.

ESSENTIAL CHARACTER.—*Calyx* 5-parted, surrounded externally by imbricated bracteæ. *Petals* 5, hypogynous. *Stamens* definite or indefinite, hypogynous, with or without an hypogynous disk; *anthers* peltate. *Ovarium* superior; *styles* several, filiform, or subulate. *Fruit* capsular, with 2 to 5 valves, and as many cells, and a loculicidal dehiscence. *Seeds* definite, villous, erect; *embryo* straight, surrounded by a small quantity of mealy *albumen*; radicle next the hilum.—*Shrubs.* Leaves fleshy, scale-like, or small, alternate, without stipulæ. *Flowers* solitary.

AFFINITIES. Dr. Ehrenberg suggests (*Ann. des Sc.* 12. 78.) that Reaumuria and Hololachna, both of which have, according to him, hypogynous stamens, may constitute a little group, to be called Reaumurieæ. To me the order appears more nearly related to Hypericineæ than to either Ficoideæ or Tamariscineæ. From the former it chiefly differs in its succulent habit, and definite villous seeds, agreeing, in Reaumuria at least, even in the obliquity of the veins of the petals, and in the leaves being dotted. From Ficoideæ its hypogynous stamens and seeds distinguish it; from Tamariscineæ its plurilocular ovarium and distinct styles; from Nitrariaceæ its erect villous seeds, distinct styles, and hypogynous stamens.

GEOGRAPHY. Natives of the Mediterranean and the milder parts of Northern Asia.

PROPERTIES. None except the presence of saline matter in great abundance.

EXAMPLES. Reaumuria, Hololachna.

XXXVIII. SAXIFRAGEÆ. The Saxifrage Tribe.

SAXIFRAGÆ, *Juss. Gen.* 308. (1789); *Vent. Tabl.* 2. 277. (1799).—SAXIFRAGEÆ, *Dec. and Duby*, 207. (1828); *Lindl. Synops.* 66. (1829.)

DIAGNOSIS. Polypetalous dicotyledons, with perigynous definite stamens, (2) ovaria adhering more or less to the calyx and to each other, indefinite seeds, and no stipulæ.

ANOMALIES. Parnassia has 4 parietal placentæ opposite the lobes of the stigma. Petals sometimes absent. Adoxa is a doubtful genus of the order, with a berry of several cells. In Heuchera the flowers are irregular.

ESSENTIAL CHARACTER.—*Calyx* either superior or inferior, of 4 or 5 sepals, which cohere more or less at their base. *Petals* 5, or none, inserted between the lobes of the calyx. *Stamens* 5-10, inserted either into the calyx (perigynous), or beneath the ovarium (hypogynous); *anthers* 2-celled, bursting longitudinally. *Disk* either hypogynous or perigynous, sometimes nearly obsolete, sometimes annular and notched, rarely consisting of 5 scales. *Ovarium* inferior, or nearly superior, usually consisting of 2 carpella, cohering more or less by their face, but distinct at the apex; sometimes 2-celled with a central placenta; sometimes 1-celled with parietal placentæ; rarely 4- or 5-celled. *Styles* none. *Stigmata* sessile on the tips of the lobes of the ovarium. *Fruit* generally a membranous 1- or 2-celled capsule with 2 bracteæ; rarely a 4-celled 4-valved capsule; sometimes a 4-celled berry. *Seeds* numerous, very minute; usually with long hexagonal reticulations on the side of a transparent testa. *Embryo* taper, in the axis of fleshy albumen, with the radicle next the hilum.—*Herbaceous* plants, often growing in patches. *Leaves* simple, either divided or entire, alternate, without stipulæ. *Flower-stems* simple, often naked.

AFFINITIES. Most nearly allied to Rosaceæ, with the herbaceous part of which they agree in habit, and from which they differ in their polyspermous partially concrete carpella, albuminous seeds, and want of stipulæ. From Cunoniaceæ they are divided rather by their habit, and by the want of stipulæ, than by any thing very positive in their fructification; the principal characteristic feature of which consists in the more perfect concretion of the carpella. Baueraceæ are known by their habit, indefinite stamens, and peculiar dehiscence of the anthers. To Caryophylleæ their habit allies them; but they differ in the insertion of their stamens, their placentation, the situation of their embryo, and otherwise. Portulaceæ, which may be compared with them, particularly on account of the situation of their stamens, want of stipulæ, and albuminous seeds, differ essentially in the structure of the embryo, in the want of symmetry in the parts of the flower, and in placentation. Grossulaceæ, however different they are in habit, agree very much in the general structure of the flowers; they differ in the ovarium being completely concrete and inferior, with two parietal placentæ, in the seeds being attached to long umbilical cords, in the albumen being corneous, and the embryo extremely minute. Chrysosplenium and Adoxa are both remarkable for the want of petals; and Parnassia, which I think, upon the whole, is a genuine genus of this order, exhibits the singular anomaly of placentæ being opposite the lobes of the stigma, an unilocular ovarium, the shell of which consists of two distinct plates connected by an intervening loose substance, and a peculiar development of an hypogynous disk, which assumes the form of 5 fringed scales, alternate with the stamens, and of a highly curious structure. Adoxa, which has a berry of several cells, and which is always referred here, appears to me far more anomalous than Parnassia. Drummondia has the stamens equal in number to the petals and opposite them, thus indicating some analogy with the monopetalous Primulaceæ.

GEOGRAPHY. Little elegant herbaceous plants, usually with white flowers, cæspitose leaves, and glandular stems : some of the species have yellow flowers, others have red, but none blue. They are natives of mountainous tracts in Europe and the northern parts of the world, frequently forming the chief beauty of that rich turf which is found near the snow in high Alpine stations. Some grow on rocks and old walls, and in hedge-rows, or near rivulets, or in groves.

PROPERTIES. According to Decandolle, the whole order is more or less astringent. The root of Heuchera americana is a powerful astringent, whence it is called in North America Alum root. *Barton*, 2. 162. Otherwise they possess no known properties; for the old idea of their being lithontriptic appears to have been derived from their name rather than their virtues.

EXAMPLES. Saxifraga, Robertsonia, Adoxa, Parnassia.

XXXIX. CUNONIACEÆ.

CUNONIACEÆ, *R. Br. in Flinders*, 548. (1814).

DIAGNOSIS. Polypetalous dicotyledons, with definite perigynous stamens, separate carpella, a more or less inferior ovarium, shrubby stem, and interpetiolar stipulæ.

ANOMALIES. Petals sometimes wanting.

ESSENTIAL CHARACTER.—*Calyx* 4 or 5 cleft, half superior or nearly inferior. *Petals* 4 or 5, occasionally wanting. *Stamens* perigynous, definite, 8-10. *Ovarium* 2-celled; the cells having 2 or many seeds; *styles* 1 or 2. *Fruit* 2-celled, capsular, or indehiscent. *Embryo* in the axis of fleshy albumen.—*Trees* or *shrubs*. *Leaves* opposite, compound or simple, usually with interpetiolar *stipulæ*.

AFFINITIES. More readily distinguished from Saxifrageæ by their widely different habit than by any very important characters in the fructification. *Brown in Flinders*, 548. The shrubby habit and remarkable interpetiolar stipules are their principal character. Baueraceæ are known by their indefinite stamens, porous anthers, and want of stipulæ.

GEOGRAPHY. Natives of the Cape, South America, and the East Indies.

PROPERTIES. A Weinmannia is used in Peru for tanning leather, and its astringent bark is employed to adulterate the Peruvian bark. The Indian Weinmannias appear to possess similar astringent qualities. *Dec.*

EXAMPLES. Cunonia, Weinmannia.

XL. BAUERACEÆ.

A section of Cunoniaceæ, *R. Brown in Flinders*, 548. (1814).

DIAGNOSIS. Polypetalous dicotyledons, with indefinite perigynous stamens, ovaria adhering more or less to the calyx and each other, anthers bursting by two pores, indefinite seeds, and no stipulæ.

ANOMALIES.

ESSENTIAL CHARACTER.—*Sepals* 8, foliaceous, inferior. *Petals* the same number, alternate with them, arising from the base of the calyx. *Stamens* indefinite, obscurely perigynous; *anthers* oblong, bursting by two pores at the apex. *Carpella* 2, a little inferior, coherent, each 1-celled, with numerous ovula attached to a common central axis; *style* one, filiform, to each ovarium. *Fruit* capsular, opening at the apex. *Seeds* indefinite, attached to a central placenta ; *embryo* in the axis of fleshy albumen, with a long taper radicle, pointing to the hilum.—*Shrubs.* *Leaves* toothed, ternate, opposite, without stipulæ. *Flowers* solitary, axillary.

AFFINITIES. I distinguish this small order both from Saxifrageæ and Cunoniaceæ by its indefinite stamens, anthers dehiscing by pores, and by its peculiar habit. It has always been considered an anomaly, with whichsoever of those two orders it has been combined, and is now conveniently separated from them. The origin of the petals and stamens appears at first sight to be hypogynous. But if a flower be carefully cut through vertically, it will be found that the ovarium coheres slightly with the calyx, and that the petals and stamens take their origin from above the point of cohesion. They are consequently perigynous, and not hypogynous.

GEOGRAPHY. Native of New Holland.

PROPERTIES. None that are known, except beauty.

EXAMPLE. Bauera only.

XLI. BRUNIACEÆ.

BRUNIACEÆ, *R. Brown in Abel's China* (1818); *Dec. Prodr.* 2. 43. (1825); *Ad. Brongniart in Ann. des Sc. Nat.* (1826).

DIAGNOSIS. Polypetalous dicotyledons, with perigynous stamens equal in number to the petals, concrete carpella, an inferior ovarium of from 1 to 3 cells, containing definite pendulous ovules, imbricated sepals, and embryo in the axis of albumen.

ANOMALIES. Berzelia has a single carpellum. Raspailia has the ovarium superior.

ESSENTIAL CHARACTER.—*Calyx* superior, 5-cleft, imbricated, occasionally nearly inferior. *Petals* alternate with the segments of the calyx, arising from its throat, imbricated. *Stamens* alternate with the petals, arising from the same point, or from a disk surrounding the ovarium ; *anthers* turned outwards, 2-celled, bursting longitudinally. *Ovarium* half inferior, with from 1 to 3 cells, in each of which there is from 1 to 2 suspended collateral ovula ; *style* simple or bifid ; *stigma* simple. *Fruit* dioecious or indehiscent, 2- or 1-celled, crowned by the persistent calyx. *Seeds* solitary or in pairs, suspended, sometimes with a short arillus ; *albumen* fleshy ; *embryo* minute at the base of the seed, with a conical superior radicle, and short fleshy cotyledons.—Branched, heath-like *shrubs*. *Leaves* small, imbricated, rigid, entire, with a callous point. *Flowers* small, capitate, or panicled, or even terminal, and solitary ; either naked, or with large involucrating bracteæ.

AFFINITIES. Nearly allied to Hamamelideæ, which are known by their habit, stipules, and deciduous valves of the anthers, and also by their valvate sepals and petals. Brongniart indicates an affinity with Myrtaceæ through Imbricaria, which is very nearly constructed as true Bruniaceæ, but has the stamens opposite the petals, and dotted leaves. The genus Raspailia is remarkable for having the stamens arising from the top of a superior ovarium ! and Thamnea is perhaps a solitary instance of a 1-celled ovarium with the ovules adhering to a central columnar axis. This order appears to me to approach Penæaceæ in several points.

GEOGRAPHY. All found at the Cape of Good Hope, with the exception of a single species inhabiting Madagascar.

PROPERTIES. Unknown.

EXAMPLES. Brunia, Linconia, Raspailia.

XLII. HAMAMELIDEÆ. THE WITCH-HAZEL TRIBE.

HAMAMELIDEÆ, *R. Br. in Abel's Voyage to China*, (1818); *A. Richard Nouv. Elém.* 532. (1828.)

DIAGNOSIS. Polypetalous dicotyledons, with perigynous stamens twice the number of the petals, concrete carpella, an inferior ovarium of 2 cells with solitary pendulous ovules, alternate leaves, deciduous stipulæ, valvate calyx, linear valvate-involute petals, and deciduous valves to the anthers.

ANOMALIES. Fothergilla is apetalous.

ESSENTIAL CHARACTER.—*Calyx* superior, in 4 pieces. *Petals* 4, linear, with a valvular æstivation. *Stamens* 8, of which 4 are alternate with the petals ; their *anthers* turned inwards, 2-celled, each cell opening by a valve which is finally deciduous, and 4 are sterile, and placed at the base of the petals. *Ovarium* 2-celled, inferior ; *ovules* solitary, pendulous or suspended ; *styles* 2. *Fruit* half inferior, capsular, usually opening with two septiferous valves. *Seeds* pendulous ; *embryo* in the midst of fleshy albumen ; *radicle* superior.—*Shrubs. Leaves* alternate, deciduous, toothed, with veins running from the midrib straight to the margin. *Stipulæ* deciduous. *Flowers* small, axillary.

AFFINITIES. Distinguished from Saxifrageæ by the deciduous valves of the anthers, definite seeds, and shrubby stem bearing alternate leaves and deciduous stipulæ. In the latter respect related to Cupuliferæ, from which the petals and calyx divide them. According to Mr. Brown, their affinity is on the one hand with Bruniaceæ, from which they are distinguished by the insertion and dehiscence of the anthers, the monospermous cells of the ovarium, the dehiscence of the capsule, the quadrifid calyx and habit ; and on the other with Cornus, Marlea, and the neighbouring genera ; in some respects also with Araliaceæ, but differing in their capsular fruit, the structure of the anthers, and other marks. See *Abel's Voyage, Appendix.*

GEOGRAPHY. Natives of North America and Japan, or the north of China.

PROPERTIES. Unknown.

EXAMPLES. Hamamelis, Fothergilla.

XLIII. PHILADELPHEÆ. THE SYRINGA TRIBE.

PHILADELPHEÆ, *Don in Jameson's Journal*, 133. (*April* 1826); *Dec. Prodr.* 3. 205. (1828.)

DIAGNOSIS. Polypetalous dicotyledons, with indefinite perigynous stamens, concrete carpella, an inferior ovarium of several cells, round anthers, indefinite ovula, and albuminous seeds.

ANOMALIES.

ESSENTIAL CHARACTER.—*Calyx* superior, with a persistent limb, having from 4 to 10 divisions. *Petals* alternate with the segments of the calyx, and equal to them in number, with a convolute-imbricate æstivation. *Stamens* indefinite, arising in 1 or 2 rows from the orifice of the calyx. *Styles* either distinct, or consolidated into one ; *stigmas* several. *Capsule* half inferior, with from 4 to 10 cells, many-seeded. *Seeds* scobiform, subulate, smooth, heaped in the angles of the cells upon an angular placenta ; *arillus ?* loose, membranous. *Albumen* fleshy ; *embryo* inverted, about as long as the albumen ; *cotyledons* oval, obtuse, flattish ; *radicle* longer than the cotyledons, superior, straight, obtuse.—*Shrubs. Leaves* deciduous, opposite, toothed, without dots or stipulæ. *Peduncles* axillary or terminal, in trichotomous cymes. *Flowers* always white.

AFFINITIES. The genera of this order were formerly referred to Myrtaceæ ; and I think there is a dissertation by the late President of the Linnean Society, in which he endeavoured to shew the difficulty of distinguishing Leptospermum even *generically* from Philadelphus,—so little did his school at that time know of the method of pursuing botanical inquiries. The affinity of the order has, however, been very properly shewn by Mr. Don to be not so much with Myrtaceæ as with Saxifrageæ, to which latter Philadelpheæ do in fact closely approach, differing widely in habit, but in fructification distinguished chiefly by the numerous cells of the fruit and the indefinite stamens. Decandolle points out an approach to Hydrangea; and if that génus does not actually belong to this order, it is at least probable that it is a link connecting it with Viburnum, agreeing almost equally with Philadelpheæ and Viburneæ in habit and fructification. Deutzia of Thunberg, which is not included in the order by Decandolle, certainly belongs to it; as I first learned from Mr. Brown's notes in Dr. Wallich's Herbarium, and as I since find stated by Mr. Don.

GEOGRAPHY. Deciduous shrubs, inhabiting thickets in Europe, North America, the north of India, and Japan.

PROPERTIES. Unknown.

EXAMPLES. Philadelphus, Deutzia.

XLIV. ESCALLONIEÆ.

ESCALLONIEÆ, *R. Brown in Franklin's Voyage*, 766. (1824.)

DIAGNOSIS. Polypetalous dicotyledons, with definite perigynous stamens, concrete carpella, an inferior ovarium of several cells with indefinite ovula, 5 sepals, and petals cohering in a tube.

ANOMALIES.

ESSENTIAL CHARACTER.—*Calyx* superior, 5-toothed. *Corolla* consisting of 5 petals, alternate with the segments of the calyx, from within which they arise, forming by their cohesion a tube, but finally separating from each other; æstivation imbricated. *Stamens* arising from the calyx, alternate with the petals; *anthers* bursting longitudinally. *Disk* conical, epigynous, plaited, surrounding the base of the style. *Ovarium* inferior, 2-celled, with two large polyspermous placentæ in the axis; *style* simple; *stigma* 2-lobed. *Fruit* capsular, 2-celled, surmounted by the persistent style and calyx, splitting by the separation of the cells at their base. *Seeds* very numerous and minute; with a transparent membranous integument; *embryo* minute, in the apex of oily *albumen*, its *radicle* at the opposite extremity of the hilum.—*Shrubs* with alternate, toothed, resinously glandular, exstipulate *leaves*, and axillary conspicuous *flowers*.

AFFINITIES. Distinguished from Grossulaceæ by the cohering petals, and by the radicle of the embryo being at the extremity most remote from the hilum; the albumen is also oily, not horny, and the placentæ are not parietal. From Philadelpheæ they are known by their glandular leaves and minute embryo; from Vaccinieæ by the final separation of the petals, and by the anthers.

GEOGRAPHY. All found in the temperate parts of South America, particularly Chile.

PROPERTIES. Unknown. Handsome shrubs, with evergreen leaves.

EXAMPLE. Escallonia.

XLV. GROSSULACEÆ. THE CURRANT TRIBE.

GROSSULARIEÆ, *Dec. Fl. Fr.* 4. 406. (1804); *Kunth Nov. G. et Sp.* 6. 58. (1823); *Dec. Prodr.* 3. 477. (1828). — RIBESIÆ, *Ach. Rich. Bot. Med.* 2. 487. (1823).— GROSSULACEÆ, *Mirb. Elém.* 2. 897. (1815); *Lindl. Synops.* 106. (1829.)

DIAGNOSIS. Polypetalous dicotyledons, with 5 perigynous fertile sta-mens, concrete carpella, an inferior ovarium with one cell and parietal pla-centæ, baccate fruit, and distinct petals and sepals.

ANOMALIES.

ESSENTIAL CHARACTER.—*Calyx* superior, 4- or 5-parted, regular, coloured. *Petals* 5, minute, inserted in the throat of the calyx. *Stamens* 5, inserted alternately with the petals, very short. *Ovarium* 1-celled, with 2 opposite parietal placentæ; *ovules* numerous; *style* 2-3-4-cleft. *Berry* crowned with the remains of the flower, 1-celled; the cell filled with pulp. *Seeds* numerous, suspended among the pulp by long filiform funiculi; *testa* exter-nally gelatinous, adhering firmly to the *albumen*, which is horny; *embryo* minute, excen-trical, with the *radicle* next the hilum. — *Shrubs*, either unarmed or spiny. *Leaves* alternate, lobed, with a plaited vernation. *Flowers* in axillary racemes, with bracteæ at their base, very rarely unisexual.

AFFINITIES. Formerly confounded with Cacteæ, to which, notwith-standing the dissimilarity of their appearance, they are indeed most closely related; the principal differences between the two orders are, that in Cacteæ the stamens are indefinite, the seeds without albumen, and the calyx and corolla undistinguishable; while in Grossulaceæ the stamens are definite, the seeds albuminous, and the calyx and corolla distinct. There are spines in both orders, and some of the Cacteæ have distinct leaves. From Ona-grariæ, Grossulaceæ are distinguished by the minute embryo, parietal placentæ, and the quinary divisions of the floral envelopes; from Homalineæ by the want of glands at the base of the sepals and petals, which are also undistin-guishable from one another in the latter; and from Loaseæ by habit, number of stamens and petals, and various other characters.

GEOGRAPHY. Natives of the mountains, hills, woods, and thickets, of the temperate parts of Europe, Asia, and America, but unknown in Africa, the tropics of either hemisphere, or the South Sea Islands. In North America they are particularly abundant, and on the mountains of Northern India they contribute to give a European character to that remarkable region.

PROPERTIES. The properties of the Gooseberry and Currant are those of the generality of the order, except that in other species a mawkish or extremely acid taste is substituted for the refreshing and agreeable flavour of the former. Some are emetic. The black Currant, which is tonic and stimulant, has fragrant glands upon its leaves and flowers; these reservoirs are also found upon some other species. Malic acid exists in Currants and Gooseberries. *Turner*, 634.

EXAMPLE. Ribes.

XLVI. CACTEÆ. THE INDIAN-FIG TRIBE.

CACTI, *Juss. Gen.* 310. (1789) *in part.* — CACTOIDEÆ, *Vent. Tabl.* 3. 289. (1799).— OPUNTIACEÆ, *Juss. Dict. Sc.* 35. 144. (1825) *in part.*; *Kunth Nov. G. et Sp.* 6. 65. (1823).—NOPALEÆ, *Dec. Théorie Elém.* 216. (1819).—CACTEÆ, *Dec. Prodr.* 3. 457. (1828); *Mem. Mus.* (1829).

DIAGNOSIS. Polypetalous succulent dicotyledons, with indefinite perigy-nous fertile stamens, concrete carpella, an inferior ovarium with one cell and parietal placentæ, baccate fruit, and imbricate petals and sepals.

ANOMALIES. The calyx and corolla are distinguishable in Rhipsalis, which is also said to have its seeds attached to a central placenta.

ESSENTIAL CHARACTER.—*Sepals* numerous, usually indefinite, and confounded with the petals, either crowning the ovarium, or covering its whole surface. *Petals* numerous, usually indefinite, arising from the orifice of the calyx, sometimes irregular. *Stamens* indefinite, more or less cohering with the petals and sepals; *filaments* long, filiform; *anthers* ovate, versatile. *Ovarium* fleshy, inferior, 1-celled, with numerous ovula arranged upon parietal placentæ, equal in number to the lobes of the stigma; *style* filiform; *stigmata* numerous, collected in a cluster. *Fruit* succulent, 1-celled, many-seeded, either smooth, or covered with scales, scars, or tubercles. *Seeds* parietal, or, having lost their adhesion, nestling in pulp, ovate or obovate, without albumen; *embryo* either straight, curved, or spiral, with a short thick radicle; *cotyledons* flat, thick, foliaceous, sometimes almost obsolete (in the leafless species). — Succulent *shrubs*, very variable in form. *Stems* usually angular, or two-edged, or foliaceous. *Leaves* almost always wanting; when present, fleshy, smooth, and entire or spine-like. *Flowers* either showy or minute, usually lasting only one day or night, always sessile.

AFFINITIES. It has been already remarked, on more than one occasion, in this work, that the state of anamorphosis, or, in other words, that remarkable distension or increase of the cellular tissue of vegetables, from which the name of succulent is derived, is no indication of natural affinity, but rather to be considered a modification of structure which may be common to all tribes. Hence the immediate relationship of Cacteæ is neither with Euphorbiaceæ, nor Laurineæ, nor Asclepiadeæ, nor Ficoideæ, nor Portulaceæ, nor Asphodeleæ, all of which contain a greater or less number of succulent genera; but with Grossulaceæ, in which no tendency whatever to anamorphosis exists. The distinction between the two orders is mentioned under Grossulaceæ. Through Rhipsalis, which is said to have a central placenta, Cacteæ are connected with Portulaceæ, to which also the curved embryo of the section of Opuntiaceæ probably indicates an approach. Decandolle further traces an affinity between these plants and Ficoideæ. For an elaborate account of this order, see his memoir above quoted.

GEOGRAPHY. America is the station of the order; no species appearing to be natives of any other part of the world; in that country they are abundant in the tropics, extending a short distance beyond them, both to the north and the south. Decandolle states that 32° or 33° north latitude is the northern limit of the order; but it is certain that a species is either wild or naturalised in Long Island, in latitude 42° north, and that there is another somewhere about 49°, in the Rocky mountains. The species which are said to be wild or naturalised in Europe, Mauritius, and Arabia, have been introduced from America, and having found themselves in situations suitable to their habits, have taken possession of the soil like actual natives: in Europe this does not extend beyond the town of Final, in 44° north latitude. There is no reason for supposing that the modern Opuntia is described in Theophrastus, as Sprengel asserts; the description of the former writer applying, as far as it applies to any thing now known, rather to some tree like Ficus religiosa. Hot, dry, exposed places are the favourite stations of Cacteæ, for which they are peculiarly adapted, in consequence of the small quantity of evaporating pores which they possess, as compared with other plants; a circumstance which, as Decandolle has satisfactorily shewn, will account for the excessively succulent state of their tissue.

PROPERTIES. The fruit is very similar in its properties to that of Grossulaceæ, some being refreshing and agreeable to the taste, others mucilaginous and insipid; they are all, however, destitute of the excessive acidity of some gooseberries and currants. The fruit of Cactus opuntia has the property of staining red the urine of those who eat it. The juice of Cactus mammillaris is remarkable for being slightly milky, and at the same time sweet and insipid.

Decandolle has the two following sections, the characters of the last of which are not, however, very certainly ascertained to be correct:—

I. OPUNTIACEÆ.

Ovula and seeds parietal.

EXAMPLES. Cactus, Opuntia, Mammillaria.

II. RHIPSALIDEÆ.

Ovula and seeds attached to a central axis.

EXAMPLE. Rhipsalis.

XLVII. ONAGRARIÆ. THE EVENING PRIMROSE TRIBE.

ONAGRÆ, *Juss. Gen.* 317. (1789). — EPILOBIACEÆ, *Vent. Tabl.* 3. 307. (1799).— ONAGRARIÆ, *Juss. Ann. Mus.* 3. 315. (1804) *in part.; Dec. Prodr.* 3. 35. (1828); *Lindl. Synops.* 107. (1829.)

DIAGNOSIS. Polypetalous dicotyledons, with definite perigynous stamens, concrete carpella, an inferior ovarium of several cells, with indefinite ovula, 4 divisions of the calyx, and roundish anthers erect in æstivation.

ANOMALIES.

ESSENTIAL CHARACTER.—*Calyx* superior, tubular, with the limb usually 4-lobed; the lobes cohering in various degrees, with a valvate æstivation. *Petals* generally equal in number to the lobes of the calyx, into the throat of which they are inserted, regular, with a twisted æstivation. *Stamens* definite, inserted into the calyx; *filaments* distinct; *pollen* triangular, usually cohering by threads. *Ovarium* of several cells, generally crowned by a disk; *style* filiform; *stigma* either capitate or 4-lobed. *Fruit* baccate or capsular, many-seeded, with from 2 to 4 cells. *Seeds* numerous, without *albumen; embryo* straight; *radicle* long and taper; *cotyledons* very short.—*Herbaceous* plants or *shrubs. Leaves* alternate or opposite, simple, entire, or toothed. *Flowers* red, purple, white, blue, or yellow, axillary, or terminal.

AFFINITIES. Onagrariæ differ from all the orders allied to them in the length of the radicle; they are particularly distinguished from Salicariæ by their inferior calyx; from Halorageæ by their filiform style, and by their exalbuminous seeds not being pendulous; from Myrtaceæ by the want of pellucid dots, and by the definite number of their stamens. *Dec.* For the distinctions between them and Hydrocaryes, Callitrichineæ, and Circæaceæ, see those orders.

The following sections of Decandolle appear worthy of being adopted:—

1. MONTINIEÆ.

Fruit capsular. Seeds with a membranous wing, imbricated, erect.— Trees or shrubs, with alternate leaves.

2. FUCHSIEÆ.

Fruit baccate. Tube of the calyx elongated beyond the ovarium.— Chiefly American trees or shrubs, with opposite leaves.

3. ONAGREÆ.

Fruit capsular, with many-seeded cells, and seeds without wings. Tube of the calyx extended beyond the ovarium. Stamens twice as many as the petals.—Herbaceous plants, sometimes slightly shrubby at the base.

4. JUSSIEÆ.

Fruit capsular, with many-seeded cells. Calyx persistent, but not tubular.—Herbaceous plants, rarely under-shrubs.

GEOGRAPHY. Chiefly natives of the temperate parts of the world, and especially of America: a good many are found in India, and a large number in Europe. In Africa they are scarcer, being mostly confined to the Cape, and to a few Jussiæas inhabiting other parts of that continent.

PROPERTIES. Few, or unknown. Œnothera biennis is cultivated for the sake of its eatable roots; and the leaves of Jussiæa peruviana form an emollient poultice. *Dec.*

EXAMPLES. Œnothera, Epilobium, Jussiæa, Fuchsia.

XLVIII. HALORAGEÆ.

HALORAGEÆ, *R. Brown in Flinders*, 17. (1814); *Dec. Prodr.* 3. 65. (1828); *Lindl. Synops.* 110. (1829). — HYGROBIEÆ, *Rich. Anal. Fr.* (1808). — HIPPURIDEÆ, *Link Enum.* 1. 5. (1821) *handb.* 1. 288. (1829.) — CERCODIANÆ, *Juss. Dict. Sc. Nat.* (1817.)

DIAGNOSIS. Polypetalous dicotyledons, with definite perigynous stamens, concrete carpella, an inferior ovarium with pendulous definite ovula, a depauperated calyx, and embryo in the midst of fleshy albumen.

ANOMALIES. Petals often wanting. Hippuris has the habit of an Equisetum.

ESSENTIAL CHARACTER. — *Calyx* superior, with a minute limb. *Petals* minute, inserted into the summit of the calyx, or wanting. *Stamens* inserted in the same place, equal in number to the petals, or occasionally fewer. *Ovarium* adhering inseparably to the calyx, with 1 or more cells; *style* none; *stigmata* equal in number to the cells, papulose, or pencil-formed; *ovula* pendulous. *Fruit* dry, indehiscent, membranous, or bony, with 1 or more cells. *Seeds* solitary, pendulous; *albumen* fleshy; *embryo* straight, in the axis; *radicle* superior, long and taper; *cotyledons* minute.—*Herbaceous* plants or *under-shrubs*, often growing in wet places. *Leaves* either alternate, opposite, or whorled. *Flowers* axillary, sessile, occasionally monœcious or diœcious.

AFFINITIES. Placed by Link among Monocotyledons, but inseparable from Dicotyledons, and especially related to Onagrariæ, from which the minute calyx and albuminous solitary pendulous seeds chiefly distinguish them. Very closely akin also to Circæaceæ and Hydrocaryes, both which see. The affinity of Callitrichineæ is probably not very great, although M. Decandolle has considered it a mere section of the order.

GEOGRAPHY. Damp places, ditches, and slow streams, in Europe, North America, Southern Africa, Japan, China, New Holland, and the South Sea Islands, are the favourite resort of this order.

PROPERTIES. Of no importance. Many are troublesome weeds.

EXAMPLES. Haloragis, Hippuris, Myriophyllum.

XLIX. CIRCÆACEÆ. THE ENCHANTER'S NIGHTSHADE TRIBE.

CIRCÆACEÆ, *Lindl. Synops.* p. 109. (1829.)

DIAGNOSIS. Polypetalous dicotyledons, with definite perigynous stamens, concrete carpella, an inferior ovarium of 2 cells, with definite erect ovula.

ANOMALIES.

ESSENTIAL CHARACTER. — *Calyx* superior, deciduous, tubular, with a two-parted limb. *Petals* 2, alternate with the lobes of the calyx. *Stamens* 2, alternate with the petals, inserted into the calyx. *Disk* large, cup-shaped, filling up the whole of the tube of the calyx, and projecting beyond it. *Ovarium* 2-celled, with an erect ovulum in each cell ; *style* simple, arising out of the disk ; *stigma* emarginate. *Fruit* 2-celled, 2-valved, 2-seeded. *Seeds* solitary, erect ; *albumen* none ; *embryo* erect ; *radicle* short, inferior.—*Herbaceous* plants; *Leaves* opposite, toothed, stalked. *Flowers* in terminal and lateral racemes, covered with uncinate hairs.

AFFINITIES. This order differs from Onagrariæ in its large fleshy disk, which fills up the tube of the calyx, in its solitary erect ovula, and in the binary division of the flower. It is connected with that order through Lopezia, with which it cannot, however, be absolutely associated ; and bears about the same relation to Onagrariæ as is borne by Halorageæ.

GEOGRAPHY. Natives of the northern parts of the world, inhabiting groves and thickets.

PROPERTIES. Unknown.

EXAMPLE. Circæa.

L. HYDROCARYES. THE WATER CHESTNUT TRIBE.

HYDROCARYES, *Link Enum. Hort. Ber.* 1. 141. (1821).—ONAGRARIÆ, § Hydrocaryes, *Dec. Prodr.* 3. 63. (1828.)

DIAGNOSIS. Polypetalous dicotyledons, with definite perigynous stamens, concrete carpella, an inferior ovarium with definite pendulous ovules, no albumen, and very unequal cotyledons.

ANOMALIES.

ESSENTIAL CHARACTER. — *Calyx* superior, 4-parted. *Petals* 4, arising from the throat of the calyx. *Stamens* 4, alternate with the last. *Ovarium* 2-celled ; *ovules* solitary, pendulous ; *style* filiform, thickened at the base ; *stigma* capitate. *Fruit* hard, indehiscent, 1-celled, 1-seeded, crowned by the indurated segments of the calyx. *Seed* solitary, large, pendulous ; *albumen* none ; *cotyledons* 2, very unequal.—*Floating* plants. Lower *leaves* opposite, upper alternate ; those under water cut into capillary segments ; *petioles* tumid in the middle. *Flowers* small, axillary.

AFFINITIES. Closely akin to Onagrariæ, from which they are distinguished by their solitary pendulous ovules ; more closely allied to Haloregeæ, from which they are divided only by their very large seeds with unequal cotyledons, developed calyx, and want of albumen ; agreeing with them, especially with Myriophyllum, in habit.

GEOGRAPHY. Found in the south of Europe, the East Indies, and China.

PROPERTIES. The great seeds are sweet and eatable.

EXAMPLE. Trapa.

LI. LOASEÆ.

LOASEÆ, *Juss. Ann. Mus.* 5. 18. (1804) ; *Dict. Sc. Nat.* 27. 93. (1823) ; *Kunth in Nov. Gen. et Sp.* 6. 115. (1823) ; *Dec. Prodr.* 3. 339. (1828.)

DIAGNOSIS. Polypetalous dicotyledons, with perigynous stamens, part of which are sterile, concrete carpella, an inferior 1-celled ovarium with parietal placentæ, and dissimilar petals and sepals.

ANOMALIES. Ovarium sometimes almost superior. Seeds definite in Mentzelia and Klaprothia.

ESSENTIAL CHARACTER. — *Calyx* superior or inferior, 5-parted, persistent, spreading in æstivation. *Petals* 5 or 10, arising from within the recesses of the calyx, cucullate, with an inflexed valvate æstivation; the interior often, when present, much smaller than the outer, and truncate at· the apex. *Stamens* indefinite, in several rows, arising from within the petals, either distinct or adhering in bundles before each petal, within the cavity of which they lie in æstivation; *filaments* subulate, unequal, the outer ones frequently destitute of anthers. *Ovarium* inferior, or nearly superior, 1-celled, with several parietal placentæ, or with 1 free central lobed one; *style* single; *stigma* 1, or several. *Fruit* capsular or succulent, inferior or superior, 1-celled, with parietal placentæ originating at the sutures. *Seeds* numerous, without arillus; *embryo* lying in the axis of fleshy albumen, with the radicle pointing to the hilum, and flat small cotyledons. — *Herbaceous* plants, hispid, with pungent hairs secreting an acrid juice. *Leaves* opposite or alternate, without stipulæ, usually more or less divided. *Peduncles* axillary, 1-flowered.

AFFINITIES. Distinguished from Onagrariæ by their unilocular ovaria and indefinite stamens, part of which are sterile; and perhaps by the latter character, and the additional 5 petals, connected with Passifloreæ, with which they also sometimes accord in habit. Their rigid stinging hairs, climbing habit, and lobed leaves, resemble those of some Urticeæ, with which, however, they have nothing more of importance in common. On the same account they may be compared with Cucurbitaceæ, with which they further agree in their inferior unilocular fruit, with parietal placentæ, and in the very generally yellow colour of their flowers. This, indeed, is the order with which, upon the whole, Loaseæ must be considered to have the closest affinity. Eschscholtzia, referred here by Decandolle, belongs to Papaveraceæ.

GEOGRAPHY. All American, and chiefly from the more temperate regions, or the tropics, of either hemisphere.

PROPERTIES. Except the stinging property which resides in the hairs of some species, nothing is known of the qualities of these plants.

EXAMPLES. Loasa, Mentzelia.

LII. SALICARIÆ. THE LOOSESTRIFE TRIBE.

SALICARIÆ, *Juss. Gen.* 330. (1789); *Lindl. Synops.* 71. (1829). — CALYCANTHEMÆ, *Vent. Tabl.* 3. 298. (1799). — SALICARINÆ, *Link Enum.* 1. 142. (1821). — LYTHRARIÆ, *Juss. Dict. Sc. Nat.* 27. 453. (1823); *Dec. Prodr.* 3. 75. (1828.)

DIAGNOSIS. Polypetalous dicotyledons, with perigynous stamens, concrete carpella, a superior ovarium with several cells, and a tubular short-toothed calyx, which covers the capsule.

ANOMALIES. Occasionally apetalous.

ESSENTIAL CHARACTER. — *Calyx* monosepalous, the lobes with a valvate or separate æstivation, their sinuses sometimes lengthened into other lobes. *Petals* inserted between the lobes of the calyx, very deciduous, sometimes wanting. *Stamens* inserted into the tube of the calyx below the petals, to which they are sometimes equal in number; sometimes they are twice, or even thrice, and four times as numerous; they are seldom four; *anthers* adnate, 2-celled, opening longitudinally. *Ovarium* superior, 2- or 4-celled; *style* filiform; *stigma* usually capitate. *Capsule* membranous, covered by the calyx, usually 1-celled, dehiscing either longitudinally or in an irregular manner. *Seeds* numerous, small, without albumen, adhering to a central placenta; *embryo* straight; *radicle* turned towards the hilum; *cotyledons* flat and leafy. — *Herbs*, rarely *shrubs*. *Branches* frequently 4-cornered. *Leaves* opposite, seldom alternate, entire, without either stipulæ or glands. *Flowers* axillary, or in terminal spikes or racemes, in consequence of the depauperation of the upper leaves.

AFFINITIES. Very near Onagrariæ, from which their superior ovarium and many-ribbed calyx distinguish them; also Melastomaceæ, from which their superior ovarium, the veining of their leaves, and the æstivation of the stamens divide them. With Labiatæ they have often a striking resemblance in habit, but this goes no further.

M. Decandolle admits the two following tribes:—

1. § Salicarieæ, *Mem. Soc. H. N. Genev.* 3. p. 2. 71.; *Prodr.* 3. 75. (1828.)

Lobes of the calyx more or less distant in æstivation, or somewhat valvate. Petals several, alternate with the lobes of the calyx, and arising from between them at the orifice of the tube; sometimes wanting. Stamens arising from lower down the tube. Seeds apterous.—Shrubs or herbaceous plants. *Dec.*

2. § Lagerströmieæ, *Dec.* l. c. p. 70.; *Prodr.* 3. 92. (1828.)

Lobes of the calyx exactly valvate in æstivation. Petals several, alternate with the lobes of the calyx, and arising from between them in the apex of the tube. Stamens two or three times as numerous, and arising from lower down the tube. Seeds with a membranous wing.—Shrubs or trees. *Dec.*

GEOGRAPHY. The Lagerströmias are all Indian or South American. The true Salicariæ are European, North American, and natives of the tropics of both hemispheres. Lythrum Salicaria, a common European plant, is singular for being found in New Holland, and for also being the only species of that order yet described from that country.

PROPERTIES. Astringency is a property of the Lythrum Salicaria, which is reputed to have been found useful in inveterate diarrhœas: another species of the same genus is accounted in Mexico astringent and vulnerary. The flowers of Lythrum? Hunteri are employed in India, mixed with Morinda, for dyeing, under the name of Dhawry. *Hunter, As. Res.* 4. 42. Heimia salicifolia, a plant remarkable, in an order with red or purple flowers, for its yellow corolla, is said to excite violently perspiration and the urinary secretion. The Mexicans consider it a potent remedy for venereal diseases, and call it Hanchinol. *Dec.* Lawsonia inermis is the plant from which the Henné of Egypt is obtained. Women in that country stain their fingers and feet with it. It is also used for dyeing skins and maroquins reddish yellow, and for many other purposes. It contains no tannin. *Ed. P. J.* 12. 416. The leaves of Ammannia vesicatoria have a strong muriatic smell; they are extremely acrid, and are used by the native practitioners of India to raise blisters, in rheumatism, &c.: bruised and applied to the part intended to be blistered, they perform their office in half an hour, and most effectually. *Ainslie,* 2. 93.

EXAMPLES. Lythrum, Lagerströmia, Ammannia.

LIII. RHIZOPHOREÆ. THE MANGROVE TRIBE.

RHIZOPHOREÆ, *R. Brown Gen. Rem. in Flinders,* p. 17. (1814); *in Congo,* p. 18. (1818); *Dec. Prodr.* 3. 31. (1828.) — PALETUVIERS, *Savigny in Lam. Dict.* 4. 696. (1796.)

DIAGNOSIS. Polypetalous dicotyledons, with perigynous stamens twice the number of the petals, concrete carpella, an inferior ovarium of 2 cells with pendulous ovules, and opposite leaves with interpetiolar stipulæ.

ANOMALIES. The leaves of Baraldeia have pellucid dots. In Cassipouna the ovarium is superior, and the seeds have albumen.

ESSENTIAL CHARACTER. — *Calyx* superior, very rarely nearly inferior, with the lobes varying in number from 4 to 13, occasionally all cohering in a calyptra. *Petals* arising from the calyx, alternate with the lobes, and equal to them in number. *Stamens* arising from the same point as the petals, and twice or thrice their number; *filaments* distinct; *anthers* erect, innate. *Ovarium* 2-celled, each cell containing 2 or more pendulous ovules. *Fruit* indehiscent, crowned by the calyx, 1-celled, 1-seeded. *Seed* pendulous, without albumen; *radicle* long; *cotyledons* 2, flat. —Coast *trees* or *shrubs*. *Leaves* simple, opposite, entire or toothed, with stipulæ between the petioles. *Peduncles* axillary.

AFFINITIES. From a consideration of the structure of Carallia and Legnotis, Mr. Brown has been led to conclude that we have a series of structures connecting Rhizophora, on the one hand, with certain genera of Salicariæ, particularly with Antherylium, though that genus wants the intermediate stipules; and, on the other, with Cunoniaceæ, especially with the simple-leaved species of Ceratopetalum. *Congo*, 437. This order agrees with Cunoniaceæ in its opposite leaves and intermediate stipulæ, and with great part of them in the æstivation of its calyx, and in the structure and cohesion of ovarium. *R. Brown, Flinders*, 549. Decandolle points out its relation to Vochyaceæ and Combretaceæ, and even to Memecyleæ through the genus Olisbea. The genera were comprehended in Lorantheæ by Jussieu. Cassipouna, mentioned as an anomalous plant, is probably the type of a distinct order.

GEOGRAPHY. Natives of the shores of the tropics, where they root in the mud, and form a dense thicket down to the verge of the ocean.

PROPERTIES. The bark is usually astringent; that of Rhizophora gymnorhiza is used in India for dyeing black. *Dec.*

EXAMPLES. Rhizophora, Bruguiera.

LIV. MELASTOMACEÆ.

MELASTOMÆ, *Juss. Gen.* p. 328. (1789); *Dict. Sc. Nat.* 29. 507. (1823). —MELASTOMACEÆ, *Don in Mem. Wern. Soc.* 4. 281. (1823); *Dec. Prodr.* 3. 99. (1828); *Mémoire* (1828).

DIAGNOSIS. Polypetalous dicotyledons, with definite perigynous stamens, concrete carpella, an inferior ovarium of several cells, long inflexed anthers, indefinite seeds, and opposite ribbed leaves without dots.

ANOMALIES. Traces of pellucid dots in Diplogenea. Ovarium more or less superior in several. Leaves sometimes not ribbed in Sonerila.

ESSENTIAL CHARACTER. — *Calyx* divided into 4, 5, or 6 lobes, cohering more or less with the angles of the ovarium, but distinct from the surface between the angles, and thus forming a number of cavities, within which the young anthers are curved downwards. *Petals* equal to the segments of the calyx, arising from their base, or from the edge of a disk that lines the calyx; twisted in æstivation. *Stamens* usually twice as many as the petals, sometimes equal to them in number; in the former case, those which are opposite the segments of the calyx are alone fertile; *filaments* curved downwards in æstivation; *anthers* long, 2-celled, usually bursting by two pores at the apex, which is rostrate, and elongated in various ways beyond the insertion of the filament; sometimes bursting longitudinally; before flowering, contained within the cases between the ovarium and sides of the calyx. *Ovarium* more or less coherent with the calyx, with several cells, and indefinite ovules; *style* 1; *stigma* simple, either capitate or minute; a cup often present upon the apex of the ovarium, surrounding the style. *Pericarpium* either dry and distinct from the calyx, or succulent and combined with the calyx, with several cells; if dehiscent, bursting through the valves, which therefore bear the septa in the middle; *placentæ* attached to a central column. *Seeds* innumerable, minute, with a brittle *testa*, and no *albumen*; usually with appendages of some kind; *embryo* straight, or curved, with equal or unequal *cotyledons*. — *Trees, shrubs*, or *herbaceous* plants. *Leaves* opposite, undivided, usually entire, without dots, with several ribs. *Flowers* terminal, usually thyrsoid.

AFFINITIES. "The family of Melastomaceæ," remarks M. Decandolle, in an excellent memoir upon the subject, "although composed entirely of exotic plants, and established at a period when but few species were known, is so well characterised, that no one has ever thought of putting any part of it in any other group, or even of introducing into it genera that do not rightly belong to it." These distinct characters are, the opposite leaves, with several great veins or ribs running from the base to the apex, something as in Monocotyledonous plants, and the long beaked anthers, to which combined there is nothing to be compared in other families. Permanent, however, as this character undoubtedly is, yet the cause of no uncertainty having been yet found in fixing the limits of the order, is rather to be attributed to the small number of species that have been examined, than to the want of connecting links: thus Diplogenea has traces of the dots of Myrtaceæ, which were not known to exist in Melastomaceæ until that genus was described; and several genera are now described with superior ovarium, a structure which was at one time supposed not to exist in the order; and, finally, in the remarkable genus Sonerila, the leaves are sometimes not ribbed.

The greatest affinity of Melastomaceæ is on the one hand with Salicariæ, on the other with Myrtaceæ; from the former they differ in the æstivation of their calyx not being valvate, from the latter in having the petals twisted before expansion and no dots on the leaves, and from both, and all others to which they can be compared, in their long anthers bent down parallel to the filaments in the flower, and lying in niches between the calyx and ovarium; with the exception of Memecyleæ, in which, however, the union between the calyx and ovarium is complete, and which have leaves destitute of the lateral ribs that so strongly point out Melastomaceæ. The structure of the seeds of Memecyleæ is also different.

From differences in the dehiscence of the anthers, Decandolle forms two sub-orders, viz.:—

1. TRUE MELASTOMAS.
Anthers opening by pores at the apex.
EXAMPLES. Melastoma, Rhexia.

2. CHARIANTHEÆ.
Anthers opening by 2 longitudinal fissures.
EXAMPLES. Charianthus, Astronia.

GEOGRAPHY. Found neither in Europe nor Asia in the temperate zone, nor in Africa north of the desert of Zahara, nor south of Brazil in South America, nor in extra-tropical Africa to the south. Beyond the tropics, 8 are found in the United States, 3 in China, and 3 in New Holland. Of the remainder, it appears that 78 are described from India or the Indian Archipelago, 12 from Africa and the adjacent islands, and 620 from America. *Dec.*

PROPERTIES. A slight degree of astringency is the prevailing character of the order, which is, although one of the most extensive known, entirely destitute of any unwholesome species. The succulent fruit of many is eatable, some of which dye the mouth black, whence the name of Melastoma. Blakea triplinervia produces a yellow fruit, which is pleasant and eatable, in the woods of Guiana. *Hamilt. Prodr.* 42.

LV. MEMECYLEÆ.

MEMECYLEÆ, *Dec. Prodr.* 3. 5. (1828.)

DIAGNOSIS. Polypetalous dicotyledons, with definite perigynous stamens, concrete carpella, an inferior ovarium with several cells, 1-ribbed leaves without dots, a few seeds, an exalbuminous embryo with convolute cotyledons, and long inflexed anthers.

ANOMALIES.

ESSENTIAL CHARACTER. — *Calyx* superior, 4- or 5-lobed, or 4-5-toothed. *Petals* 4-5, inserted into the calyx, and alternate with its lobes. *Stamens* 8-10; *filaments* distinct *anthers* incurved, 2-celled. *Style* filiform; *stigma* simple. *Berry* crowned by the limb of the calyx, 2-4-celled. *Seeds* few, pendulous, without *albumen; cotyledons* foliaceous, convolute; *radicle* straight.—*Shrubs. Leaves* opposite, simple, entire, without stipulæ or dots, almost always without more than one central rib. *Flowers* axillary, pedicellate.

AFFINITIES. Very near Myrtaceæ and Melastomaceæ, and in some respects almost intermediate between them. They agree with the former in the single rib of their leaves, and with the latter in the want of dots and in the peculiar form of the anthers; their cotyledons are those of Punica among Myrtaceæ.

GEOGRAPHY. All natives of the hottest parts of the East Indies and of the Mauritanian Islands, with the exception of the Mouririas, which are West Indian, if they belong to the order; but this is uncertain.

PROPERTIES. Unknown.

EXAMPLES. Memecylon, Mouriri.

LVI. MYRTACEÆ. THE MYRTLE TRIBE.

MYRTI, *Juss. Gen.* 323. (1789).—MYRTEÆ, *Juss. Dict. Sc. Nat.* 34. 79. (1825).— MYRTOIDEÆ, *Vent. Tabl.* (1799). — MYRTINEÆ, *Dec. Théorie, Elem.* (1819). — MYRTACEÆ, *R. Brown in Flinders*, p. 14. (1814); *Dec. Dict. Class.* v. 11. (1826); *Prodr.* 3. 207. (1829). — GRANATEÆ, *Don in Ed. Phil. Journ.* p. 134. (1826); *Dec. Prodr.* 3. 3. (1829); *Von Martius H. Reg. Monac.* (1829.)

DIAGNOSIS. Polypetalous dicotyledons, with indefinite perigynous stamens, concrete carpella, an inferior ovarium with several cells, and opposite entire leaves with pellucid dots.

ANOMALIES. Chamælauciæ have a 1-celled fruit, with erect ovula. A species of Sonneratia is apetalous. The leaves of Barringtonia are alternate and not dotted.

ESSENTIAL CHARACTER. — *Calyx* superior, 4- or 5-cleft, sometimes falling off like a cap, in consequence of the cohesion of the apex. *Petals* equal in number to the segments of the calyx, with a quincuncial æstivation; rarely none. *Stamens* either twice as many as the petals, or indefinite; *filaments* either all distinct, or connected in several parcels, curved inwards before flowering; *anthers* ovate, 2-celled, small, bursting lengthwise. *Ovarium* inferior, 2- 4- 5- or 6-celled; *style* simple; *stigma* simple. *Fruit* either dry or fleshy, dehiscent or indehiscent. *Seeds* usually indefinite, variable in form; *embryo* without *albumen*, straight or curved, with its *cotyledons* and *radicle* distinguishable or conferruminated into a solid mass.—*Trees* or *shrubs. Leaves* opposite, entire, with transparent dots, and with a vein running parallel with their margin. *Inflorescence* variable, usually axillary. *Flowers* red, white, occasionally yellow, never blue.

AFFINITIES. One of the most natural among the tribes of plants, and the most easily recognised. Its opposite exstipulate dotted entire leaves with a marginal vein, are a certain indication of it, with the exception of a few plants, which probably do not belong to the order, although at present placed in it. It is closely allied to Rosaceæ, Salicariæ, Onagrariæ, Combretaceæ, and Melastomaceæ, but cannot well be confounded either with them or any other tribe. It offers a curious instance of the facility with which the calyx and corolla can take upon themselves the same functions and transformations. In Eucalyptus, as is well known, the sepals are consolidated into a cup-like lid, called the operculum. In Eudesmia, a nearly-related genus, the calyx remains in its normal state, while the petals are consolidated into an operculum. Punica is usually referred to this order; but the descriptions that have been published of it have been founded upon so imperfect a view of its structure, that I may be permitted to dwell upon it at some length, especially as I hope to shew that it not only does not differ from the order essentially, but that it does not require to be distinguished from true Myrtaceæ even as a section. A consideration of the real structure of this plant comes the more properly within the scope of the present publication, because the genus has been considered the type of a particular order (Granateæ) by Mr. Don, in which he is supported by the high authority of Decandolle and Von Martius. The fruit of the Pomegranate is described by Gærtner and Decandolle as being divided into two unequal divisions by a horizontal diaphragm, the upper half of which consists of from 5 to 9 cells, and the lower of 3; the cells of both being separated by membranous dissepiments; the placentæ of the upper half proceeding from the back to the centre, and of the lower irregularly from their bottom; and by Mr. Don as a fleshy receptacle formed by the tube of the calyx into a unilocular berry, filled with a spongy placenta, which is hollowed out into a number of irregular cells. In fact, if a Pomegranate is examined, it will be found to agree more or less perfectly with both these descriptions. But it is clear that a fruit as thus described is at variance with all the known laws upon which compound fruits are formed. Nothing, however, is more common than that the primitive construction of fruits is obscured by the additions, or suppressions, or alterations, which its parts undergo during their progress to maturity. Hence it is always desirable to obtain a clear idea of the structure of the ovarium of all fruits which do not obviously agree with the ordinary laws of carpological composition. Now, a section of the ovarium of the Pomegranate in various directions, if made about the time of the expansion of the flowers before impregnation takes place, shews that it is in fact composed of two rows of carpella, of which three or four surround the axis, and are placed in the bottom of the tube of the calyx, and a number, varying from five to ten, surround these, and adhere to the upper part of the tube of the calyx. The placentæ of these carpella contract an irregular kind of adhesion with the back and front of their cells, and thus give the position ultimately acquired by the seeds that anomalous appearance which it assumes in the ripe fruit. If this view of the structure of the Pomegranate be correct, its peculiarity consists in this, that, in an order the carpella of which occupy but a single row around the axis, it possesses carpella in two rows, the one placed above the other, in consequence of the contraction of the tube of the calyx, from which they arise. Now, there are many instances of a similar anomaly among genera of the same order, and they exist even among species of the same genus. Examples of the latter are, Nicotiana multivalvis and Nolana paradoxa, and of the former Malope among Malvaceæ; polycarpous Ranunculaceæ as compared with Nigella, and polycarpous Rosaceæ as compared with Spiræa. In Prunus I have seen a monstrous flower producing a number of carpella around the central one, and also, in consequence of the situation, upon the calyx above it; and, finally, in the *Revue Encyclopédique* (43. 762.), a permanent variety of the Apple is described, which is exactly to Pomaceæ what Punica is to Myrtaceæ. This plant has regularly 14 styles and 14 cells, arranged in two horizontal parallel planes, namely, 5 in the middle, and 9 on the outside, smaller and nearer the top; a circumstance which is evidently to be explained by the presence of an outer series of carpella, and not upon the extravagant hypo-

65

thesis of M. Tillette de Clermont, who fancies that it is due to the cohesion of 3 flowers. The anomaly of the structure of the fruit of Punica being thus explained, nothing remains to distinguish it from Myrtaceæ but its leaves without a marginal vein, its convolute cotyledons, and pulpy seeds. There are, however, distinct traces of dots in the leaves, and the union of the venæ arcuatæ, which gives the appearance of a marginal vein to Myrtaceæ, takes place, although less regularly, in Punica; the convolute cotyledons of Punica are only in Myrtaceæ what those of Chamæmeles are in Pomaceæ, a curious but unimportant exception to the general structure; and the solitary character of the pulpy coat of the seeds will hardly be deemed by itself sufficient to characterise Granateæ. The place of Punica in the order will be probably near Sonneratia. There is no instance of a blue flower in the order.

GEOGRAPHY. Natives of hot countries both within and without the tropics; great numbers are found in South America and the East Indies, not many in Africa, and a considerable proportion of the order in New Holland and the South Sea Islands; but the genera of those countries are mostly peculiar to them. Myrtus communis, the most northern species of the order, is native of the south of Europe.

PROPERTIES. The pellucid dotting of the leaves and other parts indicates the presence of a fragrant aromatic or pungent volatile oil, which gives the principal quality to the products of the order. To this are due the grateful perfume of the Guava fruit, the powerful aroma of the flower-buds of Caryophyllus aromaticus, called by the English Cloves, and the balsamic odour of the eastern fruits called the Jamrosade and the Rose Apple. Along with this is frequently mixed an astringent principle, which sometimes predominates, to the suppression of any other property. The following are some of the less known instances of the existence of these and other qualities. The fruit of various Eugenias are found by travellers in the forests of Brazil to bear very agreeable fruit. *Pr. Max. Trav.* 75. A fruit of Brazil, called *Jaboticabeiras*, brought from the forests to the towns of St. Paul and Tejuco, belongs to this order; it is said to be delicious. *Pl. Usuelles*, 29. The young flower-buds of Calyptranthus aromatica have the flavour and quality of Cloves, for which they might be advantageously substituted, according to M. Auguste St. Hilaire. *Ibid.* no. 14. The volatile oil of Cajeputi is distilled from the leaves of Melaleuca leucadendron, and is well known as a powerful sudorific, and useful external application in chronic rheumatism. *Ainslie*, 1. 260. It is considered carminative, cephalic, and emmenagogue, and is, no doubt, a highly diffusible stimulant, antispasmodic, and diaphoretic. It has also the power of dissolving caoutchouc. *Ibid.* The root of Eugenia racemosa (Stravadium) has a slightly bitter, but not unpleasant taste. It is considered by the Hindoo doctors valuable on account of its aperient, deobstruent, and cooling properties; the bark is supposed to possess properties similar to Cinchona. *Ibid.* 2. 65. A kind of gum *Kino* is yielded by Eucalyptus resinifera, which is occasionally sold in the medicine bazars of India. *Ibid.* 1. 185. Other species of Eucalyptus yield a large quantity of tannin, which has been even extracted from the trees in New Holland, and sent to the English market. The efficacy of the bark of the root of the Pomegranate as a remedy for tape-worm is well established in India. *Ibid.* 2. 175. The leaves of Glaphyria nitida, called by the Malays *The Tree of Long Life*, (*Kayo Umur Panjang*), " probably from its maintaining itself at elevations where the other denizens of the forest have ceased to exist," afford at Bencoolen a substitute for tea; and it is known to the natives by the name of the Tea Plant. *Linn. Trans.* 14. 129.

F

66

The following are the sections of this order :—

1. CHAMÆLAUCIEÆ.

Dec. Dict. Class. v. 11. (1826); *Prodr.* 3. 208. (1829.)
Lobes of the calyx 5.. Petals the same number. Stamens in a single
row, distinct or somewhat polyadelphous, sometimes partly sterile. Fruit
dry, 1-celled; ovula numerous, erect, attached to the centre, or a central
placenta.—Heath-like New Holland shrubs. Bracteolæ 2, under the flower,
distinct, or combined, or even operculiform.
EXAMPLES. Chamælaucium, Calytrix.

2. LEPTOSPERMEÆ.

Leptospermeæ, *Dec. Dict. Class.* 11. (1826); *Prodr.* 3. 209. (1829.)
Lobes of the calyx 4 or 6. Petals the same number. Stamens distinct,
or polyadelphous. Fruit dry, many-celled.—Shrubs or trees, natives of
New Holland and the neighbouring countries. Leaves opposite or alternate. Inflorescence various; the flowers sometimes almost immersed in
the stem.
EXAMPLES. Leptospermum, Melaleuca, Eucalyptus.

3. MYRTEÆ.

Myrteæ, *Dec. Dict. Class.* 11. (1826); *Prodr.* 3. 230. (1829.)
Sepals 4 or 5. Petals the same number. Stamens distinct. Fruit
fleshy, many-celled.—Trees or shrubs, mostly intratropical, very few from
New Holland.
EXAMPLES. Myrtus, Eugenia.

4. BARRINGTONIEÆ.

Barringtonieæ, *Dec. Dict. Class.* 11. (1826); *Prodr.* 3. 288. (1829.)
Lobes of the calyx from 4 to 6. Petals as many. Stamens very nume-
rous, in several rows, equally and shortly monadelphous. Fruit berried, or
dry, indehiscent, with several cells. Cotyledons large, fleshy. — Trees.
Leaves not dotted, alternate, or almost opposite or whorled, entire or serrate.
Flowers in racemes or panicles. Probably not belonging to the order.
EXAMPLES. Barringtonia, Stravadium.

LVII. COMBRETACEÆ. THE MYROBALAN TRIBE.

COMBRETACEÆ, *R. Brown Prodr.* 351. (1810), *incidentally without a character ; A. Rich.*
Dict. Class. 4. 353. (1823); *Dec. Prodr.* 3. 9. (1828); *Mémoire* (1828).—MYRO-
BOLANEÆ, *Juss. Dict. Sc. Nat.* 31. 458. (1824.)

DIAGNOSIS. Polypetalous dicotyledons, with perigynous stamens double
the number of the petals, concrete carpella, an inferior ovarium of one cell,
with pendulous ovules hanging from the apex of the cavity, no stipulæ, oblong
petals, and convolute cotyledons.
ANOMALIES. Often apetalous.

ESSENTIAL CHARACTER.— *Calyx* superior, with a 4- or 5-lobed deciduous limb.
Petals arising from the orifice of the calyx, alternate with the lobes ; sometimes wanting. *Stamens* arising from the same part, twice as many as the segments of the calyx,
very rarely equal to them in number, or 3 times as many ; *filaments* distinct, subulate ;
anthers 2-celled, bursting longitudinally. *Ovarium* 1-celled, with from 2 to 4 ovules,
hanging from the apex of the cavity; *style* 1 ; *stigma* simple. *Fruit* drupaceous,
baccate, or nut-like, 1-celled, by abortion 1-seeded, indehiscent, often winged. *Seed*

pendulous, without albumen; *embryo* with the radicle turned towards the hilum; *plumula* inconspicuous; *cotyledons* leafy, usually convolute, occasionally plaited. — *Trees* or *shrubs*. *Leaves* alternate or opposite, without stipulæ, entire. *Spikes* axillary or terminal.

AFFINITIES. " These may be placed indifferently in the vicinity of Santalaceæ and Elæagneæ, or of Onagrariæ and Myrtaceæ, approaching the former by the apetalous genera, and the latter by those which have petals." *Dec.* To Myrtaceæ and Melastomaceæ they are related through Memecyleæ, and especially to the former, by Punica, with which they agree in the structure of their embryo. In the latter respect they also accord with Rhizophoreæ and Vochyaceæ; and with Alangieæ and Onagrariæ in the general structure of the flower. With Santalaceæ and Elæagneæ the apetalous genera agree in many important particulars.

Decandolle has two sections :—

1. TERMINALIEÆ.

Embryo cylindrical, elliptical. Cotyledons rolled spirally. Calyx 5-cleft. Petals often wanting. Stamens 10.

2. COMBRETEÆ.

Embryo cylindrical, elliptical, or angular. Cotyledons thick, plaited irregularly and longitudinally. Calyx 4-6-cleft. Petals 4-5. Stamens 8-10.

GEOGRAPHY. All natives of the tropics of India, Africa, and America. No species is extra-tropical.

PROPERTIES. Mostly astringents. Bucida Buceras yields a bark used for tanning. Terminalia Vernix is said to furnish the Chinese varnish, the juice and exhalation of which are poisonous; but this is at least doubtful. The bark of Conocarpus racemosa, one of the plants called Mangroves in Brazil, is used greatly at Rio Janeiro for tanning. *Pr. Max. Trav.* 206. The fruit of the Terminalia bellerica, or the Belleric Myrobalan, is an astringent, tonic, and attenuant. *Ainslie*, 1. 236. That of the Terminalia Chebula is much more astringent. The bark of Terminalia alata is astringent and antifebrile. *Ibid.* 2. 193. The fruit of Terminalia Chebula, as well as the galls of the same plant, are very astringent, and highly valued by dyers: with alum they give a durable yellow, and with a ferruginous mud an excellent black. *Ibid.* 2. 128. The root of T. latifolia is given in Jamaica in diarrhœa. *Ibid.*

EXAMPLES. Combretum, Bucida, Terminalia.

LVIII. ALANGIEÆ.

ALANGIEÆ, *Dec. Prodr.* 3. 203. (1828.)

DIAGNOSIS. Polypetalous dicotyledons, with numerous perigynous stamens, concrete carpella, an inferior ovarium with several cells, definite pendulous ovula, exstipulate leaves, flat cotyledons, and linear petals.

ANOMALIES. None.

ESSENTIAL CHARACTER.—*Calyx* superior, campanulate, 5-10-toothed. *Petals* 5-10, linear, reflexed. *Stamens* long, exserted, 2 or 4 times as numerous as the petals; *filaments* distinct, villous at the base; *anthers* adnate, linear, 2-celled, turned inwards, often empty. *Disk* fleshy at the base of the limb of the calyx. *Drupe* oval, somewhat crowned by the calyx, fleshy, slightly ribbed, and downy; *nucleus* 1-celled, bony, with a foramen at the apex. *Seed* 1, or according to Rheede 3, inverted, ovate; *albumen* fleshy, brittle; *embryo* straight; *radicle* long, ascending; *cotyledons* flat, foliaceous, cordate-ovate.— Large *Trees*. *Branches* often spiny. *Leaves* alternate, without stipulæ, entire, without dots. *Flowers* fascicled, axillary. *Fruit* eatable.

AFFINITIES. " Differ from Myrtaceæ in their more numerous petals, adnate anthers, 1-celled fruit, and pendulous albuminous seeds. Agree with Combretaceæ in the contracted tube of the calyx, 1-celled fruit, and pendulous seeds; but differ in the number of the petals, adnate anthers, albuminous seeds, and flat cotyledons. The order disagrees entirely with Melastomaceæ and Onagrariæ, in the form of the anthers, and 1-celled fruit. It in some measure approaches Halorageæ in the structure of the seed, but recedes from them in habit, 1-celled fruit, and single style." *Dec. Prodr.* 3. 203.

GEOGRAPHY. Natives of the East Indies.

PROPERTIES. Alangium decapetalum and hexapetalum are said by the Malays to have a purgative hydragogic property. Their roots are aromatic.

EXAMPLE. Alangium.

LIX. ELÆAGNEÆ. THE OLEASTER TRIBE.

ELÆAGNI, *Juss. Gen.* 75. (1789).—ELÆAGNEÆ, *Ach. Rich. Monogr.* (1823); *Lindl. Synopsis,* 208. (1829.)

DIAGNOSIS. Apetalous dicotyledons, with definite erect ovula, a tubular inferior calyx with the stamens alternate with its segments, and leprous leaves.

ANOMALIES. None.

ESSENTIAL CHARACTER. — *Flowers* diœcious, rarely hermaphrodite. *Male: Calyx* 4-parted; *stamens* 3, 4, or 8, sessile; *anthers* 2-celled. *Female: Calyx* inferior, tubular, persistent; the *limb* entire, or 2-4-toothed. *Ovarium* superior, simple, 1-celled; *ovulum* solitary, ascending, stalked; *stigma* simple, subulate, glandular. *Fruit* crustaceous, enclosed within the calyx become succulent. *Seed* erect; *embryo* straight, surrounded by very thin fleshy *albumen;* *radicle* short, inferior; *cotyledons* fleshy.—*Trees* or *shrubs,* covered with leprous scales. *Leaves* alternate, or opposite, entire, without stipulæ. *Flowers* axillary, often fragrant.

AFFINITIES. Its leprous leaves, superior fruit, and apetalous flowers, will at all times distinguish the Oleaster tribe, which touches at one point Thymeleæ, from which it is known by the position of its ovulum; at another Proteaceæ, known by their valvate irregular calyxes, and dehiscent fruit; at a third Santalaceæ, which have the ovarium inferior; and also at a fourth Combretaceæ, which have petals, convolute cotyledons, and a superior calyx.

GEOGRAPHY. The whole of the northern hemisphere, as far as the equator, is occupied more or less by this family, from Canada and Japan to Guiana and Java: they are not known south of the line.

PROPERTIES. The berries of Hippophae rhamnoides are occasionally eaten; the fruit of Elæagnus orientalis is almost as large as a Jujube, and is known in Persia as an article of the dessert, under the name of Zinzeyd; that of E. arborea and conferta is eaten in Nipal.

EXAMPLES. Elæagnus, Hippophae, Shepherdia, Conuleum.

LX. PROTEACEÆ.

PROTEACEÆ, *Juss. Gen.* (1789); *R. Brown in Linn. Trans.* 10. 15. (1809); *Prodr.* 363. (1810.)

DIAGNOSIS. Apetalous dicotyledons, with definite erect ovula, dehiscent fruit, a tubular inferior calyx with the stamens opposite its segments, and a valvate æstivation.

ANOMALIES. The æstivation of Franklandia is induplicate, according to Mr. Brown.

ESSENTIAL CHARACTER.—*Calyx* 4-leaved, or 4-cleft, with a valvular æstivation. *Stamens* 4, sometimes in part sterile, opposite the segments of the calyx. *Ovarium* simple, superior; *style* simple; *stigma* undivided. *Fruit* dehiscent or indehiscent. *Seed* without albumen; *embryo* with two, or occasionally several *cotyledons*, straight; *radicle* inferior.— *Shrubs* or small *trees*. *Branches* usually umbellate. *Leaves* hard, dry, divided or undivided, opposite or alternate, without *stipulæ*.

AFFINITIES. There is no difficulty in distinguishing this order; the hard woody texture of whose leaves, and irregular tubular calyxes having a valvate æstivation, stamens placed upon the lobes, along with a dehiscent fruit, at once characterise it. By these characters it is known from Elæagneæ, and all other orders. The most complete systematic monograph that has ever been written in Botany, is Mr. Brown's upon these, in the Linnæan Society's Transactions, from which I find much to extract. According to this botanist, "the radicula pointing towards the base of the fruit in all Proteaceæ, is a circumstance of the greatest importance, in distinguishing the order from the most nearly related tribes; and its constancy is more remarkable, as it is not accompanied by the usual position or even uniformity in the situation of the external umbilicus." *Linn. Trans.* 10. 36. Mr. Brown has also remarked, with his usual acuteness, that in consequence of the presence of hypogynous squamæ, we may expect to find octandrous genera belonging to this family. See *Flinders*, 2. 606. The same writer observes (*Flinders*, 568), that there is a peculiarity in the structure of the stamina of certain genera of Proteaceæ, namely Simsia, Conospermum, and Synaphea, in all of which these organs are connected in such a manner that the cohering lobes of two different anthers form only one cell. Another anomaly equally remarkable exists in Synaphea, the divisions of whose barren filament so intimately cohere with the stigma, as to be absolutely lost in its substance, while the style and undivided part of the filament remain perfectly distinct. In another place he remarks: "A circumstance occurs in some species of Persoonia, to which I have met with nothing similar in any other plant: the ovarium in this genus, whether it contain one or two ovula, has never more than one cell; but in several of the 2-seeded species, a cellular substance is, after fecundation, interposed between the ovula, and this gradually indurating, acquires in the ripe fruit the same consistence as the putamen itself, from whose substance it cannot be distinguished; and thus, a fruit originally of one cell becomes bilocular; the cells, however, are not parallel, as in all those cases where they exist in the unimpregnated ovarium, but diverge more or less upwards." *Brown in Linn. Trans.* 10. 35. This is subsequently explained, by the same author (*King's Appendix*), by the cohesion of the outer membranes of the two collateral ovula, originally distinct, but finally constituting this anomalous dissepiment, the inner membrane of the ovulum consequently forming the outer coat of the seed.

GEOGRAPHY. "The favourite station of Proteaceæ is in dry, stony, exposed places, especially near the shore, where they occur also, though more rarely, in loose sand. Scarcely any of them require shelter, and none a good soil. A few are found in wet bogs, or even in shallow pools of fresh water; and one, the Embothrium ferrugineum of Cavanilles, grows, according to him, in salt marshes. Respecting the height to which plants of this order ascend, a few facts are already known. The authors of the Flora Peruviana mention, in general terms, several species as being alpine; and Humboldt, in his valuable Chart of Equinoctial Botany, has given the mean height of

Embothrium emarginatum about 9300 feet, assigning it a range of only 300 feet. On the summits of the mountains of Van Diemen's Island, in about 43° south lat., at the computed height of about 4000 feet, I have found species of Embothrium, as well as other genera, hitherto observed in no other situation. Embothrium, however, as it is the most southern genus of any extent, so it is also, as might have been presumed, the most alpine of the family. Two genera only of this order are found in more than one continent: Rhopala, the most northern genus, though chiefly occurring in America, is to be met with also in Cochin China, and in the Malay archipelago; and Embothrium, the most southern genus of any extent, is common to New Holland and America. It is remarkable, that Proteaceæ are almost entirely confined to the southern hemisphere. This observation originated with Mr. Dryander; and the few exceptions hitherto known to it, occur considerably within the tropic. The fact is the more deserving of notice, as their diffusion is very extensive in the southern hemisphere, not merely in latitude and longitude, but also in elevation; for they are not only found to exist in all the great southern continents, but seem to be generally, though very unequally, spread over their different regions: they have been observed also in the larger islands of New Zealand and New Caledonia; but hitherto neither in any of the lesser ones, nor in Madagascar. As in America they have been found in Terra del Fuego, in Chile, Peru, and even Guiana, it is reasonable to conclude that the intermediate regions are not entirely destitute of them. But with respect to this continent, it may be observed, that the number of species seems to be comparatively small; their organisation but little varied; and further, that they have a much greater affinity with those of New Holland than of Africa. Of the botany of South Africa scarce any thing is known, except that of the Cape of Good Hope, where this family occurs in the greatest abundance and variety; but even from the single fact of a genuine species of Protea having been found in Abyssinia by Bruce, it may be presumed that in some degree they are also spread over this continent. With the shores, at least, of New Holland, under which I include Van Diemen's Island, we are now somewhat better acquainted; and in every known part of these, Proteaceæ have been met with. But it appears, that both in Africa and New Holland the great mass of the order exists about the latitude of the Cape of Good Hope, in which parallel it forms a striking feature in the vegetation of both continents. What I am about to advance respecting the probable distribution of this family in New Holland must be very cautiously received, as it is in fact chiefly deduced from the remarks I have myself made in Captain Flinders's Voyage, and subsequently during my short stay in the settlements of New South Wales and Van Diemen's Island, aided by what was long ago ascertained by Sir Joseph Banks, and by a very transitory inspection of an herbarium collected on the west coast, chiefly in the neighbourhood of Shark's Bay, by the botanists attached to the expedition of Captain Baudin. From knowledge so acquired, I am inclined to hazard the following observations: — The mass of the order, though extending through the whole of the parallel already mentioned, is by no means equal in every part of it; but on the south-west coast forms a more decided feature in the vegetation of the country, and contains a far greater number of species, than on the east; and in that part of the south coast which was first examined by Captain Flinders, it seems to be more scanty than at either of the extremes. On the west coast also, the species, upon the whole, are more similar to those of Africa than on the east, where they bear a somewhat greater resemblance to the American portion of the order. From the parallel of the map, the order diminishes in both directions; but the diminution towards the north is probably more rapid on the east than on the west coast. Within the tropic, on the east coast, no genera have hitherto been observed, which are not also found beyond it; unless that section of Grevillea, which I have called Cycloptera, be considered as a genus: whereas, at the southern limit of the order several genera make their appearance, which do not occur in its chief parallel. The most numerous genera are also the most widely diffused. Thus Grevillea, Hakea, Banksia, and Persoonia, extensive in species in the order in which they are here mentioned, are spread nearly in the same proportion; and they are likewise the only genera that have as yet been observed within the tropic. Of such of the remaining genera as consist of several species, some, as Isopogon, Petrophila, Conospermum, and Lambertia, are found in every part of the principal parallel, but hardly exist beyond it. Others, as Josephea and Synaphea, equally limited to this parallel, have been observed only towards its western extremity; while Embothrium (comprehending, for the present, under this name all the many-seeded plants of the order), which is chiefly found on the east coast, and makes very little progress towards the west, advances to the utmost limit of south latitude, and there ascends to the summits of the highest mountains. Genera consisting of one or very few species, and which exhibit generally the most remarkable deviations from the usual structure of the order, are the most local, and are found either in the principal parallel, or in the highest latitude. The range of species in the whole of the order seems to be very limited; and the few cases which may be considered as exceptions to this, occur in the most extensive genera, and in such of their species as are most strictly natives of the shores. Thus Banksia integrifolia, which grows more within the influence of the sea than any plant of the order, is probably also the most widely extended, at least in one direction, being found within the tropic, and in as high a latitude as 40°. It is remarkable, however, that with so considerable a range in latitude, its extension in longitude is comparatively small: and it is still more worthy of notice, that no species of this family has been found common to the eastern and western shores of New Holland." *Brown in Linn. Trans.* 10.

PROPERTIES. Handsome evergreen shrubs, much prized by gardeners for the neatness of their appearance, and beauty or singularity of their flowers; but of no known use, except as fire-wood, for which they are commonly employed at the Cape of Good Hope.

EXAMPLES. Protea, Banksia, Dryandra, Grevillea.

LXI. PENÆACEÆ.

PENÆACEÆ, *R. Brown, verbally* (1820); *Guillemin in Dict. Class,* 13. 171. (1828); *Martius Hort. Monac.* (1829.)

DIAGNOSIS. Apetalous dicotyledons, with definite ovula, a 4-celled ovarium, and a solid homogeneous embryo.

ANOMALIES.

ESSENTIAL CHARACTER.—*Calyx* inferior, with 2 or more bracteæ at its base, hypocrateriform, with a 4-lobed limb valvate in æstivation, or deeply 4-parted imbricated in æstivation. *Stamens* either 4, arising from below the recesses of the limb, with which they alternate, or 8, arising from near the base of the calyx; *anthers* 2-celled, turned inwards, usually with membranous valves lying on the face of a thick fleshy connectivum, sometimes with fleshy valves, and an obliterated connectivum. *Ovarium* superior, 4-celled, with a simple *style* and 4 *stigmas*; *ovules* either ascending, collateral, in pairs, or solitary and suspended; the foramen always next the placenta. *Fruit* capsular, 4-celled, dehiscent, or indehiscent? *Seed* erect or inverted; *testa* brittle; *nucleus* a solid fleshy mass, with no distinction of *albumen* or *embryo; radicular end* next the hilum?; *hilum* fungous.—*Shrubs. Leaves* opposite, imbricated, without stipulæ. *Flowers* terminal and axillary, usually red.

AFFINITIES. According to an observation of Jussieu, this order is allied to Epacrideæ; but I confess I am unable to perceive on what account. To me it appears related in the first degree to some apetalous dicotyledons, such as Proteaceæ, with some of which the species agree in habit, and in the case of Penæa fruticulosa even in the thickened connectivum and the structure of the lobes of the stigma, each of which is strikingly like that of a Grevillea. To Bruniaceæ they must be compared, notwithstanding the presence of petals in that order, for the sake of Linconia, in which the pendulous ovula agree with P. marginata (Geissoloma *m.*), and the thickened connectivum of the anthers, which is common to several species, although not present in Geissoloma. The fungous hilum of the seed is similar to that of Polygaleæ, with which, however, Penæaceæ have no other apparent relation.

This order exhibits a singular instance of two distinct kinds of æstivation and attachment of ovula among species which it is impossible to separate from each other. In true Penæa the æstivation is valvate, and the ovula ascending, while in Geissoloma the former is imbricate, and the latter suspended. Penæa has also tetrandrous flowers, with peculiarly fleshy anthers, while Geissoloma has octandrous flowers, with no peculiar fleshiness in the anthers.

GEOGRAPHY. Evergreen shrubs, natives of the Cape of Good Hope.

PROPERTIES. A subviscid, sweetish, somewhat nauseous gum-resin, called Sarcocolla, is produced by Penæa mucronata (and others). It was supposed by the Arabians to possess, as its name indicates, the power of agglutinating wounds. *Ainslie,* 1. 380. It contains a peculiar principle, named *Sarcocollim,* which has never been detected in any other vegetable matter, and having the property of forming oxalic acid, being treated with nitric acid. *Dec.*

EXAMPLES. Penæa, Geissoloma.

LXII. ARISTOLOCHIÆ. THE BIRTHWORT TRIBE.

ARISTOLOCHIÆ, *Juss. Gen.* (1789); *R. Brown Prodr.* 349. (1810); *Lindley's Synopsis,* 224. (1829).—PISTOLOCHINÆ and ASARINÆ, *Link Handb.* 1. 367. (1829.)

DIAGNOSIS. Apetalous dicotyledons, with indefinite ovules, a many-celled ovarium, and a valvate calyx.

ANOMALIES.

ESSENTIAL CHARACTER.—*Flowers* hermaphrodite. *Calyx* superior, tubular, with 3 segments, which are valvate in æstivation, sometimes regular, sometimes very unequal. *Stamens* 6 to 10, epigynous, distinct, or adhering to the style and stigmas. *Ovarium* inferior, 3- or 6-celled; *ovules* numerous, horizontally attached to the axis; *style* simple, *stigmas* radiating, as numerous as the cells of the ovarium. *Fruit* dry or succulent, 3- or 6-celled, many-seeded. *Seeds* with a very minute embryo placed in the base of fleshy albumen.— *Herbaceous* plants or *shrubs,* the latter often climbing. *Leaves* alternate, simple, stalked, often with leafy stipulæ. *Flowers* axillary, solitary, brown or some dull colour.

AFFINITIES. These are usually stationed upon the limits of monocotyledons and dicotyledons, agreeing with the former in the ternary division of the flower, and in some respects in habit; with the latter in the more essential points of their structure. Their affinity to Cytineæ, an order itself upon the limits of the vascular and cellular divisions of vegetables, is undoubtedly very intimate. Decandolle, in the *Botanicon Gallicum*, places them between Elæagneæ and Euphorbiaceæ, to the former of which they approach through Asarum, but with the latter of which their relation is not obvious. To Passifloreæ they may be compared, on account of the twining habit, alternate leaves, and leafy habit of many species; and to Cucurbitaceæ, on account of their twining habit, and inferior ovarium.

GEOGRAPHY. Very common in the equinoctial parts of South America, and rare in other countries; found sparingly in North America, Europe, and Siberia; more frequently in the Basin of the Mediterranean, and in small numbers in India.

PROPERTIES. These are in general tonic and stimulating; Aristolochia is, as its name implies, reputed emmenagogue, especially the European species rotunda, longa, and Clematitis. An infusion of the dried leaves of Aristolochia bracteata is given by native Indian practitioners as an anthelmintic; fresh bruised and mixed with castor oil, they are considered as a valuable remedy in obstinate psora. The root of Aristol. indica is supposed by the Hindoos to possess emmenagogue and antarthritic virtues; it is very bitter. Arist. odoratissima, a native of the West Indies, is a valuable bitter, and alexipharmic. *Ainslie*, 2. 5. The Aristolochia fragrantissima, called in Peru Bejuca de la Estrella, or Star Reed, is highly esteemed in Peru as a remedy against dysenteries, malignant inflammatory fevers, colds, rheumatic pains, &c. The root is the part used. See *Lambert's Illustration of Cinchona*, p. 150, &c. The power of the root of Aristolochia serpentaria in arresting the progress of the worst forms of typhus, is highly spoken of by Barton, 2. 51. It has an aromatic smell, approaching that of Valerian, with a warm, bitterish, pungent taste. Asarum canadense, called Wild Ginger in the United States, is nearly allied in medical properties to the Aristolochia serpentaria. *Barton*, 2. 88. The root of Asarum europæum, or Asarabacca, is used by native practitioners in India as a powerful evacuant: they also employ the bruised and moistened leaves as an external application round the eyes in certain cases of ophthalmia. *Ainslie*, 1. 24. The leaves and roots of the same plant are emetic; but this quality is lost, according to Decandolle, by keeping or by steeping in vinegar.

EXAMPLES. Aristolochia, Asarum, Trichopus.

LXIII. CYTINEÆ.

CYTINEÆ, *Adolphe Brongn. in Ann. des Sc. Nat.* 1. 29. (1824).—PISTIACEÆ, *Agardh. Aphor. Bot.* p. 240. (1826).—RHIZANTHEÆ, *Blume in Batav. Zeitung,* (1825); *Flora Javæ,* (1829).—ARISTOLOCHIÆ, § Cytineæ, *Link Handb.* 1. 368. (1829.)

DIAGNOSIS. Apetalous leafless dicotyledons, with indefinite ovules, a 1-celled ovarium with parietal placentæ and indehiscent fruit.

ANOMALIES. No spiral vessels exist in these plants.

ESSENTIAL CHARACTER.—*Flowers* diœcious, monœcious, or hermaphrodite. *Calyx* superior, with a limb divided into several divisions, which are imbricated in æstivation. *Stamens* cohering in a solid central column, from the apex of which arise some horned processes; *anthers* adnate, either bursting longitudinally and externally, or having their inside cellular, and discharging their pollen by orifices at the apex. *Ovarium* inferior,

1- or many-celled, with broad parietal placentæ, which are covered with an indefinite number of minute ovules. *Fruit* an inferior pulpy berry. *Seeds* extremely minute, (their nucleus consisting of a mass of grumous matter. *Blume.*)—*Parasitical* brown or colourless plants, without spiral vessels. *Stem* simple, covered with a few leaves in the form of scales. *Flowers* in spikes or heads, or solitary.

AFFINITIES. These very curious plants are all parasitical, with scales in room of leaves. Among them is the very remarkable plant described by Mr. Brown in the 13th vol. of the Linnean Society's Transactions, under the name of Rafflesia, to which I refer those who are desirous either of knowing what is the structure of one of the most anomalous of vegetables, or of finding a model of botanical investigation and sagacity, or of consulting one of the most beautiful specimens of botanical analysis which Mr. Bauer has ever made. The affinity of these plants appears to be greater with Aristolochiæ than any other phænogamous tribe. But the most interesting circumstance of their organisation is, that they exhibit in some degree the structure both of flowering and flowerless, or of vascular and cellular plants. Like flowering or vascular plants, they have a distinct floral envelope, and distinct sexual organs, not essentially, or in fact very, different from those of ordinary vegetables. Like flowerless or cellular plants, they are destitute of all trace of spiral vessels, and their seeds appear to be composed of a homogeneous mass of grumous matter, in which no radicle or cotyledons, no ascending or descending extremity, no definite points of vegetation, can be distinguished.

GEOGRAPHY. Natives of the south of Europe, and the East Indies.

PROPERTIES. Probably all astringents. Cytinus contains Gallic acid; and, according to M. Pelletier (*Bull. Pharm.* 1813. p. 290.) it has the singular property of precipitating gelatine, although it does not contain tannin. Rafflesia is used in Java as a powerful astringent, for certain purposes.

EXAMPLE. Cytinus.

LXIV. SANTALACEÆ. THE SANDERS-WOOD TRIBE.

SANTALACEÆ, *R. Brown Prodr.* 350. (1810); *Juss. Dict. des Sc. Nat.* 47. 287. (1827); *Lind. Synops.* 207. (1829).—OSYRIDEÆ, *Juss. in Ann. Mus. vol.* 5. (1802).—NYSSACEÆ, *Juss. in Dict. des Sciences*, 35. 267. (1825).—OSYRINÆ, *Link Handb.* 1. 371. (1829.)

DIAGNOSIS. Apetalous dicotyledons, with definite pendulous ovules, solitary flowers, and a 1-celled ovarium, with a tubular superior calyx.

ANOMALIES. Osyris differs in its diœcious flowers, in having a trifid calyx with only three stamens, and, according to the younger Gærtner, an erect seed with an embryo curved and lying a little out of the axis of the albumen, with its radicle superior, and therefore turned away from the hilum.

ESSENTIAL CHARACTER.—*Calyx* superior, 4- or 5-cleft, half-coloured, with valvate æstivation. *Stamens* 4 or 5, opposite the segments of the calyx, and inserted into their bases. *Ovarium* 1-celled, with from 1 to 4 *ovules*, fixed to the top of a central placenta near the summit; *style* 1; *stigma* often lobed. *Fruit* 1-seeded, hard and dry, and drupaceous. *Albumen* fleshy, of the same form as the seed; *embryo* in the axis, inverted, taper.—*Trees* or *shrubs*, sometimes *under-shrubs* or *herbaceous* plants. *Leaves* alternate, or nearly opposite, undivided, sometimes minute, and resembling stipulæ. *Flowers* in spikes, seldom in umbels, or solitary, small. *R. Br.*

AFFINITIES. Closely allied to Elæagneæ and Thymelæeæ. Mr. Brown observes (*Flinders*, 569.) that one of the most remarkable characters of this tribe consists in its unilocular ovarium containing more than one, but always a determinate number of ovula, which are pendulous, and attached to the apex of a central receptacle. This receptacle varies in its figure in the different genera, in some being filiform, in others nearly filling the cavity of the ovarium. It appears, from the botanical Appendix to Captain Flinders's Voyage, that there is a very remarkable species of Exocarpus (a genus belonging to this tribe), which bears its flowers upon the margins of dilated foliaceous branches, analogous to those of Xylophylla. I refer Nyssaceæ to this, without any doubt. According to Jussieu, who is the only botanist that has noticed that tribe, it contains but the single genus Nyssa, differing from Elæagneæ in its inferior ovarium, albuminous pendulous seed, and superior radicle. It is more nearly allied to Santalaceæ; but its ovarium contains, instead of three ovules adhering to a central placenta, one only, which is pendulous, and its embryo is not cylindrical, but has enlarged foliaceous cotyledons. It has been long since remarked by Mr. Brown, that Anthobolus and Exocarpus differ from Santalaceæ in having a superior ovarium : Jussieu, in his last observations upon this tribe, does not absolutely separate those genera, but he suggests the possibility of their forming a new family along with Cervantesia of the *Flora Peruviana.*

GEOGRAPHY. Found in Europe and North America, in the form of little obscure weeds; in New Holland, the East Indies, and the South Sea Islands, as large shrubs, or small trees.

PROPERTIES. Sanders-wood is the produce of Santalum album. In India it is esteemed by the native doctors as possessing sedative and cooling qualities, and as a valuable medicine in gonorrhœa. It is also employed as a perfume. *Ainslie,* 1. 377. The Thesiums are scentless and slightly astringent. *Dec.*

EXAMPLES. Santalum, Nyssa, Thesium.

LXV. THYMELÆÆ. THE MEZEREUM TRIBE.

THYMELÆÆ, *Juss. Gen.* 76. (1789); *R. Br. Prodr.* 358. (1810); *Lindley's Synopsis,* 208. (1829.)

DIAGNOSIS. Apetalous dicotyledons, with definite pendulous ovula, a single 1-celled superior ovarium, indehiscent fruit, and exstipulate leaves.

ANOMALIES.

ESSENTIAL CHARACTER.—*Calyx* inferior, tubular, coloured; the limb 4-cleft, seldom 5-cleft, with an imbricated æstivation. *Corolla* 0, or sometimes scale-like petals in the orifice of the calyx. *Stamens* definite, inserted in the tube or its orifice, often 8, sometimes 4, less frequently 2; when equal in number to the segments of the calyx or fewer, opposite to them; *anthers* 2-celled, dehiscing lengthwise in the middle. *Ovarium* solitary, with one solitary pendulous ovulum; *style* 1; *stigma* undivided. *Fruit* hard, dry, and nut-like, or drupaceous. *Albumen* none, or thin and fleshy; *embryo* straight, inverted; *cotyledons* plano-convex; *radicle* short, superior; *plumula* inconspicuous.—*Stem* shrubby, very seldom herbaceous, with tenacious bark. *Leaves* without stipulæ, alternate or opposite, entire. *Flowers* capitate or spiked, terminal or axillary, occasionally solitary. *R. Br.*

AFFINITIES. Closely akin to Santalaceæ, Elæagneæ, and Proteaceæ from all which they are readily known by obvious characters; especially from the two latter by the pendulous ovula, and from the former by the inferior calyx. Aquilarineæ, placed by Decandolle near Chailletiaceæ, among polypetalous orders, differ from Thymeleæ chiefly in their 2-valved fruit;

the scales in the throat of several genera of Thymelææ being of the same nature as the bodies wrongly called petals in Aquilarineæ.

GEOGRAPHY. Natives sparingly of Europe, and the northern parts of the world, common in the cooler parts of India and South America, and abundant at the Cape of Good Hope and in New Holland.

PROPERTIES. The great feature of this order is the causticity of the bark, which acts upon the skin as a vesicatory, and causes excessive pain in the mouth if chewed. A decoction of it is said to have been found useful in venereal complaints. The berries of D. Laureola are poisonous to all animals except birds. *Dec.* The bark is composed of interlaced fibres, which are extremely tough, but which are easily separable; in Jamaica a species is found which is called the Lace Bark Tree, in consequence of the beautifully reticulated appearance of the inner bark: cordage has been manufactured from several species. A very soft kind of paper is made from the inner bark of Daphne Bholua, in Nipal. *Dec. Prodr.* 68. Daphne Gnidium and Passerina tinctoria are used in the south of Europe to dye wool yellow.

EXAMPLES. Daphne, Passerina, Struthiola.

LXVI. HERNANDIEÆ.

HERNANDIEÆ, *Blume Bijdr.* 550. (1825.)

DIAGNOSIS. Apetalous dicotyledons, with an inferior tubular deciduous calyx, a single pendulous ovulum, no albumen, lobed cotyledons, and a calycine involucellum to the female or hermaphrodite flowers.

ANOMALIES.

ESSENTIAL CHARACTER.—*Flowers* monœcious or hermaphrodite, with a calycine involucellum to the females or hermaphrodites. *Calyx* petaloid, inferior, tubular, 4-8-parted, deciduous. *Stamens* definite, inserted into the calyx in two rows, of which the outer is often sterile; *anthers* bursting longitudinally. *Ovarium* superior, 1-celled; *ovulum* pendulous; *style* 1, or none; *stigma* peltate. *Drupe* fibrous, 1-seeded. *Seed* solitary, pendulous; *embryo* without albumen, inverted; *cotyledons* somewhat lobed, shrivelled, oily. — *Trees. Leaves* alternate, entire. *Spikes* or *corymbs* axillary or terminal.

AFFINITIES. Adopted from Blume. It appears very near Thymelææ, differing almost solely in the fibrous drupaceous fruit, lobed cotyledons, and the presence of a sort of involucrum to the female or hermaphrodite flowers. Hernandia has been hitherto referred to Laurineæ or Myristiceæ, from both of which it is obviously very different. Blume refers Inocarpus to the same order; but this measure appears questionable.

GEOGRAPHY. Natives of the Indian archipelago and Guiana.

PROPERTIES. The bark, seed, and young leaves, are all slightly purgative. According to Rumphius, the fibrous roots of Hernandia sonora, chewed and applied to wounds caused by the Macassar poison, form an effectual cure. The juice of its leaves is a powerful depilatory; it destroys hair wherever it is applied, without pain. The wood appears to be very light. According to Aublet, that of H. guianensis takes fire readily from a flint and steel, and is used as amadou.

EXAMPLE. Hernandia.

LXVII. AQUILARINEÆ. THE AGALLOCHUM TRIBE.

AQUILARINEÆ, *R. Brown Cong.* p. 25. (1818); *Dec. Prodr.* 2. 59. (1825.)

DIAGNOSIS. Apetalous dicotyledons, with definite suspended ovula, a solitary superior 1-celled ovarium, tubular calyx, and stamina alternately fertile and scale-like, arising from the throat.

ANOMALIES.

ESSENTIAL CHARACTER. — *Calyx* turbinate, coriaceous, 5-lobed. *Petals* 0. *Stamens* monadelphous, 10 fertile, 10 sterile; the former inserted between the latter, which are petaloid or scale-like; *anthers* innate, 2-celled, bursting longitudinally. *Ovarium* superior, 1-celled, ovate, crowned by a short simple *stigma ; ovules* 2, parietal, suspended, with their foramen in their apex, which is tapering and turned to the bottom of the cell. *Capsule* pyriform, 2-valved, 1-celled, with the valves bearing the seed. *Seeds* solitary, with an arillus or tail, (probably suspended, with the same form as the ovulum, and with the radicle at the opposite extremity to the hilum.) — *Trees. Leaves* alternate, entire.

AFFINITIES. M. Decandolle places this order between Chailletiaceæ, but with indications of doubt, and an erroneous character; and Mr. Brown seems willing (*Congo*, 444.) to consider the order a section of Chailletiaceæ, adding, that it would not be difficult to shew its affinity to Thymeleæ. In this I fully concur, after an examination of a specimen of Aquilaria Agallochum, for which I am indebted to the East India Company; in fact, Aquilarineæ chiefly differ from Thymeleæ in their dehiscent fruit, and probably also in the direction of their radicle. In both orders the ovarium is superior and 1-celled, both have similar scale-like bodies at the orifice of the calyx, and no petals, both suspended ovula, a single style, and capitate stigma.

GEOGRAPHY. Natives of the East Indies.

PROPERTIES. Aloes wood, a fragrant resinous substance, of a dark colour, is the inside of the trunk of the Aquilaria ovata and A. Agallochum. It is considered a cordial by some Asiatic nations, and has been prescribed in Europe in gout and rheumatism. *Ainslie*, 1. 479.

EXAMPLE. Aquilaria.

LXVIII. OLACINEÆ.

OLACINEÆ, *Mirb. Bull. Philom.* n. 75. 377. (1813); *Dec. Prodr.* 1. 531. (1824.)

DIAGNOSIS. Polypetalous dicotyledons, with hypogynous definite stamens, concrete carpella, an ovarium of 1 cell with a columnar placenta in the axis, an imbricated calyx, unsymmetrical flowers, definite (3) pendulous ovules, and bifid petals with appendages.

ANOMALIES. According to Decandolle and others, the ovarium of some consists of several cells, but this is doubtful. Ximenia has entire petals, but it is not certain that it belongs to the order.

ESSENTIAL CHARACTER. — *Calyx* small, entire, or slightly toothed, finally becoming, in many cases, enlarged. *Petals* definite, hypogynous, valvate in æstivation, either altogether separate, or cohering in pairs by the intervention of stamina. *Stamens* definite, part fertile, part sterile; the former varying in number from 3 to 10, hypogynous, usually cohering with the petals, and alternate with them; the latter opposite the petals, to which they in part adhere, their upper end resembling an appendage; *filaments* compressed; *anthers* innate, oblong, 2-celled, bursting longitudinally. *Ovarium* superior, 1-celled, with 3 ovules pendulous from the top of a central column or placenta. *R. Br. Style* filiform ;

stigma simple. *Fruit* somewhat drupaceous, indehiscent, frequently surrounded by the enlarged calyx, 1-celled, 1-seeded. *Seed* erect; *albumen* large, fleshy; *embryo* small, in the base of albumen, its *radicle* near the hilum. — *Trees* or *shrubs*. *Leaves* simple, alternate, entire, without stipulæ; occasionally altogether wanting. *Flowers* small, axillary.

AFFINITIES. M. Decandolle places this order near Aurantiaceæ, with which it agrees in many respects, differing, however, in the structure of the ovarium, the want of a disk, the unsymmetrical flowers, &c. Jussieu, on the contrary, regards the affinity as strongest with Sapoteæ, considering the corolla as monopetalous. But the obvious affinity of Olax with Aquilarineæ and Samydeæ induces me to concur with Mr. Brown in considering the order nearly akin to Santalaceæ, among Monochlamydeæ. In the mean while its artificial characters place it among Thalamifloræ.

GEOGRAPHY. A small order, consisting of tropical or nearly tropical shrubs, chiefly found in the East Indies, New Holland, and Africa. One only is known in the West Indies. None have been described from any part of South America, south of Dutch Guiana.

PROPERTIES. The wood of Heisteria coccinea is the Partridge wood of the cabinet-makers.

EXAMPLES. Olax, Fissilia.

LXIX. CHAILLETIACEÆ.

CHAILLETIÆ, *R. Brown Cong.* p. 23. (1818). — CHAILLETIACEÆ, *Dec. Prodr.* 2. 57. (1825.)

DIAGNOSIS. Polypetalous dicotyledons, with definite perigynous stamens, concrete carpella, a superior ovarium with 2 or 3 cells and 5 hypogynous glands, and alternate stipulate leaves.

ANOMALIES.

ESSENTIAL CHARACTER. — *Sepals* 5, with an incurved valvate æstivation. *Petals* 5, alternate with the sepals, and arising from the base of the calyx, usually 2-lobed. *Stamens* 5, alternate with the petals, and combined with them at the base; *anthers* ovate, versatile. *Glands* usually 5, hypogynous, opposite the petals. *Ovarium* superior, 2- or 3-celled; *ovules* twin, pendulous; *style* simple; *stigma* obsoletely 3-lobed. *Fruit* drupaceous, rather dry, 1- 2- or 3-celled. *Seeds* solitary, pendulous, without albumen; *embryo* thick, with a thick superior *radicle* and fleshy *cotyledons*. — *Trees* or *shrubs*. *Leaves* alternate, with two stipulæ, entire. *Flowers* small, axillary, their peduncle often connate with the petiole.

AFFINITIES. Whether what are here called petals are not rather abortive stamina is doubted by botanists, and hence the station of the order is by one referred to Dichlamydeæ, and by another to Monochlamydeæ, and is compared, on the one hand, with Terebintaceæ or Rosaceæ, and, on the other, with Samydeæ and Amentaceæ. To me it seems that what appear to be petals are so; a fact which it is difficult to doubt, when it is remembered that both organs are mere transformations of one common type, and that it is in appearance and position only that they differ. Decandolle stations it between Homalineæ and Aquilarineæ, to the latter of which it has probably most affinity; it agrees with the former in the presence of glands round the ovarium, but differs in its superior ovarium with the placentæ in the axis, and many other characters.

GEOGRAPHY. Of the few known species belonging to this order, 2 are found in Sierra Leone, 2 in Madagascar, 2 in equinoctial America, and 1 in Timor.

PROPERTIES. The fruit of Chailletia toxicaria is said to be poisonous.

EXAMPLES. Chailletia, Leucosia, Tapura.

LXX. HOMALINEÆ.

HOMALINEÆ, R. Brown in Congo, (1818); Dec. Prodr. 2. 53. (1825.)

DIAGNOSIS.   Polypetalous dicotyledons, with perigynous stamens, concrete carpella, an inferior ovarium of 1 cell with parietal placentæ, and petals and sepals resembling each other, with glands at their base.

ANOMALIES.   It is said there are no glands in Napimoga.   Astranthus is said to have a superior ovarium; but this requires confirmation.

ESSENTIAL CHARACTER.—*Calyx* funnel-shaped, superior, with from 5 to 15 divisions. *Petals* alternate with the segments of the calyx, and equal to them in number. *Glands* present in front of the segments of the calyx. *Stamens* arising from the base of the petals, either singly or in threes or sixes; *anthers* 2-celled, opening longitudinally. *Ovarium* half inferior, 1-celled, with numerous ovula; *styles* from 3 to 5, simple, filiform, or subulate; *ovules* attached to as many parietal placentæ as there are styles. *Fruit* berried or capsular. *Seeds* small, ovate, or angular, with an embryo in the middle of fleshy albumen. — *Trees* or *shrubs*. *Leaves* alternate, with deciduous stipulæ, toothed or entire. *Flowers* in spikes, racemes, or panicles.

AFFINITIES.   According to Mr. Brown, related to Passifloreæ, especially to Smeathmannia, from which, however, their inferior ovarium distinguishes them, to say nothing of their general want of stipulæ and glands on the leaves, of the presence of glands at the base of the floral envelopes, and of their erect and very different habit.   With Malesherbiaceæ they agree and disagree much, as with Passifloreæ.   From Rosaceæ, Bixineæ, and Flacourtianeæ, to all which they have a greater or less degree of affinity, they differ in many obvious particulars.   Decandolle places them between Samydeæ and Chailletiaceæ, describing them as apetalous, but classing them with his Dichlamydeæ; Mr. Brown also understands them as without petals; but I confess I cannot comprehend what petals are, if the inner series of the floral envelopes of these plants are not so; an opinion which their supposed affinity with Passifloreæ would confirm, if analogy could be admitted as evidence in cases which can be decided without it.   I may remark, that the statement of M. Decandolle, that the stamens are opposite the sepals (*Prodr.* 3. 53.) is inaccurate; they are, as Mr. Brown describes them (*Congo*), opposite the petals.

GEOGRAPHY.   All tropical, and chiefly African or Indian.   Four or five species are described from the West Indies and South America.

PROPERTIES.   Unknown.

EXAMPLES.   Astranthus, Blackwellia, Homalium.

LXXI. SAMYDEÆ.

SAMYDEÆ, Vent. Mem. Inst. 2. 142. (1807); Gærtn. fil. Carp. 3. 238. 242. (1805); Kunth. Nov. Gen. 5. 360. (1821); Dec. Prodr. 2. 47. (1825.)

DIAGNOSIS.   Apetalous dicotyledons, with indefinite ovules, a 1-celled ovarium with parietal placentæ, dehiscent fruit, hermaphrodite flowers, perigynous monadelphous stamens, and leaves with a mixture of round and oblong dots.

ANOMALIES.

ESSENTIAL CHARACTER.—*Sepals* 3, 5, or 7, more or less cohering at the base, usually coloured inside; æstivation somewhat imbricated, very seldom completely valvate. *Petals* 0. *Stamens* arising from the tube of the calyx, 2, 3, or 4 times as many as the sepals; *filaments* monadelphous, either all bearing anthers, or alternately shorter, villous or ciliated, and alternately bearing ovate 2-celled erect *anthers*. *Ovarium* superior, 1-celled; *style* 1, filiform; *stigma* capitate, or slightly lobed; *ovula* indefinite, attached to parietal placentæ. *Capsule* coriaceous, with 1 cell and from 3 to 5 valves, many-seeded, the valves dehiscing imperfectly, often somewhat pulpy inside, and coloured. *Seeds* fixed to the valves, without order, on the papillose or pulpy part, with a fleshy arillus and excavated hilum; *albumen* fleshy; *embryo* inverted, minute; *cotyledons* ovate, foliaceous; *radicle* pointing to the extremity remote from the hilum. — *Trees* or *shrubs*. *Leaves* alternate, often somewhat distichous, simple, entire or toothed, evergreen, with stipulæ, usually with pellucid dots, which are most frequently oblong. *Peduncles* axillary, solitary, or numerous.

AFFINITIES. Placed in Dichlamydeæ by Decandolle, who, however, describes them as apetalous, " unless the petaloid layer covering the inner surface of the sepals be considered a corolla," a proposition which it is impossible to admit. This order appears to be of very uncertain affinity. Its fruit approximates it to Bixineæ, its dotted leaves to Terebintaceæ, near which Decandolle stations it, and its perigynous stamens to Rosaceæ, with which its alternate stipulate leaves also ally it. Mr. Brown observes, that Samydeæ are especially distinguished by their leaves having a mixture of *round* and *linear* pellucid dots, which distinguish them from all the other families with which they are likely to be confounded. *Congo*, 444.

GEOGRAPHY. Chiefly natives of the West Indies and South America; a very few only are described from India.

PROPERTIES. Unknown. The bark and leaves are said to be astringent in a slight degree. *Dec.*

EXAMPLES. Samyda, Casearia.

LXXII. SANGUISORBEÆ. THE BURNET TRIBE.

ROSACEÆ, § Sanguisorbeæ, *Juss. Gen.* 336. (1789); *Dec. Prodr.* 2. 588. (1828); *Lindl. Synops.* 102. (1829.)

DIAGNOSIS. Apetalous dicotyledons, with definite suspended ovula, an inferior tubular indurated calyx, with perigynous stamens, indehiscent fruit, and alternate stipulate leaves.

ANOMALIES. The stipulæ of Cliffortia cohere with the leaves. Alchemilla arvensis has simple 1-celled anthers bursting transversely, and ascending ovula.

ESSENTIAL CHARACTER. — *Flowers* often unisexual. *Calyx* with a thickened tube and a 3- 4- or 5-lobed limb, its tube lined with a disk. *Petals* none. *Stamens* definite, sometimes fewer than the segments of the calyx, with which they are alternate, arising from the orifice of the calyx; *anthers* 2-celled, innate, bursting longitudinally, occasionally 1-celled, bursting transversely. *Ovarium* solitary, simple, with a style proceeding from the apex or the base; *ovulum* solitary, always attached to that part of the ovarium which is next the base of the style; *stigma* compound or simple. *Nut* solitary, enclosed in the often indurated tube of the calyx. *Seed* solitary, suspended or ascending; *embryo* without albumen; *radicle* superior; *cotyledons* large, plano-convex. — *Herbaceous* plants or *undershrubs*, occasionally spiny. *Leaves* simple and lobed, or compound, alternate, with stipulæ. *Flowers* small, often capitate.

AFFINITIES. This order, usually combined with Rosaceæ, appears to me to demand a distinct station, on account of its constantly apetalous flowers, its indurated calyx, and the reduction of carpella to one only; it is, however, not, as far as I know, distinguishable by any other characters. The presence of petals, a character assigned to Acæna, I have shewn, in the *Botanical Register*, to have no existence. Usually the ovulum is suspended, the style arising from below the apex of the carpellum; but when the style proceeds from the base of the carpellum, the ovulum is ascending, in all cases adhering to the ovarium immediately over against the origin of the style. A genus usually referred to this order, the Cephalotus of Labillardiére, offers a remarkable exception to the usual characters, in having a coloured calyx, in the senary division of its flower, and in the presence of ascidia, or pitchers, among its leaves, resembling those of Nepenthes. It is, however, by no means well ascertained that this is the station of Cephalotus, its seeds being unknown. Various kinds of adhesion between the leaves and the stipules take place in the genus Cliffortia, and have given rise to a number of errors; for an explanation of which, see M. Decandolle's remarks in the *Annales des Sciences Naturelles*, 1. 447.

GEOGRAPHY. Natives of heaths, hedges, and exposed places in Europe, North and South America beyond the tropics, and the Cape of Good Hope; in which latter country they represent the Rosaceæ of Europe.

PROPERTIES. Their general character is astringency. A decoction of Alchemilla vulgaris is slightly tonic. This is asserted, by Frederick Hoffmann, and others, to have the effect of restoring the faded beauty of ladies to its earliest freshness. Sanguisorba officinalis, or common Burnet, is a useful fodder. *A. R.*

EXAMPLES. Acæna, Sanguisorba, Margyricarpus.

LXXIII. ROSACEÆ. THE ROSE TRIBE.

ROSACEÆ, *Juss. Gen.* 334. *in part* (1789); *Dec. Prodr.* 2. 525. *in part* (1825); *Dec. and Duby Botan. Gall. in part* (1828); *Lindl. Synops.* p. 88. (1829.)

DIAGNOSIS. Polypetalous dicotyledons, with lateral styles, superior simple ovaria, regular perigynous stamens, exalbuminous definite seeds, and alternate stipulate leaves.

ANOMALIES. Stipulæ absent in Lowea. Albumen present in Neillia, according to Don. The fruit of Spiræa sorbifolia (Schizonotus m.) is capsular.

ESSENTIAL CHARACTER.—*Calyx* 4- or 5-lobed, with a disk either lining the tube or surrounding the orifice; the fifth lobe next the axis. *Petals* 5, perigynous, equal. *Stamens* indefinite, arising from the calyx, just within the petals, in æstivation curved inwards; *anthers* innate, 2-celled, bursting longitudinally. *Ovaries* superior, either solitary or several, 1-celled, sometimes cohering into a plurilocular pistillum; *ovula* 2, or more, suspended, very rarely erect; *styles* lateral; *stigmata* usually simple, and emarginate on one side. *Fruit* either 1-seeded nuts, or acini, or follicles containing several seeds. *Seeds* suspended, rarely ascending. *Embryo* straight, with a taper short radicle pointing to the hilum, and flat cotyledons. *Albumen* usually almost obliterated when the seeds are ripe; if present, fleshy.—*Herbaceous* plants or *shrubs*. *Leaves* simple or compound, alternate, with 2 stipulæ at their base.

AFFINITIES. The genera of this order are uniform in their structure and sensible qualities. Neuradeæ, at present included, will probably be hereafter removed to a more appropriate station. Distinguished from Pomaceæ by their superior fruit and usually suspended seeds; from Leguminosæ by their regular petals and stamens, and especially by the odd segment of the 5-lobed calyx of that order being anterior, not posterior, as in Rosaceæ; from Chrysobalaneæ by their styles proceeding from the side of the ovarium near the apex, and not from the base, by their regular petals and stamens, and by their fruit not being a drupe. Amygdaleæ, often combined with Rosaceæ, are particularly characterised by their terminal styles, drupaceous fruit, and hydrocyanic juice, along with which is a formation of gum. Sanguisorbeæ are apetalous, with definite stamens alternate with the segments of the calyx. Related in many points to Saxifrageæ.

GEOGRAPHY. Natives chiefly of the temperate or cold climates of the northern hemisphere; a very few are found on high land within the tropics, and an inconsiderable number in the southern hemisphere. Only one species is found in the West Indies, viz. Rubus jamaicensis; thirteen are natives of high land in the East Indies, within the tropics, viz. Potentilla Leschenaultiana, and twelve species of Rubus; the South American species chiefly consist of a few kinds of Rubus; at the Cape of Good Hope the order is unknown.

PROPERTIES. No Rosaceous plants are unwholesome; they are chiefly remarkable for the presence of an astringent principle, which has caused some of them to be reckoned febrifuges. The root of Tormentilla is used for tanning in the Feroe Isles. *Dec.* Potentilla anserina has been used by tanners; P. reptans as a febrifuge. *Ibid.* Geum urbanum and rivale have been compared, for efficacy, to Cinchona. *Ibid.* The fruits of many species of Fragaria (Strawberry) and Rubus (Raspberry and Blackberry) are valuable articles of the dessert. The leaves of Rubus arcticus and Rosa rubiginosa have been employed as substitutes for Tea. *Ibid.* The roots of Gillenia trifoliata and stipulacea are emetic, and perhaps tonic. *Barton,* 1. 69. They are used in the United States as Ipecacuanha. *Dec.* The root of Spiræa ulmaria has been used as a tonic. *A. R.* Agrimonia eupatoria yields a decoction useful as a gargle. *Ibid.* The root of Rubus villosus is a popular astringent medicine in North America. Two or three teaspoonsful of the decoction, administered three or four times a-day, has been found useful in cholera infantum. *Barton,* 2. 157. One of the most powerful anthelmintics in the world belongs to this family. It is an Abyssinian plant, known to botanists by the name of Brayera anthelmintica. Upon the authority of Dr. Brayer, after whom it is named, two or three doses of the infusion are sufficient to cure the most obstinate case of tænia. See *Brayer's Notice upon the subject.* The various species of Rosa form some of the greatest beauties of the garden. The fruit of R. canina and other allied species is astringent, and employed in medicine against chronic diarrhœa and other maladies. The petals of R. damascena yield a highly fragrant essential oil, called Attar of Roses; those of R. gallica are astringent when dried with rapidity, and are sometimes found useful in cases of debility, such as leucorrhœa, diarrhœa, &c. *A. R.*

The following divisions have been established among Rosaceous plants:

1. § POTENTILLEÆ. *Cinquefoils.*

§ Potentillæ, *Juss. Gen.* 337. (1789.) — § Dryadeæ, *Vent. Tabl.* 3. 349. (1799); *Dec. Prodr.* 2. 549. (1825.) — Fragariaceæ, *Rich. in Nestl. Potentill.* (1816); *Lindl. Synops.* 90. (1829.)

Fruit consisting either of small nuts or acini, arising from a common receptacle, and invested with a dry permanent calyx. *Calyx* either 4- or 5-cleft, sometimes bearing *bracteolæ* on its tube equal in number to the segments, and alternate with them. *Petals* 5. *Seed* solitary, erect, or inverted. — Mostly *herbaceous* plants, very seldom *shrubs; leaves* usually compound; *stipulæ* adhering to the petiole.

EXAMPLES. Potentilla, Fragaria, Geum.

2. § Roseæ. *True Roses.*

§ Rosæ, *Juss. Gen.* 335. (1789.) — § Rosæ, *Dec. Prodr.* 2. 596. (1825); *Lindl. Synops.* 99. (1829.)

Nuts numerous, hairy, terminated by the persistent lateral style, and enclosed within the fleshy tube of the calyx, which is contracted at its orifice, where it is surrounded by a fleshy disk. *Seed* suspended. *Sepals* 5. *Petals* 5. *Stamens* indefinite. — *Shrubs*, with prickly or naked stems. *Leaves* pinnate. *Flowers* red, white, or yellow, usually fragrant.

EXAMPLES. Rosa, Lowea.

3. § Spiræaceæ. *Spiræas.*

§ Spiræeæ, *Juss. Gen.* 339. (1789.) — § Ulmariæ, *Vent. Tabl.* 3. 351. (1799.) — § Spiræaceæ, *Dec. Prodr.* 2. 541. (1825); *Lindl. Synops.* 89. (1829.)

Follicles several, invested by the calyx. *Seeds* from 1 to 6, suspended from the inner edges of the follicle. — *Shrubs* or *herbaceous* plants.

EXAMPLES. Spiræa, Gillenia, Schizonotus.

? 4. § Neuradeæ. *Neuradas.*

§ Neuradeæ, *Dec. Prodr.* 2. 548. (1825.)

Calyx 5-cleft, with a short tube adhering to the ovarium, the lobes somewhat incumbent or valvate in æstivation. *Petals* 5. *Stamens* 10. *Carpella* 10, combined in a 10-celled compressed capsule. *Seeds* solitary, obliquely pendulous. — *Herbaceous* plants, native of sandy plains, suffrutescent at the base, and usually decumbent. *Leaves* with 2 stipulæ, downy, sinuate-pinnatifid, or bipinnatifid. *Seeds* germinating in the capsule.

EXAMPLE. Neurada.

Is not this rather a tribe of Ficoideæ, as has been suggested by M. de Jussieu? to which, however, the want of albumen, the form of the embryo, and the texture of the leaves, are objections. *Dec. Prodr.* 2. 548.

LXXIV. POMACEÆ. THE APPLE TRIBE.

ROSACEÆ, § Pomaceæ, *Juss. Gen.* 334. (1789); *Dec. Prodr.* 2. 626. (1825.) — POMACEÆ, *Lindl. in Linn. Trans.* 13. 93. (1821); *Synops.* 103. (1829.)

DIAGNOSIS. Polypetalous dicotyledons, with perigynous indefinite stamens, ovaria adhering more or less to the calyx, and alternate stipulate leaves.

ANOMALIES. In Amelanchier the simple ovaria are spuriously 2-celled. In Cratægus the ovaria are very rarely solitary.

ESSENTIAL CHARACTER. — *Calyx* superior, 5-toothed; the odd segment posterior. *Petals* 5, unguiculate, inserted in the throat of the calyx; the odd one anterior. *Stamens* indefinite, inserted in a ring in the throat of the calyx. *Disk* thin, clothing the sides of the limb of the calyx. *Ovaria* from 1 to 5, adhering more or less to the sides of the calyx and each other; *ovules* usually 2, collateral, ascending, very rarely solitary; *styles* from 1 to 5; *stigmata* simple. *Fruit* a pome, 1- to 5-celled, seldom spuriously 10-celled; the endocarpium either cartilaginous, spongy, or bony. *Seeds* ascending, solitary. *Albumen* none; *embryo* erect, with flat *cotyledons*, or convolute ones in Chamæmeles, and a short conical *radicle.* — *Trees* or *shrubs. Leaves* alternate, stipulate, simple, or compound. *Flowers* in terminal cymes, white or pink.

AFFINITIES. Closely allied to Rosaceæ, from which they differ in the adhesion of the ovaria with the sides of the calyx, and more or less with each other. Their fruit is always a pome; that is, it is made up of a fleshy calyx adhering to fleshy or bony ovaria, containing a definite number of seeds. Pomaceæ are peculiarly distinguished by their ovula being in pairs, and side by side; while Rosaceæ, when they have 2 or more ascending ovules, always have them placed one above the other. Cultivated plants of the order are very apt to produce monstrous flowers, which depart sometimes in a most remarkable degree from their normal state. No order can be more instructively studied with a view to morphological inquiries; particularly the common Pear when in blossom. A remarkable permanent monster of this kind, with 14 styles, 14 ovaria, and a calyx with 10 divisions in two rows, is described in the *Revue Encyclopédique*, (43. 762.); it exhibits a tendency, on the part of Pomaceæ, to assume the indefinite ovaria and double calyx of Rosaceæ. I have seen a Prunus in a similar state. Amygdaleæ are known by their superior solitary ovarium and drupaceous fruit, and by the presence of Prussic acid, which, however, exists in Cotoneaster microphylla, a plant of the order Pomaceæ.

GEOGRAPHY. Found plentifully in Europe, Northern Asia, the mountains of India, and North America; rare in Mexico, unknown in Africa, except on its northern shore, and in Madeira, and entirely absent from the southern hemisphere; a solitary species is found in the Sandwich Islands.

PROPERTIES. The fruit as an article of food, and the flowers for their beauty, are the chief peculiarities of this order, which consists exclusively of trees and bushes, without any herbaceous plant. The Apple, the Pear, the Medlar, the Quince, the Service, the Rowan Tree or Mountain Ash, are all well known, either for their beauty or their use. The wood of the Pear is almost as hard as Box, for which it is even substituted by wood engravers; the timber of the Beam Tree (Pyrus Aria) is invaluable for axletrees. The bark of Photinia dubia is used in Nipal for dyeing scarlet. *Dec. Prodr.* 238. Malic acid is contained, in considerable quantity, in apples; it is also almost the sole acidifying principle of the berries of the Mountain Ash (Pyrus aucuparia). *Turner*, 634.

EXAMPLES. Pyrus, Cratægus, Cydonia.

LXXV. AMYGDALEÆ. THE ALMOND TRIBE.

AMYGDALEÆ, *Juss. Gen.* 340. *a* § *of* Rosaceæ (1789). — DRUPACEÆ, *Dec. Fl. Française*, 4. 479. (1815); *Prodr.* 2. 529. (1825) *a* § *of* Rosaceæ; *Lindl. Synops.* 89. (1829) *a* § *of* Rosaceæ.

DIAGNOSIS. Polypetalous dicotyledons, with a superior solitary simple ovarium having a terminal style, regular perigynous indefinite stamens, a drupaceous fruit, an exalbuminous suspended seed, and alternate stipulate simple leaves yielding hydrocyanic acid.

ANOMALIES.

ESSENTIAL CHARACTER.— *Calyx* 5-toothed, deciduous, lined with a disk; the fifth lobe next the axis. *Petals* 5, perigynous. *Stamens* 20, or thereabouts, arising from the throat of the calyx, in æstivation curved inwards; *anthers* innate, 2-celled, bursting longitudinally. *Ovary* superior, solitary, simple, 1-celled; *ovula* 2, suspended; *styles* terminal, with a furrow on one side, terminating in a reniform *stigma*. *Fruit* a drupe, with the putamen sometimes separating spontaneously from the sarcocarp. *Seeds* mostly solitary, suspended, in consequence of the cohesion of a funiculus umbilicalis, arising from the base of the cavity of the ovarium, with its side. *Embryo* straight, with the radicle pointing to the hilum; *cotyledons* thick; *albumen* none.— *Trees* or *shrubs*. *Leaves* simple, alternate, usually glandular towards the base; *stipulæ* simple, mostly glandular. *Flowers* white or pink. *Hydrocyanic acid* present in the leaves and kernel.

AFFINITIES. Distinguished from Rosaceæ and Pomaceæ by their fruit being a drupe, their bark yielding gum, and by the presence of hydrocyanic acid; from Leguminosæ by the latter character, and also by their regular petals and stamens, and especially by the odd segment of the 5-lobed calyx of that order being inferior, not superior; from Chrysobalaneæ by their hydrocyanic acid, terminal styles, and regular petals and stamens. I have seen a monstrous Plum with an indefinite number of ovaria arising irregularly from the tube of the calyx, and therefore exhibiting a tendency, on the part of this order, to assume one of the distinguishing characters of Rosaceæ.

GEOGRAPHY. Natives exclusively of the northern hemisphere, where they are found in cold or temperate climates. One species, Cerasus occidentalis, is a native of the West Indies; a kind of Almond, Amygdalus microphylla, inhabits hot arid plains in Mexico; and another, A. cochinchinensis, is reputed to grow in the woods of Cochinchina.

PROPERTIES. The astringent febrifugal properties of Rosaceæ, with which order these are usually combined, are also found in Amygdaleæ; as in the bark of Cerasus virginiana, which is prescribed in the United States, and of the C. capollim of Mexico. They are, however, better known for yielding an abundance of prussic, or hydrocyanic, acid, a deadly principle residing in the leaves and kernel; in consequence of which some of the species are poisonous to cattle which feed upon them: as, for example, the Cerasus capricida, which kills the goats of Nipal; and the C. virginiana, which is known in North America to be dangerous. They all of them, also, yield a gum, analogous to gum tragacanth. Notwithstanding, however, the poisonous principle that is present in them, their fruit is, in many cases, a favourite food; that of the Amygdalus (peach and nectarine), Prunus (plum and apricot), and Cerasus (cherry), are among the most delicious with which we are acquainted; the seed of Amygdalus is familiar to us under the name of almonds, and its oil under the name of oil of almonds. The bark of the root of Cerasus capollim is used in Mexico against dysentery. *Dec.* The leaves of Prunus spinosa (sloe), and Cerasus avium (wild cherry), have been employed as a substitute for tea. *Ibid.* The former are well known to afford one of the means used in Europe for adulterating the black tea of China. Prunus domestica, or the common plum, yields those fruits sold in the shops under the name of prunes, which are chiefly prepared in France, from the varieties called the St. Catherine and the green-gage; and in Portugal from a sort which derives its name from the village of Guimaraens, where they are principally dried. They contain so large a quantity of sugar, that brandy is distilled from them when fermented; and it has even been proposed to manufacture sugar from them. *A. R.* The kernel of Prunus brigantiaca yields a fixed oil, called *Huile des Marmottes,* which is used instead of olive or almond oil. *Ibid.* The bark of Prunus spinosa is one of the substances that has been reported to resemble Jesuits' bark in its effects. *Ibid.* Prunus cocomilia yields a bark, the febrifugal properties of which are spoken of very highly. According to M. Tenore, it is a specific for the cure of the dangerous intermittent fevers of Calabria, where it grows. A variety of Cerasus avium is used for the preparation, in the Vosges and the Black Forest, of the liqueur known under the name of Kirschenwasser. The flowers of Amygdalus persica (peach) are gently laxative, and are used advantageously for children. The kernel of Cerasus occidentalis is used for flavouring the liqueur Noyau.

EXAMPLES. Prunus, Amygdalus, Cerasus.

LXXVI. CHRYSOBALANEÆ. The Cocoa-Plum Tribe.

CHRYSOBALANEÆ, *R. Brown, in Tuckey's Voyage to the Congo, App.* (1818); *Dec. Prodr.* 2. 525. *a sect. of* Rosaceæ (1825); *Reichenb. Conspectus,* 171. *a sect. of* Onagrariæ, (1828.)

DIAGNOSIS. Polypetalous dicotyledons, with a superior solitary ovarium, having a style proceeding from its base, irregular perigynous petals and stamens, a drupaceous fruit adhering obliquely to the calyx, exalbuminous definite erect seeds, and alternate stipulate simple leaves.

ANOMALIES. Hirtella has fleshy albumen and leafy cotyledons, according to Gærtner; and one species of the same genus is described as apetalous. Cycnia has a semipetaloid irregular calyx and no petals.

ESSENTIAL CHARACTER. — *Calyx* 5-lobed, sometimes bracteolate at the base. *Petals* more or less irregular, either 5 or none. *Stamens* either definite or indefinite, usually irregular either in size or position. *Ovarium* superior, solitary, 1- or 2-celled, cohering more or less on one side with the calyx; *ovula* twin, erect; *style* single, arising from the base; *stigma* simple. *Fruit* a drupe of 1 or 2 cells. *Seed* usually solitary, erect. *Embryo* with fleshy cotyledons, and no albumen. — *Trees* or *shrubs. Leaves* simple, alternate, stipulate, with no glands, and veins that run parallel with each other from the midrib to the margin. *Flowers* in racemes, or panicles, or corymbs.

AFFINITIES. The obvious affinity of this order is with Amygdaleæ, from which it differs in having irregular stamens and petals, and a style proceeding from the base of the ovarium. With Rosaceæ, to which Chrysobalaneæ have a strict relation, they agree in the same manner as Amygdaleæ, excepting the characters just pointed out. To Leguminosæ, with drupaceous fruit, they approach closely in the irregularity of their stamens and corolla, and especially in the cohesion which takes place between the stalk of the ovarium and the sides of the calyx; a character found, as M. Decandolle well remarks, in Jonesia and Bauhinia, undoubted leguminous plants: they are distinguished from this latter order by the position of their style and ovula, and by the relation which is borne to the axis of inflorescence by the odd lobe of the calyx being the same as is found in Rosaceæ. Brown remarks (*Congo,* 434), that the greater part of the order has the flowers more or less irregular, and that the simple ovarium of Parinarium has a dissepiment in some degree analogous to the movable dissepiment of Banksia and Dryandra; but we now know, from the more recent observations of this learned botanist upon the ovulum, that this dissepiment arises differently. The analogy of structure, as to the dissepiment of Parinarium, is to be sought in Amelanchier.

GEOGRAPHY. These plants are principally found in the tropical regions of Africa and America: none are recorded as natives of Asia; but there is reason to believe, from specimens of large trees seen in the forests of India, without flowers or fruit, by Dr. Wallich, that one or two species of Parinarium are indigenous in equinoctial Asia; and my genus Cycnia, founded upon a spiny plant from Nipal (*Wall. Cat. Herb. Ind.*), is apparently referable to this order. One species of Chrysobalanus is found as far to the north as the pine-barrens of Georgia in North America; a climate, however, as in all the regions bounding the Gulf of Mexico on the north, much more heated than that of most other countries in the same parallel of latitude.

PROPERTIES. No medicinal properties have been ascribed to Chrysobalaneæ. The fruit of Chrysobalanus Icaco is eaten in the West Indies, under the name of the cocoa-plum; another is brought to market in Sierra Leone (C. luteus); and the Rough-skinned, or Gray, plum of the same colony is the produce of Parinarium excelsum. The kernel of Parinarium campestre and montanum is said by Aublet to be sweet and good to eat.

EXAMPLES. Chrysobalanus, Parinarium, Hirtella.

LXXVII. LEGUMINOSÆ. THE PEA TRIBE.

LEGUMINOSÆ, *Juss. Gen.* 345. (1789); *Brown Diss.* (1822); *Dec. Prodr.* 2. 93. (1825); *Lindl. Synops.* 75. (1829.)

DIAGNOSIS. Polypetalous dicotyledons, with a terminal style and solitary simple superior ovarium, perigynous definite stamens, exalbuminous seeds, peritropal ovula, leguminous fruit, and alternate stipulate leaves.

ANOMALIES. The Detariums are apetalous and drupaceous. Ceratonia, Copaifera, and five or six other genera, are also apetalous. Some Mimoseæ are monopetalous; the latter section and Swartzieæ have usually also hypogynous stamens. Diphaca and a species of Cæsalpinia have regularly 2 ovaria. Ormosia has 2 stigmas. *Dec.* Sophora, Myrospermum, and some others, have no stipulæ. Some have opposite leaves.

ESSENTIAL CHARACTER. — *Calyx* 5-parted, toothed, or cleft, inferior, with the odd segment anterior; the segments often unequal, and variously combined. *Petals* 5, or by abortion 4, 3, 2, 1, or none, inserted into the base of the calyx, either papilionaceous or regularly spreading; the odd petal posterior. *Stamens* definite or indefinite, perigynous, either distinct or monadelphous, or diadelphous; very seldom triadelphous; *anthers* versatile. *Ovarium* simple, superior, 1-celled, 1- or many-seeded; *style* simple, proceeding from the upper margin; *stigma* simple. *Fruit* either a legume or a drupe. *Seeds* attached to the upper suture, solitary or several, occasionally with an arillus; *embryo* destitute of *albumen*, either straight or with the radicle bent upon the cotyledons; *cotyledons* either remaining under ground in germination, or elevated above the ground, and becoming green like leaves. — *Herbaceous* plants, *shrubs*, or vast *trees*, extremely variable in appearance. *Leaves* alternate, most commonly compressed; *petiole* tumid at the base. *Stipulæ* 2 at the base of the petiole, and 2 at the base of each leaflet. *Pedicels* usually articulated, with 2 bracteolæ under the flower.

AFFINITIES. The most common feature is, to have what are called papilionaceous flowers; and when these exist, no difficulty is experienced in recognising the order, for papilionaceous flowers are found no where else. Another and a more invariable character is to have a leguminous fruit; and by one of these two characters all the plants of the family are known. It is remarkable, however, for the complete obliteration of one or other of these distinctions in many cases. Mimosa and its allies have, instead of the irregular arrangement which characterises a papilionaceous flower, its parts of fructification disposed with the utmost symmetry; and Detarium, instead of a legumen, bears a fruit not distinguishable from a drupe. This last circumstance is easily to be understood, if we bear in mind that a legume and a drupe differ more in name than reality, the latter being formed upon precisely the same plan as the former, but with this modification, that its pericarpium is thickened, more or less fleshy on the outside and stony on the inside, 1-seeded, and indehiscent. Hence some of the regular-flowered genera with distinct stamens may be said to be Rosaceous in flower, and Leguminous in fruit. Simple, therefore, as the diagnosis of the order usually is, Mr. Brown is perfectly correct in asserting that, until he indicated the difference of the position of the odd lobe of the calyx in Leguminosæ and Rosaceæ (Amygdaleæ), no positive character had been discovered to distinguish the one order from the other. The presence of stipulæ at the base of the leaflets of the compound leaves of Leguminosæ is a character in the vegetation by which they may be known from Rosaceæ. Myroxylon agrees with Samydeæ in the remarkable glandular marking of the leaves, in which the pellucid spaces are both round and linear—a very singular and uncommon character, which was first pointed out by Mr. Brown. *Congo,* 444. Very few double flowers are known in this order; those of Spartium junceum and Ulex europæus are the most remarkable : the nature of the latter I have described in detail in the *Trans. of the Hort. Soc.* vol. 7. p. 237. Two ovaria are common in Wisteria sinensis; and the same phenomenon is to be seen, according to Decandolle, in Gleditschia : it appears also to be normal in Diphaca and Cæsalpinia digyna. M. Aug. de St. Hilaire is said (*Dec. Mém.* 52) to have found a Mimosa in Brazil with 5 carpella : on account of these, and other circumstances, M. Decandolle assumes the carpellum of Leguminosæ to be solitary by abortion, and that a whorl of 5 is that which is necessary to complete the symmetry of the flowers. Of the accuracy of this view I am satisfied; but I think it might have been proved as satisfactorily from analogy, without the aid of such instances. In consequence of the highly irritable nature of the leaves of many of the plants of this order, and of the tendency to irritability discoverable in them all, some botanists have placed them at the extremity of their system, in contact with the limits of the animal kingdom. See *Agardh Classes,* p. 4, and *Martius, H. R. M.* p. 176. For observations upon the nature of this irritability, see *Dutrochet sur la Motilité, Paris,* 1824, in which the author endeavours to shew that the motion is the effect of galvanic agency; and the same writer's *Nouvelles Recherches sur l'Exosmose, &c.,* in which he alters the explanation of the manner in which galvanism produces the motion, adhering, however, to his opinion of that subtle principle being the real agent. This ingenious naturalist might have been satisfied with attributing the phenomenon to an inherent vital action, without puzzling himself with a vain search after first causes, which always leaves the most successful inquirer exactly where he set out. For remarks upon the order in general, see M. Decandolle's valuable Mémoire, published at Paris in 1825-6, in one thick volume 4to. The relation that is borne by this order to Chrysobalaneæ and Amygdaleæ has been already explained under those orders. To the tribes formerly included under the name of Terebintaceæ, Leguminosæ are nearly allied in many important circumstances, but are distinguished by their stipules, which nevertheless exist in Canarium among Burseraceæ, and which do not exist in Sophora, a genuine, and Myrospermum, a spurious Leguminous genus. The affinity of the latter to Amyrideæ is, however, so great, that it appears to me very questionable whether it ought not to be absolutely referred to that order rather than to Leguminosæ. With Xanthoxyleæ they are allied through Ailanthus. The monadelphous stamens, irregular flowers, occasional simple ovarium, style, and stigma of Polygaleæ, are all so many points of affinity with Leguminosæ.

In many respects this order is one of the most important which the botanist can study, but especially as it serves to shew how little real importance ought to be attached to dehiscence of fruit in determining the limits of natural orders. What may be called the normal fruit of Leguminosæ is a legume, that is to say, a dry simple ovarium, with a suture running along both its margins, so that at maturity it separates through the middle of each suture into two valves; but every conceivable degree of deviation from this type occurs : the Arachis and many more are indehiscent; Detarium is drupaceous; in Carmichælia the valves separate from the suture, which remains entire, like the replum of Cruciferæ; in all Lomentaceous genera, such as Ornithopus, the valves are indehiscent in the line of the suture, but separate transversely; in Entada a combination of the peculiarities of Carmichælia and Lomentaceæ occurs; and, finally, in Hæmatoxylon the valves adhere by the suture and split along the axis. The divisions which have been proposed in this extensive order are of unequal value; it is possible that two of them, namely, Mimoseæ and Cæsalpinieæ may deserve, as Mr. Brown seems to think, the rank of independent orders; for they really appear to be of the same importance with reference to Papilionaceæ, as Amygdaleæ and Pomaceæ are with respect to Rosaceæ, or as Amyrideæ, Connaraceæ, Anacardiaceæ, and Burseraceæ, with respect to each other. I give them, however, as I find them in Decandolle.

His first and most important division depends upon the form of the embryo, out of which arise the divisions called Curvembriæ and Rectembriæ; viz. —

CURVEMBRIÆ.

Radicle bent back upon the cotyledons.

These are distinguished into two tribes by the structure of their flowers, viz. —

Tribe 1. SWARTZIEÆ.

Calyx bladdery, with indistinct lobes. Stamens hypogynous. Corolla none, or petals only 1 or 2.

EXAMPLES. Swartzia, Baphia.

Tribe 2. PAPILIONACEÆ.

Calyx with distinct lobes. Stamens perigynous. Corolla papilionaceous.

EXAMPLES. Vicia, Pisum, Sophora.

The germination of this tribe varies thus: — some of the species push their cotyledons above ground, which become green, resembling leaves; and of these none bear seeds which are eaten by man or animals: others germinate with their cotyledons under ground, and it is among these only that all the kinds which bear what we call pulse are found: the former Decandolle calls *Phyllolobeæ*, and they are divided by him into sections, viz. 1. § Sophoreæ, 2. § Loteæ, 3. § Hedysareæ; the latter he designates as *Sarcolobeæ*, which comprehend, 4. § Vicieæ, 5. § Phaseoleæ, 6. § Dalbergieæ.

RECTEMBRIÆ.

Radicle of the embryo straight.

The tribes are known by the position of their stamens and the æstivation of their petals.

Tribe 3. MIMOSEÆ.

Sepals and petals valvate in æstivation. Stamens hypogynous.

EXAMPLES. Acacia, Mimosa, Inga.

Tribe 4. CÆSALPINIEÆ.

Petals imbricated in æstivation, and stamens perigynous.

EXAMPLES. Arachis, Cæsalpinia, Cassia.

Of the genera comprehended in this tribe, those which have petals, and their stamens variously combined, are called § Geoffrieæ; such as have petals, the stamens being distinct, are § Cassieæ; and a couple of genera, with drupaceous fruit and no petals, constitute § Detarieæ.

The reader is referred to the 2d volume of Decandolle's *Prodromus* for further information upon these divisions.

GEOGRAPHY. The geographical distribution of this order has been considered with great care by Decandolle, from whom I take the substance of what follows.

One of the first things that strikes the observer is, that if a number of genera of Leguminosæ have as extensive a range as those of other orders, there is a very considerable number of which the geographical limits are clearly defined. Thus the genera of New Holland are in most cases unknown beyond that vast island; the same may be said of North and South America, and the Cape of Good Hope; and there are between 14 and 15 genera unknown beyond the limits of Europe and the neighbouring borders of Asia and Africa. About 92 genera out of 280 are what are called sporadic, or dispersed over different and widely separated regions, such as Tephrosia, Acacia, Glycine, and Sophora. The species are found more or less in every part of the known world, with the exception, perhaps, of the island of Tristan d'Acugna and St. Helena, neither of which do they inhabit; but they are distributed in extremely unequal proportions; in general they diminish sensibly in approaching the pole, especially the Rectembriæ, which are unknown in northern regions. This will be apparent from the following table:—

	Curvembr.	Rectembr.
Europe, with the exception of the Mediterranean	184	0
Siberia	128	1
United States	167	16
China, Japan, and Cochinchina	64	13
Levant	247	3
Basin of the Mediterranean	466	2
Canaries	21	0
Arabia and Egypt	78	9
Mexico	90	62
West Indies	134	87
East Indies	330	122
Equinoctial America	246	359
Equinoctial Africa	81	49
New Holland	154	75
Isles of Southern Africa	29	13
South America beyond the tropics	18	11
Cape of Good Hope	334	19
South Sea Islands	11	2

This distribution, if condensed, will give the following results:—

Equinoctial zone	910	692
Beyond the tropics to the north	1277	35
——————————— south	417	107

PROPERTIES. This order is not only among the most extensive that are known, but also one of the most important to man, with reference to the objects either of ornament, of utility, or of nutriment, which it comprehends. When we reflect that the Cercis, which renders the gardens of Turkey resplendent with its myriads of purple flowers; the Acacia, not less valued for its airy foliage and elegant blossoms than for its hard and durable wood; the Braziletto, Logwood, and Rosewoods of commerce; the Laburnum; the classical Cytisus; the Furze and the Broom, both the pride of the otherwise dreary heaths of Europe; the Bean, the Pea, the Vetch, the Clove, the Trefoil, the Lucerne, all staple articles of culture by the farmer, are all species of Leguminosæ; and that the Gums Arabic and Kino, and various precious medicinal drugs, not to mention Indigo, the most useful of all dyes, are products of other species,—it will be perceived that it would be difficult to point out an order with greater claims upon the attention. It would be in vain to attempt to enumerate all its useful plants or products, in lieu of which I shall speak of the most remarkable, and of those which are least known.

The beauty of Dr. Wallich's Amherstia nobilis, a large tree bearing pendulous racemes of deep scarlet flowers, is unequalled in the vegetable kingdom. The general character of the order is to be eminently wholesome; but there are some singular exceptions to this. The seeds of Lathyrus

Aphaca are said to produce intense headach if eaten abundantly: the seeds of the Laburnum are poisonous; they contain a principle called Cytisine. The root of a species of Mimosa, called Spongia, is accounted a poison in Brazil. *Ed. P. J.* 14. 267. The leaves and branches of Tephrosia are used for intoxicating fish; the leaves of Ornithopus scorpioides are capable of being employed as vesicatories. The juice of Coronilla varia is poisonous. *Dec.* The powerful purgative effects of Senna are possessed also by other species, even by Colutea arborescens and Coronilla emerus. Cassia marilandica is found in North America a useful substitute for the Alexandrian Senna. *Barton,* 1. 143. The Senna of the shops consists, according to M. Delile, of Cassia acutifolia, Cassia Senna, and Cynanchum Argel. He says the Cassia lanceolata of Arabia does not yield the Senna of commerce. The active principle of Senna is called Cathartine. It was discovered by MM. Lassaigne and Fenuelle. *Ed. P. J.* 7. 389. Purgative properties are also found in the pulp within the fruit of Cathartocarpus fistula and Ceratonia siliqua, of Mimosa fagifolia, and also of the Tamarind, the preserved pulp of which is so well known as a delicious confection. Malic acid exists in the Tamarind, mixed with tartaric and citric acids. *Turner,* 634. The same may be said of Inga fæculifera, or the *Pois doux,* of St. Domingo, that bears pods filled with a sweet pulp, which the natives use. *Hamilt. Prodr.* 62. The roots of the liquorice contain an abundance of a sweet subacrid mucilaginous juice, which is much esteemed as a pectoral; similar qualities are ascribed to Trifolium alpinum roots. The root of Abrus precatorius possesses exactly the properties of the liquorice root of the shops. *Ainslie,* 2. 79. In Java it is found demulcent. The seeds are considered by some as ophthalmic and cephalic, externally applied. The roots of Beans, Genistas, Ononis, Guilandina Nuga and Moringa, Anthyllis cretica, &c. are diuretic. *Dec.* Those of Dolichos tuberosus and bulbosus, and Lathyrus tuberosus, are wholesome food. Some are reported to produce powerfully bitter and tonic effects. Various species of Geoffræa, the bark of Æschinomene grandiflora and of Cæsalpinia Bonduccella are of this class. The kernels of Guilandina Bonduccella are very bitter, and are supposed by the native doctors of India to possess powerful tonic virtues. When pounded small and mixed with castor oil, they form a valuable external application in incipient hydrocele. *Ainslie,* 2. 136. The leaves are a valuable discutient, fried with a little castor oil, in cases of hernia humoralis. *Ibid.* The bark of Acacia Arabica is considered in India a powerful tonic; a decoction of its pods is used as a substitute for that of the seeds of Mimosa saponaria for washing. *Ibid.* 2. 142. The root of Hedysarum sennoides is accounted in India tonic and stimulant. *Ibid.* 2. 53. These powers are probably connected with the astringent and tanning properties of several others. Some of the Algarobas or Prosopises of the western part of South America bear fruit, the pericarp of which consists almost wholly of tannin. The bark of some of the species of Acacia abound to such a degree in tanning principles as to have become objects of commercial importance. In 1824 some tons of the extract of Acacia bark were imported from New South Wales for the use of tanners. *Ed. P. J.* 11. 266. The pods of Cassia Sabak and Acacia nilotica are used in Nubia for tanning. *Delile Cent.* 10. The valuable astringent substance, called Catechu, or Terra Japonica, is procured by boiling and evaporating the brown heart-wood of Acacia Catechu, or Khair Tree: it is obtained by simply boiling the chips in water until the inspissated juice has acquired a proper consistency; the liquor is then strained, and soon coagulates into a mass. *Brewster,* 5. 349. Gum Kino is the produce of Pterocarpus erinacea *R. Br.,* Gum Dragon and Sandalwood of Pterocarpus Draco and Santalinus, Gum Lac of Erythrina monosperma,

Gum Anime of Hymenæa Courbaril *Dec.*, Gum Arabic is yielded by Acacia senegalensis and some others, Gum Tragacanth by Astragalus creticus and similar species. According to Mr. Don (*Prodr.* no. 247.), the Manna of Arabia is produced by several species of Hedysarum, related to H. Alhagi. The Dalbergia monetaria of Linnæus yields a resin very similar to Dragon's Blood. *Ainslie*, 1. 115. A similar juice is yielded by Butea frondosa and superba. *Dec.* Among the woods of trees of this order, the most important is that of the Locust Tree, Robinia pseudacacia, which is a light bright yellow, hard and durable, but brittle. The Brazil wood of commerce is obtained from Cæsalpinia Braziliensis. The fine Jacaranda, or Rosewood of commerce, so called because when fresh it has a faint but agreeable smell of roses, is produced by a species of Mimosa in the forests of Brazil. *Pr. Max. Trav.* 69. Among dyes are Indigo, produced by all Indigoferas and some Galegas, Logwood, the wood of Hæmatoxylon campeachianum, and the red dye yielded by several Cæsalpinias. The colouring matter of Logwood is a peculiar principle, called Hæmatin. The wood of Pterocarpus santalinus yields a deep red colouring matter; it is known in commerce under the name of Saunders Wood. *Ainslie*, 1. 386. All the species of the genus Copaifera, and 16 are known, yield the Balsam of Copaiva; but it is not in all of them of equal quality. C. multijuga is said by Von Martius to afford the greatest abundance. *Hayne in Linnæa*, 1826. 418. The Balsam is known in Venezuela under the name of Tacamahaca. *Dec. Prodr.* 2. 508. Myroxylon peruiferum, the Quinquino of Peru, produces a fragrant resin, in much use both for burning as a perfume, and for medicinal purposes, called the Balsam of Tolu. *Lambert's Illustration*, 95. Both it and the Balsam of Peru are also yielded, according to Ach. Richard, by M. toluiferum. *Ann. des Sc.* 2. 172. The root of Clitoria Ternatea is emetic. *Ainslie*, 2. 140. The seed of Psoralea corylifolia is considered by the native practitioners of India stomachic and deobstruent. *Ibid.* 141. According to Dr. Horsfield, the Acacia scandens of Java is classed among the emetics. *Ibid.* 2. 108. The roots and herbage of Baptisia tinctoria have been found to possess antiseptic and subastringent properties. They have also a cathartic and emetic effect. *Barton*, 2. 57. The seeds of Cassia auriculata are considered by the Indian doctors as refrigerant and attenuant. *Ainslie*, 2. 32. The leaves of Coronilla picta are highly esteemed among the Hindoos, on account of the virtues they are said to possess in hastening suppuration when applied in the form of a poultice, that is, simply made warm, and moistened with a little castor oil. *Ibid.* 2. 64. The seeds of Parkia africana are roasted as we roast coffee, then bruised, and allowed to ferment in water. When they begin to become putrid, they are well washed and pounded; the powder is made into cakes, somewhat in the fashion of our chocolate; they form an excellent sauce for all kinds of meat. The farinaceous matter surrounding the seeds forms a pleasant drink, and they also make it into a sweetmeat. *Brown in Denham*, 29. The irritating effects of the hairs which clothe the pods of Dolichos pruriens, or Cowhage, are well known. A strong infusion of the root of the same plant, sweetened with honey, is used by the native practitioners of India in cases of cholera morbus. *Ainslie*, 1. 93. The native practitioners in India prescribe the dried buds and young flowers of Bauhinia tomentosa in certain dysenteric affections. *Ibid.* 2. 48. A decoction of the bitter root of Galega purpurea (Tephrosia) is prescribed by the Indian doctors in cases of dyspepsia, lientery, and tympanitis. *Ibid.* 2. 49. The powdered leaf of Indigofera Anil is used in hepatitis. *Ibid.* 1. 179. The volatile oil of the Coumarouma odorata, or Tonka Bean, has been ascertained to be a peculiar principle called Coumarin. It was mistaken by M. Vogel for Benzoic acid. *Turner*, 660. It may be found in a crystallised state between the skin and the kernel, and exists abundantly in the flowers of Melilotus officinalis. *Ed. P. J.* 3. 407. It has been found that a peculiar acid, called Carbazotic, is formed by the action of nitric acid upon Indigo. *Turner*, 641. Sulphur exists in combination with different bases in peas and beans. *Ed. P. J.* 14. 172. The leaves of the Phaseolus trilobus (called Sem, or Simbi) are considered by Indian practitioners cooling, sedative, antibilious, and tonic, and useful as an application to weak eyes. *Trans. M. and P. Soc. Calc.* 2. 406.

LXXVIII. URTICEÆ. The Nettle Tribe.

URTICEÆ, *Juss. Gen.* 400. (1789); *Lindley's Synopsis*, 218. (1829). — CÆNOSANTHEÆ and CANNABINÆ, *Blume Bijdr.* (1825.) *both sections of* Urticeæ.

DIAGNOSIS. Apetalous dicotyledons, with definite erect ovula, an inferior calyx, distinct stipulæ, and an embryo with the radicle remote from the hilum.

ANOMALIES.

ESSENTIAL CHARACTER. — *Flowers* monœcious or diœcious, scattered or clustered. *Calyx* membranous, lobed, persistent. *Stamens* definite, distinct, inserted into the base of the calyx, and opposite its lobes; *anthers* curved inwards in æstivation, curving backwards with elasticity when bursting. *Ovarium* superior, simple; *ovule* solitary, erect; *stigma* simple. *Fruit* a simple indehiscent nut, surrounded either by the membranous or fleshy calyx. *Embryo* straight, curved, or spiral, with or without albumen; *radicle* superior, and therefore remote from the hilum; *cotyledons* lying face to face. — *Trees*, or *shrubs*, or *herbs*. *Leaves* alternate, with stipulæ, hispid or scabrous, often covered with pungent hairs.

AFFINITIES. The position of the ovulum, the want of milk, the flowers being arranged in loose racemes or panicles, not in fleshy heads, and their habit, distinguish Urticeæ from Artocarpeæ. From Polygoneæ they are known by their want of stipulæ, from Chenopodeæ and Scleranthceæ by their stinging or scabrous surface, the position of the radicle, and their elastic stamens; and from Euphorbiaceæ by the simplicity of their ovarium; from Betulineæ by the presence of a calyx, and from Cupuliferæ by their superior simple ovarium. They agree with the two latter orders remarkably in stipulation.

GEOGRAPHY. Widely dispersed over every part of the world; appearing in the most northern regions, and in the hottest climates of the tropics; growing now upon dry walls, where there is scarcely nutriment for a moss or a lichen, and inhabiting the dampest recesses of the forest.

PROPERTIES. The tenacity of the fibres of many species is such that cordage has been successfully manufactured from them. The leaves of Hemp are powerfully narcotic. The Turks know its stupifying qualities under the name of Malach. Linnæus speaks of its vis narcotica, phantastica, dementens, anodyna, and repellens. Even the Hottentots use it to get drunk with, and call it Dacha. The Arabians name it Hashish. *Ainslie*, 2. 189. A most powerfully narcotic gum-resin, called in Nipal Cheris or Cherris, is supposed to be obtained from a variety of Cannabis sativa. *Ibid.* 2. 73. The effects of the venomous sting of the common nettles, Urtica dioica, urens, and pilulifera of Europe, are too well known. Their effects are, however, not to be compared for an instant with those of some Indian species. M. Leschenault (*Mém. Mus.* 6. 362.) thus describes the effect of gathering Urtica crenulata in the Botanic Garden at Calcutta:—" One of

the leaves slightly touched the first three fingers of my left hand : at the time
I only perceived a slight pricking, to which I paid no attention. This was
at seven in the morning. The pain continued to increase; in an hour it had
become intolerable : it seemed as if some one was rubbing my fingers with a
hot iron. Nevertheless, there was no remarkable appearance; neither swelling, nor pustule, nor inflammation. The pain rapidly spread along the arm,
as far as the armpit. I was then seized with frequent sneezing and with a
copious running at the nose, as if I had caught a violent cold in the head.
About noon I experienced a painful contraction of the back of the jaws,
which made me fear an attack of tetanus. I then went to bed, hoping that
repose would alleviate my suffering; but it did not abate; on the contrary,
it continued during nearly the whole of the following night; but I lost the
contraction of the jaws about seven in the evening. The next morning the
pain began to leave me, and I fell asleep. I continued to suffer for two
days; and the pain returned in full force when I put my hand into water. I
did not finally lose it for nine days." A similar circumstance occurred, with
precisely the same symptoms, to a workman in the Calcutta Garden. This
man described the sensation, when water was applied to the stung part, as if
boiling oil was poured over him. Another dangerous species was found by
the same botanist in Java (U. stimulans), but its effects were less violent.
Both these seem to be surpassed in virulence by a nettle called *daoun
setan*, or devil's leaf, in Timor; the effects of which are said, by the natives,
to last for a year, and even to cause death.

The common Hop, Humulus lupulus, is a rather anomalous genus of this
order, remarkable, as is well known, for its bitterness; the active principle of
it is called by chemists Lupulin.

EXAMPLES. Urtica, Parietaria, Böhmeria.

LXXIX. ULMACEÆ. The Elm Tribe.

ULMACEÆ, *Mirbel Elém.* 905. (1815); *Lindl. Synops.* 225. (1829.)—CELTIDEÆ, *Rich.*

DIAGNOSIS. Apetalous dicotyledons, with definite suspended ovula,
solitary or loosely clustered flowers, a 2-celled indehiscent fruit, and alternate
stipulate scabrous leaves.

ANOMALIES.

ESSENTIAL CHARACTER. — *Flowers* hermaphrodite or polygamous. *Calyx* divided,
campanulate, inferior. *Stamens* definite, inserted into the base of the calyx; erect in æstivation. *Ovarium* superior, 2-celled; *ovules* solitary, pendulous; *stigmas* 2, distinct. *Fruit*
1- or 2-celled, indehiscent, membranous or drupaceous. *Seed* solitary, pendulous; *albumen*
none, or in very small quantity; *embryo* with foliaceous cotyledons; *radicle* superior. —
Trees or *shrubs*, with scabrous, alternate, simple, deciduous leaves, and stipulæ.

AFFINITIES. Nearly related to Urticeæ, from which they are only
distinguishable by the 2-celled fruit, pendulous seeds, and radicle turned
towards the hilum; from Artocarpeæ they are known by their inflorescence, dry fruit, and double ovarium.

GEOGRAPHY. Natives of the north of Asia, the mountains of India,
China, North America, and Europe; in the latter of which countries they
form valuable timber-trees.

PROPERTIES. The inner bark of the Elm is slightly bitter and astringent, but it does not appear to possess any important quality. The sub-

stance which exudes spontaneously from it is called Ulmin; it is also found in the Oak, Chestnut, and other trees, and, according to Berzelius, is a constituent of most kinds of bark. *Turner*, 700.

EXAMPLES. Ulmus, Celtis.

LXXX. ARTOCARPEÆ. THE BREAD-FRUIT TRIBE.

ARTOCARPEÆ, *R. Brown in Congo* (1818); *Blume Bijdr.* 479; *and* PHOLEOSANTHEÆ, 435, *both sections of* Urticeæ (1825.)—SYCOIDEÆ, *Link Handb.* 1. 292. (1829.)

DIAGNOSIS. Apetalous lactescent dicotyledons, with flowers in fleshy heads, definite suspended ovula, alternate stipulate leaves, and radicle turned towards the hilum.

ANOMALIES. Antiaris has solitary flowers, and the ovarium cohering with the involucrum.

ESSENTIAL CHARACTER. — *Flowers* monœcious, in heads or catkins. *Calyx* with an uncertain number of divisions, which are often membranous ; sometimes tubular, or entire. *Stamens* uncertain in number, either solitary or several, straight. *Ovarium* 1- or 2-celled, superior, rarely inferior ; *ovulum* suspended ; *style* single, filiform ; *stigma* bifid. *Fruit* usually a fleshy receptacle, either covered by numerous nuts, lying among the persistent fleshy calyxes, or enclosing them within its cavity ; occasionally consisting of a single nut, covered by a succulent involucrum. *Seed* suspended solitary, ; *embryo* inverted, with its *radicle* pointing to the *hilum,* straight or curved, with or without *albumen.* — *Trees, shrubs,* or *herbs. Leaves* alternate, toothed or lobed, or entire, smooth or covered with asperities ; *stipulæ* membranous, deciduous, convolute in vernation.

AFFINITIES. The Fig may be taken as the type of this order, which agrees with Urticeæ in its apetalous flowers, scabrous alternate leaves, and membranous stipulæ ; but which differs in its habit and milky juice, and in the position of the ovulum, which is constantly suspended, not erect. Mr. Brown, indeed, in his Appendix to the Congo Expedition, says that in Artocarpeæ " the ovulum, which is always solitary, is erect, while the embryo is inverted or pendulous." But this statement must be an oversight : I have constantly found the ovulum suspended in Artocarpus incisa, Maclura aurantiaca, Ficus Carica, and other species, and in all the Dorstenias, in the whole of which there is a very conspicuous foramen immediately against the point of attachment of the ovulum.

GEOGRAPHY. Natives of all parts of the tropics, particularly of the East Indies ; a few species, in the form of Morus and Maclura, and the cultivated Fig, straggle northwards as far as Canada and Persia. Dorstenias are remarkable for being herbaceous Brazilian weeds, in an order composed otherwise of trees or shrubs.

PROPERTIES. The Fig, the Bread-fruit, the Jack, and the Mulberry, are all found here, and are a curious instance of wholesome or harmless plants in an order which contains the most deadly poison in the world, the Upas of Java ; the juice, however, of even those which have wholesome fruit, is acrid and suspicious ; and in a species of Fig, Ficus toxicaria, is absolutely venomous. The juice of all of them contains a greater or less abundance of caoutchouc, and the Cecropia peltata is reported to yield American caoutchouc. But Humboldt doubts whether this is the fact, as its juice is difficult to inspissate. *Cinch. For.* p. 44. The seeds of a plant nearly allied to Cecropia, called Musanga by the Africans of the Gold Coast, as well as those of Artocarpus, are eatable as nuts. The famous Cow Tree, or Palo de Vacca, of South America, which yields a copious supply of a rich and wholesome milk, belongs to this order; it is supposed to be related to Brosimum. Brosimum alicastrum abounds in a tenacious gummy milk; its leaves and young shoots are much eaten by cattle, but when they become old they cease to be innocuous. The roasted nuts are used instead of bread, and have much the taste of Hazel nuts. *Swartz*, 1. 19. A kind of paper is manufactured from Broussonetia papyrifera. The bark of the Morus alba contains moroxylic acid in combination with lime. *Turner*, 640. Fustick, a yellow dye, is the wood of Morus tinctoria. The seeds of Ficus religiosa are supposed by the doctors of India to be cooling and alterative. *Ainslie*, 2. 25. The leaves of Ficus septica are emetic. *Ibid.* The Cochin-chinese consider that plant caustic and anthelmintic. The bark of Ficus racemosa is slightly astringent, and has particular virtues in hæmaturia and menorrhagia. The juice of its root is considered a powerful tonic. *Ibid.* 2. 31. The white glutinous juice of Ficus indica is applied to the teeth and gums, to ease the toothache; it is also considered a valuable application to the soles of the feet when cracked and inflamed. The bark is supposed to be a powerful tonic, and is administered by the Hindoos in diabetes. *Ibid.* 2. 11. Gum lac is obtained from the Ficus indica in great abundance. The tena-city of life in some plants of this family is remarkable. A specimen of Ficus australis lived and grew suspended in the air, without earth, in one of the hothouses in the Botanic Garden, Edinburgh, for eight months, without experiencing any apparent inconvenience. *Ed. P. J.* 3. 80. The celebrated Banyan Tree of India is Ficus religiosa. Prince Maximilian, of Wied Neuwied, says that the colossal wild Fig-trees " are one of the most grate-ful presents of nature to hot countries: the shade of such a magnificent tree refreshes the traveller when he reposes under its incredibly wide-spreading branches, with their dark green shining foliage. The Fig-trees of all hot countries have generally very thick trunks, with extremely strong boughs, and a prodigious crown." *Travels*, p. 104. Is it possible that the Indian poison with which the Nagas tip their arrows, of the tree that produces which nothing is known, can belong to this tribe? See, for an account of the effect of this poison, *Brewster's Journal*, 9. 219. The poisonous pro-perty of the Upas has been found to depend upon the presence of that most virulent of all principles, called strychnia. *Turner*, 650.

Examples. Artocarpus, Morus, Maclura.

LXXXI. STILAGINEÆ.

Stilagineæ, *Agardh's Classes*, 199. (1824); *Von Martius Hort. Reg. Monac.* (1829.)

Diagnosis. Apetalous dicotyledons, with unisexual spiked flowers, collateral pendulous ovules, solitary ovaria, 2-lobed anthers bursting verti-cally, and 1-seeded fruit with an albuminous seed.

Anomalies.

Essential Character. — *Flowers* unisexual. *Calyx* 3- or 5-parted. *Corolla* 0. *Stamens* 2, or more, arising from a tumid receptacle; *filaments* capillary; *anthers* innate, 2-lobed, with a fleshy connectivum and vertical cells opening transversely. *Ovarium* supe-rior; *stigma* sessile, 3-4-toothed. *Fruit* drupaceous, with 1 seed and the remains of ano-ther. *Seed* suspended; *embryo* green, with foliaceous cotyledons, lying in the midst of copious fleshy albumen. — *Trees* or *shrubs*. *Leaves* alternate, simple, with deciduous stipulæ.

Affinities. An obscure order, of the limits of which nothing has been well made out. Judging from the genera Stilago and Antidesma, it is very near Cupuliferæ, from which it differs chiefly in its superior ovarium and copious fleshy albumen.

GEOGRAPHY. Natives of the East Indies.

PROPERTIES.

EXAMPLES. Stilago, Antidesma.

LXXXII. CUPULIFERÆ. THE OAK TRIBE.

CUPULIFERÆ, *Rich. Anal. du Fr.* (1808); *Lindl. Synops.* 239. (1829); *Blume Flora Javæ,* (1829).— CORYLACEÆ, *Mirb. Elém.* 906. (1815).— QUERCINEÆ, *Juss. in Dict. Sc. Nat.* vol. 2. *Suppl.* 12. (1816.)

DIAGNOSIS. Apetalous dicotyledons, with definite pendulous ovules, 2 or more in each cell, amentaceous flowers, single inferior ovaria enclosed in a cupule, and alternate stipulate leaves with veins proceeding straight from the midrib to the margin.

ANOMALIES.

ESSENTIAL CHARACTER.— *Flowers* unisexual; males amentaceous, females aggregate or amentaceous. *Males: Stamens* 5 to 20, inserted into the base of the scales or of a membranous calyx, generally distinct. *Females : Ovaries* crowned by the rudiments of a superior calyx, seated within a coriaceous involucrum (*cupule*) of various figure, and with several cells and several ovules, the greater part of which are abortive; *ovules* twin or solitary, pendulous; *stigmata* several, sub-sessile, distinct. *Fruit* a bony or coriaceous 1-celled nut, more or less enclosed in the involucrum. *Seeds* solitary, 2 or 3, pendulous; *embryo* large, with plano-convex fleshy cotyledons, and a minute superior radicle. — *Trees* or *shrubs*. *Leaves* with stipulæ, alternate, simple, with veins proceeding straight from the midrib to the margin.

AFFINITIES. These are known among European trees by their amentaceous flowers and peculiarly veined leaves; from all other plants they are distinguished by their apetalous superior rudimentary calyx, fruit enclosed in a peculiar husk or cup, and nuts containing but 1 cell and 1 or 2 seeds, in consequence of the abortion of the remainder. They are nearly akin to Salicineæ and Betulineæ, from which the presence of a calyx, and, in the former case, the veining of their leaves, distinguish them. To Urticeæ they are nearly allied, but differ in their many-celled ovarium, pendulous ovula, and superior calyx.

GEOGRAPHY. Inhabitants of the forests of all the temperate parts of the continent both of the Old and New World; extremely common in Europe, Asia, and North America; more rare in Barbary and Chile, and the southern parts of South America; and unknown at the Cape. The species which are found within the tropics of either hemisphere are chiefly Oaks, which abound in the high lands, but are unknown in the valleys of equatorial regions.

PROPERTIES. An order which comprehends the Oak, the Hazel Nut, the Beech, and the Spanish Chestnut, can scarcely require much to be said to a European reader of its properties, which are of too common a use to be unknown even to the most ignorant. Gallic acid exists abundantly in the Oak. The leaves of Quercus falcata are employed, on account of their astringency, externally in cases of gangrene; and the same astringent principle, which pervades all the order, has caused them to be employed even as febrifuges, tonics, and stomachics. Cork is the bark of Quercus suber; it contains a peculiar principle called Suberin (*Turner,* 700), and an acid called the Suberic (*Ibid.* 641)   The galls that writing ink is prepared from are the produce of the Oak, from which they derive their astringency.   The acorns of a species known in the Levant under the name of Velonia (Quercus ægilops) are imported for the use of dyers.

EXAMPLES.   Quercus, Corylus, Fagus.

LXXXIII. BETULINEÆ.   THE BIRCH TRIBE.

AMENTACEÆ, *Juss. Gen.* 407. (1789) *in part*; *Lindl. Synops.* § 228. (1829). — BETULINEÆ, *L. C. Richard MSS. A. Richard, Elém. de la Bot. ed.* 4. 562. (1828.)

DIAGNOSIS.   Achlamydeous dicotyledons, with a 2-celled ovarium, definite pendulous seeds, and amentaceous flowers.

ANOMALIES.   The male flowers have occasionally a distinct calyx.

ESSENTIAL CHARACTER. — *Flowers* unisexual, monœcious, amentaceous; the males sometimes having a membranous lobed calyx. *Stamens* distinct, scarcely ever monadelphous; *anthers* 2-celled.   *Ovarium* superior, 2-celled; *ovules* definite, pendulous; *style* single, or none; *stigmas* 2.   *Fruit* membranous, indehiscent, by abortion 1-celled.   *Seeds* pendulous, naked; *albumen* none; *embryo* straight; *radicle* superior. — *Trees* or *shrubs*. *Leaves* alternate, simple, with the venæ primariæ running straight from the midrib to the margin; *stipulæ* deciduous.

AFFINITIES.   This order approaches more near to Urticeæ and Cupuliferæ than either Plataneæ or Salicineæ, which may be considered dismemberments of it.   In the male flowers of several species there is a distinct membranous calyx, very like that of Ulmus; the seeds are definite and pendulous, and the leaves have the same venation as Cupuliferæ.   It is distinguished by the 2 distinct cells of the fruit, by the want of a calyx to the female flowers, and by its solitary pendulous seeds.

GEOGRAPHY.   Inhabitants of the woods of Europe, Northern Asia, and North America, and even making their appearance on the mountains of Peru and Columbia.

PROPERTIES.   Fine timber-trees, usually with deciduous leaves; their bark astringent, and sometimes employed as a febrifuge; but chiefly valued for their importance as ornaments of a landscape.   Their wood is often light, and of inferior quality, but that of the Black Birch of North America is one of the hardest and most valuable we know.

EXAMPLES.   Betula, Alnus.

LXXXIV. SALICINEÆ.   THE WILLOW TRIBE.

AMENTACEÆ, *Juss. Gen.* 407. (1789) *in part*; *Lindl. Synops.* § 229. (1829). — SALICINEÆ, *L. C. Richard MSS.; Ach. Richard. Elém. de la Bot. ed.* 4. 560. (1828.)

DIAGNOSIS.   Achlamydeous dicotyledons, with a 1- or 2-celled ovarium, indefinite comose seeds, and amentaceous flowers.

ANOMALIES.

ESSENTIAL CHARACTER. — *Flowers* unisexual, either monœcious or diœcious, amentaceous.   *Stamens* distinct or monadelphous; *anthers* 2-celled.   *Ovarium* superior, 1- or 2-celled; *ovules* numerous, erect, at the base of the cell or adhering to the lower part of the sides; *style* 1 or 0; *stigmas* 2. *Fruit* coriaceous, 1- or 2-celled, 2-valved, many-seeded. *Seeds* either adhering to the lower part of the axis of each valve, or to the base of the cell, comose; *albumen* 0; *embryo* erect; *radicle* inferior.—*Trees* or *shrubs*. *Leaves* alternate, simple, with deliquescent venæ primariæ, and frequently with glands; *stipulæ* deciduous or persistent.

AFFINITIES. The hairy seeds, and polyspermous 2-valved fruit, distinguish this from Betulineæ, the only order with which it is likely to be confounded. It is usually combined with that order and Cupuliferæ, under the name of Amentaceæ; but it is more consonant with modern views of division to keep them all separate.

GEOGRAPHY. Natives, generally, of the same localities as Betulineæ, but extending further to the north than the species of that order. The most northern woody plant that is known is a kind of Willow, Salix arctica. They are found sparingly in Barbary, and there is a species of Willow even in Senegal.

PROPERTIES. Valuable trees, either for their timber or for economical purposes; the Willow, the Sallow, and the Poplar, being the representatives. Their bark is usually astringent, tonic, and stomachic; that of Populus tremuloides is known as a febrifuge in the United States; the leaves of Salix herbacea, soaked in water, are employed in Iceland for tanning leather. Willow bark has been found by Sir H. Davy to contain as much tanning principle as that of the Oak. *Ed. P. J.* 1. 320. It has lately acquired a great reputation in France as a febrifuge.

EXAMPLES. Populus, Salix.

LXXXV. PLATANEÆ. THE PLANE TRIBE.

PLATANEÆ, *Lestiboudois* according to *Von Martius. Hort. Reg. Monacensis*, p. 46. (1829.)

DIAGNOSIS. Achlamydeous dicotyledons, with a 1-celled ovarium, pendulous ovules, alternate leaves, and amentaceous flowers.

ANOMALIES.

ESSENTIAL CHARACTER. — *Flowers* amentaceous, naked; the sexes in distinct amenta. *Stamens* single, without any floral envelope, but with several small scales and appendages mixed among them; *anthers* linear, 2-celled. *Ovaria* terminated by a thick style, having the stigmatic surface on one side; *ovules* solitary, or two, one above the other, and suspended. *Nuts*, in consequence of mutual compression, clavate, with a persistent recurved style. *Seeds* solitary, or rarely in pairs, pendulous, elongated; *testa* thick; *embryo* long, taper, lying in the axis of fleshy albumen, with the radicle turned to the extremity next (opposite, *A. Rich.*) the hilum.— *Trees* or *shrubs*. *Leaves* alternate, palmate, or toothed, with scarious sheathing stipulæ. *Amenta* round, pendulous.

AFFINITIES. Formerly comprehended in the tribe called Amentaceæ, this order is particularly known by its round heads of flowers, its 1-celled ovarium, containing 1 or 2 pendulous ovula, and its embryo lying in fleshy albumen, by which it is distinguishable from both Betulineæ, Myriceæ, and Artocarpeæ, with all which, especially the latter, it has a close affinity. From the latter, indeed, it is chiefly known by the want of calyx, by the presence of albumen, and the absence of milk; the habit of the two orders being much the same. According to Gærtner, the radicle is next the hilum; according to Achille Richard (*Dict. Class.* 14. 23.), it is at the other extremity.

GEOGRAPHY. Natives of Barbary, the Levant, and North America.

PROPERTIES. Noble timber-trees, the wood of which is extremely valuable; the bark of Platanus is remarkable for falling off in hard irregular patches,—a circumstance which arises from the rigidity of its tissue, on account of which it is incapable of stretching as the wood beneath it increases in diameter.

EXAMPLE. Platanus.

LXXXVI. MYRICEÆ. THE GALE TRIBE.

MYRICEÆ, *Rich. Anal. du Fr.* (1808); *Ach. Rich. Elém. de la Bot. ed* 4. 561. (1828); *Lindl. Synops.* 242. (1829).—CASUARINEÆ, *Mirbel in Ann. Mus.* 16. 451. (1810); *R. Brown in Flinders,* 2. 571. (1814.)

DIAGNOSIS. Achlamydeous dicotyledons, with a 1-celled ovarium, erect ovules, a naked embryo, and amentaceous flowers.

ANOMALIES. Casuarina is leafless.

ESSENTIAL CHARACTER.—*Flowers* unisexual, amentaceous. *Males: Stamens* 1 or several, each with a hypogynous scale. *Anthers* 2- or 4-celled, opening lengthwise. *Females: Ovarium* 1-celled, surrounded by several hypogynous scales; *ovulum* solitary, erect, with a foramen in its apex; *stigmas* 2, subulate. *Fruit* drupaceous, often covered with waxy secretions, formed of the hypogynous scales of the ovarium, become fleshy and adherent; or dry and dehiscent, with the scales distinct. *Seed* solitary, erect; *embryo* without albumen; *cotyledons* 2, plano-convex; *radicle* short, superior.—*Leafy shrubs,* with resinous glands and dots, the leaves alternate, simple, with or without stipulæ; or *leafless shrubs* or *trees,* with filiform branches bearing membranous toothed sheaths at the articulations.

AFFINITIES. The nearest approach made by these plants is probably to Ulmaceæ and Betulineæ, from the former of which they are readily known by their amentaceous flowers and want of a perianthium; from the latter they are distinguished by their erect ovula, aromatic leaves, and 1-celled ovarium. In the latter respect they resemble Piperaceæ, from which, however, they differ materially in other points. The only anomalous genus is Casuarina, which has the habit of a gigantic Equisetum, and which can scarcely be compared with any other dicotyledonous tree. Mr. Brown, in the Appendix to *Flinders's Voyage,* has the following observations on the structure of this remarkable genus, from which it will be seen that he does not consider it achlamydeous, as I do.

" In the male flowers of all the species of Casuarina, I find an envelope of four valves, as Labillardiére has already observed in one species, which he has therefore named C. quadrivalvis. *Plant. Nov. Holl.* 2. p. 67. t. 218. But as the two lateral valves of this envelope cover the others in the unexpanded state, and appear to belong to a distinct series, I am inclined to consider them as bracteæ. On this supposition, which, however, I do not advance with much confidence, the perianthium would consist merely of the anterior and posterior valves; and these, firmly cohering at their apices, are carried up by the anthera, as soon as the filament begins to be produced, while the lateral valves or bracteæ are persistent; it follows from it, also, that there is no visible perianthium in the female flower; and the remarkable economy of its lateral bracteæ may, perhaps, be considered as not only affording an additional argument in support of the view now taken of the nature of the parts, but also as in some degree again approximating Casuarina to Coniferæ, with which it was formerly associated. The outer coat of the seed or caryopsis of Casuarina consists of a very fine membrane, of which the terminal wing is entirely composed; between this membrane and the crustaceous integument of the seed, there exists a stratum of spiral vessels, which Labillardiére, not having distinctly seen, has described as an ' integumentum arachnoideum;' and within the crustaceous integument there is a thin proper membrane, closely applied to the embryo, which the same author has entirely overlooked. The existence of spiral vessels, particularly in such quantity, and, as far as can be determined in the dried specimens, unaccompanied by other vessels, is a structure at least very unusual in the integuments of a seed or caryopsis, in which they are very seldom at all visible; and have never, I believe, been observed in such abundance as in this genus, in all whose species they are equally obvious."

GEOGRAPHY. Found in the cold parts of Europe and North America, the tropics of South America, the Cape of Good Hope, India, and New Holland; in the latter country the order is chiefly represented by Casuarina.

PROPERTIES. Aromatic shrubs, or trees of considerable size. Comptonia asplenifolia possesses astringent and tonic properties, and is much used in the domestic medicine of the United States, in cases of diarrhœa. *Barton,* 1. 224. The root of Myrica cerifera is a powerful astringent, and wax is obtained in great abundance from its berries. The fruit of Myrica sapida is about as large as a cherry, and, according to Buchanan, is a pleasant acid and eatable in Nipal. *Don,* p. 56. It has a pleasant, refreshing, acidulous taste. *Wall. Tent.* 60.

EXAMPLES. Myrica, Nageia, Casuarina.

LXXXVII. JUGLANDEÆ. THE WALNUT TRIBE.

JUGLANDEÆ, *Dec. Théorie,* 215. (1813); *Kunth in Ann. des Sc. Nat.* 2. 343. (1824.)

DIAGNOSIS. Apetalous dicotyledons, with ascending definite ovules, amentaceous flowers, and a superior calyx.

ANOMALIES.

ESSENTIAL CHARACTER. — *Flowers* unisexual. *Calyx* in the males oblique, membranous, irregularly divided, attached to a single bractea; in the females superior, with 4 divisions. *Petals* in the males 0; in the females occasionally present, and 4 in number, arising from between the calyx and the styles, and cohering at the base. *Stamens* indefinite, (3-36), hypogynous; *filaments* very short, distinct; *anthers* thick, 2-celled, innate, bursting longitudinally. *Disk* 0. *Ovarium* inferior, 1-celled; *ovulum* solitary, erect; *styles* 1 or 2, and very short, or none; *stigmas* much dilated, either 2 and lacerated, or discoid and 4-lobed. *Fruit* drupaceous, 1-celled, with 4 imperfect partitions. *Seed* 4-lobed; *embryo* shaped like the seed; *albumen* 0; *cotyledons* fleshy, 2-lobed, wrinkled; *radisle* superior. — *Trees.* *Leaves* alternate, unequally pinnated, without pellucid dots or stipulæ. *Flowers* amentaceous.

AFFINITIES. These have usually been mixed with Terebintaceæ, to which they, however, do not appear so closely allied as to Corylaceæ, with which they accord in their amentaceous unisexual flowers, and superior calyx. Among apetalous orders, their pinnated resinous undotted leaves particularly distinguish them.

GEOGRAPHY. Chiefly found in North America; one species, the common Walnut, is a native of the Levant and Persia; another, of Caucasus; and a third, of the West India Islands.

PROPERTIES. The fruit of the Walnut is esteemed for its sweetness and wholesome qualities. It abounds in a kind of oil, of a very drying nature. The rind of the fruit, and even the skin of the kernel, are extremely astringent. Juglans cathartica and cinerea are esteemed anthelmintic and cathartic; the fruit of several kinds of Hickory is eaten in America. The timber of all is valuable; that of J. regia for its rich deep brown colour when polished, and that of Carya alba for its elasticity and toughness.

EXAMPLES. Juglans, Carya.

LXXXVIII. EUPHORBIACEÆ. The Euphorbium Tribe.

Euphorbiæ, *Juss. Gen.* 385. (1789). — Euphorbiaceæ, *Ad. de Juss. Monogr.* (1824);
Lindl. Synops. 220. (1829.)

Diagnosis. Apetalous dicotyledons, with definite suspended ovules,
a 3-celled ovarium, unisexual flowers, and embryo in the midst of oily
albumen.

Anomalies. Carpella occasionally 2, or more than 3.

Essential Character. — *Flowers* monœcious or diœcious. *Calyx* lobed, inferior,
with various glandular or scaly internal appendages; (sometimes wanting). *Males: Stamens* definite or indefinite, distinct or monadelphous; *anthers* 2-celled. *Females: Ovarium* superior, sessile, or stalked, 2- 3- or more celled; *ovules* solitary or twin, suspended
from the inner angle of the cell; *styles* equal in number to the cells, sometimes distinct,
sometimes combined, sometimes none; *stigma* compound, or single with several lobes.
Fruit consisting of 2, 3, or more dehiscent cells, separating with elasticity from their
common axis. *Seeds* solitary or twin, suspended, with an arillus; *embryo* enclosed in
fleshy albumen; *cotyledons* flat; *radicle* superior. — *Trees, shrubs,* or *herbaceous plants,*
often abounding in acrid milk. *Leaves* opposite or alternate, simple, rarely compound,
usually with stipulæ. *Flowers* axillary or terminal, usually with bracteæ, sometimes
enclosed within an involucrum.

Affinities. If the group of apetalous orders be considered a natural
one, Euphorbiaceæ will stand by the side, or in the vicinity, of Urticeæ, with
which, however, they have few points in common, except the want of a
corolla; or near Myristiceæ, with which the columnar stamens of many
species, and the acridity of their juice, may be said to accord. But it is
probable that the real relationship of the order is of a very different kind.
Jussieu long ago perceived a resemblance between Euphorbiaceæ and
Rhamneæ, a resemblance which A. Brongniart has since adverted to (*Monogr.
des Rhamn.* p. 35); and which chiefly depends upon a similarity in habit,
an embryo with flat foliaceous cotyledons, solitary seeds, a great reduction in
size of the petals of Rhamneæ, as if the order was tending towards an
apetalous state, and a frequent division of the fruit into three parts.
Auguste St. Hilaire (*Pl. Usuelles*, no. 18.) inquires whether they are not
intermediate between Menispermeæ and Malvaceæ. There can be no
doubt of their relation to the latter, that is to say, to the orders of polypetalous dicotyledons with hypogynous stamens and a valvate calyx, if we
consider their general habit, especially that of the Crotons, the presence of
abundance of stellate hairs, and their definite seeds; but these points are
not sufficient to approximate the orders very nearly: in fact, the true
affinities of Euphorbiaceæ cannot be said to be at present well understood.
Ach. Richard suggests some affinity with Terebintaceæ, as well as Rhamneæ. *Elémens, ed.* 4. 558.

Geography. This extensive order, which probably does not contain
fewer than 1500 species, either described or undescribed, exists in the greatest
abundance in equinoctial America, where about 3-8ths of the whole number
have been found; sometimes in the form of large trees, frequently of bushes,
still more usually of diminutive weeds, and occasionally of deformed, leafless,
succulent plants, resembling the Cacti in their port, but differing from them
in every other particular. In the Western world they gradually diminish as
they recede from the equator, so that not above 50 species are known in
North America, of which a very small number reaches as far as Canada.
In the Old World the known tropical proportion is much smaller, arising
probably from the species of India and equinoctial Africa not having been
described with the same care as those of America; not above an eighth

having been found in tropical Africa, including the islands, and a sixth being perhaps about the proportion in India. A good many species inhabit the Cape, where they generally assume a succulent habit; and there are almost 120 species from Europe, including the basin of the Mediterranean : of these, 16 only are found in Great Britain, and 7 in Sweden.

PROPERTIES. The excellent monograph of M. Adrien de Jussieu contains the best information that exists upon this subject; and I accordingly avail myself of it, making a few additions to his facts. The general property is that of excitement, which varies greatly in degree, and consequently in effect. This principle resides chiefly in the milky secretion of the order, and is most powerful in proportion as that secretion is abundant. The smell and taste of a few are aromatic; but in the greater part the former is strong and nauseous, the latter acrid and pungent. The hairs of some species are stinging. The bark of various species of Croton is aromatic, as Cascarilla; and the flowers of some, such as Caturus spiciflorus, give a tone to the stomach. Many of them act upon the kidneys, as several species of Phyllanthus, the leaves of Mercurialis annua, and the root of Ricinus communis. Several are asserted by authors to be useful in cases of dropsy; some Phyllanthuses are emmenagogue. The bark of several Crotons, the wood of Croton Tiglium and common Box, the leaves of the latter, of Cicca disticha, and of several Euphorbias, are sudorific, and used against syphilis; the root of various Euphorbias, the juice of Commia, Anda, Mercurialis perennis, and others, are emetic; and the leaves of Box and Mercurialis, the juice of Euphorbia, Commia, and Hura, the seeds of Ricinus, Croton Tiglium, &c. &c., are purgative. Many of them are also dangerous, even in small doses, and so fatal in some cases, that no practitioner would dare to prescribe them; as, for example, Manchineel. In fact, there is a gradual and insensible transition, in this order, from mere stimulants to the most dangerous poisons. The latter have usually an acrid character, but some of them are also narcotic, as those Phyllanthuses, the leaves of which are thrown into water to intoxicate fish. Whatever the stimulating principle of Euphorbiaceæ may be, it seems to be of a very volatile nature, because application of heat is sufficient to dissipate it. Thus the root of the Jatropha Manihot or Cassava, which when raw is one of the most violent of poisons, becomes a wholesome nutritious article of food when roasted. In the seeds the albumen is harmless and eatable, but the embryo itself is acrid and dangerous. Independently of this volatile principle, there are two others belonging to the order, which require to be noticed : the first of these is Caoutchouc, that most innocuous of all substances, produced by the most poisonous of all families, which may be almost said to have given a new arm to surgery, and which has become an indispensable necessary of life; it exists in Artocarpeæ and elsewhere, but is chiefly the produce of species of Euphorbiaceæ. The other is the preparation called Turnsol, which, although chiefly obtained from Crozophora (Croton) tinctoria, is to be procured equally abundantly from many other plants of the order.

The properties of Euphorbiaceæ are so important, that I do not think I should fulfil the object of this work, if I did not, in addition to the foregoing general view of the order, add a detailed list of the qualities of the most important species named by writers.

Acalypha Cupameni, an Indian herb, has a root which, bruised in hot water, is cathartic; a decoction of its leaves is also laxative. *Rheede*, 10. 161. The nut of Aleurites ambinux is eatable and aphrodisiac, but rather indigestible. *Commers. according to Ad. de J.* The nuts of another species are eaten in Java and the Moluccas; but they are intoxicating, unless they are roasted. *Rumph.* The Anda of Brazil is famous for the purgative qualities of its seeds, which are fully as powerful as those of the Palma Christi. The Brazilians make use of them in cases of indigestion, in liver complaints, the jaundice, and dropsy. The rind, roasted on the fire, passes as a certain remedy for diarrhœa brought on by cold. According to Marcgraaf, the fresh rind steeped in water communicates to it a narcotic property which is sufficient to stupify fish. *Martius Amœn. Monac.* p. 3. The seeds are either eaten raw, or are prepared as an electuary; they yield an oil, which is said, by M. Auguste St. Hilaire, to be drying and excellent for painting; in short, much better than nut oil. *Pl. Usuelles,* 54. The bark of Briedelia spinosa, an Indian shrub, is, according to Roxburgh, a powerful astringent; the leaves are greedily eaten by cattle, which by their means free themselves of intestinal worms. The leaves of common Box are sudorific and purgative; according to Hanway, camels eat them in Persia, but they die in consequence. *Ad. de J.* The flowers of Caturus spiciflorus are spoken of as a specific in diarrhœa, either taken in decoction or in conserve. *Burm. Ind.* 303. The succulent fruit of Cicca disticha and racemosa is sub-acid, cooling, and wholesome. Its leaves are sudorific, and its seeds cathartic. The capsules of Cluytia collina are poisonous, according to Roxburgh. The root and bark of Codiæum variegatum are acrid, and excite a burning sensation in the mouth if chewed; but the leaves are sweet and cooling. *Rumphius.* The juice of Commia cochinchinensis is white, tenacious, emetic, purgative, and deobstruent. Cautiously administered, it is a good medicine in obstinate dropsy and obstructions. *Lour.* 743. The Quina Blanca of Vera Cruz is produced by the Croton Eluteria of Swartz, and is probably the Cascarilla of Europe. *Schiede in Ann. des Sc.* 18. 217. The drastic oil of Tiglium is expressed from the seeds of Croton Tiglium, formerly known in Europe under the name of *Grana molucca.* It is said, by Dr. Ainslie, to have proved in a singular manner emmenagogue. *Mat. Med.* 1. 108. A decoction of Croton perdicipes, called Pe de Perdis, Alcamphora, and Cocallera, in different provinces of Brazil, is much esteemed as a cure for syphilis, and as a useful diuretic. *Pl. Us.* 59. The root of another species, called Velame do Campo, C. campestris, has a purgative root, also employed against syphilitic disorders. *Ib.* 60. The leaves of a species of Croton (C. gratissimum, *Burchell,*) are so fragrant as to be used by the Koras of the Cape of Good Hope as a perfume. *Burch.* 2. 263. Crozophora tinctoria yields the preparation called Turnsol; the plant itself is acrid, emetic, and drastic. An abundance of useful oil is obtained from two species of Elæococca; it is, however, only fit for burning and painting, on account of its acridity. *Ad. de J.* Six sorts of European Euphorbias are named, by Deslongchamps, as fit substitutes for Ipecacuanha, the best of which he states to be E. Gerardiana, the powdered root of which vomits easily in doses of 18 or 20 grains. *Ainslie,* 1. 123. The root of Euphorbia Ipecacuanha is said, by Barton, to be equal to the true Ipecacuanha, and in some respects superior; it is not unpleasant either in taste or smell. *Barton,* 1. 218. Various species of fleshy Euphorbia, especially the Euph. antiquorum and canariensis, produce the drug Euphorbium of the shops, which is the inspissated milky juice of such plants. In India it is mixed with the oil expressed from the seeds of Sesamum orientale, and used externally in rheumatic affections, and internally in cases of obstinate constipation. It is little used in Europe. Orfila places it among his poisons. *Ainslie,* 1. 121. Euphorbia papillosa is administered, in Brazil, as a purgative; but is apt, if given in too strong a dose, to cause dangerous superpurgations. *Pl. Usuelles,* 18. The juice of the leaves of Euphorbia nereifolia is prescribed by the native practitioners of India, internally as a purge and deobstruent, and externally, mixed with Margosa oil, in such cases of contracted limb as are induced by ill-treated rheumatic affections. The leaves have, no doubt, a diuretic quality. *Ainslie*, 2. 98. The leaves and seeds of Euphorbia thymifolia are given, by the Tamool doctors of India, in worm cases, and in certain bowel affections of children. *Ib*. 2. 76. The same persons give the fresh juice of Euphorbia pilulifera in aphthous affections. The fresh acrid juice of Euphorbia Tirucalli is used in India as a vesicatory. *Ib*. 2. 133. The Ethiopians are said, by Virey, to form a mortal poison for their arrows from the juice of Euphorbia heptagona. *Hist. des Médic*. 299. The juice of Excæcaria Agallocha, and even its smoke when burnt, affects the eyes with intolerable pain, as has been experienced occasionally by sailors sent ashore to cut fuel, who, according to Rumphius (2. 238.), having accidentally rubbed their eyes with the juice, became blinded, and ran about like distracted men, and some of them finally lost their sight. The famous Manchineel tree, Hippomane Mancinella, is said to be so poisonous, that persons have died from merely sleeping beneath its shade. This is doubted, indeed, by Jacquin, who, however, admits its extremely venomous qualities; but it is by no means improbable that the story has some foundation in truth, particularly if, as Ad. de Jussieu truly remarks, the volatile nature of the poisonous principle of these plants is considered. The juice of Hura crepitans is stated to be of the same fatal nature as that of Excæcaria; its seeds are said to have been administered to negro slaves as purgatives, in number not exceeding 1 or 2, with fatal consequences. *Ad. de J*. The powdered fruit of Hyænanche globosa is used in the colony of the Cape of Good Hope to poison hyænas, as nux vomica to poison stray dogs in Europe. From the seeds of Jatropha glauca the Hindoos prepare, by careful expression, an oil which, from its stimulating quality, they recommend as an external application in cases of chronic rheumatism and paralytic affections. *Ainslie*, 2. 6. The seeds of Jatropha Curcas are purgative and occasionally emetic; an expressed oil is obtained from them, which is reckoned a valuable external application in itch and herpes; it is also used, a little diluted, in chronic rheumatism. The varnish used by the Chinese for covering boxes is made by boiling this oil with oxide of iron. The leaves are considered as rubefacient and discutient; the milky juice is supposed to have a detergent and healing quality, and dyes linen black. *Ibid*. 2. 46. The roots of the Jatropha Manihot, or Mandiocca, yield a flour of immense importance in South America: this is obtained by crushing the roots, after the bark has been removed, and then straining off the water; after which the mass is gradually dried in pans over a fire. The seeds of several species of Jatropha are purgative, but they sometimes act so dangerously as to require extreme caution in administering them. Mercurialis perennis is purgative and dangerous. According to Sloane, it has sometimes produced violent vomiting, incessant diarrhœa, a burning heat in the head, a deep and long stupor, convulsions, and even death; yet this very plant, when boiled, has been eaten as a potherb. The leaves of Maprounea brasiliensis, or the Marmeleiro do Campo of Brazil, yield a black dye, which is, however, fugitive. A decoction of its root is also administered in derangement of the stomach; — a most remarkable circumstance, if we consider the close relation that is borne by it to Manchineel and other most poisonous trees. According to M. Auguste St. Hilaire, the Maprounea is destitute of the milky juice of Sapium, Excæcaria, Hippomane, and other dangerous genera. *Pl. Us*. 65. The seeds of Omphalea are eaten safely, if the embryo is first removed; if this is not done, they are cathartic. Both Pedilanthus tithymaloides and padifolius are used medicinally in the West Indies: the former, known under the name of Ipecacuanha, is used for the same purposes as that drug; the latter, called the Jew Bush, or Milk plant, is used in decoction of the recent plant as an antisyphilitic, and in cases of suppression of the menses. *Hamilt. Prodr. Fl. Ind.* 43. The root, leaves, and young shoots of Phyllanthus Niruri are considered, in India, deobstruent, diuretic, and healing. The leaves are very bitter, and a good stomachic. *Ainslie,* 2. 151. Some other species, particularly Ph. urinaria, are powerful diuretics. The fruit of Phyllanthus Emblica is frequently made into pickle; it is acid, and, when dry, very astringent. *Ibid.* 1. 240. The bruised leaves of Phyllanthus Conami are used for inebriating fishes. *Aubl.* 928. The boiled leaves of Plukenetia corniculata are said to be an excellent potherb, for which purpose it is cultivated in Amboyna. *Rumph.* The purgative quality of Ricinus, the Castor oil plant, is well known; the root is said to be diuretic. The juice of Sapium aucuparium is reputed poisonous. A case is mentioned by Tussac (*Journ. Bot.* 1813. 1. 117.) of a gardener whose nostrils became swollen and seized with erysipetalous phlegmasis, in consequence of the fumes only of this plant. The root of Tragia involucrata is reckoned, by the Hindoo doctors, among those medicines which they conceive to possess virtues in altering and correcting the habit in cases of cachexia, and in old venereal complaints attended with anomalous symptoms. *Ainslie,* 2. 62. There is reason to believe that the timber imported from the coast of Africa, under the name of African Teak, belongs to some tree of this order. From a species of a tree, stated by Mr. Brown to be of an unpublished genus, it is said that a substance resembling caoutchouc is procured in Sierra Leone. *Congo,* 444.

EXAMPLES. Euphorbia, Croton, Buxus, Jatropha.

LXXXIX. RESEDACEÆ. THE MIGNONETTE TRIBE.

RESEDACEÆ, *Dec. Théor. ed.* 1. 214. (1813); *Lindl. Synops.* 219. (1829).

DIAGNOSIS. Apetalous dicotyledons, with indefinite ovules, a 1-celled ovarium with parietal placentæ, dehiscent fruit, irregular flowers partly sterile, and a reniform embryo.

ANOMALIES.

ESSENTIAL CHARACTER.—*Florets* included within a many-parted involucrum, neuter on the outside, hermaphrodite in the centre. *Calyx* 1-sided, undivided, glandular. *Stamens* of the sterile florets linear, petaloid. *Stamens* of the fertile florets perigynous, definite; *filaments* erect; *anthers* 2-celled, opening longitudinally. *Ovarium* sessile, 3-lobed, 1-celled, many-seeded, with 3 parietal placentæ. *Stigmata* 3, glandular, sessile. *Fruit* dry and membranous, or succulent, opening at the apex. *Seeds* several, reniform, attached to 3 parietal placentæ; *embryo* taper, arcuate, without albumen; *radicle* superior.—*Herbaceous* plants, with alternate *leaves,* the surface of which is minutely papillose; and minute, gland-like *stipulæ.*

AFFINITIES. The character which is here assigned to Resedaceæ is in conformity with an opinion I published some years ago, that the part called calyx by botanists is an involucrum, the supposed petals neutral florets, and the disk or nectary a calyx surrounding a fertile floret in the middle. The reasons I assigned for this opinion were, firstly, " That there is a difference in the time of expansion of the neutral florets and of the stamens of the fertile one; the former being quite open in very many capitula, before one anther of the latter has burst in a single flower. Secondly, That there is an evident analogy between the appendages of the neutral florets and the stamens of the perfect florets; inasmuch as in Reseda odorata those of the upper sterile florets are of nearly the same number as the real stamens ; because in Reseda alba, and some others, in which a union of filaments takes place in the perfect floret, there is a corresponding but more complete union of the sterile appendages ; and because occasionally in Reseda odorata, stamens are changed into bodies altogether similar to the sterile appendages; and in Reseda Phyteuma the same appearance is always assumed by the perfect stamens after the anthers have performed their functions. Thirdly, That there is an equal analogy between the calyx of the neutral florets and that of the perfect floret; because both have a peculiar glandular margin, the same form, both produce their stamens from their surface ; and because the upper edge of the calyx in the sterile florets has the same relation to the axis of each particular head as that of the perfect floret has to the axis of the whole inflorescence. In Reseda Phyteuma, which has the margin of its neutral florets rolled back, the same thing occurs in the perfect floret. Fourthly, That there is no instance of the same analogy existing between the disk and petals of other plants." *Coll. Bot.* no. 22. Hence I inferred that the genus must be excluded from even the vicinity of Capparideæ, with which it is usually placed. This view of the structure of Reseda, however paradoxical it may appear, has been adopted by M. Decandolle; but Mr. Brown, in the Appendix to *Major Denham's Narrative*, has advanced various arguments in opposition to it. By these I was at first induced to believe that I was mistaken in my theory ; but upon reflection, and a subsequent repetition of the observations I originally made, I have been led to decide that Mr. Brown's arguments, strong as they undoubtedly are, do not carry conviction with them, and are, in fact, less weighty than they seem to be. In the first place, this learned botanist does not attempt to invalidate some of the arguments upon which I was led to my original conclusion ; and secondly, those which he has advanced in support of the contrary opinion appear to me to be open to objection. Mr. Brown's arguments in favour of the popular mode of understanding the structure of Reseda are :

1st. That the presence and appearance of the hypogynous disk, the anomalous structure of the petals, and the æstivation of the flower, all occur in a greater or less degree in Capparideæ, and have been found united in no other family of plants ; and,

2d. That the appendages (which I consider abortive stamens), being formed before the part upon which they rest (and which I have called calyx), are consequently to be referred to the corolla rather than the stamens : this, at least, is how I understand the chief argument employed by Mr. Brown. I hope I do not misunderstand.

3d. That the processes of the supposed petals are analogous to those of Dianthus, Lychnis, and Silene.

To the first of these arguments I reply, that, without meaning in the slightest degree to doubt the accuracy of Mr. Brown's observations, which I know are beyond question, I have not been able to discover any Capparideous plants which are in my judgment analogous in the conformation of their parts to Reseda ; and that, even presuming appearances of analogy to exist more unequivocally than Mr. Brown states that they do, such a fact would not by itself shake the evidence I have produced to the contrary. To this I may add, that analogical evidence in support of my position, fully as powerful as that said to exist against it in Capparideæ, is furnished by Datisca, a genus I think evidently very near Reseda, which is unquestionably apetalous, and of which the calyx of the female flowers may without difficulty be compared with that of Reseda, except that it is adherent to the ovarium.

To the second objection it may be answered, that in organs of so anomalous a structure as those of Reseda, there can be no difficulty in supposing that anomaly to overcome the ordinary laws of successive formation ; that, moreover, the argument is founded upon an assumption that the petals are always formed before the stamens ; a point with no proof of which am I acquainted, and which I think open to considerable doubt : for instance, are the petals of Illecebreæ developed before the stamens, or subsequent to them ? and how is the existence of apetalous species in polypetalous genera to be reconciled with such a theory ? Besides this, is not the circumstance described by Mr. Brown of the stamina not being covered by the supposed petals in the slightest degree in any stage of development, an admission that in Reseda itself the formation of the stamens is anterior to that of the corolla? and if this is true of perfect stamens, why should it not be true of sterile ones? Mr. Brown also states that, at the period when what he calls the unguis of the petals (but what I call the calyx of the neutral florets) is scarcely to be detected, that part which is commonly called the disk (but which I consider the calyx of a sterile floret) is hardly visible also : — is not this a proof of the identity of the two parts ? and if so, they must be either all disks, which is absurd, or all calyxes, which is that for which I contend.

With regard to the third objection, that the processes of the supposed petals of Reseda are analogous to those of Silene, Lychnis, &c., I entertain a different opinion, for the following reasons : — The coronal processes of Silene consist of cellular tissue only, without any trace of vessels, and are analogous to the crests or lamellæ upon the labellum of Orchideæ, the anomalous subulate processes of Gilliesia, the scales of the orifice of some Boragineæ, the hump on the calyx of Scutellaria, and perhaps also the ligula of grasses. But in Reseda each of the processes has a central vascular axis, and is anatomically undistinguishable from the filament of the fertile stamens ; being thus analogous to the ligulate or subulate processes of Büttneriaceæ, or the coronal processes of Schwenkia, Brodiæa, and Leucocoryne, all of which are notoriously abortive stamina. I know of no instance of mere processes arising from the surface of a petal having a vascular axis; for Polygala, after the explanation that has been given of its structure by Auguste St. Hilaire, will hardly be considered an instance : neither am I acquainted with any case of sterile stamens being destitute of such an axis, unless they are in a very rudimentary state, which those of Reseda are not.

To conclude, I would beg those who still entertain doubts upon this subject to examine Reseda Phyteuma, and to set out in their inquiry from that species, in which, according to Mr. Don (*Ed. New Phil. Journ. Oct.* 1828), one of the sterile stamens occasionally bears an anther; a statement which, if there is no mistake, sets the question at rest for ever. Viewing the structure of Reseda in the usual way, its affinity would be obviously with Capparideæ, with which it entirely agrees in its seeds ; but in the light in which I see it, its proximity will be to Euphorbiaceæ and Datisceæ, particularly to the latter ; and if to them, also to Corylaceæ and Ulmaceæ, with the calyx of which, especially that of the male flowers of Fagus, the calyx of Reseda has much in common. I consider that Resedaceæ bear about the same relation to Euphorbiaceæ as Campanulaceæ to Compositæ, as Cinchonaceæ to Stellatæ, or as Hydrangeaceæ to Viburnum.

GEOGRAPHY. Weeds inhabiting exclusively Europe, the adjoining parts of Asia, the basin of the Mediterranean, and the adjacent islands.

PROPERTIES. Nothing further is known of them than that Reseda luteola yields a yellow dye, and that the Mignonette (R. odorata) is among the most fragrant of plants.

EXAMPLES. Reseda, Ochradenus.

XC. DATISCEÆ.

DATISCEÆ, *R. Brown in Denham,* 25. (1826.)

DIAGNOSIS. Apetalous dicotyledons, with indefinite ovules, a 1-celled ovarium with parietal placentæ, dehiscent fruit, regular unisexual flowers, and a straight embryo.

ANOMALIES.

ESSENTIAL CHARACTER. — *Flowers* unisexual. *Calyx* of the males divided into several pieces; of the females superior, toothed. *Stamens* several; *anthers* 2-celled, membranous, linear, bursting longitudinally. *Ovarium* 1-celled, with polyspermous parietal placentæ; *stigmas* equal in number to the placentæ, recurved. *Fruit* capsular, opening at the vertex, 1-celled, with polyspermous parietal placentæ. *Seeds* enveloped in a membranous finely reticulated integument; *embryo* straight, without *albumen,* its *radicle* turned towards the *hilum.*—*Herbaceous* branched *plants.* *Leaves* alternate, cut, compound, without stipulæ. *Flowers* in axillary racemes.

AFFINITIES. Mr. Brown is of opinion that this order differs widely from Reseda; but it strikes me that there is no group of plants to which it bears a greater affinity, if the flowers of Reseda are considered apetalous, which Mr. Brown, however, does not admit. Their habit is very similar. The structure of the fruit is absolutely the same, except that the calyx of one is superior, and of the other inferior; both are destitute of albumen; their anthers are also essentially alike. I consider Datisceæ a connecting link between Resedaceæ and Urticeæ.

GEOGRAPHY. The very few species of which this order consists are scattered over North America, Siberia, northern India, the Indian archipelago, and the south-eastern corner of Europe.

PROPERTIES. Datisca is bitter.

EXAMPLES. Datisca, Tetrameles.

XCI. EMPETREÆ. THE CROWBERRY TRIBE.

EMPETREÆ, *Nutt. Gen.* 2. 233.; *Don in Edinb. New Phil. Journ.* (1826); *Lindley's Synopsis,* 224. (1829.)

DIAGNOSIS. Apetalous dicotyledons, with definite ascending ovules, inferior distinct imbricated sepals, distinct stamens, and an embryo in the axis of fleshy albumen.

ANOMALIES.

ESSENTIAL CHARACTER.—*Flowers* unisexual. *Sepals* hypogynous imbricated scales. *Stamens* equal in number to the sepals, and alternate with them; *anthers* roundish, 2-celled, the cells distinct, bursting longitudinally. *Ovarium* superior, seated in a fleshy disk, 3- 6- or 9-celled; *ovules* solitary, ascending; *style* 1; *stigma* radiating, the number of its rays corresponding with the cells of the ovarium. *Fruit* fleshy, seated in the persistent calyx, 3- 6- or 9-celled; the coating of the cells bony. *Seeds* solitary, ascending; *embryo* taper, in the axis of fleshy watery albumen; *radicle* inferior.— Small acrid *shrubs* with heath-like evergreen *leaves* without stipulæ; and minute *flowers* in their axillæ.

AFFINITIES. Although the institution of this order is attributable to Mr. Nuttall, the final determination and characterising it is due to the exactness of Mr. Don, who has made numerous remarks upon it in the work above quoted. According to this gentleman, the order holds a kind of intermediate place between Euphorbiaceæ and Celastrineæ, agreeing in habit with the former, especially with Micranthea, and some species of Phyllanthus, more than with the latter.

GEOGRAPHY. A very small group, comprising a few species from North America, the south of Europe, and Straits of Magellan.

PROPERTIES. Unknown.

EXAMPLES. Empetrum, Corema, Ceratiola.

XCII. STACKHOUSEÆ.

STACKHOUSEÆ, *R. Br. in Flinders*, 555. (1814.)

DIAGNOSIS. Polypetalous dicotyledons, with 5 perigynous stamens, concrete carpella, a superior deeply lobed ovarium with several cells and lateral styles, and regular flowers.

ANOMALIES.

ESSENTIAL CHARACTER.— *Calyx* 1-leaved, 5-cleft, equal, with an inflated tube. *Petals* 5, equal, arising from the top of the tube of the calyx; their claws combined in a tube longer than the calyx; their limb narrow, stellate. *Stamens* 5, distinct, unequal (2 alternately shorter), arising from the throat of the calyx. *Ovarium* superior, 3- or 5-lobed, the lobes distinct, each with a single erect ovulum; *styles* from 3 to 5, sometimes combined at the base; *stigmas* simple. *Fruit* of from 3 to 5, indehiscent, winged, or wingless pieces; *column* central, persistent. *Embryo* erect, in the axis of, and almost as long as, the fleshy *albumen.*— *Herbaceous* plants. *Leaves* simple, entire, alternate, sometimes minute. *Stipulæ* lateral, very minute. *Spike* terminal, each flower with 3 bracteæ.

AFFINITIES. Between Celastrineæ and Euphorbiaceæ, according to Mr. Brown; from the latter of which they differ in the presence of petals, in the structure of their fruit, and in the position of their seeds, besides other characters; from the former in the presence of stipulæ, in the cohesion of the petals in a tube, in the want of a fleshy disk, in the deeply lobed ovarium, and so on.

GEOGRAPHY. A few New Holland shrubs compose all that is known of the order.

PROPERTIES. Unknown.

EXAMPLE. Stackhousia.

XCIII. CELASTRINEÆ.

CELASTRINEÆ, *R. Brown in Flinders*, 22. (1814); *Dec. Prodr.* 2. 2. (1825); *Ad. Brongniart Mémoire sur les Rhamnées*, 16. (1826); *Lindl. Synops.* 74. (1829.)

DIAGNOSIS. Polypetalous dicotyledons, with 4 or 5 perigynous stamens alternate with the petals, concrete carpella, a superior ovarium with several cells surrounded by a large fleshy disk, ascending ovules, and alternate simple leaves without stipulæ.

ANOMALIES. Flowers unisexual in Maytenus. Petals none in Alzatea.

ESSENTIAL CHARACTER.— *Sepals* 4 or 5, imbricated, inserted into the margin of an expanded torus. *Petals* inserted by a broad base, under the margin of the disk, with an imbricate æstivation. *Stamens* alternate with the petals, inserted into the disk, either at the margin or within it; *anthers* innate. *Disk* large, expanded, flat, closely surrounding the ovarium, covering the flat expanded torus. *Ovarium* superior, immersed in the disk and adhering to it, with 3 or 4 cells; *cells* 1- or many-seeded; *ovules* ascending from the axis, attached to a short funiculus. *Fruit* superior; either a 3- or 4-celled capsule, with 3 or 4 septiferous valves; or a dry drupe, with a 1- or 2-celled nut, the cells of which are 1- or many-seeded. *Seeds* ascending, seldom inverted by resupination, either provided with an arillus, or without one; *albumen* fleshy; *embryo* straight; *cotyledons* flat and thick, with a short inferior radicle. — *Shrubs*. *Leaves* simple, alternate or opposite. *Flowers* in axillary cymes.

AFFINITIES. Formerly confounded with Rhamneæ, this order was first separated by Mr. Brown, who distinguished it particularly by the relation which its stamens bear to the petals. It also differs in its imbricated calyx, and in its disk being hypogynous. According to Brongniart, Celastrineæ have more relation to several orders with hypogynous stamens than to any with perigynous ones, especially to Malpighiaceæ, to which they are related through Hippocrateaceæ, which are in fact, according to Mr. Brown, scarcely distinct from Celastrineæ. *Brongn. Mém.* p. 15. Related to Euphorbiaceæ.

GEOGRAPHY. Natives of the warmer parts of Europe, North America, and Asia, but far more abundant beyond the tropics than within them; a great number of species inhabit the Cape of Good Hope. Some are found in Chile and Peru, and a few in New Holland.

PROPERTIES. I find nothing recorded about the properties of the species of this order, except a remark by Decandolle, that a decoction of the young branches of Maytenus are employed in Chile as a wash for swellings produced by the poisonous shade of the tree Lithi. *Essai*, 123. ed. 2.

EXAMPLES. Euonymus, Celastrus, Alzatea.

XCIV. HIPPOCRATEACEÆ.

HIPPOCRATICEÆ, *Juss. Ann. Mus.* 18. 483. (1811.) — HIPPOCRATEACEÆ, *Kunth in Humb. N. G. Am.* 5. 136. (1821); *Dec. Prodr.* 1. 567. (1829.)

DIAGNOSIS. Polypetalous dicotyledons, with definite hypogynous stamens (3) cohering at the base in a fleshy cup, concrete carpella, an ovarium of several cells with the placentæ in the axis, an imbricated calyx, unsymmetrical flowers, erect ovules, undivided petals without appendages, and indehiscent apterous fruit.

ANOMALIES.

ESSENTIAL CHARACTER.—*Sepals* 5, very seldom 4 or 6, very small, combined as far as the middle, persistent. *Petals* 5, very seldom 4 or 6, equal, hypogynous? somewhat imbricated in æstivation. *Stamens* 3, very seldom 4 or 5; *filaments* cohering almost as far as the apex into a tube dilated at the base, and forming about the ovarium a thick disk-like cup; *anthers* 1-celled, opening transversely at the apex, 2- or even 4-celled. *Ovarium* concealed by the tube, 3-cornered, distinct; *style* 1; *stigmas* 1-3; *ovula* erect. *Fruit* either consisting of 3 samaroid carpella, or berried with from 1 to 3 cells. *Seeds* in each cell 4, or more, but definite, attached to the axis in pairs, some of them occasionally abortive, erect, without albumen; *embryo* straight; *radicle* pointing towards the base; *cotyledons* flat, elliptical oblong, somewhat fleshy, cohering when dried.—Arborescent or climbing *shrubs*, which are almost always smooth. *Leaves* opposite, simple, entire or toothed, somewhat coriaceous. *Racemes* axillary, in corymbs or fascicles. *Flowers* small, not shewy.

AFFINITIES. The ternary number of the stamens, along with the quinary number of the petals and sepals, is the prominent characteristic of this order, which was formerly included in Acerineæ by M. de Jussieu, which is placed between Erythroxyleæ and Marcgraaviaceæ by Decandolle, but which is, to all appearance, much more nearly related to Celastrineæ, as Mr. Brown has remarked; for " the insertion of the ovula is either towards the base, or is central; the direction of the radicle is always inferior." *Brown, Congo*, 427. In Hippocratea ovata the testa and cotyledons are furnished in the inside with innumerable trachea-like threads; the same economy has been remarked by Du Petit Thouars in the pericarp of Calypso. *Dec. Prodr.* 1. 567. The only similar cases of this curious structure with which I am acquainted are in Collomia, in which I have detected it (*Bot. Reg.* fol. 1166.), and in Casuarina, in which it has been described by Mr. Brown; plants having no apparent affinity with Hippocrateaceæ.

GEOGRAPHY. The principal part are South American, about 1-seventh are natives of Africa or the Mauritian Islands, and the same number has been recorded as East Indian.

PROPERTIES. The fruit of Tonsella pyriformis, a native of Sierra Leone, is eatable. It is about the size of a Bergamot Pear; its flavour is rich and sweet. *Hort. Trans.* The nuts of Hippocratea comosa are oily and sweet. *Swartz.* 1. 78.

EXAMPLES. Hippocratea, Anthodon, Salacia.

XCV. BREXIACEÆ.

DIAGNOSIS. Polypetalous dicotyledons, with definite hypogynous stamens, a hypogynous disk, concrete carpella, an ovarium of several cells with the placentæ in the axis, an imbricated calyx, symmetrical flowers, indefinite exalbuminous seeds with a straight embryo, and drupaceous fruit and arborescent stems.

ANOMALIES.

ESSENTIAL CHARACTER. — *Calyx* inferior, small, persistent, 5-parted; æstivation imbricated. *Petals* 5, hypogynous, imbricated in æstivation. *Stamens* 5, hypogynous, alternate with the petals, arising from a narrow cup, which is toothed between each stamen; *anthers* oval, innate, 2-celled, bursting longitudinally, fleshy at the apex; *pollen* triangular, cohering by means of fine threads. *Ovarium* superior, 5-celled, with numerous ovules attached in two rows to placentæ in the axis; *style* 1, continuous; *stigma* simple. *Fruit* drupaceous, 5-celled, many-seeded. *Seeds* indefinite, attached to the axis, with a double integument, the inner of which is membranous; *albumen* 0; *cotyledons* ovate, obtuse; *radicle* cylindrical, centripetal.—*Trees*, with nearly simple trunks. *Leaves* coriaceous, alternate, simple, not dotted, with deciduous minute stipulæ. *Flowers* green, in axillary umbels, surrounded by bracteæ on the outside.

AFFINITIES. The solitary genus upon which this order is founded does not exhibit any very obvious affinities, for which reason it is probable that other genera remain to be discovered which will establish the connexion that is at present wanting. Its habit is that of some Myrsineæ, especially Theophrasta, from which it differs in being polypetalous, in the stamens being alternate with the petals, and in many other circumstances. With Rhamneæ and Celastrineæ its relation is no doubt strong, but its stamens are hypogynous, not perigynous, and its seeds indefinite. Some resemblance may be traced between it and Anacardiaceæ, especially in the resinous appearances visible upon the young shoots, and also in habit; but its fructification is entirely at variance with that order. With Pittosporeæ it agrees in its hypogynous definite stamens, its polyspermous fruit, its alternate undivided leaves, and habit; but it disagrees in a number of important particulars. Upon the whole, however, I think it approaches more nearly to

Celastrineæ than to any other order. The fruit is well described by Dr. Wallich in the *Flora Indica.*

GEOGRAPHY. Madagascar trees.

PROPERTIES. Unknown.

EXAMPLE. Brexia.

XCVI. RHAMNEÆ. THE BUCKTHORN TRIBE.

RHAMNI, *Juss. Gen.* 376. (1789).— RHAMNEÆ, *Dec. Prodr.* 2. 19. (1825); *Brongniart Mémoire sur les Rhamnées,* (1826); *Lindl. Synops.* 72. (1829.)

DIAGNOSIS. Polypetalous dicotyledons, with perigynous definite stamens opposite the cucullate petals, concrete carpella, a superior ovarium with several cells surrounded by a fleshy disk, solitary erect ovula, valvate calyx, and alternate simple leaves with minute stipulæ.

ANOMALIES. Sometimes the ovarium is inferior. Leaves opposite in Colletia and Retanilla. Stipules and petals often wanting.

ESSENTIAL CHARACTER. — *Calyx* monophyllous, 4-5-cleft, with a valvate æstivation. *Petals* distinct, cucullate, or convolute, inserted into the orifice of the calyx, occasionally wanting. *Stamens* definite, opposite the petals. *Disk* fleshy. *Ovarium* superior, or half superior, 2- 3- or 4-celled; *ovules* solitary, erect. *Fruit* fleshy and indehiscent, or dry and separating in 3 divisions. *Seeds* erect; *albumen* fleshy, seldom wanting; *embryo* almost as long as the seed, with large flat *cotyledons*, and a short inferior *radicle.* — *Trees* or *shrubs,* often spiny. *Leaves* simple, alternate, very seldom opposite, with minute *stipulæ. Flowers* axillary or terminal.

AFFINITIES. Under this name have been for a long time confounded four orders, very different in characters, and even in natural affinities, the peculiarities of three of which have been pointed out by M. Ad. Brongniart in his memoir upon the subject, and a fourth has been distinguished by myself. These orders are Rhamneæ properly so called, Celastrineæ, Ilicineæ, and Staphyleaceæ, the respective affinities of which will be found under each. M. Brongniart indicates the relation that Rhamneæ bear, thus: if we take the insertion of stamens as the most important distinction of plants, it will be found that among polypetalous orders with perigynous stamens, Pomaceæ are those to which Rhamneæ have the closest relation, agreeing with them in the ovarium, the cells of which are determinate in number, in the ascending ovules, and in their alternate leaves usually having two stipulæ at their base; the number and position of their stamens, and the structure of their seeds, separate them widely. But if the insertion of the stamens is left out of consideration, they will be found to have many characters in common with Büttneriaceæ (*Brown in Flinders*, 22.); such as, the æstivation of the calyx, the form of the petals, the position of the stamens in front of those petals, the structure of the ovarium and seeds in many important points; the principal differences between them are, in fact, the stamens being turned outwards in Büttneriaceæ, which are also destitute of a disk, have hypogynous stamens, and always 2 or more ovules. Euphorbiaceæ are allied to Rhamneæ; but the constant separation of sexes in the former family, their hypogynous stamens and suspended ovules, are all important marks of distinction. Nitrariaceæ may be compared with Rhamneæ in several points.

GEOGRAPHY. Found over nearly all the world, except in the arctic zone; the maximum of species is said to be dispersed through the hottest parts of the United States, the south of Europe, the north of Africa, Persia and India in the northern hemisphere, and the Cape of Good Hope and

New Holland in the southern. Some of the genera appear to be confined to particular countries, as all the true Ceanothuses to North America, Phylicas to the Cape, Cryptandra and Pomaderris to New Holland.

PROPERTIES. The berries of various species of Rhamnus are violent purgatives, and have been highly spoken of in dropsy. They also yield a dye, varying in tint from yellow to green; the ripe berries of R. catharticus, mixed with gum arabic and lime-water form the green colour known under the name of Bladder-green. The French berries of the shops (Graines d'Avignon, Fr.) are the fruit of Rh. infectorius and saxatilis, and amygdalinus. The fruit of Zizyphus is destitute of these purgative qualities, and, on the contrary, is often wholesome and pleasant to eat, as in the case of the Jujube and the Lote, the latter of which is now known to have given their name to the classical Lotophagi. The peduncles of Hovenia dulcis become extremely enlarged and succulent, and are in China a fruit in much esteem, resembling in flavour, as it is said, a ripe Pear. Some species are astringent. Sageretia theezans is used for tea by the poorer classes in China; an infusion of the twigs of Ceanothus americanus has been named as useful, on account of its astringency, to stop gonorrhœal discharges; antisyphilitic virtues are ascribed to the root of the same, and also of Berchemia volubilis; and it is said, by Rumphius, that in the Moluccas the bark of Zizyphus Jujuba is employed as a remedy for diarrhœa. *Brongn.*

EXAMPLES. Rhamnus, Phylica, Hovenia.

XCVII. STAPHYLEACEÆ. THE BLADDER-NUT TRIBE.

CELASTRINEÆ, § Staphyleaceæ, *Dec. Prodr.* 2. 2. (1825).—STAPHYLEACEÆ, *Lindl. Synops.* 75. (1829.)

DIAGNOSIS. Polypetalous dicotyledons, with 5 perigynous stamens alternate with the petals, concrete carpella, a superior ovarium of several cells surrounded by a fleshy disk, erect ovules, and opposite pinnated leaves with common and partial stipulæ.

ANOMALIES. Flowers unisexual in Turpinia.

ESSENTIAL CHARACTER. — *Sepals* 5, connected at the base, coloured, with an imbricated æstivation. *Petals* 5, alternate, with an imbricated æstivation. *Stamens* 5, alternate with the petals, perigynous. *Disk* large, urceolate. *Ovarium* 2- or 3-celled, superior; *ovula* erect; *styles* 2 or 3, cohering at the base. *Fruit* membranous or fleshy, indehiscent or opening internally, often deformed by the abortion of some of the parts. *Seeds* ascending, roundish, with a bony testa; *hilum* large, truncate; *albumen* none; *cotyledons* thick. — *Shrubs. Leaves* opposite, pinnate, with both common and partial stipulæ. *Flowers* in terminal, stalked racemes.

AFFINITIES. Combined with Celastrineæ by Decandolle, but distinguished by Ad. Brongniart (*Mém. sur les Rhamnées*, p. 16.), this order appears to me to be essentially characterised by its opposite pinnated stipulate leaves, and to indicate an affinity between Celastrineæ and Sapindaceæ.

GEOGRAPHY. The very few species which belong here are irregularly scattered over the face of the globe. Of the genus Staphylea, 1 is found in Europe, 1 in North America, 1 in Japan, 2 in Jamaica, 1 in Peru; and of Turpinia, 1 is Mexican, and 1 East Indian.

PROPERTIES. Unknown.

EXAMPLES. Staphylea, Turpinia.

XCVIII. HIPPOCASTANEÆ. The Horse-Chestnut Tribe,

HIPPOCASTANEÆ, *Dec. Théorie, ed.* 2. 244. (1819); *Prodr.* 1. 597. (1824).—
CASTANEACEÆ, *Link Enum.* 1. 354. (1821.)

DIAGNOSIS. Polypetalous dicotyledons, with hypogynous definite stamens, concrete carpella, an ovarium of several cells with the placentæ in the
axis, an imbricated calyx, unsymmetrical flowers, definite erect ovules, undivided petals without appendages, dehiscent fruit, and compound palmate
leaves.

ANOMALIES.

ESSENTIAL CHARACTER.—*Calyx* campanulate, 5-lobed. *Petals* 5, or 4 by the abortion of one of them, unequal, hypogynous. *Stamens* 7-8, distinct, unequal, inserted upon
a hypogynous disk; *anthers* somewhat incumbent. *Ovarium* roundish, 3-cornered,
3-celled; *style* 1, filiform, conical, acute; *ovula* 2 in each cell. *Fruit* coriaceous, 1- 2- or
3-valved, 1- 2- or 3-celled, 1- 2- or 3-seeded. *Seeds* large, roundish, with a smooth shining
coat, and a broad pale hilum; *albumen* none; *embryo* curved, inverted, with fleshy, very
thick, gibbous, cohering cotyledons, germinating under ground; *plumula* unusually large,
2-leaved; *radicle* conical, curved, turned towards the hilum.—*Trees* or *shrubs*, Leaves
opposite, without *stipulæ*, compound, quinate or septenate. *Racemes* terminal, somewhat
panicled; the pedicels with an articulation.

AFFINITIES. The want of symmetry in the parts of the flower, and their
compound leaves, approximate Hippocastaneæ to Sapindaceæ; the same
character brings them near Acerineæ, from both which they are distinguished
by the structure of their fruit and seeds. They also approach Rhizoboleæ,
as is stated in speaking of that order.

GEOGRAPHY. The north of India and North America contain the few
species that belong to this order.

PROPERTIES. Handsome trees or small bushes, chiefly remarkable for
their large seeds, with an extensive hilum. These seeds contain a great
quantity of starch, which renders them nutritive for man and many other
animals. They also contain a sufficient proportion of potash to be useful
as cosmetics, or as a substitute for soap; they are bitter, and have been employed as a sternutatory. The bark of the common Horse Chestnut is bitter,
astringent, and febrifugal.

EXAMPLES. Æsculus, Pávia.

XCIX. RHIZOBOLEÆ.

RHIZOBOLEÆ, *Dec. Prodr.* 1. 599. (1824); *Cambessédes in Aug. St. Hil. Fl. Bras. Merid.*
1. 322. (1827.)

DIAGNOSIS. Polypetalous dicotyledons, with hypogynous indefinite
stamens, concrete carpella, an ovarium of several cells: with solitary peltate
ovules, an imbricated calyx, exstipulate compound leaves, and round anthers
bursting longitudinally.

ANOMALIES.

ESSENTIAL CHARACTER.—*Sepals* 5, more or less combined, imbricated in æstivation.
Petals 5, thickish, unequal, arising along with the stamens from a hypogynous disk.
Stamens extremely numerous, slightly monadelphous, arising in a double row from a disk,
the innermost being shorter and often abortive; *anthers* roundish. *Ovarium* superior,
4-celled, 4-seeded; *styles* 4; *stigmas* simple; *ovula* peritropal. *Fruit* formed of 4 combined nuts, part of which are sometimes abortive; each nut indehiscent, 1-seeded, 1-celled,
with a thick double putamen. *Seed* reniform, without albumen, with a funiculus which

is dilated into a spongy excrescence; *embryo* very large, constituting nearly the whole of the almond-like substance of the nut, with a long 2-edged cauliculus, having two small cotyledons at the top, and lying in a furrow of the radicle.— *Trees.* *Leaves* opposite, stalked, compound, without stipulæ. *Flowers* racemose.

AFFINITIES. A very distinct order, related on the one hand to Anacardiaceæ, and particularly to Mangifera, but perhaps rather to be associated with Sapindaceæ, in consideration of its hypogynous flowers and its fruit; in some measure also related to Hippocastaneæ on account of its opposite compound palmate leaves; but in Hippocastaneæ the radicle is small, and the cotyledons very large, while in Rhizoboleæ the radicle is enlarged, and the cotyledons small. In both orders the albumen seems to be absorbed by the various parts of the embryo. *Decand. Prodr.* 1. 599.

GEOGRAPHY. Six large trees found in the forests of the hottest parts of South America constitute the whole of the order.

PROPERTIES. Some of them are known for producing the Souari (vulgò Suwarrow) Nuts, or Brazil Nuts of the shops, the kernel of which is one of the most delicious fruits of the nut kind that is known. An oil is extracted from them not inferior to that of the Olive.

EXAMPLE. Caryocar.

C. SAPINDACEÆ. THE SOAP-TREE TRIBE.

SAPINDI, *Juss. Gen.* 246. (1789).—SAPINDACEÆ, *Juss. Ann. Muss.* 18. 476. (1811); *Dec. Prodr.* 1. 601. (1824.)

DIAGNOSIS. Polypetalous dicotyledons, with hypogynous definite stamens irregularly arranged upon a disk, concrete carpella, an ovarium of several cells with the placentæ in the axis, an imbricated calyx, unsymmetrical flowers, petals usually with some interior appendage, and very unequal sepals.

ANOMALIES. In Tina the flowers appear to be symmetrical. Stadmannia, Amirola, and Dodonæa, have no petals.

ESSENTIAL CHARACTER. — *Sepals* 4 or 5, either distinct or cohering at the base; *æstivation* imbricate. *Petals* generally equal in number to the sepals, occasionally one less, very rarely none, hypogynous; sometimes naked, sometimes villous or glandular in the middle, sometimes with an interior petaloid scale. *Stamens* irregularly arranged, distinct, double the number of the petals, inserted on a hypogynous glandular disk. *Ovarium* roundish; *style* 1 or 3; *ovula* arising from the middle of the axis, definite (collateral), ascending. *Fruit* drupaceous or capsular, 3-celled, or by abortion 1- or 2-celled. *Seeds* solitary, attached to the axis, without albumen; *embryo* with the radicle pointing towards the base of the cell; *cotyledons* more or less curved upon the radicle, occasionally straight.— Erect or climbing *trees* or *shrubs*, very seldom *herbaceous* plants. *Leaves* alternate, often compound, having frequently pellucid lines or dots.

AFFINITIES. Very near Meliaceæ, which agree in habit and in their pinnated leaves, but which are known by their monadelphous stamens and symmetrical flowers. To Polygaleæ they are no doubt akin in the singular combination of 8 stamens with 5 unequal sepals, and· an uncertain number of petals; and also in their arillus, which may be compared to the caruncula of Polygaleæ, although somewhat different in its origin. The dried leaves resemble, as Decandolle remarks, those of Connaraceæ. Their climbing habit and tendency to produce tendrils indicate a relation to Vites, which, however, is not very near. Mr. Brown remarks, that although in the far greater part of this family the ovulum is erect and the radicle of the embryo inferior, yet it includes more than one genus in which both the seeds and embryo are inverted. *Congo,* 427. (1818.)

GEOGRAPHY. Natives of most parts of the tropics, but especially of South America and India; the tribe called Paullinieæ is most abundant in the former, and Sapindeæ in the latter region. Africa knows many of them, but they are wanting in the cold regions of the north. None are found in Europe or the United States of America. Dodonæas represent the order in New Holland.

PROPERTIES. It is singular that while the leaves and branches of many of these plants are unquestionably poisonous, the fruit of others is valuable as an article of the dessert. Thus the Longan, the Litchi, and the Rambutan, fruits among the most delicious of the Indian archipelago, are the produce of different species of Euphoria. The fruit of Schmidelia edulis is known at desserts in Brazil under the name of Fruta de paraô; it is said to have a sweet and pleasant taste. *Pl. Us.* 67. That of Sapindus esculentus is very fleshy, and much esteemed by the inhabitants of Certaô, by whom it is called Pittomba. *Ibid.* 68. Some species of Paullinia are stated, upon various authorities, to be poisonous, especially the P. australis, to which principally M. Auguste de St. Hilaire attributes the poisonous quality of the Lecheguana honey. *Ed. P. J.* 14. 269. The arillus of Paullinia subrotunda and of Blighia sapida is eatable. The leaves of Magonia pubescens and glabrata, called Tinguy in Brazil, are used for stupifying fishes; their bark is employed for healing sores in horses caused by the stings of insects. *A. St. Hil. Hist. des Pl.* 238. The fruit of Sapindus saponaria is saponaceous. The root of Cardiospermum halicacabum is aperient. *Ainslie,* 2. 204.

EXAMPLES. Sapindus, Blighia, Paullinia.

CI. ACERINEÆ. THE SYCAMORE TRIBE.

ACERA, *Juss. Gen.* 50. (1789); *Ann. Mus.* 18. 477. (1811).—ACERINEÆ, *Dec. Théorie,* ed. 2. 244. (1819); *Prodr.* 1. 593. (1824); *Lindl. Synops.* 55. (1829).

DIAGNOSIS. Polypetalous dicotyledons, with distinct hypogynous definite stamens, concrete carpella, an ovarium of several cells with the placentæ in the axis, an imbricated calyx, unsymmetrical flowers, definite erect ovules, undivided petals without appendages, and indehiscent winged fruit.

ANOMALIES. The leaves of Negundium are compound.

ESSENTIAL CHARACTER.—*Calyx* divided into 5, or occasionally from 4 to 9 parts, with an imbricate æstivation. *Petals* equal in number to the lobes of the calyx, inserted round a hypogynous disk. *Stamens* inserted upon a hypogynous disk, generally 8, not often any other number, always definite. *Ovarium* 2-lobed; *style* 1; *stigmas* 2. *Fruit* formed of two parts, which are indehiscent and winged; each 1-celled, with 1 or 2 seeds. *Seeds* erect, with a thickened lining to the testa; *albumen* none; *embryo* curved, with foliaceous wrinkled *cotyledons,* and an inferior *radicle.*— *Trees. Leaves* opposite, simple, rarely pinnate, without stipulæ. *Flowers* often polygamous, sometimes apetalous, in axillary corymbs or racemes.

AFFINITIES. Related closely to Malpighiaceæ in their winged fruit, to Sapindaceæ in the pinnate leaves of two species, and the unsymmetrical flowers of the whole.

GEOGRAPHY. Europe, the north of India, and North America, are the stations of this order, which is unknown in Africa and the southern hemisphere.

PROPERTIES. They are only known for the sugary sap of Acer saccharinum and other species, from which sugar is extracted in abundance.
EXAMPLES. Acer, Negundium.

CII. ERYTHROXYLEÆ.

ERYTHROXYLEÆ, *Kunth in Humb. N. G. Am.* 5. 175. (1821); *Dec. Prodr.* 1. 573. (1824.)

DIAGNOSIS. Polypetalous dicotyledons, with definite hypogynous stamens, concrete carpella, an entire ovarium of 1 cell, an imbricated calyx, symmetrical flowers, definite pendulous ovules, distinct sessile stigmas, and drupaceous fruit.
ANOMALIES.

ESSENTIAL CHARACTER.—*Sepals* 5, combined at the base, persistent. *Petals* 5, hypogynous, broad at the base, with a plaited scale there, equal, the margins lying upon each other in æstivation. *Stamens* 10; *filaments* combined at the base into a cup; *anthers* innate, erect, 2-celled, dehiscing lengthwise. *Ovarium* 1-celled, or 3-celled, with 2 cells spurious; *styles* 2, distinct; *stigmas* 3, somewhat capitate, or united almost to the point; *ovulum* solitary, pendulous. *Fruit* drupaceous, 1-seeded. *Seed* angular; *albumen* corneous; *embryo* linear, straight, central; *cotyledons* linear, flat, leafy; *radicle* superior, taper, straight; *plumula* inconspicuous.—*Shrubs* or *trees ;* young shoots often compressed and covered with acute imbricated scales. *Leaves* alternate, seldom opposite, usually smooth; *stipules* axillary. *Flowers* small, whitish or greenish. *Peduncles* with bracteæ at the base.

AFFINITIES. Separated from Malpighiaceæ by Kunth on account of the appendages of the petals, the presence of albumen, the fruit being often 1-celled by abortion, and their peculiar habit. *Dec.* Mr. Brown suggests that Erythroxylon belongs to Malpighiaceæ, or at least that it approximates very closely to that family. *Congo,* 426.
GEOGRAPHY. Chiefly West Indian and South American. A few are found in the East Indies, and several in the Mauritius and Madagascar.
PROPERTIES. The wood of some is bright red; that of E. hypericifolium is called in the Isle of France Bois d'huile. A permanent reddish brown dye is obtained from the bark of Erythroxylum suberosum, called in Brazil Gallinha choca and Mercurio do campo. *Pl. Us.* 69.
EXAMPLES. Erythroxylum, Sethia.

CIII. MALPIGHIACEÆ. THE BARBADOES CHERRY TRIBE.

MALPIGHIACEÆ, *Juss. Gen.* 252. (1789); *Ann. Mus.* 18. 479. (1811); *Dec. Prodr.* 1. 577. (1824.)

DIAGNOSIS. Polypetalous dicotyledons, with definite hypogynous stamens, concrete carpella, a nearly entire ovarium of 3 cells, a glandular imbricated calyx, symmetrical flowers, definite pendulous ovules, a single style, exalbuminous seeds, fruit without a woody axis, unguiculate petals, and leaves without pellucid dots.
ANOMALIES. Styles sometimes distinct. Leaves in an African species alternate. Petals occasionally wanting.

ESSENTIAL CHARACTER. — *Sepals* 5, slightly combined, persistent. *Petals* 5, unguiculate, inserted in a hypogynous disk, occasionally rather unequal, very seldom wanting. *Stamens* 10, alternate with the petals, seldom fewer, occasionally solitary ; *filaments* either distinct, or partly monadelphous; *anthers* roundish. *Ovarium* 1, usually 3-lobed, formed of 3 carpella, more or less combined; *styles* 3, distinct or combined; *ovula* suspended. *Fruit* dry or berried, 3-celled or 3-lobed, occasionally 1- or 2-celled by abortion. *Seeds* solitary, pendulous, without albumen; *embryo* more or less curved, or straight; *radicle* short; *lobes* leafy or thickish.—Small *trees* or *shrubs*, sometimes climbing. *Leaves* opposite, scarcely ever alternate, simple, without dots, with stipulæ mostly. *Flowers* in racemes or corymbs. *Pedicels* articulated in the middle, with 2 minute bracteæ.

AFFINITIES. Distinguished from Erythroxyleæ by the structure of the ovarium; and from Acerineæ by the unguiculate petals, the glandular calyx, and the symmetrical flowers. Mr. Brown remarks, that the insertion of the ovulum is always towards its apex, or considerably above its middle; and the radicle of the embryo is uniformly superior, in which point Banisteria offers no exception to the general structure, although Gærtner has described its radicle as inferior. *Congo*, 426.

GEOGRAPHY. Almost exclusively found in the equinoctial parts of America; of 180 species enumerated by Decandolle, only 5 are East Indian, 1 is found at the Cape, 1 in Arabia, and 5 in equinoctial Africa, or the contiguous islands.

PROPERTIES. Little is known of this subject. The wood of some kinds is bright red. The fruit of many is eaten in the West Indies; the hairs of a few species are painfully pungent. The bark of Malp. Moureila, according to Aublet, is employed in Cayenne as a febrifuge.

The following sections are employed by Decandolle:—

1. MALPIGHIEÆ.

Styles 3, distinct or cohering in 1. Fruit fleshy, indehiscent.—Leaves opposite.

EXAMPLES. Malpighia, Bunchozia.

2. HIPTAGEÆ.

Style 1, or 3 combined in 1. Carpella of the fruit dry, indehiscent, 1-seeded, often variously expanded into wings.—Leaves opposite or verticillate.

EXAMPLES. Hiptage, Thryallis, Aspicarpa.

3. BANISTERIEÆ.

Styles 3, distinct. Carpella of the fruit dry, indehiscent, monospermous, variously expanded into wings.—Leaves opposite, rarely whorled.

EXAMPLES. Hiræa, Banisteria.

CIV. VITES. THE VINE TRIBE.

VITES, *Juss. Gen.* 267. (1789).—SARMENTACEÆ, *Vent. Tabl.* 3. 167. (1799).—VINIFERÆ, *Juss. Mem. Mus.* 3. 444. (1817).—AMPELIDEÆ, *Kunth in Humboldt, N. G. et Sp.* 5. 223. (1821); *Dec. Prodr.* 1. 627. (1824.)

DIAGNOSIS. Polypetalous dicotyledons, with definite hypogynous stamens, concrete carpella, an entire ovarium of 2 cells, a small almost entire open calyx, symmetrical flowers, definite erect ovules, baccate fruit, tumid joints, and a climbing habit.

ANOMALIES. Leea and Lasianthera are monopetalous; but it is doubtful whether they belong to the order.

ESSENTIAL CHARACTER.—*Calyx* small, nearly entire at the edge. *Petals* 4 or 5, inserted on the outside of a disk surrounding the ovarium; in æstivation turned inwards at the edge, in a valvate manner. *Stamens* equal in number to the petals, inserted upon the disk, sometimes sterile by abortion; *filaments* distinct, or slightly cohering at the base;

anthers ovate, versatile. *Ovarium* superior, 2-celled; *style* 1, very short; *stigma* simple; *ovula* erect, definite. *Berry* round, often by abortion 1-celled, pulpy. *Seeds* 4 or 5, or fewer by abortion, bony, erect; *albumen* hard; *embryo* erect, about one half the length of the albumen; *radicle* taper; *cotyledons* lanceolate, plano-convex. — Scrambling, climbing *shrubs*, with tumid separable joints. *Leaves* with stipulæ at the base, the lower opposite, the upper alternate, simple or compound. *Peduncles* racemose, sometimes by abortion changing to tendrils. *Flowers* small, green.

AFFINITIES. The tumid joints, which separate from each other by an articulation, along with the many other points of agreement in their fructification, approximate them to Geraniaceæ. Their compound leaves, and their evident relation to Leea, which is itself possibly Meliaceous, indicates their affinity to the latter order; and their habit and inflorescence to Caprifoliaceæ, through Hedera. The tendrils of the order are the branches of inflorescence, the flowers of which are abortive.

GEOGRAPHY. Inhabitants of woods in the milder and hotter parts of both hemispheres, especially in the East Indies.

PROPERTIES. Acid leaves, and a fruit like that of the common grape, is the usual character of the order. The sap or tears of the vine are a popular remedy in France for chronic ophthalmia, but they are of little value. The leaves, on account of their astringency, are sometimes used in diarrhœa. But the dried fruit and wine are the really important products of the grape; products which are, however, yielded by no other of the order, if we except the Fox-grapes of North America, which scarcely deserve to be excepted. The acid of the grape is chiefly the tartaric; malic acid, however, exists in them. The sugar contained in grapes differs slightly from common sugar in composition, containing a smaller quantity of carbon. *Turner*, 682.

M. Decandolle has 2 tribes, the last of which is doubtful.

Tribe 1. VINIFERÆ, or SARMENTACEÆ.

Corolla polypetalous. Stamens opposite the petals. Peduncles often with tendrils.

EXAMPLES. Cissus, Vitis.

Tribe 2. LEEACEÆ.

Corolla monopetalous. Stamens alternate? with the petals, often monadelphous. Fruit and seeds scarcely known. Tendrils wanting.

EXAMPLES. Leea, Lasianthera.

CV. MELIACEÆ. THE BEAD-TREE TRIBE.

MELIÆ, *Juss. Gen.* 263. (1789); *Mem. Mus.* 3. 436. (1817); *Dec. Prodr.* 1. 619. (1824.)

DIAGNOSIS. Polypetalous dicotyledons, with definite hypogynous stamens combined in a long tube, concrete carpella, an ovarium of several cells with the placentæ in the axis, an imbricated calyx, symmetrical flowers, definite exalbuminous apterous seeds with straight embryo, and sub-sessile anthers.

ANOMALIES.

ESSENTIAL CHARACTER.— *Sepals* 4 or 5, more or less united. *Petals* the same number, hypogynous, conniving at the base, or even cohering, usually having a valvate æstivation. *Stamens* twice as many as the petals (occasionally equal in number, sometimes 3 or 4 times as many); *filaments* cohering in a long tube; *anthers* sessile within the orifice of the tube. *Disk* frequently highly developed, surrounding the ovarium like a cup. *Ovarium* single, with several cells; *style* 1; *stigmas* distinct or combined; *ovules* 1 or 2 in each cell. *Fruit* berried, drupaceous or capsular, many-celled, often, in consequence of abortion, 1-celled, the valves, if present, having the dissepiments in their middle. *Seeds* without albumen, not winged; *embryo* inverted. — *Trees or shrubs.* *Leaves* alternate, without *stipulæ*, simple or compound.

AFFINITIES. This order is not well understood. It is apparently akin to Sapindaceæ, with which it agrees in habit, but from which it is distinguished by its stamens and symmetrical flowers. To Cedreleæ it is most closely allied, and therefore connected with Rutaceæ through Flindersia. Humiriaceæ are principally distinguished by their highly developed connectivum and partially united stamens. Styraceæ are very nearly akin to Meliaceæ, but they are monopetalous.

GEOGRAPHY. Found principally in the hotter parts of the East and West Indies, South America, and Africa. The common Bead-tree, Melia Azedarach, has the most northern position, in Syria.

PROPERTIES. The false Winter's Bark, a good tonic and stimulant, not much known, is yielded by Canella alba; it is aromatic, and used as a condiment in the West Indies. The bark of Guarea Trichilioides is, according to Aublet, purgative and emetic. The root of Melia Azedarach is bitter and nauseous, and is used in North America as anthelmintic; the pulp that surrounds the seeds is said to be deleterious; but this is denied by M. Turpin, who asserts that dogs which he has seen eat it experienced no inconvenience; and children in Carolina eat them with impunity. *Ach. R.* It is supposed that the Melia Azedarachta, or Neemtree of India, possesses febrifuge properties. See *Trans. of the M. and Ph. Soc. of Calcutta*, 3. 430. A kind of Toddy, which the Hindoo doctors consider a stomachic, is obtained by tapping this, which is also called the Margosa-tree. *Ainslie*, 1. 453. From the fruit of the same plant an oil is obtained, which is fit for burning and for other domestic purposes, and, as Ach. Richard well observes (*Bot. Méd.* 708.), is another instance, after the Olive, of the pericarp yielding that substance which is usually obtained from the seed. This oil is said to possess antispasmodic qualities. *Dec.* A warm pleasant-smelling oil is prepared from the fruit of Trichilia speciosa, which the Indian doctors consider a valuable external remedy in chronic rheumatism and paralytic affections. *Ainslie*, 2. 71. Some delicious fruits of the Indian archipelago, called Langsat, or Lanséh, and Ayer Ayer, are species of the genus Lansium; they have a watery pulp, with a cooling pleasant taste. Milnea edulis is another plant of the order, with eatable fruit.

M. Decandolle has the following sections (*Prodr.* 1. 619.):—

1. MELIEÆ.

Cotyledons flat and leafy.
EXAMPLES. Melia, Turræa.

2. TRICHILIEÆ.

Cotyledons very thick.
EXAMPLES. Ekebergia, Guarea.

CVI. CEDRELEÆ.

CEDRELEÆ, *Brown in Flinders*, 64. (1814). — MELIACEÆ, § Cedreleæ, *Dec. Prodr.* 1. 624. (1824.)

DIAGNOSIS. Polypetalous dicotyledons, with definite hypogynous stamens combined in a tube, concrete carpella, an ovarium of several cells with the placentæ in the axis, an imbricated calyx, symmetrical flowers, indefinite exalbuminous winged seeds with a straight embryo, and sub-sessile anthers.

ANOMALIES. Flindersia has dotted leaves.

ESSENTIAL CHARACTER. — *Calyx* 5-cleft, persistent. *Petals* 5, sessile, inserted at the base of a staminiferous disk, imbricated in æstivation. *Stamens* 10, inserted on the outside, below the apex of a hypogynous disk ; those which are opposite the petals sterile ; *anthers* acuminate, attached near the base ; their cells side by side, bursting longitudinally. *Disk* hypogynous, cup-shaped, with 10 plaits. *Ovarium* superior, 5-celled ; *style* simple ; *stigma* deeply 5-lobed, peltate. *Capsule* separable into 5 pieces, which are combined at the base, before bursting, with a short central axis, which is finally distinct and persistent. *Placenta* central, with 5 longitudinal lobes, which occupy the cavities of the capsule, and therefore alternate with the pieces, dividing each cavity in two ; finally becoming loose, and having 2 (or more) seeds on each side. *Seeds* erect, or ascending, with their apex terminated in a wing ; *testa* coriaceous, thickened at the base and sides ; *albumen* 0, (a little, *Dec.*) ; *cotyledons* flat, transverse ; *radicle* transverse, very short, distant from the hilum, (embryo erect, *Dec.*) — *Leaves* alternate, without stipulæ, compound. *Inflorescence* terminal, panicled. *R. Br.*

AFFINITIES. Nearly related to Meliaceæ, in whose affinities they participate. Chiefly distinguished by their winged and indefinite seeds. Flindersia, a genus established by Mr. Brown in the Appendix to Captain Flinders' Voyage, differs from Cedreleæ both in the insertion of its seeds, which are erect, in the dehiscence of its capsules, and also in having movable dissepiments : these last, however, Mr. Brown considers as segments of a common placenta, having a peculiar form. Flindersia is also distinct from the whole order, in having its leaves dotted with pellucid glands, in which respect it serves to connect Cedreleæ with Hesperideæ (Aurantiaceæ), and, notwithstanding the absence of albumen, even with Diosmeæ. See the *Appendix and Atlas to Flinders' Voyage.*

GEOGRAPHY. These are common to America and India, but have not yet been found on the continent of Africa, nor in any of the adjoining islands. *Brown Congo*, 465.

PROPERTIES. The bark of Cedrela is fragrant and resinous ; that of C. Toona, and of Swietenia Mahagoni, is also accounted febrifugal. The mahogany wood used by cabinet-makers is the produce of the last-mentioned plant. The bark of Swietenia febrifuga, called on the Coromandel coast the Red Wood Tree, is a useful tonic in India in intermittent fevers ; but Dr. Ainslie found that if given beyond the extent of 4 or 5 drachms in the 24 hours, it deranged the nervous system, occasioning vertigo and subsequent stupor. Oxleya xanthoxyla, a large tree, is the Yellow-wood of New South Wales.

EXAMPLES. Cedrela, Flindersia, Oxleya.

CVII. HUMIRIACEÆ.

HUMIRIACEÆ, *Adrien de Jussieu in Aug. de St. Hil. Flora Bras. Merid.* 2. 87. (1829.)

DIAGNOSIS. Polypetalous dicotyledons, with definite hypogynous stamens, concrete carpella, an entire ovarium of 5 cells, an imbricated calyx, symmetrical flowers, definite pendulous ovules, a single style, albuminous seeds, fruit without a woody axis, a dilated connectivum, and leaves without pellucid dots.

ANOMALIES.

ESSENTIAL CHARACTER. — *Calyx* in 5 divisions. *Petals* alternate with the lobes of the calyx, and equal to them. *Stamens* hypogynous, 2-celled, 4 or many times as numerous as the petals, monadelphous ; *anthers* with a fleshy connectivum, extended beyond the 2 lobes. *Ovarium* superior, usually surrounded by an annular or toothed disk, 5-celled, with from 1 to 2 suspended ovules in each cell; *style* simple; *stigma* lobed. *Fruit* drupaceous, with 5 or fewer cells. *Seed* with a membranous integument; *embryo* straight, oblong, lying in fleshy albumen; *radicle* superior.—*Trees* or *shrubs.* Leaves alternate, simple, coriaceous, without stipulæ. *Flowers* somewhat cymose.

AFFINITIES. These are not well made out: they differ from Meliaceæ very much in habit, and in many respects in fructification, especially in having the æstivation of the corolla quincuncial, not valvate, and the stamens sometimes indefinite; the anthers also of Humiriaceæ, as Von Martius observes (*Nov. Gen. &c.* 2. 147.), are very different from Meliaceæ in the great dilatation of their connectivum; their albuminous seeds and slender embryo are at variance with Meliaceæ. In the latter respect, and in their balsamic wood, they agree better with Styracineæ, as also in the variable direction of the embryo. Besides these points of affinity, Von Martius compares Humiriaceæ with Chlenaceæ, on account of both orders containing definite and indefinite monadelphous stamens, several stigmas, partially abortive cells, inverted albuminous seeds, and a singular complicated vernation, by which two longitudinal lines are impressed upon each leaf. To me it appears, that the real affinity is with Aurantiaceæ; an affinity indicated by their inflorescence, the texture of their stamens, their disk, their winged petioles, and their balsamic juices.

GEOGRAPHY. All Brazilian trees.

PROPERTIES. Humirium floribundum, when the trunk is wounded, yields a fragrant liquid yellow balsam, called Balsam of Umiri, resembling the properties of Copaiva and Balsam of Peru. *Martius.*

EXAMPLE. Humirium.

CVIII. AURANTIACEÆ. THE ORANGE TRIBE.

AURANTIACEÆ, *Corr. Ann. Mus.* 6. 376. (1805); *Mirb. Bull. Philom.* 379. (1813); *Dec. Prodr.* 1. 535. (1824.)

DIAGNOSIS. Polypetalous dicotyledons, with definite hypogynous stamens, concrete carpella, an entire ovarium of several cells, an open calyx, symmetrical flowers, definite pendulous ovules, a single style, a pulpy fruit without a woody axis, exalbuminous seeds, and compound dotted leaves.

ANOMALIES.

ESSENTIAL CHARACTER.—*Calyx* urceolate or campanulate, somewhat adhering to the disk, short, 3- or 5-toothed, withering. *Petals* 3 to 5, broad at the base, sometimes distinct, sometimes slightly combined, inserted upon the outside of a hypogynous disk, slightly imbricated at the edges. *Stamens* equal in number to the petals, or twice as many, or some multiple of their number, inserted upon a hypogynous disk; *filaments* flattened at the base, sometimes distinct, sometimes combined in one or several parcels; *anthers* terminal, innate. *Ovarium* many-celled; *style* 1, taper; *stigma* slightly divided, thickish; *Fruit* pulpy, many-celled, with a leathery rind replete with receptacles of volatile oil, and sometimes separable from the cells; *cells* often filled with pulp. *Seeds* attached to the axis, sometimes numerous, sometimes solitary, usually pendulous, occasionally containing more embryos than one; *raphe* and *chalaza* usually very distinctly marked; *embryo* straight; *cotyledons* thick, fleshy; *plumula* conspicuous.—*Trees* or *shrubs,* almost always smooth, and filled every where with little transparent receptacles of volatile oil. *Leaves* alternate, often compound, always articulated with the petiole, which is frequently winged. *Spines,* if present, axillary.

AFFINITIES. Readily known by the abundance of oily receptacles which are dispersed over all parts of them, by their deciduous petals, and compound leaves with a winged petiole. They are nearly related to Amy-rideæ and Connaraceæ on the one hand, and to various genera of Diosmeæ on the other, but are distinguished from them all by a variety of obvious characters. The raphe and chalaza are usually distinctly marked upon the testa, and sometimes beautifully. Decandolle considers the rind of the Orange to be of a different origin and nature from the pericarpium of other fruit, and more analogous to the torus or disk of Nelumboneæ; but if the ovarium and ripe fruit are compared, it will be readily seen that this hypo-thesis is untenable, and that there is no difference between the rind of an orange and an ordinary pericarpium.

GEOGRAPHY. Almost exclusively found in the East Indies, whence they have in some cases spread over the rest of the tropics. Two or three species are natives of Madagascar; one is described as found wild in the woods of Essequebo; and Prince Maximilian of Wied Neuwied speaks of a wild Orange of Brazil, called Caranja da terra, which has by no means the delicious refreshing qualities of the cultivated kind, but a mawkish sweet taste. *Travels*, 76.

PROPERTIES. The wood is universally hard and compact; they abound in a volatile, fragrant, bitter, exciting oil; the pulp of the fruit is always more or less acid. *Dec.* The Orange, the Lemon, the Lime, and the Citron, fruits which, although natives of India, have now become so com-mon in other countries as to give a tropical character to a European dessert, are the most remarkable products of this order. If to this be added the excellence of their wood, and the fragrance and beauty of their flowers, I know not if an order more interesting to man can be pointed out. The fruits just mentioned are not, however, its only produce. The Wam-pee, a fruit highly esteemed in China and the Indian archipelago, is the produce of Cookia punctata. The berries of Glycosmis citrifolia are deli-cious; those of Triphasia trifoliata are extremely agreeable. The produc-tiveness of the common Orange is enormous. A single tree at St. Mi-chael's has been known to produce 20,000 oranges fit for packing, exclu-sively of the damaged fruit and the waste, which may be calculated at one-third more. The juice of the Lime and the Lemon contains a large quantity of citric acid. *Turner*, 632. Oranges contain malic acid. *Ib.* 634. A decoction of the root and bark of Ægle Marmelos is supposed, on the Malabar coast, to be a sovereign remedy in hypochondriasis, melancholia, and palpitation of the heart; the leaves in decoction are used in asthmatic complaints, and the fruit a little unripe is given in diarrhœa and dysentery. Roxburgh adds, that the Dutch in Ceylon prepare a perfume from the rind; the fruit is most delicious to the taste, and exquisitely fragrant and nu-tritious, but laxative; the mucus of the seed is a good cement for some purposes. *Ainslie*, 2. 87. The leaves of Bergera Königii are considered by the Hindoos stomachic and tonic; an infusion of them toasted stops vomiting. The green leaves are used raw in dysentery; the bark and root internally as stimuli. *Ibid.* 2. 139. The young leaves of Feronia elephantum have, when bruised, a most delightful smell, very much resembling anise. The native practitioners of India consider them stomachic and carminative. Its gum is very like gum arabic. *Ibid.* 2. 83.

EXAMPLES. Citrus, Limonia, Bergera.

CIX. SPONDIACEÆ. The Hogplum Tribe.

Spondiaceæ, *Kunth in Ann. Sc. Nat.* 2. 362. (1824). — Terebintaceæ, trib. 3.
Dec. Prodr. 2. 74. (1825.)

Diagnosis. Polypetalous dicotyledons, with 10 perigynous stamens, concrete carpella, a superior ovarium of several cells, regular flowers, an annular disk, solitary pendulous ovula, and alternate pinnated leaves with pellucid dots.

Anomalies.

Essential Character. — *Flowers* sometimes unisexual. *Calyx* 5-cleft, regular, persistent or deciduous. *Petals* 5, inserted below a disk surrounding the ovarium, somewhat valvate or imbricate in æstivation. *Stamens* 10, perigynous, arising from the same part as the petals. *Disk* annular, in the males orbicular, with 10 indentations. *Ovarium* superior, sessile, from 2- to 5-celled; *styles* 5, very short; *stigmas* obtuse; *ovulum* 1 in each cell, pendulous. *Fruit* drupaceous, 2-5-celled. *Seeds* without albumen; *cotyledons* plano-convex; *radicle* superior, pointing to the hilum (inferior in Spondias, according to *Gærtner*). — *Trees* without spines. *Leaves* alternate, unequally pinnate, without pellucid dots, a few simple leaves occasionally intermixed. *Stipules* 0. *Inflorescence* axillary and terminal in panicles or racemes.

Affinities. Very near Anacardiaceæ in the structure of their fruit, which is almost that of Mangifera, except that it is compound and not simple; destitute, however, of the resinous juice of that order. They are remarkable for the great development of their disk.

Geography. Natives of the West Indies, the Society Islands, and the Isle of Bourbon.

Properties. The fruit of several species of Spondias is eatable in the West Indies, where they are called Hog Plums.

Example. Spondias.

CX. CONNARACEÆ.

Terebintaceæ, *Juss. Gen.* 368. (1789) *in part.* — Connaraceæ, *R. Brown in Congo*, 431. (1818); *Kunth in Ann. Sc. Nat.* 2. 359. (1824.) — Terebintaceæ, trib. 7. Dec. Prodr. 2. 84. (1825.)

Diagnosis. Polypetalous dicotyledons, with definite hypogynous stamens, anthers bursting by longitudinal slits, distinct simple carpella, exstipulate leaves without pellucid dots, no albumen, and terminal stigmas.

Anomalies.

Essential Character. — *Flowers* hermaphrodite, rarely unisexual. *Calyx* 5-parted, regular, persistent; *æstivation* either imbricate or valvular. *Petals* 5, inserted on the calyx, imbricated, rarely valvate in æstivation. *Stamens* twice the number of petals, hypogynous, those opposite the petals shorter than the others; *filaments* usually monadelphous. *Ovarium* solitary and simple, or several, each with a separate style and stigma; *ovula* 2, collateral, ascending; *styles* terminal; *stigmas* usually dilated. *Fruit* dehiscent, single, or several together, splitting lengthwise internally. *Seeds* erect, in pairs or solitary, with or without albumen, often with an arillus; *radicle* superior, at the extremity opposite the hilum; *cotyledons* thick in the species without albumen, foliaceous in those with albumen. — *Trees* or *shrubs*. *Leaves* compound, not dotted, alternate, without stipulæ. *Flowers* terminal and axillary, in racemes or panicles, with bracteæ.

Affinities. Connarus can only be distinguished from Leguminosæ by the relation the parts of its embryo have to the umbilicus of the seed, (*Brown in Congo*, 432.); that is to say, by the radicle being at the extremity most remote from the hilum. This observation must, however, be understood to refer only to some particular cases in Leguminosæ, and also to the fructification; the want of stipulæ and regular flowers being usually sufficient to point them out. From Anacardiaceæ and other Terebintaceous orders they are at once known by the total want of resinous juice.

GEOGRAPHY. All found in the tropics of Asia, Africa, and America.

PROPERTIES. Unknown.

EXAMPLES. Connarus, Omphalobium.

CXI. AMYRIDEÆ.

TEREBINTACEÆ, *Juss. Gen.* 368. (1789) *in part.*—AMYRIDEÆ, *R. Brown in Congo*, 431. (1818); *Kunth in Ann. Sc. Nat.* 2. 353. (1824).—TEREBINTACEÆ, trib. 5. *Dec. Prodr.* 2. 81. (1825.)

DIAGNOSIS. Polypetalous dicotyledons, with definite hypogynous stamens, anthers bursting by longitudinal slits, distinct simple carpella, exstipulate dotted leaves, and no albumen.

ANOMALIES.

ESSENTIAL CHARACTER.—*Calyx* small, regular, persistent in 4 divisions. *Petals* 4, hypogynous, with imbricated æstivation. *Stamens* double the number of the petals, hypogynous. *Ovarium* superior, 1-celled, seated on a thickened disk; *stigma* sessile, capitate; *ovules* 2, pendulous. *Fruit* indehiscent, sub-drupaceous, 1-seeded, glandular. *Seed* without albumen; *cotyledons* fleshy; *radicle* superior, very short.— *Trees* or *shrubs*, abounding in resin. *Leaves* opposite, compound, with pellucid dots. *Inflorescence* axillary and terminal, panicled. *Pericarpium* covered with granular glands, filled with an aromatic oil.

AFFINITIES. The general structure of this order is that of Anacardiaceæ, but in qualities it more nearly resembles Burseraceæ. M. Kunth suggests its relation to Aurantiaceæ, to which its dotted leaves, capitate stigmas, and pericarpia filled with reservoirs of oil, appear to approximate it.

GEOGRAPHY. Natives exclusively of the tropics of India and America, with the exception of one species found in Florida.

PROPERTIES. Fragrant resinous shrubs. The Gum Elemi Tree of Nevis is, according to Dr. Hamilton, a plant related to the genus Amyris, which he calls A.? hexandra. *Prodr. Fl. Ind.* 35. The gum-resin, called Bdellium, is probably produced by a species of Amyris, the Niouttout of Adanson, according to Virey. *Hist. Nat. des Méd.* 291. The layers of the liber of a species of Amyris were found by M. Cailliaud to be used by the Nubian Mahometans as paper, on which they write their legends. *Delile Cent.* 13. Amyris toxifera is said to be poisonous. *Dec.* Resin of Coumia is produced by A. ambrosiaca. *Ibid.*

EXAMPLE. Amyris.

CXII. BURSERACEÆ.

TEREBINTACEÆ, *Juss. Gen.* 368. (1789) *in part.*—BURSERACEÆ, *Kunth in Ann. Sc. Nat.* 2. 333. (1824).—TEREBINTACEÆ, trib. 4. *Dec. Prodr.* 2. 75. (1825.)

DIAGNOSIS. Polypetalous dicotyledons, with 2 or 4 times as many perigynous stamens as petals, concrete carpella, a superior ovarium of several cells, regular flowers, an annular disk, collateral ovules, and pinnated alternate leaves without pellucid dots.

ANOMALIES.

ESSENTIAL CHARACTER.— *Flowers* hermaphrodite, occasionally unisexual. *Calyx* persistent, somewhat regular, with from 2 to 5 divisions. *Petals* 3-5, inserted below a disk arising from the calyx; *æstivation* usually valvate. *Stamens* 2 or 4 times as many as the petals, perigynous, all fertile. *Disk* orbicular or annular. *Ovarium* 2-5-celled, superior, sessile; *style* 1 or 0; *stigmas* equal in number to the cells; *ovula* in pairs, attached to the axis, collateral. *Fruit* drupaceous, 2-5-celled, with its outer part often splitting into valves. *Seeds* without albumen; *cotyledons* either wrinkled and plaited, or fleshy; *radicle* superior, straight, turned towards the hilum.— *Trees* or *shrubs*, abounding in balsam, resin, or gum. *Leaves* alternate, unequally pinnate, occasionally with *stipulæ*, usually without pellucid dots. *Flowers* axillary or terminal, in racemes or panicles.

AFFINITIES. Differ from Anacardiaceæ, to which they are closely allied in their compound ovarium and pinnated leaves, and also in the very generally valvate æstivation of the calyx.

GEOGRAPHY. Exclusively natives of tropical India, Africa, and America.

PROPERTIES. They have all an abundance of fragrant resinous juice, which is, however, destitute of the acridity and staining property of Anacardiaceæ. The resin of Boswellia is used in India as frankincense, and also as pitch. It is hard and brittle, and, according to Dr. Roxburgh, is boiled with some low-priced oil to render it soft and fit for use. The native doctors prescribe it, mixed with ghee (clarified butter), in cases of gonorrhœa, and also in what they call Ritta Kaddapoo, which signifies flux accompanied with blood. The wood is heavy, hard, and durable. *Ainslie*, 1. 137. The Boswellia serrata, called Libanus thurifera by Colebrooke, produces the gum-resin Olibanum, a substance chiefly used as a grateful incense, but which also possesses stimulant, astringent, and diaphoretic properties. *Ibid*. 1. 267. A kind of coarse resin is obtained from Boswellia glabra, and is used boiled with oil for pitching the bottom of ships. *Ibid*. The Bursera paniculata, called Bois de Colophane in the Isle of France, gives out, from the slightest wound in the bark, a copious flow of limpid oil of a pungent turpentine odour, which soon congeals to the consistence of butter, assuming the appearance of camphor. *Brewster*, 2. 182. The gum of Canarium commune has the same properties as those of the Balsam of Copaiva; the three-cornered nuts are eaten in Java both raw and dressed, and an oil is expressed from them, which is used at table when fresh, and for burning when stale. The raw nuts are, however, apt to bring on diarrhœa. *Ainslie*, 2. 60. Balsam of Acouchi is produced by Icica acuchini, Gum elemi by Icica heptophylla, Balm of Gilead by Balsamodendron Gileadense, Opobalsamum or Balsam of Mecca by B. opobalsamum, a substance like Gum elemi by Icica Icicariba, and Carana, and a yellow concrete essential oil by Bursera acuminata.

EXAMPLES. Boswellia, Bursera, Balsamodendrum.

CXIII. ANACARDIACEÆ. THE CASHEW TRIBE.

TEREBINTACEÆ, *Juss. Gen.* 368. (1789) *in part.*— CASSUVIEÆ or ANACARDIEÆ, *Brown in Congo*, 431. (1818).— TEREBINTACEÆ, *Kunth in Ann. des Sc. Nat.* 2. 333. (1824.) Trib. 1 and 2. *Dec. Prodr.* 2. 62. &c. (1825); *Juss. Dict. des Sc. Nat.* v. 53. (1828.)

DIAGNOSIS. Polypetalous dicotyledons, with perigynous stamens, a superior simple ovarium, solitary exalbuminous seeds, and alternate exstipulate leaves without pellucid dots.

ANOMALIES. There is, according to Mr. Brown (*Congo*, 431.), an unpublished genus of this order, with ovarium inferior. The stamens of Melanorhæa are indefinite and hypogynous.

ESSENTIAL CHARACTER. — *Flowers* usually unisexual. *Calyx* usually small and persistent, with 5, or occasionally 3-4, or 7 divisions. *Petals* equal in number to the segments of the calyx, perigynous, (occasionally wanting,) imbricated in æstivation. *Stamens* equal in number to the petals and alternate with them, or twice as many or even more, equal or alternately shorter, or partly sterile ; *filaments* distinct, or in the genera without a disk cohering at the base. *Disk* fleshy, annular or cup-shaped, hypogynous, occasionally wanting. *Ovarium* single, very rarely 5 or 6, of which 4 or 5 are abortive, superior, (very rarely inferior), 1-celled ; *styles* 1 or 3, occasionally 4, sometimes none ; *stigmas* as many ; *ovulum* solitary, attached by a cord to the bottom of the cell. *Fruit* indehiscent, most commonly drupaceous. *Seed* without albumen ; *radicle* either superior or inferior, but always directed towards the hilum, sometimes curved suddenly back ; *cotyledons* thick and fleshy, or leafy. — *Trees* or *shrubs*, with a resinous, gummy, caustic, or even milky juice. *Leaves* alternate, simple, or ternate or unequally pinnate, without pellucid dots. *Flowers* terminal or axillary, with bracteæ.

AFFINITIES. The order called Terebintaceæ by Jussieu and many other botanists has been broken up into several by Brown and Kunth, but preserved entire by Decandolle, who does not, however, appear to have devoted particular attention to the subject. I follow the former botanists, abandoning altogether the name Terebintaceæ, which is about equally applicable to either Anacardiaceæ, Burseraceæ, Connaraceæ, Spondiaceæ, or Amyrideæ, the five orders which have been formed at its expense. All these are nearly related to each other, and whatever affinity is borne by one of them will be participated in by them all in a greater or less degree. They are distinguished from Rhamneæ by their resinous juice, superior ovarium, imbricated calyx, and stamens not opposite the petals ; from Celastrineæ by several of the same characters, and want of albumen; from Rosaceæ and Leguminosæ by their definite stamens, dotted leaves, very minute stipulæ if any, resinous juice, dotted leaves, solitary ovula, or by some one or other of these characters. To Diosmeæ they approach very nearly, and also to Xanthoxyleæ, from which some of them differ in their perigynous stamens. Melanorhæa is remarkable for its indefinite stamens, and especially for its hypogynous petals becoming enlarged, foliaceous, and deep red as the fruit advances to maturity.

GEOGRAPHY. Chiefly natives of tropical America, Africa, and India ; a few are found beyond the tropics, both to the north and the south. Pistacias and some species of Rhus inhabit the south of Europe; many of the latter genus occupy stations in North America and Northern India, and also at the Cape of Good Hope; Duvaua and Schinus inhabit exclusively Chile and the adjacent districts.

PROPERTIES. Large trees, with inconspicuous flowers, abounding in a resinous, sometimes acrid, highly poisonous juice, are the ordinary representatives of this order, to which belong the Cashew Nut, the Pistacia Nut, and the Mango fruit. Some trees are celebrated for yielding a clammy juice, which afterwards turns black, and is used for varnishing in India. One kind is from the common Cashew nut. The varnish of Sylhet is chiefly procured from Semecarpus anacardium, the marking nut-tree of commerce; and the varnish of Martaban from a plant called by Dr. Wallich Melanorhæa usitatissima. All these varnishes are extremely dangerous to some constitutions; the skin, if rubbed with them, inflames and becomes covered with pimples that are difficult to heal; the fumes have been known to produce a painful swelling and inflammation of the skin, which, in a case recorded by Dr. Brewster, extended from the hands as far as the face and eyes, which became swelled to an alarming degree. I have known an instance of similar effects having been produced by roasting the nuts of Anacardium occidentale. But there are some constitutions that are not affected in any degree by such poisons. These varnishes are at first white, and afterwards become black. This has been ascertained by Dr. Brewster to arise from the recent varnish being an organised substance, consisting of an immense congeries of small parts, which disperse the sun's rays in all directions, like a thin film of unmelted tallow; while the varnish which has been exposed to the air loses its organised structure, becomes homogeneous, and then transmits the sun's rays of a rich, deep, uniform red colour. *Brewster*, 8. 100. The same is probably the substance mentioned by Dr. Ainslie (1. 190) as the Black Lac of the Burmah country, with which the natives lacker various kinds of ware. A valuable black hard varnish is obtained from Stagmaria verniciflua in the Indian archipelago : this resin is extremely acrid, causing excoriations and blisters if applied to the skin. *Ed. P. J.* 6. 400. A black varnish well known in India is manufactured from the nuts of Semecarpus anacardium and the berries of Holigarna longifolia. *Ibid.* 4. 450. The leaves of some species of Schinus are so filled with a resinous fluid, that the least degree of unusual repletion of the tissue causes it to be discharged; thus some of them fill the air with fragrance after rain; and S. Molle and some others expel their resin with such violence when immersed in water as to have the appearance of spontaneous motion, in consequence of the recoil. Schinus Arroeira is said by M. Auguste St. Hilaire to cause swellings in those who sleep under its shade. *Ibid.* 14. 267. The fresh juicy bark of the Arueira shrub (Schinus Molle) is used in Brazil for rubbing newly-made ropes with, which it covers with a very durable bright dark-brown coating. The juice of the same plant is applied by the Indians in diseases of the eyes. *Pr. Maxim. Trav.* 270. This last plant, and also Rhus coriaria, possess acid qualities. The fruit of Cassuvium occidentale and Anacardium orientale is said to exercise a singular effect upon the brain. *Virey Bull. Pharm.* 1814. p. 271. Mastich is the produce of Pistacia atlantica and Lentiscus; Scio turpentine is yielded by Pistacia Terebinthus; a substance like mastich is exuded by Schinus Molle, and the Peruvians use it for strengthening their gums. The juice of many species of Rhus is milky, stains black, and is sometimes extremely poisonous. Rhus coriaria is used by tanners. The bark of Rhus glabrum is considered a febrifuge, and is also employed as a mordant for red colours. Several Comocladias stain the skin black. *Dec.*

Decandolle distinguishes 2 sections of this order (*Prodr.* 2. 62.), viz.

1. ANACARDIEÆ.
Cotyledons thick, folded back upon the radicle.
EXAMPLES. Anacardium, Holigarna, Mangifera.

2. SUMACHINEÆ.
Cotyledons foliaceous. Radicles bent back upon their line of union.
EXAMPLES. Rhus, Mauria.

CXIV. XANTHOXYLEÆ.

TEREBINTACEÆ, *Juss. Gen.* 368. (1789) *in part.* — XANTHOXYLEÆ, *Nees and Martius in Nov. Act. Bonn.* 11. (1823); *Adrien de Jussieu Rutacées,* p. 114. (1825). — PTELEACEÆ, *Kunth Ann. des Sc.* 2. 354. (1824). — TEREBINTACEÆ, trib. 6. *Dec. Prodr.* 2. 82. (1825.)

DIAGNOSIS. Polypetalous dicotyledons, with definite hypogynous stamens, partially concrete carpella, an imbricated calyx, symmetrical unisexual flowers, definite pendulous ovules, capsular or drupaceous fruit, and exstipulate dotted leaves.

ANOMALIES. Many species have distinct carpella.

ESSENTIAL CHARACTER. — *Flowers* unisexual, regular. *Calyx* in 3, or more commonly in 4 or 5 divisions. *Petals* the same number, very rarely none, usually longer than the calyx; *æstivation* generally twisted, convolute. *Stamens* equal to the petals in number, or twice as many, arising from around the base of the stalk of the abortive carpella; in the female flowers wanting or imperfect. *Ovarium* made up of the same number of pieces as there are petals, or of a smaller number, either altogether combined, or more or less distinct; *ovules* in each cell 2, collateral, or one above the other, very seldom 4; *styles* more or less combined, according to the degree of cohesion of the carpella. *Fruit* either berried or membranous, sometimes of from 2 to 5 cells, sometimes consisting of several drupes or 2-valved capsules, of which the sarcocarp is fleshy and partly separable from the endocarp. *Seeds* solitary or twin, pendulous, usually smooth and shining, with a testaceous integument; *embryo* lying within fleshy albumen; *radicle* superior; *cotyledons* ovate, flat. — *Trees* or *shrubs*. *Leaves* without stipulæ, alternate or opposite, either simple, or more commonly abruptly or unequally pinnate, with pellucid dots. *Flowers* axillary or terminal, gray, green, or pink. The various parts bitter or aromatic.

AFFINITIES. This is one of the families which comprehend genera with both distinct and concrete carpella; the latter are often entirely distinct, even in the ovarium; but most frequently there is a union, or at least a cohesion, of the styles, by which their tendency to concretion may be recognised. In a few instances the carpella are absolutely solitary. " The place originally assigned, and for a long time preserved, for most of the genera of Xanthoxyleæ, proves sufficiently how near the affinity is between them and Terebintaceæ. If, with Messrs. Brown and Kunth, the latter are divided into several orders, Xanthoxyleæ will be most immediately allied to Burseraceæ and Connaraceæ, agreeing with the former in the genera with a simple fruit, and with the latter in those with a compound one. Notwithstanding the distance which usually intervenes in classifications between Aurantiaceæ and Terebintaceæ, there are nevertheless many points of resemblance between them; Correa has pointed out a passage from one to the other through Cookia; Kunth, in new-modelling the genus Amyris, and in considering it the type of a distinct order, suspects its near affinity with Aurantiaceæ; we cannot, therefore, be surprised at the existence also of relations between the latter and Zanthoxyleæ. A mixture of bitter and aromatic principles, the presence of receptacles of oil that are scattered over every part, which give a pellucid dotted appearance to the leaves, and which cover the rind of the fruit with opaque spaces, — all these characters give the two families a considerable degree of analogy. This has already been indicated by M. de Jussieu in speaking of Toddalia, and in his remarks upon the families of Aurantiaceæ and Terebintaceæ; and it is confirmed by the continual mixture, in all large herbaria, of unexamined plants of Terebintaceæ, Xanthoxyleæ, and Aurantiaceæ. The fruit of the latter is, however, extremely different; their seeds resembling, as they do, Terebintaceæ, are on that very account at variance with Xanthoxyleæ, but at the same time establish a further point of affinity between them and some Rutaceous plants which are destitute of albumen. Unisexual flowers, fruit separating into distinct cocci, seeds solitary or twin in these cocci, enclosing a usually smooth and blackish integument, which is even sometimes hollowed out on its inner edge, a fleshy albumen surrounding an embryo the radicle of which is superior, are all points of analogy between Xanthoxyleæ and Euphorbiaceæ, particularly between those which have in their male flowers from 4 to 8 stamens inserted round the rudiment of a pistil, and in the female flowers cells with 2 suspended, usually collateral, ovules.

Finally, several Xanthoxyleous plants have in their habit, and especially in their foliage, a marked resemblance to the Ash. The diœcious flowers of Fraxinus, its ovarium, the two cells of which are compressed, having a single style, 2 ovules in the inside, and scales on the outside, and which finally changes into a samara which is 1-celled and 1-seeded by abortion, all establish certain points of contact between Ptelea and Fraxinus." *Ad. de Juss.*

GEOGRAPHY. Most of the species belong to America, especially to the tropical parts; some are found in temperate regions; they are rare in Africa; some exist in the Isles of France and Madagascar, many are natives of India and China, and 1 is found in New Holland.

PROPERTIES. Nearly all aromatic and pungent. The Fagaras are popularly called Peppers in the countries where they are found. Xanthoxylum Clava and fraxineum are powerful sudorifics and diaphoretics; they are remarkable, according to Barton, for their extraordinary power in exciting salivation, whether applied immediately to the gums or taken internally: these two plants are reputed to have been used successfully in paralysis of the muscles of the mouth and in rheumatic affections. Xanthoxylum caribæum is held to be a febrifuge. *Dec.* A plant called Coentrilho in Brazil (Xanthoxylum hiemale) is employed as a remedy for pain in the ear, for which purpose the powder of its bark is made use of. Its wood is very hard, and valuable for building. *Pl. Usuelles*, 37. The fruit of Ptelea has a strong, bitter, aromatic taste, and is said to have been used with some success as a substitute for hops. *Dec.* The bark of a species of Brucea is stated by Dr. Horsfield to be of a bitter nature, and to possess properties similar to those of Quassia Simarouba. *Ainslie*, 2. 105. The Brucea antidysenterica contains a poisonous principle called Brucia, which is similar in its effects to Strychnia, but 12 or 16 times less energetic than that alkali. *Turner*, 652.

EXAMPLES. Xanthoxylum, Toddalia, Blackburnia.

CXV. DIOSMEÆ. THE BUCKU TRIBE.

DIOSMEÆ, *R. Brown in Flinders*, (1814).—RUTACEÆ, *Dec. Prodr.* 1. 709. (1824) *chiefly.* — DIOSMEÆ, *Ad. de Jussieu Rutacées*, 1. 83. (1825). — FRAXINELLEÆ, *Nees and Martius Nov. Act. Bonn.* 11. 149. (1823). — CUSPARIEÆ, *Dec. Mem. Mus.* 9. 141. (1822); *Prodr.* 1. 729. (1824), *a § of* Rutaceæ.

DIAGNOSIS. Polypetalous dicotyledons, with definite hypogynous stamens, concrete carpella, an entire ovarium of several cells, an imbricated calyx, symmetrical hermaphrodite flowers, 2 ovules, endocarp separable from the sarcocarp as a 2-valved coccus, and exstipulate dotted leaves.

ANOMALIES. Some of the genera are monopetalous, others have the carpella in great part distinct. Empleurum has no petals. Dictamnus and some others have irregular flowers and more ovules than 2. According to Mr. Brown, there is a New Holland genus, with perigynous stamens, 10 segments of the calyx, 10 petals, and indefinite stamens.

ESSENTIAL CHARACTER. — *Flowers* hermaphrodite, regular or irregular. *Calyx* in 4 or 5 divisions. *Petals* either as many as the divisions of the calyx, distinct, or combined into a kind of spurious monopetalous corolla, or occasionally wanting; *æstivation* for the most part twisted-convolute, very rarely somewhat valvular. *Stamens* equal in number to the petals, or twice as many, or even fewer in consequence of abortion, hypogynous, very rarely perigynous, placed on the outside of a disk or cup surrounding the ovarium, and either free or combined with the base of the calyx, or sometimes obsolete. *Ovarium* sessile or stalked, its lobes equal to the number of petals, or fewer; *ovules* twin and collateral, or one above the other, very rarely 4; *style* single, occasionally divided towards the base into as many parts as there are lobes of the ovarium; *stigma* simple or dilated. *Fruit* consisting of several capsules, either cohering firmly or more or less distinct; the endocarp separating entirely from the sarcocarp, which is 2-valved; the former 2-valved also, the valves dividing at the base, but connected by a membrane which bears the seeds. *Seeds* twin or solitary, with a testaceous integument; *embryo* with a superior *radicle*, which is either straight or oblique, and *cotyledons* of variable form; *albumen* none.— *Trees* or *shrubs*, very rarely *herbaceous* plants. *Leaves* without stipulæ, opposite or alternate, simple or pinnate, covered with pellucid resinous dots. *Flowers* axillary or terminal. All the parts aromatic.

AFFINITIES. M. A. de Jussieu, from whose excellent memoir upon Rutaceæ I have borrowed the greater part of my remarks upon Rutaceæ, Zygophylleæ, Xanthoxyleæ, and Simarubaceæ, speaks thus of Diosmeæ (*Mém.* p. 19.): —

" Diosmeæ are the group to which Mr. Brown gives that name, with the exception, however, of some of the genera which he refers to it; and they are that by the characters of which botanists have generally defined Rutaceæ. It is not necessary to describe the floral envelopes, the stamens, the disk, or the structure of the seed, because these parts vary according to the sections, which are in part characterised by their differences, and they will be better examined in their respective places. But it is important to understand the ovaria, and especially the pericarp, the structure of which is very characteristic. The ovaria, whether combined by their central axis, or more or less distinct, always contain 2 ovula; if 4, or sometimes but 1 are found, this occurs only in genera stationed at the extreme limits of the group. They are collateral, or more frequently placed one above the other, and then one is usually ascending, and the other suspended. This position, which at first sight appears singular, is very natural ; for the ovary is usually pierced by the vessels of the style only in the middle, and it is at that point that the two ovules are inserted, both at nearly the same height. If, therefore, they are placed one above the other, it is indispensable that one should ascend, and the other descend. These ovules may be considered peritropal, rather than either ascending or suspended, or, in other terms, attached by their middle rather than by either extremity." —
" If the ovarium of a Diosmea is divided across, its coat will be found to consist of two layers, the outer rather the most fleshy, and the inner thin or almost absent on the side next the axis, the side which is traversed from bottom to top by the vessels of the peduncle. These vessels, at a certain height, meet those of the style, either at the point of its insertion or below it; united to these, they penetrate the cavity of the cell, the shell of which they pierce, and there form funiculi, to which the ovula are attached. Thus far the structure of Diosmeæ is little different from that of other Rutaceous plants. But this becomes modified as the ovarium advances towards the state of fruit. The endocarp hardens by degrees, and at the same time separates from the sarcocarp. Its form resembles that of a bivalve shell, and may be more especially compared to that of a mussel; it presents two extremities, one superior, the other inferior, two lateral faces which are more or less convex, and two edges more or less acute, which unite them, the one external, the other internal. The two valves are woody and touch at the edges, except perhaps at a part of their inside where they are separated; this space is filled by a membrane which passes from one to the other : it is either slightly fleshy, or, which is more common, extremely thin, thickened in the middle by the passage of the vessels of the seed which penetrate it; and as, after having pierced it, they are almost immediately inserted into the seed, the latter appears to be actually borne by the membrane itself. When the fruit is perfectly ripe,

the sarcocarp of each cell opens from above inwards, following a longitudinal furrow, which had become visible some time previously. Its inner surface is seen to be covered by projecting lignified vessels, which are directed obliquely from the inner edge towards the outer, and are indicated externally by some transverse projections. The endocarp is loose in the inside of the shell, unless at its membrane, by means of which it continues to preserve some degree of adhesion with the other parts; but it soon opens, the two valves separate in different directions, and force out the seeds. When this separation takes place, the membrane is torn all round, and either falls away or sticks to the seed. In the latter case it is found attached to the hilum, if one seed only has ripened; but then in removing it, the remains of the abortive ovule may be found on one side. If both seeds have arrived at maturity, they are usually seen one resting on the other by their contiguous flattened extremities, and the membrane extends along their inner edge, being enlarged at their point of contact, where two transverse prolongations are perceptible."

M. A. de Jussieu then proceeds to point out the inaccuracy of calling, with some, this endocarp an arillus,—a name which, as Auguste St. Hilaire somewhere remarks, has been applied to as many different things as the Linnean term nectarium; or, with others, applying the same name to the persistent membrane.

Diosmeæ are nearly related to Rutaceæ, from which they differ in the remarkable structure of their fruit, and in having two ovula in each cell; with Humiriaceæ they have an analogy through the tribe called Cuspariæ, some of which have monadelphous stamens; with Aurantiaceæ they agree in their dotted leaves, definite stamens, occasional production of double embryos, fleshy disk, and sometimes in habit in the tribe of Cuspariæ. Xanthoxyleæ and Simarubaceæ accord with them in a multitude of points.

GEOGRAPHY. One genus, Dictamnus, is found in the south of Europe. The Cape of Good Hope is covered with different species of Diosma and nearly allied genera; New Holland abounds in Boronias, Phebaliums, Correas, Eriostemons, and the like; great numbers inhabit the equinoctial regions of America.

PROPERTIES. The Diosmas, or Bucku plants, of the Cape, are well known for their powerful and usually offensive odour; they are recommended as antispasmodics. The American species possess, in many cases, febrifugal properties. There is an excellent bark of this nature, used by the Catalan Capuchin friars of the missions on the river Carony in South America, called the Quina de la Guayna, or de la Angostura, or Angostura bark: this, which has been successively ascribed to Brucea ferruginea and two species of Magnolia, is now known to be the produce of Cusparia febrifuga (Bonplandia trifoliata *W.*), a plant of this family. *Humb. Cinch. For.* p. 38. *Eng. ed.* Evodia febrifuga, one of the Quinas of Brazil, has a bark so powerfully febrifugal as to compete with that of Cinchona. A bark much spoken of by the miners of Brazil, under the name of Casca de larangeira da terra, and in which Cinchonine was detected by Dr. Gomez, probably belongs to this tree. *Pl. Usuelles*, no. 4. One of the Quinas of Brazil is the Ticorea febrifuga: its bark is a powerful medicine in intermittent fevers. *Ibid.* 16. Hortia Braziliana possesses similar properties, but in a less degree. *Ibid.* 17. An infusion of the leaves of Ticorea jasminiflora is drank in Brazil as a remedy for the disease called by the Brazilian Portuguese Bobas, and by the French Frambæsia. *A. St. Hil. Hist.* 141. Dictamnus abounds in volatile oil to such a degree, that the atmosphere surrounding it actually becomes inflammable in hot weather. Its root was formerly employed as a sudorific and vermifuge.

A. de Jussieu divides the species of this order geographically, and, what is very singular, he finds their fructification in accord with their geographical distribution. His sections are:

1. EUROPEAN.
One from the south of Europe.

2. CAPE.
All from the Cape of Good Hope, and scarcely extending beyond the colony.

3. AUSTRALASIAN.
Inhabitants of New Holland, within or without the tropics, and Van Diemen's Island.

4. AMERICAN.
Sect. I. South America, New Zealand, the Friendly Islands, Mexico.
Sect. II. (Cusparieæ, *Dec.* Fraxinellæ, *Nees and Martius* chiefly.) South America, West Indies.
EXAMPLES. Diosma, Adenandra, Agathosma, Monniera, Ticorea.

CXVI. RUTACEÆ. THE RUE TRIBE.

RUTÆ, *Juss. Gen.* 296. (1789) *in part.* — RUTACEÆ, *Dec. Prodr.* 1. 709. (1824) *in part.* — RUTEÆ, *Adrien de Juss. Rutacées,* 78. (1825.)

DIAGNOSIS. Polypetalous dicotyledons, with definite hypogynous stamens, concrete carpella, an entire ovarium of several cells, an imbricated calyx, symmetrical hermaphrodite flowers, capsular fruit, endocarp not separable from the sarcocarp, and exstipulate dotted leaves.

ANOMALIES. Cyminosma differs in habit from the rest.

ESSENTIAL CHARACTER. — *Flowers* hermaphrodite, regular. *Calyx* with 4 or 5 divisions. *Petals* alternate with the divisions of the calyx, with a twisted-convolute æstivation, rarely convolute, or twisted separately. *Stamens* 2 or occasionally 3 times as many as the petals, inserted round the base of the stalk of the calyx, which is sometimes disciform. *Ovarium* divided more or less deeply into 3 or 5 lobes, with from 3 to 5 cells ; *ovules* in each cell 4, or from 4 to 20, pendulous, or attached to the axis ; *style* simple, or often (in the ovaries which are deeply lobed) separated at the base ; *stigma* 3- or 5-cornered, or furrowed. *Capsule* either with 3 loculicidal valves, or with from 4 to 5 lobes, which open internally at the apex ; the sarcocarp not separable from the endocarp. *Seeds* often fewer than the ovules, pendulous or adnate, reniform, pitted, with a testaceous integument ; *embryo* lying within fleshy *albumen*, white or greenish ; *radicle* superior ; *cotyledons* flat. *Ad. J.* — *Herbaceous* plants, or *small shrubs. Leaves* without stipulæ (with one exception), alternate, simple, deeply lobed, or pinnate, commonly with pellucid dots. *Flowers* often with a centrifugal inflorescence, white, or more frequently yellow.

AFFINITIES. Allied to Zygophylleæ by Peganum, which A. de Jussieu actually places with Rutaceæ, although its stipulate leaves destitute of pellucid dots appear to determine its greatest affinity to be with Zygophylleæ. From Diosmeæ they differ in scarcely any thing except the dehiscence of their fruit.

GEOGRAPHY. Found in the south of Europe, whence they extend in our hemisphere as far as the limits of the Old World, following the southern part of the temperate zone, and very rarely advancing within the tropics. *Ad. de J.*

PROPERTIES. Their powerful odour and their bitterness characterise them; they act principally on the nerves. Common Rue, and another species, are said to be emmenagogue, anthelmintic, and sudorific.

EXAMPLES. Ruta, Peganum.

CXVII. CORIARIEÆ.

CORIARIEÆ, *Dec. Prodr.* 1. 739. (1824.)

DIAGNOSIS. Polypetalous dicotyledons, with definite hypogynous stamens, anthers bursting by longitudinal slits, 5 distinct simple carpella surrounding a fleshy axis, exstipulate leaves without pellucid dots, no albumen, filiform stigmas, and sepaloid petals.

ANOMALIES.

ESSENTIAL CHARACTER.—*Flowers* either hermaphrodite, or monœcious, or diœcious. *Calyx* campanulate, 5-parted, ovate. *Petals* 5, alternate with the lobes of the calyx, and smaller than they are, fleshy, with an elevated keel in the inside. *Stamens* 10, arising from the torus, 5 between the lobes of the calyx and the angles of the ovarium, 5 between the petals and the furrows of the ovarium; *filaments* filiform; *anthers* oblong, 2-celled. *Ovarium* seated on a thickish torus, 5-celled, 5-angled; *style* 0; *stigmas* 5, long, subulate; *ovula* solitary, pendulous. *Carpella* 5, when ripe close together but separate, indehiscent, 1-seeded, surrounded with glandular lobes. *Seed* pendulous; *albumen* none; *embryo* straight; *radicle* superior; *cotyledons* 2, fleshy. — *Shrubs*, with opposite square branches, often 3 on each side, 2 of them being secondary to an intermediate principal one. *Leaves* opposite, simple, 3-ribbed, entire, ovate, or cordate. *Buds* scaly. *Racemes* terminal, simple, leafy at the base; *pedicels* often with two little bracteæ in the middle.

AFFINITIES. Placed by M. Decandolle immediately after Ochnaceæ, with which the order no doubt agrees, in having its ovaria distinct, and surrounding a fleshy axis; but the stigmata in Coriarieæ are long, linear, and distinct, with no style, while Ochnaceæ have a single style connecting the carpella and minute stigmas; the former, therefore, are apocarpous, the latter syncarpous. Coriarieæ are also certainly allied to Rutaceæ, but they differ from them as they do from Ochnaceæ; and besides, the carpella are in Rutaceæ connate. With Connaraceæ they agree in several points, while they are different in others. Upon the whole, their exact affinity may be considered unsettled.

M. Decandolle understands Coriaria as apetalous, but I do not see upon what principle, either of structure or analogy. In his *Essai sur les Propriétés Médicales* he referred it to the vicinity of Rhamneæ, p. 350. Jussieu referred it to Malpighiaceæ.

GEOGRAPHY. 4 from Peru, 1 from the south of Europe and north of Africa, 1 from New Zealand, and 1 from Mexico.

PROPERTIES. Coriaria myrtifolia is used by dyers for staining black. Its fruit is poisonous. It is said that several soldiers of the French army in Catalonia were affected by eating it; 15 became stupified, and 3 died. *Dec.*

EXAMPLE. Coriaria.

CXVIII. OCHNACEÆ.

OCHNACEÆ, *Dec. Ann. Mus.* 17. 398. (1811); *Prodr.* 1. 735. (1824.)

DIAGNOSIS. Polypetalous dicotyledons, with hypogynous stamens, and a deeply lobed ovarium, the style arising from the base of the concrete carpella, which are seated upon a succulent disk; anthers opening by pores.

ANOMALIES. Stamens definite or indefinite.

ESSENTIAL CHARACTER. — *Sepals* 5, persistent, imbricated in æstivation. *Petals* hypogynous, definite, sometimes twice as many as the sepals, deciduous, spreading, imbricated in æstivation. *Stamens* 5, opposite the sepals, or 10, or indefinite in number, arising from a hypogynous disk; *filaments* persistent; *anthers* 2-celled, innate, opening by pores. *Carpella* equal in number to the petals, lying upon an enlarged, tumid, fleshy disk (the *gynobase*); their *styles* combined in one; *ovula* erect. *Fruit* composed of as many pieces as there were carpella, indehiscent, somewhat drupaceous, 1-seeded, articulated with the gynobase, which grows with their growth. *Seeds* without albumen; *embryo* straight; *radicle* short; *cotyledons* thick. —Very smooth *Trees* or *shrubs*, having a watery juice. *Leaves* alternate, simple, entire, or toothed, with 2 stipulæ at the base. *Flowers* usually in racemes, with an articulation in the middle of the pedicels.

AFFINITIES. Very near Rutaceæ, from which they are distinguished by their erect ovula, the dehiscence of their anthers, and many more characters. They are to Polypetalæ what Labiatæ and Boragineæ are to Monopetalæ.

GEOGRAPHY. All found in tropical India, Africa, and America.

PROPERTIES. Walkera serrata has a bitter root and leaves, and is employed in Malabar, in decoction in milk or water, as a tonic, stomachic, and anti-emetic. The bark of Ochna hexasperma is used in Brazil as a cure of the sores produced in cattle by the punctures of insects. It probably acts as an astringent. *Pl. Usuelles*, 38.

EXAMPLES. Ochna, Gomphia.

CXIX. ZYGOPHYLLEÆ. THE BEAN CAPER TRIBE.

ZYGOPHYLLEÆ, *R. Brown in Flinders*, (1814); *Dec. Prodr.* 1. 703. (1824); *Adrien de Juss. Rutacées*, 67. (1825.)

DIAGNOSIS. Polypetalous dicotyledons, with definite hypogynous stamens, concrete carpella, an entire ovarium of several cells, an imbricated calyx, symmetrical flowers, pendulous ovules, stamens arising from hypogynous scales, and opposite stipulate leaves without pellucid dots.

ANOMALIES. Ovules occasionally erect. Tribulus has the fruit separating into spiny nuts, with transverse phragmata, and no albumen. Melianthus has very irregular flowers.

ESSENTIAL CHARACTER. —*Flowers* hermaphrodite, regular. *Calyx* divided into 4 or 5 pieces, with convolute æstivation. *Petals* unguiculate, alternate with the segments of the calyx and a little longer, in æstivation, which is usually convolute, at first very short and scale-like. *Stamens* double the number of the petals, dilated at the base, sometimes naked, sometimes placed on the back of a small scale, hypogynous. *Ovarium* simple, surrounded at the base with glands or a short sinuous disk, more or less deeply 4- or 5-furrowed, with 4 or 5 cells; *ovula* in each cell 2 or more, attached to the inner angle, pendulous, or occasionally erect; *style* simple, usually with 4 or 5 furrows; *stigma* simple, or with 4 or 5 lobes. *Fruit* capsular, rarely somewhat fleshy, with 4 or 5 angles or wings, bursting by 4 or 5 valves bearing the dissepiments in the middle, or into as many close cells; the sarcocarp not separable from the endocarp. *Seeds* usually fewer than the ovules, either compressed and scabrous when dry, or ovate and smooth, with a thin herbaceous integument. *Embryo* green ; *radicle* superior ; *cotyledons* foliaceous ; *albumen* whitish, between horny and cartilaginous, in Tribulus wanting. *Ad. J.* — *Herbaceous* plants, *shrubs,* or *trees*, with a very hard wood, the branches often articulated at the joints. *Leaves* opposite, with stipulæ, very seldom simple, usually unequally pinnate, not dotted. *Flowers* solitary, or in pairs or threes, white, blue, or red, often yellow.

AFFINITIES. Nearly related to Oxalideæ, from which, however, they are distinguished by a multitude of characters. With Simarubaceæ they accord in the stamens springing from the back of a hypogynous scale; a structure well worth more attentive consideration than it has yet received. Something analogous to it will be found in Caryophylleæ. M. Adrien de Jussieu also observes that the petals are remarkable for their being, in an early state, minute and hidden by the calyx, which they only exceed about the time of flowering, while in other Rutaceous orders the petals are always larger than the calyx. The distinguishing characters in its vegetation or habit are, the leaves being constantly opposite, with lateral or intermediate stipulæ, being generally compound, and always destitute of the pellucid glands which universally exist in true Diosmeæ. *Brown in Denham*, 26. It is also a very common character of the order to have the radicle at that extremity of the seed which is most remote from the hilum ; but this, which is of great importance in many natural families, is of less value in Zygophylleæ. (See many good remarks upon this subject in Mr. Brown's *Appendix to Denham*, p. 27.)

Biebersteinia, appended to this order by A. de Jussieu, is a genus that requires further examination.

GEOGRAPHY. Guaiacum, Porlieria, and Larrea, are peculiar to America. Fagonia is distributed over the south of Europe, the Levant, Persia, and India. Zygophyllum inhabits the same regions, and also the south of Africa, and is represented in New Holland by Röpera. Tribulus is found in all the Old World within the tropics, or in countries bordering upon them. *Ad. de J.* Melianthus, a most anomalous genus, is remarkable for being found both at the Cape of Good Hope and in Nipal, without any intermediate station.

PROPERTIES. Zygophyllum Fabago is sometimes employed as an anthelmintic. The ligneous plants of the order are remarkable for the extreme hardness of their wood. All the Guaiacums are well known for their exciting properties; the bark and wood of Guaiacum sanctum and officinale have a somewhat bitter and acrid flavour, and are principally employed as sudorifics, diaphoretics, or alteratives ; they contain a particular matter often designated as resin or gum-resin, but which is now considered a distinct substance, called Guaiacine. *Dec.* The wood of Guaiacum officinale, or Lignum vitæ, is remarkable for the direction of its fibres, each layer of which crosses the preceding diagonally ; a circumstance first pointed out to me by Professor Voigt.

EXAMPLES. Zygophyllum, Tribulus.

CXX. SIMARUBACEÆ. THE QUASSIA TRIBE.

SIMARUBACEÆ, *Rich. Anal. de Fr.* 21. (1808). — SIMARUBEÆ, *Dec. Diss. Ochn. Ann. Mus.* 17. 323. (1811); *Prodr.* 1. 733. (1824); *Adrien de Juss. Rutacées*, 129. (1825.)

DIAGNOSIS. Polypetalous dicotyledons, with definite hypogynous stamens, concrete carpella, an entire ovarium of several cells, an imbricated

calyx, symmetrical flowers, solitary pendulous ovules, stamens arising from hypogynous scales, and exstipulate leaves without dots.
Anomalies.

Essential Character. — *Flowers* hermaphrodite, or occasionally unisexual. *Calyx* in 4 or 5 divisions. *Petals* the same number, longer, either spreading or combined in a tube ; *æstivation* twisted. *Stamens* twice as many as the petals, each arising from the back of a hypogynous scale. *Ovarium* 4- or 5-lobed, placed upon a stalk from the base of which the stamens arise, 4- or 5-celled, each cell with 1 suspended ovulum ; *style* simple ; *stigma* 4- or 5-lobed. *Fruit* consisting of 4 or 5 drupes arranged around a common receptacle, indehiscent. *Seeds* pendulous, with a membranous integument ; *embryo* without *albumen* ; *radicle* superior, short, drawn back within the thick cotyledons. — *Trees* or *shrubs*. *Leaves* without stipulæ, alternate, occasionally simple, most usually compound without dots. *Peduncles* axillary or terminal. *Flowers* whitish, green, or purple. The different parts bitter.

Affinities. Akin to Zygophylleæ in their stamens inserted upon hypogynous scales, and to Ochnaceæ in their deeply-lobed ovarium, or nearly separate ovaria ; from these latter they are distinguished by their want of a succulent disk, their suspended not erect ovules, and their anthers bursting by longitudinal slits, not by terminal pores. A. de Jussieu says, " They are known from all Rutaceous plants by the co-existence of three characters ; namely, ovaria with but one ovulum, indehiscent drupes, and- exalbuminous seeds, the membranous integument of the embryo and the radicle being retracted within thick cotyledons."
Geography. All natives of tropical America, India, or Africa, with the exception of 1 Nipal plant.
Properties. All intensely bitter. The wood of Quassia is well known. A plant called Paraïba in Brazil, the Simaruba versicolor of St. Hilaire, possesses such excessive bitterness that no insects will attack it. Specimens of it placed among dried plants which were entirely devoured by the larvæ of a species of Ptinus, remained untouched. The Brazilians use an infusion in brandy as a specific against the bite of serpents, and also employ it with very great success to cure the lousy diseases to which people are very subject in those countries. *Pl. Usuelles*, no. 5.
Examples. Quassia, Simaruba.

CXXI. PITTOSPOREÆ.

Pittosporeæ, *R. Brown in Flinder's Voyage*, 2. 542. (1814); *Dec. Prodr.* 1. 345. (1824); *Ach. Rich. in Dict. Class.* 13. 643. (1828.)

Diagnosis. Polypetalous dicotyledons, with definite hypogynous stamens, distinct except at the base, concrete carpella, an ovarium of several cells with the placentæ in the axis, an imbricated calyx, symmetrical flowers, indefinite seeds with a minute embryo in fleshy albumen, and simple leaves.
Anomalies.

Essential Character. — *Sepals* 5, deciduous, either distinct or partially cohering ; *æstivation* imbricated. *Petals* 5, hypogynous, sometimes slightly cohering ; *æstivation* imbricated. *Stamens* 5, hypogynous, distinct, alternate with the petals. *Ovarium* single, distinct, with the cells or the placentæ 2 or 5 in number, and many-seeded ; *style* 1 ; *stigmas* equal in number to the placentæ. *Fruit* capsular or berried, with many-seeded cells, which are sometimes incomplete. *Seeds* often covered with a glutinous or resinous pulp ; *embryo* minute, near the hilum, lying in fleshy *albumen ; radicle* rather long ; *cotyledons* very short. — *Trees* or *shrubs*. *Leaves* simple, alternate, without stipulæ, usually entire. · *Flowers* terminal or axillary, sometimes polygamous.

Affinities. Mr. Brown, in establishing these as an order, remarks that they are widely different from Rhamneæ or Celastrineæ, but without pointing out their real affinity ; Decandolle places them between Polygaleæ and Frankeniaceæ ; according to Achille Richard they are very near Rutaceæ, to which he thinks them allied by a crowd of characters.

GEOGRAPHY. Chiefly New Holland plants. A few are found in Africa and the adjacent islands, and 1 in Nipal. Mr. Brown remarks that Pittosporum itself has been found not only in New Holland, but also in New Zealand, Norfolk Island, the Society and Sandwich Islands, the Moluccas, China, Japan, and even Madeira. *Flinders, 542.*

PROPERTIES. The wood of Senacia undulata is handsomely veined, whence it is called in the Mauritius Bois de joli cœur. *Dec.* The berries of Billardiera are eatable. The bark of Pittosporum Tobira has a resinous smell. Nothing is known of the properties of any.

EXAMPLES. Billardiera, Pittosporum, Bursaria.

CXXII. GERANIACEÆ. THE GERANIUM TRIBE.

GERANIA, *Juss. Gen.* 268. (1789). — GERANIACEÆ, *Dec. Fl. Fr.* 4. 828. (1805) ; *Prodr.* 1. 637. (1824) ; *Lindl. Synops.* 56. (1829.)

DIAGNOSIS. Polypetalous dicotyledons, with definite monadelphous hypogynous stamens, concrete carpella, an entire ovarium of several cells, an imbricated calyx, symmetrical flowers, solitary pendulous ovules, and carpella adhering to a woody axis, separating with elasticity and curling back.

ANOMALIES. Petals none in Rhyncotheca, which also has albumen. Flowers sometimes irregular.

ESSENTIAL CHARACTER. — *Sepals* 5, persistent, more or less unequal, with an imbricated æstivation ; 1 sometimes saccate or spurred at the base. *Petals* 5, seldom 4 in consequence of 1 being abortive, unguiculate, equal or unequal, either hypogynous or perigynous. *Stamens* usually monadelphous, hypogynous, twice or thrice as many as the petals ; some occasionally abortive. *Ovarium* composed of 5 pieces placed round an elevated axis, each 1-celled, 1-seeded ; *ovula* pendulous ; *styles* 5, cohering round the elongated axis. *Fruit* formed of 5 pieces, cohering round a lengthened indurated axis ; each piece consisting of 1 cell, containing 1 seed, having a membranous pericarpium, and terminated by an indurated style, which finally curls back from the base upwards, carrying the pericarpium along with it. *Seeds* solitary, pendulous, without albumen. *Embryo* curved ; *radicle* pointing to the base of the cell ; *cotyledons* foliaceous, convolute, and plaited. — *Herbaceous plants* or *shrubs. Stems* tumid, and separable at the joints. *Leaves* either opposite or alternate ; in the latter case opposite the peduncles.

AFFINITIES. In many points nearly related to Oxalideæ, Balsamineæ, and Tropæoleæ, with which they are by some botanists associated. They are, however, distinguished by the peculiar dehiscence of the fruit, their stems with tumid joints, their convolute plaited cotyledons, and habit. In the arrangement of their carpella about an elevated axis they agree with all those orders formerly comprehended under the common name of Rutaceæ, from which the length of that axis, and many other characters, distinguish them. Their analogy with Vites is pointed out in speaking of that order. In many respects they border close upon Malvaceæ.

GEOGRAPHY. Very unequally distributed over various parts of the world. A great proportion is found in the Cape of Good Hope, chiefly of the genus Pelargonium ; Erodium and Geranium are principally natives of Europe, North America, and Northern Asia, and Rhyncotheca of South America. It is remarkable that Pelargonium is found in New Holland.

PROPERTIES. An astringent principle and an aromatic or resinous flavour are the characteristics of this order. The stem of Geranium spinosum burns like a torch, and gives out an agreeable odour. The root of Geranium maculatum is considered a valuable astringent in North America, where it is sometimes called Alum root. *Barton*, 1. 155. In North Wales Geranium Robertianum has acquired celebrity as a remedy for nephritic complaints. *Ibid.* Some of the Pelargoniums are acidulous, but this genus is chiefly known as an object of garden culture, for which its great beauty, and the facility with which the species or supposed species intermix, render it well adapted.

EXAMPLES. Geranium, Monsonia, Erodium.

CXXIII. OXALIDEÆ. THE WOODSORREL TRIBE.

OXALIDEÆ, *Dec. Prodr.* 1. 689. (1824) ; *Lindl. Synops.* 59. (1829.)

DIAGNOSIS. Polypetalous dicotyledons, with definite hypogynous stamens distinct except at the base, concrete carpella, an ovarium of several cells with the placentæ in the axis, an imbricated calyx, symmetrical flowers, indefinite exalbuminous seeds with a straight embryo, and compound leaves.

ANOMALIES.

ESSENTIAL CHARACTER.—*Sepals* 5, sometimes slightly cohering at the base, persistent, equal. *Petals* 5, hypogynous, equal, unguiculate, with a spirally-twisted æstivation. *Stamens* 10, usually more or less monadelphous, those opposite the petals forming an inner series, and longer than the others ; *anthers* 2-celled, innate. *Ovarium* with 5 angles and 5 cells ; *styles* 5, filiform ; *stigmata* capitate or somewhat bifid. *Fruit* capsular, membranous, with 5 cells, and from 5 to 10 valves. *Seeds* few, fixed to the axis, enclosed within a fleshy integument, which curls back at the maturity of the fruit, and expels the seeds with elasticity. *Albumen* between cartilaginous and fleshy. *Embryo* the length of the albumen, with a long radicle pointing to the hilum, and foliaceous cotyledons.— *Herbaceous plants, undershrubs,* or *trees. Leaves* alternate, compound, sometimes simple by abortion, very seldom opposite or somewhat whorled.

AFFINITIES. Formerly included in Geraniaceæ, from which, in the judgment of many, they are not sufficiently distinct. According to M. Decandolle they are rather allied to Zygophylleæ ; an opinion in which I am inclined to concur, and which their compound leaves appear to confirm. Averrhoa differs from the rest in its arborescent habit. They are generally described with an arillus; but, according to M. Auguste St. Hilaire, the part so called is nothing but the outer integument of the seed. *Pl. Us.* 43.

GEOGRAPHY. Natives of all the hotter and temperate parts of the world, most abundantly known in America and the Cape of Good Hope, and most rarely in the East Indies and equinoctial Africa.

PROPERTIES. Averrhoa Bilimbi and the pinnated Oxalis called Biophytum have sensitive leaves. Their foliage is generally acid, so that they are fit to supply the place of sorrel. Some of the species are astringent, and have been employed in spitting of blood. Oxalis acetosella contains pure oxalic acid. *Turner*, 623. Several species of Oxalis are used in Brazil against malignant fevers. *Pl. Usuelles*, 43. The fruit of Averrhoa is intensely acid. A species of Oxalis found in Columbia bears tubers like a potato, and is one of the plants called Arracacha.

EXAMPLES. Oxalis, Biophytum, Averrhoa.

CXXIV. TROPÆOLEÆ. The Nasturtium Tribe.

Tropæoleæ, *Juss. Mem. Mus.* 3. 447. (1817) ; *Dec. Prodr.* 1. 683. (1824.)

Diagnosis. Polypetalous dicotyledons, with definite hypogynous distinct stamens, concrete carpella, an ovarium of 3 cells with the placentæ in the axis, an imbricated calyx with 1 of the sepals spurred, unsymmetrical flowers, definite pendulous ovules, and indehiscent fruit.

Anomalies. Magallana has winged fruit, 1-celled and 1-seeded by abortion. In Trop. pentaphyllum, according to Aug. St. Hilaire (*Pl. Us.* 41.), the calyx is valvular, and the petals only 2.

Essential Character.—*Sepals* 5, the upper one with a long distinct spur; *æstivation* quincuncial. *Petals* 5, unequal, irregular, the 2 upper sessile and remote, arising from the throat of the calyx, the 3 lower stalked and smaller, sometimes abortive. *Stamens* 8, perigynous, distinct; *anthers* innate, erect, 2-celled. *Ovarium* 1, 3-cornered, made up of 3 carpella; *style* 1; *stigmas* 3, acute; *ovula* solitary, pendulous. *Fruit* indehiscent, separable into 3 pieces from a common elongated axis. *Seeds* large, without albumen, filling the cavity in which they lie; *embryo* large; *cotyledons* 2, straight, thick, consolidated together into a single body; *radicle* lying within projections of the cotyledons. — Smooth *herbaceous* plants, of tender texture and with an acrid taste, trailing or twining. *Leaves* alternate, without stipulæ, petiolate, with radiating ribs. *Peduncles* axillary, 1-flowered.

Affinities. Very near Geraniaceæ, with which they agree even in their spur (which in Pelargonium is often present, but adnate to the pedicel), and also Balsamineæ, and Hydrocereæ, from which they differ chiefly in the structure of their fruit.

Geography. All natives of South America, mostly upon high land.

Properties. The fleshy fruit of Tropæolum majus is acrid, and possesses the properties of Cress; and M. Decandolle remarks, that the caterpillar of the Cabbage butterfly feeds exclusively upon Cruciferæ and Tropæolum. The root of Tr. tuberosum is eaten in Peru. Tropæolum pentaphyllum is used in Brazil as an antiscorbutic, under the Portuguese name of Chagas da Miuda. *Pl. Usuelles*, 41.

Example. Tropæolum.

CXXV. HYDROCEREÆ.

Hydrocereæ, *Blume Bijdr.* 241. (1825.)

Diagnosis. Polypetalous dicotyledons, with definite hypogynous stamens, concrete carpella, an entire ovarium of several cells with placentæ in the axis, an imbricated calyx, one of the sepals of which is spurred, symmetrical flowers, definite pendulous ovules, and a drupaceous fruit.

Anomalies.

Essential Character.— *Sepals* 5, deciduous, coloured, unequal; the lowermost elongated into a spur. *Petals* 5, hypogynous, unequal; the upper arched. *Stamens* 5, hypogynous, connate at the apex; *anthers* slightly connate, 2-celled, bursting at the apex. *Ovarium* 5-celled, 5-angled, with 2 or 3 ovula in each cell; *stigmas* 5, sessile, acute. *Fruit* succulent, with 5 cells, each of which has a bony hard lining, and contains a single seed. *Seed* solitary, without albumen; *cotyledons* plano-convex; *radicle* superior. — *Herbaceous*. *Stems* angular. *Leaves* alternate, without stipulæ, serrated. *Peduncles* axillary, manyflowered.

Affinities. Closely related to Balsamineæ and Tropæoleæ, from which they are only distinguished by their symmetrical flowers and drupaceous fruit.

GEOGRAPHY. A single species, native of marshes and wet places in Java.

PROPERTIES. Unknown.

EXAMPLE. Hydrocera.

CXXVI. BALSAMINEÆ. THE BALSAM TRIBE.

BALSAMINEÆ, *Ach. Rich. Dict. Class.* 2. 173. (1822); *Dec. Prodr.* 1. 685. (1824); *Lindl. Synops.* 59. (1829.)

DIAGNOSIS. Polypetalous dicotyledons, with definite hypogynous stamens, concrete carpella, an ovarium of 5 cells with the placentæ in the axis, an imbricated calyx, unsymmetrical flowers with one of the sepals spurred, and indefinite ovules.

ANOMALIES.

ESSENTIAL CHARACTER. — *Sepals* 5, irregular, deciduous, the two inner and upper of which are connate, the lower spurred. *Petals* 4, hypogynous, united in pairs, so that apparently there are only 2 petals; the fifth wanting. *Stamens* 5, hypogynous; *filaments* subulate; *anthers* 2-celled, bursting lengthwise. *Ovarium* single; *stigma* sessile, more or less divided in 5; *cells* 5, many-seeded. *Fruit* capsular, with 5 elastic valves, and 5 cells formed by membranous projections of the placenta, which occupies the axis of the fruit, and is connected with the apex by 5 slender threads. *Seeds* numerous, suspended; *albumen* none; *embryo* straight, with a superior *radicle* and plano-convex *cotyledons*. — Succulent *herbaceous* plants. *Leaves* simple, opposite or alternate, without stipulæ. *Peduncles* axillary.

AFFINITIES. So nearly related to Geraniaceæ, of which it is, in the opinion of many, a mere section, that it is only distinguishable by the spurred calyx, polyspermous fruit, and unsymmetrical flowers. Tropæoleæ differ in their fruit, Oxalideæ in their compound leaves and symmetrical flowers. M. Kunth, in a memoir printed in 1827, was the first to point out the true structure of this family, which had been more or less misunderstood by all previous observers. I had overlooked this memoir at the time of the publication of my *Synopsis of the British Flora*, whence the old erroneous character is given in that work. The following is the substance of M. Kunth's remarks: — Linnæus attributed to the Impatiens Balsamina a calyx of 2 leaves, 5 unequal petals, a nectary, a single ovary, a sessile stigma, and a unilocular polyspermous capsule, opening in 5 valves. M. de Jussieu describes it nearly in the same way, with the exception of considering the capsule as having 5 cells, and the corolla as consisting of 4 petals, the lower of which is spurred. These erroneous characters have been reproduced by most authors. Dr. Hooker alone refers the part which has the spur to the calyx, which he consequently makes to consist of 3 pieces. M. Achille Richard has come nearest the truth in the *Dictionnaire Classique*, where he describes the calyx as consisting of 4 pieces, and the 4 petals united in pairs. The fact is, that the structure is usually this: the centre of the flower is occupied by an ovarium, surmounted by a stigma divided into 5 acute lobes. Around this stand 5 hypogynous stamens, placed in a single row and at equal distances from each other. Hence the normal number of the parts of the flower should be 5. The corolla, however, consists of 2 bifid petals placed right and left, with a wider space between their upper than their lower edges. Upon comparing the position of these with the stamens, it appears that each occupies the place of 3 stamens, whence it is impossible to doubt that they each consist of 2 soldered together. On the other hand, the space between them, which answers to 2 stamens, is an equal proof of the abortion of a fifth petal. And this view of the structure is confirmed by the sepals. Thus on the outside of each pair of petals, at their base, is found a leaflet, the situation of which is opposite a stamen; and opposite the space left by the abortion of the fifth petal is a large broad leaflet, made up by the union of 2 sepals. The position of the fifth sepal, which is that which is spurred, is between 2 petals and opposite a stamen.

GEOGRAPHY. Natives of damp places among bushes in the East Indies; 1 is found in Madagascar, 1 in Europe, 2 in North America, and 1 in Russia in Asia.

PROPERTIES. Chiefly remarkable for the elastic force with which the valves of the fruit separate at maturity, expelling the seeds. For a supposed explanation of this phenomenon, see Dutrochet *Nouvelles Recherches sur l'Exosmose et Endosmose.* According to Decandolle, they are diuretic.

EXAMPLE. Balsamina Impatiens.

CXXVII. VOCHYACEÆ.

VOCHYSIACEÆ, *Mart. Nov. Gen.* 1. 123. (1824). —VOCHYSIEÆ, *A. St. Hil. Mem. Mus.* 6. 265. (1820); *Dec. Prodr.* 3. 25. (1828.)

DIAGNOSIS. Polypetalous dicotyledons, with definite perigynous stamens, concrete carpella, and irregular flowers with a spurred calyx.

ANOMALIES. Ovarium either superior or inferior. The leaves of Salvertia have no stipulæ.

ESSENTIAL CHARACTER. — *Sepals* 4-5, combined at the base, imbricated in æstivation, the upper one calcarate. *Petals* 1, 2, 3, or 5, alternate with the segments of the calyx, and inserted into their base, unequal. *Stamens* 1-5, usually opposite the petals, rarely alternate with them, arising from the bottom of the calyx, for the most part sterile, 1 of them having an ovate fertile 4-celled anther. *Ovarium* superior, or partially inferior, 3-celled; *ovules* in each cell solitary or twin, attached to the base of the axis; *style* and *stigma* 1. *Capsule* 3-cornered, 3-celled, 3-valved, the valves bursting along their middle. *Seed* without albumen, erect; *embryo* straight, inverted; *cotyledons* large, foliaceous, convolute, plaited; *radicle* short, superior. — *Trees.* *Branches* opposite, when young 4-cornered. *Leaves* opposite, sometimes towards the extremities of the branches alternate, entire, with 2 stipulæ at the base. *Flowers* usually in terminal panicles or racemes.

AFFINITIES. "An order at present but ill understood, in habit and flower somewhat allied to Guttiferæ or Marcgraaviaceæ, but distinct from both in the stamens being inserted into the calyx; perhaps more directly connected with Combretaceæ, on account of the convolute cotyledons and inverted seeds; and even perhaps allied to some Onagrariæ, on account of the abortive solitary stamen." *Dec. Prodr.* 3. 25. Is not the order nearer Violaceæ? an affinity strongly pointed out by the irregular flowers, 3-locular ovarium, and stipulæ, but impeded by the perigynous insertion of the stamens.

GEOGRAPHY. Natives of equinoctial America, where they inhabit ancient forests, by the banks of streams, sometimes rising up mountains to a considerable elevation. They are often trees with large spreading heads.

PROPERTIES. Unknown.

EXAMPLES. Vochya, Amphilochia, Erisma.

CXXVIII. TREMANDREÆ.

TREMANDREÆ, *R. Brown in Flinders*, p. 12. (1814); *Dec. Prodr.* 1. 343. (1824.)

DIAGNOSIS. Polypetalous dicotyledons, with 8 or 10 hypogynous distinct stamens, concrete carpella, a 2-celled ovarium with a definite number of pendulous ovules, a calyx with valvate æstivation, anthers bursting by pores, and entire petals involute in æstivation.

ANOMALIES.

ESSENTIAL CHARACTER. — *Sepals* 4 or 5, equal, with a valvular æstivation, slightly cohering at the base, and deciduous. *Petals* equal in number to the sepals, with an involute æstivation, enwrapping the stamens, much larger than the calyx, and deciduous. *Stamens* hypogynous, distinct, 2 before each petal, and therefore either 8 or 10; *anthers* 2- or 4-celled, opening by a pore at the apex. *Ovarium* 2-celled; *ovules* from 1 to 3 in each cell, pendulous; *style* 1; *stigmas* 1 or 2. *Fruit* capsular, 2-celled, 2-valved; *dehiscence* loculicidal. *Seeds* pendulous, ovate, with a thickened appendage at the apex, but with no appendage about the hilum; *embryo* cylindrical, straight, in the axis of fleshy albumen; the *radicle* next the hilum. — Slender heath-like *shrubs*, with their hairs usually glandular. *Leaves* alternate or whorled, without stipulæ, entire or toothed. *Pedicels* solitary, axillary, 1-flowered.

AFFINITIES. Not very certain; many genera probably still remain to be discovered. According to Decandolle, they are related to Polygaleæ; from which they differ in a number of points, especially in their distinct stamens and regular flowers; agreeing with them in having a remarkable tumour, called a caruncula, at one end of the seeds, which are also definite and pendulous in both orders.

GEOGRAPHY. All natives of New Holland.

PROPERTIES. Unknown.

EXAMPLES. Tetratheca, Tremandra.

CXXIX. POLYGALEÆ. THE MILKWORT TRIBE.

POLYGALEÆ, *Juss. Ann. Mus.* 14. 386. (1809); *Mem. Mus.* 1. 385. (1815); *Dec. Prodr.* 1. 321. (1824); *Lindl. Synops.* 39. (1829); *Aug. de St. Hilaire and Moquin-Tandon Mem. Mus.* 17. 313. (1829.)

DIAGNOSIS. Polypetalous dicotyledons, with definite hypogynous stamens in one parcel, concrete carpella, an ovarium of 2 cells with the placentæ in the axis, an imbricated calyx, unsymmetrical flowers, definite pendulous ovules, and dehiscent fruit.

ANOMALIES. Sepals 4, and all petaloid in some Kramerias. Flowers generally monopetalous. Ovarium sometimes 1-celled by abortion. Fruit indehiscent in Mundia, Monnina, Securidaca, and Krameria. The latter has also no albumen. Stamens distinct in Krameria.

ESSENTIAL CHARACTER. — *Sepals* 5, very irregular, distinct, often glumaceous; 3 exterior, of which 1 is superior and 2 anterior; 2 interior (*the wings*) usually petaloid, and alternate with the upper and lower ones. *Petals* hypogynous, usually 3, of which 1 is anterior and larger than the rest (*the keel*), and 2 alternate with the upper outer, and lateral inner sepals, and often connate with the keel; sometimes 5, and then the 2 additional ones minute and between the wings and the lower sepals. *Keel* sometimes entire, and then either naked or crested; sometimes 3-lobed, and then destitute of a crest. *Stamens* hypogynous, 8, usually combined in a tube, unequal, and ascending; sometimes 4, and distinct; the tube split opposite the upper sepal; *anthers* clavate, innate, mostly 1-celled and opening at their apex, sometimes 2-celled; very rarely the dehiscence is longitudinal. *Disk* either absent or present, regular or irregular. *Ovarium* superior, compressed, with 2 cells, which are anterior and posterior, the upper one occasionally suppressed; *ovules* solitary, very rarely twin, pendulous; *style* simple, curved, sometimes very oblique and cucullate at the apex, which is also entire or lobed; *stigma* simple. *Fruit* usually opening through the valves; occasionally indehiscent, membranous, fleshy, coriaceous, or drupaceous, winged or apterous. *Seeds* pendulous, with a caruncula next the hilum, naked or enveloped with hairs; the outer integument crustaceous, the inner membranous; *albumen* abundant, fleshy, rarely reduced to a thin gelatinous plate, very seldom wanting; *embryo* straight, or slightly curved, with the radicle next the hilum.—*Shrubs* or *herbaceous* plants. *Leaves* generally alternate, sometimes opposite, mostly simple, and always destitute of *stipules. Flowers* usually racemose, very often small and inconspicuous, but shewy in many Polygalas. *Pedicels* with 3 bracteæ.

AFFINITIES. The structure of this order has been admirably explained by Messrs. Aug. de St. Hilaire and Moquin-Tandon, from whose memoir above quoted, the foregoing character and almost all that is said here is extracted, and to which I refer those readers who wish to study the subject more intimately. Before adverting to the affinities of this order, it will be useful to consider what is the nature of the irregularity of the flowers; an irregularity which is such as to obscure, in a great measure, the relative position of the sepals and petals. The calyx apparently consists of but three pieces, which are usually green, and like sepals in their common state; but their real number is 5, the two coloured lateral petal-like bodies, sometimes lying within the apparent sepals, being in reality part of the series of the calyx. The corolla is mostly monopetalous, and, if carefully examined, formed of 3 pieces; namely, the keel and two petals, all soldered together. We have, therefore, an abortion of two petals, according to the laws of alternation: but this is not all; there is not only an abortion of two petals, but of those two which would, if present, be found right and left of the keel. The monopetalous corolla is, therefore, formed by the cohesion of the two posterior and the one anterior petal of a pentapetalous corolla, of which the two lateral petals are suppressed. The keel has an appendage of an anomalous character, called technically a crest, and often consisting of one or even two rows of fringes or divisions, originating not from the margin but from within it, and sometimes cohering in a common membrane at their base. M. de St. Hilaire has satisfactorily shewn that this crest is nothing more than the deeply-lobed middle segment of a keel, with these lobes in such a state of cohesion that the central lobe is pushed outwards, while the lateral ones cohere by their own margins and with its back. The stamens are only 8, two therefore are suppressed; or in Krameria 4, one being suppressed. I may remark, in addition, that the relative position of the fifth sepal and petal respectively, was first indicated by Mr. Brown. *Denham*, 31.

Polygaleæ are stationed by Decandolle between Droseraceæ and Tremandreæ, and in the immediate vicinity of Violaceæ. With the latter they are related on account of their hypogynous stamens, irregular flowers, and cucullate stigma; and with Tremandreæ on account of the caruncula of their seed. To Fumariaceæ they approach in the general aspect of their flowers; but if my theory of the structure of that order be admitted, their resemblance would not be so great as it appears to be. Leguminosæ are perhaps, notwithstanding their perigynous stamens, the order with which Polygaleæ have the greatest affinity: the irregularity of corolla is of a similar nature in both; there is in Leguminosæ a tendency to suppress the upper lateral petals, in Erythrina, as in Polygala; the ascending direction of the style and a cohesion of stamens are characters common to both orders. That part of the *Mémoires du Muséum* in which the second part of the paper above referred to is to appear, not having reached this country when the present sheet is sending to press, I have no means of knowing what the views of St. Hilaire and Moquin-Tandon are of the affinities of the tribe.

GEOGRAPHY. Most of the genera are limited to one or two of the five parts of the globe; thus Salomonia is only found in Asia, Soulamea in the Moluccas, Muraltia at the Cape of Good Hope, Krameria and Securidaca in the two Americas, and finally Monnina and Badiera in South America. Comesperma is found both in Brazil and New Holland, and, what is very remarkable, there is in the former country a species of the Cape genus Mundia. Polygala itself is found in four of the five parts; under the torrid zone and in temperate climates, at Cayenne, and on the mountains of Switzerland; it is, however, very unequally distributed. This genus inhabits almost every description of station, — dry plains, deep morasses, woods, mountains, cultivated and barren soils. Comesperma is only known in Brazilian woods, and Monnina and Krameria in open places.

PROPERTIES. Bitterness in the leaves and milk in the root are their usual characteristics; but the order has not been well investigated with respect to its qualities. Polygala senega root is stimulant, diuretic, sialagogue, expectorant, purgative, emetic, and sudorific, and also emmenagogue. It has been used with great success in croup. *Barton*, 2. 116. P. sanguinea, according to the same writer, possesses similar qualities. A peculiar vegetable principle, called Senegin, has been discovered by Gehlen in the root of Polygala senega, and M. Reschier is also said to have procured a principle called Polygaline from the same plant; but it is not known whether these two substances are the same. *Stephens and Church*, no. 103. The bark of Monnina polystachya, called Yallhoy in Peru, is stated to be extremely useful in cases of dysentery. It also possesses detersive properties in a great degree. The ladies of Peru ascribe the beauty of their hair to the use of its infusion, and the silversmiths of Huanaco employ it for cleansing and polishing their wrought silver. *Lambert's Illust. Cinch.* 132, &c. Krameria, a genus of an extremely anomalous structure, which, although most likely really belonging to the order, differs from it in many important points, is also remarkable for its tonic and excessively astringent qualities. Its root is sold in Europe under the name of Ratanhia, and is one of the substances which, in conjunction with gum kino, is used for adulterating port wine in England. According to M. Cadet, this root contains gallic acid, but neither tannin nor resin.

EXAMPLES. Polygala, Krameria, Monnina, Securidaca.

CXXX. VIOLACEÆ. THE VIOLET TRIBE.

VIOLARIEÆ, *Dec. Fl. Fr.* 4. 801. (1805); *Juss. Ann. Mus.* 18. 476. (1811); *Dec. Prodr.* 1. 287. (1824). — VIOLACEÆ, *Lindl. Synops.* 35. (1829).

DIAGNOSIS. Polypetalous dicotyledons, with definite hypogynous stamens, concrete carpella, a 1-celled ovarium with narrow parietal placentæ, 5 distinct sepals, an erect embryo, stipulate leaves, and a capsule with loculicidal dehiscence.

ANOMALIES. The berry of Pentaloba is 5-lobed, but there is only 1 style. The plants called Sauvageæ, if they really belong to the order, have a septicidal dehiscence.

ESSENTIAL CHARACTER.—*Sepals* 5, persistent, with an imbricate æstivation, usually elongated at the base. *Petals* 5, hypogynous, equal or unequal, usually withering, and with an obliquely convolute æstivation. *Stamens* 5, alternate with the petals, occasionally opposite them, inserted on a hypogynous disk, often unequal; *anthers* bilocular, bursting inwards, either separate or cohering, and lying close upon the ovarium; *filaments* dilated, elongated beyond the anthers; two, in the irregular flowers, generally furnished with an appendage or gland at their base. *Ovarium* 1-celled, many-seeded, or rarely 1-seeded, with 3 parietal placentæ opposite the 3 outer sepals; *style* single, usually declinate, with an oblique hooded *stigma*. *Capsule* of 3 valves, bearing the placentæ in their axis. *Seeds* often with a tumour at their base; *embryo* straight, erect, in the axis of fleshy *albumen.*—*Herbaceous* plants or *shrubs*. *Leaves* simple, usually alternate, sometimes opposite, stipulate, entire, with an involute *vernation*. *Inflorescence* various.

AFFINITIES. Mr. Brown, in speaking of Violaceæ, mentions, in his Appendix to the Congo Voyage, a genus, at that time unpublished, called Hymenanthera, having 5 scales alternating with the petals, with a bilocular berry, in each cell of which is a single pendulous seed. It appears very paradoxical to associate such a plant with an order otherwise well defined; and Mr. Brown himself seems to think it should be placed between Violeæ and Polygaleæ. The structure of this genus points out strongly the relation of Violaceæ to Polygaleæ, to the latter of which, however, it rather appears to me to be referable. These two orders differ from each other, in the latter having a 2-celled not 1-celled ovarium, leaves without stipulæ, and 1-celled anthers. Droseraceæ are known from Violaceæ by their numerous styles, minute embryo, circinate leaves, and want of stipulæ. Passifloreæ, to which the baccate genera of Violaceæ, and especially Corynostylis (Calyptrion, *Dec.*), which has a twining stem, undoubtedly approach, are distinguished by a multitude of characters. The irregular flowers, dilated filaments and sepals, and stipulate leaves, of Violaceæ, usually indicate them at once; but the regular-flowered fruticose genera, which constitute the tribe of Alsodineæ, are not to be recognised by a combination of such characters.

GEOGRAPHY. Of these tribes, Violeæ chiefly consist of European, Siberian, and American plants; a few only being found within the tropics of Asia. They are abundant in South America, the forms of which are, however, materially different from those of the more temperate parts of the world, most of them being shrubs, while the northern Violets are uniformly herbaceous, or nearly so. Alsodineæ are exclusively South American and African, with the exception (?) of Pentaloba, which, upon the authority of Loureiro, is Cochinchinese. Sauvageæ are exclusively South American or African.

PROPERTIES. The roots of all Violaceæ appear to be more or less emetic, a property which is strongly possessed by the South American species, and in a less degree only by those of Europe. Hence they form part of the herbs known under the name of Ipecacuanha. Ionidium parviflorum is used by the Spanish Americans, and I. Poaya by the Brazilians, as a substitute for Ipecacuanha. *Pl. Us.* 9. and 20. The root of another species, called Poaya, Poaya da praia, and Poaya branca, the Ionidium Itubu of Kunth, is commonly sold as true Ipecacuanha, to which it approaches very nearly in its properties. At Pernambuco it is esteemed the very best remedy that can be employed in dysentery; and the inhabitants of Rio-Grande-do-Norte consider it a specific against gout. *Ibid.* no. 11. The foliage of the Conohoria Lobolobo is used in Brazil for the same purposes as Spinach with us. Boiled, it becomes mucilaginous. *Ibid.* 10. Viola canina is reputed a powerful agent for the removal of cutaneous affections; and Anchietea salutaris is accounted by the Brazilians not only a purgative, but also a remedy against similar maladies. M. A. St. Hilaire remarks, that this notion deserves attention, as connected with the depurative properties ascribed in Europe to Viola canina, to which, although Anchietea is botanically related, there is nothing in its appearance which would have led the Portuguese settlers to attribute the virtues of the one to the other. *Ibid.* no. 19. Sauvagesia erecta is very mucilaginous, on which account it has been used in Brazil for complaints of the eyes, in Peru in disorders of the bowels, and in the Antilles as diuretic, or rather in cases of slight inflammation of the bladder.

The sections adopted by Decandolle are these : —

1. VIOLEÆ.

Petals unequal. Sepals 3 outer and broader, 2 interior. Fruit with a loculicidal dehiscence. Stamens alternate with the petals; filaments dilated, extended beyond the anthers, distinct (approximated or contracted), or occasionally connate; cells of the anthers finally 2-valved.

EXAMPLES. Calyptrion, Viola, Glossarrhen.

2. ALSODINEÆ. *R. Brown Congo*, p. 21. (1818.)

Petals unequal. Stamens usually either connected at the base, or adhering to the inside of an elevated cup, situated between the petals and stamens.

EXAMPLES. Conohoria, Rinorea, Ceranthera.

3. SAUVAGEÆ.

Dehiscence of the capsule septicidal. Stamens 5, fertile, opposite the petals, distinct; filaments neither dilated nor extended beyond the anthers. Scales 5, petaloid, alternate with the stamens. Intermediate between Violaceæ and Frankeniaceæ.

EXAMPLES. Sauvagesia, Lavradia.

CXXXI. PASSIFLOREÆ. THE PASSION FLOWER TRIBE.

PASSIFLOREÆ, *Juss. Ann. Mus.* 6. 102. (1805); *Id. Dict. des Sciences Nat.* 38. 48. (1825); *Dec. Prodr.* 3. 321. (1828); *Achille Richard Dict. Class.* 13. 95. (1828).

DIAGNOSIS. Polypetalous dicotyledons, with definite perigynous stamens, filamentous or membranous processes upon the tube of the calyx, concrete carpella, a superior 1-celled ovarium with parietal placentæ, corolla with an imbricated æstivation, glandular leaves, arillate seeds, and embryo in the midst of fleshy albumen.

ANOMALIES. Some apetalous.

ESSENTIAL CHARACTER.— *Sepals* 5, sometimes irregular, combined in a tube of variable length, the sides and throat of which are lined by filamentous or annular processes, apparently metamorphosed petals. *Petals* 5, arising from the throat of the calyx, on the outside of the filamentous processes, occasionally wanting, sometimes irregular, imbricated in æstivation. *Stamens* 5, monadelphous, rarely indefinite, surrounding the stalk of the ovarium; *anthers* turned outwards, linear, 2-celled, bursting longitudinally. *Ovarium* seated on a long stalk, superior, 1-celled; *styles* 3, arising from the same point, clavate; *stigmas* dilated. *Fruit* surrounded by the calyx, stalked, 1-celled, with 3 parietal polyspermous placentæ, sometimes 3-valved. *Seeds* attached in several rows to the placenta, with a brittle sculptured testa surrounded by a pulpy arillus; *embryo* straight, in the midst of fleshy thin albumen; *radicle* turned towards the hilum; *cotyledons* flat, leafy.— *Herbaceous* plants or *shrubs*, usually climbing, very seldom arborescent. *Leaves* alternate, with foliaceous stipulæ, often glandular. *Flowers* axillary or terminal, often with a 3-leaved involucre.

AFFINITIES. The real nature of the floral envelopes of this remarkable order is a question upon which botanists entertain very different opinions, and their ideas of its affinities are consequently much at variance. According to Jussieu (*Dict. des Sciences*, 38. 49.), the " parts taken for petals are nothing but inner divisions of the calyx, usually in a coloured state, and wanting in several species;" and therefore, in the judgment of this venerable botanist, the order is apetalous, or monochlamydeous. Decandolle adopts the same view of the nature of the floral envelopes as Jussieu; but he nevertheless considers the order polypetalous; a conclusion which I confess myself unable to understand, upon the supposition of the inner series of floral envelopes being calyx. Other botanists, and I think with justice, consider the outer series of the floral envelopes as the calyx, and the inner as the corolla, for two principal reasons. In the first place, they have the ordinary position and appearance of calyx and corolla, the outer being green, and the inner coloured; and, in the second place, there is no essential difference between the calyx and corolla, except the one being the outer, and the other the inner of the floral envelopes. And if the real nature of these parts is to be determined by analogy, an opinion in which I do not, however, concur, the great affinity, as I think, of the order with Violaceæ would confirm the idea of its being polypetalous rather than apetalous. The nature of the filamentous appendages, or rays as they are called, which proceed from the orifice of the tube, and of the membranous or fleshy, entire or lobed, flat or plaited, annular processes which lie between the petals and the stamens, is ambiguous. I am disposed to refer them to a peculiar form of petals, rather than to the stamens, for the reasons which I have assigned in the *Hort. Trans.* vol. 6. p. 309, for understanding the normal metamorphosis of the parts of fructification to be centripetal. There can, at least, be no doubt of their being of an intermediate nature between petals and stamens. With regard to the affinity of Passifloreæ, Jussieu, swayed by the opinion he entertains of their being apetalous, and Decandolle, who partly agrees and partly disagrees with Jussieu in his view of their structure, both assign the order a place near Cucurbitaceæ; but when we consider the stipitate fruit, occasionally valvular, the parietal placentæ, the sometimes irregular flowers, the stipulate leaves, and the climbing habit of these plants, it is difficult not to admit their affinity with Capparideæ and Violaceæ, the dilated disk of the former of which is probably analogous to the innermost of the annular processes of Passiflora. That the fleshy covering of the seeds in this order is a real arillus, is clear from the seeds of a capsular species nearly related to Pass. capsularis, but apparently unpublished, a drawing of which, by M. Ferdinand Bauer, exists in the Library of the Horticultural Society. In this plant the apex of the sculptured testa is uncovered by the arillus.

GEOGRAPHY. These plants are the pride of South America and the West Indies, where the woods are filled with their species, which climb about from tree to tree, bearing at one time flowers of the most striking beauty, and of so singular an appearance, that the zealous Catholics who discovered them, adapted Christian traditions to those inhabitants of the South American wildernesses; and at other times fruit, tempting to the eye and refreshing to the palate. One or two extend northwards into North America. Several are found in Africa and the neighbouring islands; and a few in the East Indies, of which the greater part belong to the genus Modecca.

PROPERTIES. Nothing is known of the properties of this order further than that the succulent arillus and pulp that surround the seeds are fragrant, juicy, cooling, and pleasant, in several species.

EXAMPLES. Passiflora, Tacsonia, Murucuja, Smeathmanuia.

CXXXII. MALESHERBIACEÆ.

MALESHERBIACEÆ, *Don in Jameson's Journal,* 321. (1826). — PASSIFLOREÆ,
§ Malesherbieæ, *Dec. Prodr.* 3. 337. (1828).

DIAGNOSIS, Polypetalous dicotyledons, with definite perigynous stamens, a membranous ring at the mouth of the tube of the calyx, concrete
carpella, a superior 1-celled ovarium with parietal placentæ, styles widely
apart at the base, corolla with a twisted æstivation, exstipulate glandless
leaves, exarillate seeds, and an embryo in the midst of fleshy albumen.
ANOMALIES.

ESSENTIAL CHARACTER.— *Calyx* tubular, membranous, inflated, 5-lobed, the lobes
with an imbricated æstivation. *Petals* 5, alternate with the segments of the calyx,
persistent, with a convolute æstivation, arising from without a short membranous rim
or corona. *Stamens* 5 or 10, perigynous; *filaments* filiform, distinct, or connected with
the stalk of the ovarium; *anthers* versatile. *Ovarium* superior, stipitate, 1-celled, with
the placentæ at the base, from which the ovules arise by the intervention of umbilical
cords; *styles* 3, filiform, very long, arising from distinct points of the apex of the ovarium;
stigmas clavate. *Fruit* capsular, 1-celled, 3-valved, membranous more or less, many-
seeded. *Seeds* attached by umbilical cords to placentæ arising either from the axis of
the valves, or from their base; *testa* crustaceous, brittle, with a fleshy crest, and no
arillus; *embryo* taper, in the midst of fleshy *albumen*, with the radicle next the hilum.—
Herbaceous or *half-shrubby* plants. *Leaves* alternate, lobed, without stipulæ. *Flowers*
axillary or terminal, solitary, yellow or blue.

AFFINITIES. According to Mr. Don, by whom these plants were first
considered the rudiments of an order, " they agree on the one hand with
Passifloreæ, and on the other with Turneraceæ;" and I am persuaded that
this is their true position. From the former they differ in the insertion of
their styles, in their versatile anthers, in their short placentæ, membranous
fruit, taper embryo, want of arillus and of stipules, and altogether in their
habit: from Turneraceæ, to which their habit quite allies them, they differ
in the presence of a perigynous membrane, in the remarkable insertion of
the styles, and in the want of all trace of an arillus; agreeing with that
order in the æstivation of the corolla, and in the principal other points of
their structure. I have modified the essential character of the order, in
consequence of the inspection of a Chilian plant, of which specimens are
in my possession.

GEOGRAPHY. Natives of Chile.
PROPERTIES. Unknown, except as objects of great beauty.
EXAMPLE. Malesherbia.

CXXXIII. TURNERACEÆ.

LOASEÆ, § Turneraceæ, *Kunth N. G. et Sp.* 6. 123. (1823). — TURNERACEÆ, *Dec.
Prodr.* 3. 345. (1828).

DIAGNOSIS. Polypetalous dicotyledons, with 5 perigynous stamens, concrete carpella, a superior 1-celled ovarium with 3 parietal placentæ, corolla
with a twisted æstivation, and embryo in the midst of fleshy albumen.
ANOMALIES.

ESSENTIAL CHARACTER.— *Calyx* inferior, often coloured, with 5 equal lobes, imbricated in æstivation. *Petals* 5, inserted into the tube of the calyx, equal, with a twisted
æstivation. *Stamens* 5, inserted into the tube of the calyx below the petals, with which
they are alternate; *filaments* distinct; *anthers* oblong, erect, 2-celled. *Ovarium* superior,
1-celled, with 3 parietal placentæ; *ovules* indefinite; *styles* 3 or 6, cohering more or less,

and simple branched or multifid at the apex. *Capsule* 3-valved, 1-celled, opening from the point about as far as the middle, the valves bearing the placentæ in the middle. *Seeds* with a thin membranous arillus on one side, crustaceous, reticulated; *embryo* slightly curved, in the middle of fleshy albumen; *radicle* turned towards the hilum; *cotyledons* somewhat plano-convex.—*Herbaceous* plants, having sometimes a tendency to become shrubby, with a simple pubescence, which does not sting. *Leaves* alternate, simple, without stipulæ, with occasionally 2 glands at the apex of the petiole. *Flowers* axillary, their pedicel either distinct or cohering with the petiole; with 2 bracteolæ. *Petals* yellowish, rarely blue.

AFFINITIES. Placed by Decandolle between Loaseæ and Fouquieraceæ, chiefly, it should seem, on account of its manifest relation to the former, and its perigynous stamens. To me it appears that those botanists are right who place it in the vicinity of Cistineæ, from which it differs more in the insertion of the stamens, and in the approximation of the radicle to the hilum, than in any other character, agreeing with them very much in habit. With Malvaceæ they agree in the twisted æstivation of the corolla, and in habit. With Loaseæ and Passifloreæ they have also much in common; and the circumstance of their certain relationship to Cistineæ gives great weight to the ingenious approximation, by M. Du Petit Thouars, of Passifloreæ to Violaceæ. The presence of glands upon the ends of the petioles of Turneraceæ is a confirmation of their affinity to the former. They are distinguished from Loaseæ by their fruit being superior and 1-celled, with parietal placentas, and by their definite stamens; the former character is, however, weakened by the nearly superior fruit of some Loaseæ.

GEOGRAPHY. Natives exclusively of the West Indies and South America. There seems no good reason for supposing Turnera trioniflora to be a native of Japan.

PROPERTIES. Unknown.

EXAMPLES. Turnera, Piriqueta.

CXXXIV. CISTINEÆ. THE ROCK-ROSE TRIBE.

CISTI, *Juss. Gen.* 294. (1789).—CISTOIDEÆ, *Vent. Tabl.* 3. 219. (1799).—CISTINEÆ, *Dec. Prodr.* 1. 263. (1824); *Lindl. Synops.* 36. (1829).

DIAGNOSIS. Polypetalous dicotyledons, with indefinite hypogynous stamens, concrete carpella, a 1-celled ovarium with narrow parietal placentæ, 5 sepals, and an inverted embryo.

ANOMALIES.

ESSENTIAL CHARACTER.—*Sepals* 5, continuous with the pedicel, persistent, unequal, the three inner with a twisted æstivation. *Petals* 5, hypogynous, very fugitive, crumpled in æstivation, and twisted in a direction contrary to that of the sepals. *Stamens* indefinite, hypogynous, distinct; *anthers* innate. *Ovarium* distinct, 1- or many-celled; *ovula* with the foramen at their apex; *style* single; *stigma* simple. *Fruit* capsular, usually 3- or 5-valved, occasionally 10-valved, either 1-celled with parietal placentæ in the axis of the valves, or imperfectly 5- or 10-celled with dissepiments proceeding from the middle of the valves, and touching each other in the centre. *Seeds* indefinite in number. *Embryo* inverted, either spiral or curved in the midst of mealy albumen.—*Shrubs* or herbaceous plants. *Branches* often viscid. *Leaves* usually entire, opposite or alternate, stipulate or exstipulate. *Racemes* usually unilateral. *Flowers* white, yellow, or red, very fugacious.

AFFINITIES. Distinguished from Violaceæ, with which they were formerly confounded, by their indefinite stamens and inverted embryo; from Bixineæ by this last character, by their mealy albumen, habit, and not having the leaves ever dotted; from Hypericineæ by the latter character, and the structure of their fruit.

GEOGRAPHY. S. Europe and the north of Africa are the countries that Cistineæ chiefly inhabit. They are rare in North America, extremely uncommon in South America, and scarcely known in Asia.

PROPERTIES. None, except that the resinous balsamic substance, called Labdanum, is obtained from Cistus creticus.

EXAMPLES. Cistus, Helianthemum.

CXXXV. BIXINEÆ. THE ARNOTTO TRIBE.

BIXINEÆ, *Kunth Diss. Malv.* p. 17. (1822); *Dec. Prodr.* 1. 259. (1824).

DIAGNOSIS. Polypetalous dicotyledons, with indefinite hypogynous stamens, concrete carpella, a 1-celled ovarium with narrow parietal placentæ, 4-7 sepals, and an erect embryo.

ANOMALIES. Corolla often wanting.

ESSENTIAL CHARACTER.—*Sepals* 4-7, either distinct or cohering at the base, with an imbricated æstivation. *Petals* 5, like the sepals, or wanting. *Stamens* indefinite, distinct, inserted upon a receptacle at the base of the calyx; *anthers* 2-celled. *Ovarium* superior, sessile, 1-celled; *ovula* proceeding from 4 to 7 parietal placentæ; *style* single, or in 2 or 4 divisions. *Fruit* capsular, or berried, 1-celled, many-seeded. *Seeds* attached to parietal placentæ, and enveloped in pulp; *albumen* either fleshy or very thin; *embryo* included, either straightish or curved, with leafy *cotyledons*; *radicle* pointing to the hilum. — *Trees* or *shrubs.* *Leaves* alternate, simple, entire, usually with pellucid dots; *stipules* deciduous; *peduncles* axillary, 1- many-flowered, with bracteæ.

AFFINITIES. The carpological characters of this order are very much those of Cistineæ and Homalineæ; from the former, Bixineæ differ in the position of their radicle, and in many other particulars; from the latter they are distinguished by their hypogynous stamens, and consequently superior fruit, by the distinct nature of the sepals and petals, when the latter are present, &c. Their dotted leaves are remarkable among all the neighbouring orders, and would alone suffice to characterise them, if they were constant, but they are occasionally not dotted. Some of the genera were formerly referred to Rosaceæ; but the affinity of this order with that is very weak; the plants which were formerly placed in it were imperfectly known.

GEOGRAPHY. All natives of the hotter parts of America, or of the islands of the Mauritius.

PROPERTIES. Bixa yields the substance known to the English by the name of Arnotto, and to the French by that of Rocou. It is the pulp that envelopes the seeds, and which is slightly purgative and stomachic. Farmers use it to stain their cheeses, and dyers for a reddish colour. The bark of Ludia is said to be emetic: but it is uncertain whether that genus does not belong to Homalineæ.

EXAMPLES. Bixa, Prockia.

CXXXVI. SARRACENNIEÆ.

SARRACENNIEÆ, *Turpin in Dict. des Sc.* c. ic. (?); *De la Pylaie in Ann. Linn. Par.* 6. 388. t. 13. (1827); *Hooker Fl. Boreal. Am.* p. 33. (1829).

DIAGNOSIS. Polypetalous dicotyledons, with hypogynous indefinite distinct stamens, concrete carpella, an ovarium of several cells with the placentæ in the axis, a regular calyx with imbricate æstivation, and a peltate petaloid persistent stigma.

ANOMALIES.

ESSENTIAL CHARACTER.—*Sepals* 5, persistent, often having a 3-leaved involucrum on the outside; *æstivation* imbricate. *Petals* 5, hypogynous, unguiculate, concave. *Stamens* indefinite, hypogynous; *anthers* oblong, adnate, 2-celled, bursting internally and longitudinally. *Ovarium* superior, 5-celled, with polyspermous placentæ in the axis; *style* single; *stigma* much dilated, peltate, with 5 angles. *Capsule* crowned by the persistent stigma, with 5 cells and 5 loculicidal valves. *Seeds* very numerous, minute, slightly warted, covering 5 large placentæ, which project from the axis into the cavity of the cells; *albumen* abundant; *embryo* cylindrical, lying near the base of the seed, with the *radicle* turned to the *hilum.* — *Herbaceous* perennial plants, living in bogs. *Roots* fibrous. *Leaves* radical, with a hollow urn-shaped petiole, at the apex of which is articulated the lamina, which covers the petiole like a lid. *Scapes* each having one large flower, of a more or less herbaceous colour.

AFFINITIES. These are not well made out. It is usual to refer Sarracennia to the vicinity of Papaveraceæ, on account of its remarkably dilated stigma, which is compared to the radiant stigma of Papaver, its indefinite stamens and small embryo lying at the base of copious albumen; and there can be no doubt that these points of resemblance are important. But I believe it is also akin to Droseraceæ, or at least to that order, whatever it may be, which shall finally comprehend Dionæa. With this genus no one has suspected the analogy of Sarracennia; a circumstance which has arisen, I presume, chiefly from attention having been turned to the fructification rather than the vegetation of those genera. If we compare the foliage of Dionæa with that of Sarracennia, we shall find that the pitcher of the latter is represented by the dilated footstalk of the former, which only requires its margins to cohere to be identical with it, and that the lid of the pitcher of the latter is analogous to the irritable lamina of the former. In both genera the stamens are hypogynous; both have a single stigma, which in Sarracennia is petaloid, in Dionæa is merely fringed; both have an embryo lying at the base of copious albumen, and both have polyspermous placentæ. In the internal arrangement of the fruit the two genera are dissimilar; but the differences depend upon peculiar modifications of structure, which cannot be considered to affect affinities otherwise so strongly indicated. In the remarkable structure of the leaves this order agrees with Nepentheæ, which are probably not so distantly related as they are usually supposed to be, and also with a single genus of Rosaceæ (Cephalotus).

GEOGRAPHY. They are exclusively confined to the bogs of North America.

PROPERTIES. Unknown.

EXAMPLE. Sarracennia.

CXXXVII. DROSERACEÆ. THE SUNDEW TRIBE.

DROSERACEÆ, *Dec. Théorie*, 214. (1819); *Prodr.* 1. 317. (1814); *Lindl. Synops.* 38. (1829).

DIAGNOSIS. Polypetalous dicotyledons, with definite hypogynous stamens, concrete carpella, a 1-celled ovarium with narrow parietal placentæ, 5 sepals, an erect embryo, and circinate vernation.

ANOMALIES. The anthers of Byblis and Roridula open by pores. Vernation not circinate in Dionæa.

ESSENTIAL CHARACTER.— *Sepals* 5, persistent, equal, with an imbricate æstivation. *Petals* 5, hypogynous. *Stamens* distinct, withering, either equal in number to the petals and alternate with them, or 2, 3, or 4 times as many. *Ovarium* single; *styles* 3-5, either wholly distinct, or slightly connected at the base, bifid or branched. *Capsule* of 1 or 3 cells, and 3 or 5 valves, which bear the placentæ either in the middle or at their base. *Seeds* either naked or furnished with arillus. *Embryo* straight, erect, in the axis of a fleshy or cartilaginous albumen. *Cotyledons* rather thick.—Delicate *herbaceous* plants, often covered with glands. *Leaves* alternate, with *stipulary* ciliæ and a circinate vernation. *Peduncles*, when young, circinate.

AFFINITIES. Nearly allied to Violaceæ, from which their circinate vernation, several styles, minute embryo, and exstipulate leaves, distinguish them. They are also no doubt related to Saxifrageæ, to which order it is possible that one of the genera referred to Droseraceæ by Decandolle (Romanzovia), actually belongs. The most material circumstance that separates them from Saxifrageæ is their hypogynous, not perigynous stamens. But when we consider how difficult it frequently is, to determine whether the point of origin of the stamens in Saxifrageæ is from the calyx or from below the ovarium, this distinction will cease to have much value. Besides the line of origin of the stamens, these two orders are also distinguished by their vernation and placentation; but in the latter respect Parnassia among Saxifrageæ accords with Droseraceæ; and in the former Dionæa among Droseraceæ accords with Saxifrageæ. It is not, however, quite certain that this last-mentioned genus is actually referable to Droseraceæ, from which it differs remarkably in the structure of its ovarium, in its style, and in its foliage. I am persuaded that Droseraceæ are fully as nearly related to Saxifrageæ as to Violaceæ; and this fact shews how much the artificial distribution of orders is at variance with natural affinities. Droseraceæ are also allied to Sarracennieæ: see that order.

GEOGRAPHY. At the Cape of Good Hope, in South America, North America, New Holland, China, Europe, Madagascar, the East Indies, wherever there are marshes or morasses, these plants are found. Drosophyllum lusitanicum is remarkable for growing on the barren sands of Portugal.

PROPERTIES. The leaves of Dionæa muscipula are irritable, and collapse when touched. The common Droseras are rather acid, slightly acrid, and, according to some, poisonous to cattle. The Drosera communis of Brazil is said by M. A. St. Hilaire to be poisonous to sheep. *Pl. Usuelles,* no. 15.

EXAMPLES. Drosera, Drosophyllum.

CXXXVIII. NEPENTHEÆ. THE PITCHER-PLANT TRIBE.

ARISTOLOCHIÆ, § Nepenthinæ, *Link Handb.* 1. 369. (1829).

DIAGNOSIS. Apetalous dicotyledons, with a 4-celled ovarium, indefinite ovula, a regular imbricated calyx, and pitcher-shaped leaves.

ANOMALIES. The direction of the radicle uncertain.

ESSENTIAL CHARACTER.—*Flowers* diœcious. *Calyx* 4-leaved, inferior, oppositely imbricated in æstivation. *Stamens* cohering in a solid column, bearing at the apex about 16 anthers, collected in various directions in one head; *anthers* 2-celled, opening longitudinally and externally. *Ovarium* superior, 4-cornered, 4-celled, with an indefinite number of ascending ovules attached to the sides of the dissepiments; *stigma* sessile, simple. *Fruit* capsular, 4-celled, 4-valved, with the seeds sticking to the sides of the dissepiments, which proceed from the middle of the valves. *Seeds* indefinite, very minute, fusiform, with a lax outer integument; *albumen* oblong, much less than the seed, lying about the middle of the outer integument; *embryo* in the midst of fleshy albumen, with 2 cotyledons placed face to face; (*radicle* turned towards the hilum, *Ad. Brongn.* Nees, and *Esenbeck;* turned to the extremity opposite the hilum, *Richard*).—*Herbaceous* or *half-shrubby* caulescent plants. *Leaves* alternate, slightly sheathing at the base, with a dilated foliaceous petiole, pitcher-shaped at the end, which is articulated with a lid-like lamina. *Racemes* terminal, dense, many-flowered.

AFFINITIES. The relation that is borne by the highly curious plants which this order contains was not even guessed at until M. Adolphe Brongniart pointed out a resemblance between them and Cytineæ, which had not before been suspected, but which he considered so important as to justify him in placing it in the same order. While we admit the ingenuity with which this opinion is sustained, it is impossible to agree with M. Brongniart in the conclusion at which he has arrived. To say nothing of the extreme dissimilarity in habit between these plants, the structure of their fruit appears to me essentially different; and the seeds of Cytinus being unknown, the resemblance between it and Nepenthes is reduced to a similarity in the arrangement of the anthers, which cannot in the present case be considered of much importance, as it in some degree depends upon the unisexuality of the flowers of both genera. It appears to me that, in the existing state of our knowledge, there is no order to which Nepenthes can be safely approximated : it has a remote affinity with Droseraceæ, but a number of connecting links is required to fill up the space between them. The best account of the structure of Nepenthes will be found in the *Ann. des Sc.* 1. 42. and 3. 366. The structure of the pitcher-shaped leaves is analogous to that of Sarracennieæ, and Cephalotus among Rosaceæ. The water contained in the unopened pitcher of a plant which flowered in the Botanic Garden, Edinburgh, was found by Dr. Turner " to emit, while boiling, an odour like baked apples, from containing a trace of vegetable matter, and to yield minute crystals of superoxalate of potash on being slowly evaporated to dryness." *B. Mag.* 2798. There is a good account of the germination of Nepenthes, in Jameson's Journal for April 1830, from which it may be concluded that the long loose tunic of the seed is intended to act at first as a buoy, to float the seed upon the surface of the water, and afterwards as an anchor, to keep it fast upon the mud until it can have struck root.

GEOGRAPHY. All natives of swamps in the East Indies and China.

PROPERTIES. Unknown.

EXAMPLE. Nepenthes.

CXXXIX. LINEÆ. THE FLAX TRIBE.

LINEÆ, *Dec. Théorie*, ed. 1. 217. (1819) ; *Prodr.* 1. 423. (1824) ; *Lindl. Synops.* 53. (1829).

DIAGNOSIS. Polypetalous dicotyledons, with definite hypogynous stamens, concrete carpella, an entire ovarium of several cells with placentæ in the axis, an imbricated regular calyx, symmetrical flowers, definite pendulous ovules, distinct style, capitate stigmas, stamens immediately hypogynous, flat cotyledons, and a capsular many-celled fruit.

ANOMALIES.

ESSENTIAL CHARACTER.— *Sepals* 3-4-5, with an imbricated æstivation, continuous with the peduncle, persistent. *Petals* equal in number to the sepals, hypogynous, unguiculate, with a twisted æstivation. *Stamens* equal in number to the petals, and alternate with them, united at the base in a hypogynous ring, from which proceed little teeth opposite to the petals, and indicating abortive stamens ; *anthers* ovate, innate. *Ovarium* with about as many cells as sepals, seldom fewer ; *styles* equal in number to the cells ; *stigmas* capitate. *Capsule* generally pointed with the indurated base of the styles, many-celled ; each cell partially divided in two by an imperfect spurious dissepiment, and dehiscing with two valves at the apex. *Seeds* in each cell single, compressed, inverted ; *albumen* usually absent ; inner lining of the *testa* tumid ; *embryo* straight, fleshy, with the radicle pointing towards the hilum ; *cotyledons* flat.— *Herbaceous* plants, or small *shrubs*. *Leaves* entire, without stipulæ, usually alternate. *Petals* very fugitive.

AFFINITIES. It is remarked by Decandolle, that these are intermediate, as it were, between Caryophylleæ, Malvaceæ, and Geraniaceæ, from all which, however, they are obviously distinguished.

GEOGRAPHY. Europe and the north of Africa are the principal stations of this order, which is, however, scattered more or less over most parts of the globe. Several are natives of North and South America, 2 only are found in India, 1 in New Zealand, and none in New Holland; for the L. angustifolium mentioned by Decandolle as having been sent him from that country, had probably, as he suggests, been introduced from Europe. It is stated by Dr. Richardson, that the most northern limit of this order in North America is 54° N. *Ed. P. J.* 12. 209.

PROPERTIES. The tenacity of their fibre, and the mucilage of their diuretic seeds, are the striking characters of Lineæ, which are also usually remarkable for the beauty of their flowers. The leaves of L. catharticum are purgative. Linum selaginoides is considered in Peru bitter and aperient. *Dec.*

EXAMPLES. Linum, Radiola.

CXL. CARYOPHYLLEÆ. THE CHICKWEED TRIBE.

CARYOPHYLLEÆ, *Juss. Gen.* 299. (1789); *Dec. Prodr.* 1. 351. (1824); *Lindl. Synops.* p. 43. (1829).

DIAGNOSIS. Polypetalous dicotyledons, with definite hypogynous stamens, concrete carpella, an ovarium of 1 or several cells with placentæ in the axis, an imbricated calyx, symmetrical flowers, an embryo coiled round mealy albumen, and opposite entire leaves with herbaceous stems.

ANOMALIES. Some are apetalous; others are accidently unsymmetrical in their fructification.

ESSENTIAL CHARACTER.—*Sepals* 4-5, continuous with the peduncle; either distinct, or cohering in a tube, persistent. *Petals* 4-5, hypogynous, unguiculate, inserted upon the pedicel of the ovarium; occasionally wanting. *Stamens* twice as many as the petals, inserted upon the pedicel of the ovarium along with the petals; *filaments* subulate, sometimes monadelphous; *anthers* innate. *Ovarium* stipitate on the apex of a pedicel (called the gynophorus); *stigmata* 2-5, sessile, filiform, papillose on the inner surface. *Capsule* 2-5-valved, either 1-celled or 2-5-celled, in the latter case with a loculicidal dehiscence. *Placenta* central, in the 1-celled capsules distinct, in the 2-5-celled capsules adhering to the edge of the dissepiments. *Seeds* indefinite in number, rarely definite; *albumen* mealy; *embryo* curved round the albumen; *radicle* pointing to the hilum.—*Herbaceous* plants, occasionally becoming *suffrutescent. Stems* tumid at the articulations. *Leaves* always opposite and entire, often connate at the base.

AFFINITIES. On the one hand these plants are allied to Frankeniaceæ, with which they agree in their unguiculate petals, bearing processes at their orifice, and in some measure in habit; and on the other to Lineæ, from which they are principally distinguished by their unilocular, or, if plurilocular, several-seeded capsules, and albuminous seeds. Geraniaceæ, Oxalideæ, Violaceæ, and Portulaceæ, are all also allied in many particulars, but they are readily distinguished. Elatineæ differ in their exalbuminous seeds and capitate stigmas. Bartling combines in one order Caryophylleæ, Paronychiæ, Amarantaceæ, Phytolacceæ, and Chenopodeæ; and all these orders, although artificially separated widely, do in fact concur in a number of essential points; but the rest may be readily known from Caryophylleæ by their want of petals; their combining character is the embryo curved round the albumen, in which particular Polygoneæ also agrees with them. Macræa, a genus of mine, which Mr. Don states to be the same as Viviania, a neglected genus of Cavanilles (see *Jameson's Journal, Jan.* 1830, p. 170.), if really belonging to the order, differs remarkably in the curved embryo lying, according to Dr. Hooker, in the midst of fleshy albumen, in its dry persistent petals, and in the vernation of both the calyx and petals; but I incline to think that this remarkable genus indicates the existence of an order allied to Frankeniaceæ or Geraniaceæ more closely than to Caryophylleæ. Hydropityon, doubtfully referred here by Decandolle, belongs to Scrophularineæ, as I learn from Mr. Bentham.

GEOGRAPHY. Natives principally of the temperate and frigid parts of the world, where they inhabit mountains, hedges, rocks, and waste places. Those which are found within the tropics are usually natives of high elevations and mountainous tracts, almost always reaching the limits of eternal snow, where many of them exclusively vegetate. The Mollugos are the most tropical form of the order. A little plant, called Physa, is found in Madagascar; and some Silenes are scattered in many different parts of the globe. According to the calculations of Humboldt, Caryophylleæ constitute $\frac{1}{22}$ of the flowering plants of France, $\frac{1}{27}$ of Germany, $\frac{1}{17}$ of Lapland, $\frac{1}{72}$ of North America.

PROPERTIES. Remarkable for little except their uniform insipidity. A few, such as the Dianthuses and Lychnises, are handsome flowers; but the greater part are mere weeds. Saponaria officinalis, Gypsophila Ostruthium, Lychnis dioica, and L. chalcedonica, have saponaceous properties: Saponaria has been used in syphilis. Arenaria peploides, having been fermented, is used in Iceland as a sort of food. A decoction of the root of Silene virginica is said to have been employed in North America as anthelmintic. *Dec.*

Decandolle admits two sections (*Prodr.* 1.)

1. SILENEÆ.
Sepals united in a cylindrical tube.
EXAMPLES. Lychnis, Dianthus.
2. ALSINEÆ. *Dec. Fl. Franc.* 4. 766.
Sepals distinct, or only cohering at the base.
EXAMPLES. Stellaria, Alsine.

CXLI. FRANKENIACEÆ.

FRANKENIACEÆ, *Aug. St. Hilaire Mém. Plac. Centr.* 39. (1815); *Dec. Prodr.* 1. 349. (1824); *Lindl. Synops.* 38. (1829).

DIAGNOSIS. Polypetalous dicotyledons, with definite hypogynous stamens, concrete carpella, a 1-celled ovarium with narrow parietal placentæ, 5 connate sepals, an erect embryo, exstipulate leaves, and a capsule with septicidal dehiscence.

ANOMALIES. None, if Luxemburgia be excluded.

ESSENTIAL CHARACTER.— *Sepals* 4-5, united in a furrowed tube, persistent, equal. *Petals* alternate with the sepals, hypogynous, unguiculate, with appendages at the base of the limb. *Stamens* hypogynous, either equal in number to the petals, and alternate with them, or having a tendency to double the number; *anthers* roundish, versatile. *Ovarium* superior; *style* filiform, 2-fid or 3-fid. *Capsule* 1-celled, enclosed in the calyx, 2- 3- or 4-valved, many-seeded; *dehiscence* septicidal. *Seeds* attached to the margins of the valves, very minute; *embryo* straight, erect, in the midst of albumen (divided into two plates, *Gærtn. fil.*)—*Herbaceous* plants or *under-shrubs.* *Stems* very much branched. *Leaves* opposite, exstipulate, with a membranous sheathing base; often revolute at the edge. *Flowers* sessile in the divisions of the branches, and terminal, embosomed in leaves, usually pink.

AFFINITIES. Allied on the one hand to Caryophylleæ, from which they are distinguished by their different placentation, and by the form of their embryo; to Lineæ, from which they are known by their unilocular fruit; and on the other to Violaceæ, which differ in having a loculicidal, not septicidal, dehiscence. Their habit is that of Amarantaceæ and Illecebreæ, from which their petals and compound fruit divide them.

GEOGRAPHY. This order is chiefly found in the north of Africa and south of Europe. Two species are natives of the Cape of Good Hope, 1 of South America, 4 of New Holland, and 3 of temperate Asia. None have been found in tropical India or North America.

PROPERTIES. Unknown.

EXAMPLE. Frankenia.

CXLII. TAMARISCINEÆ. THE TAMARISK TRIBE.

TAMARISCINEÆ, *Desvaux, in a Dissertation read before the French Institute* (in 1815), according to the *Ann. Sc. Nat.* 4. 344. (1825); *A. St. Hil. Mem. Mus.* 2. 205. (1816); *Ehrenb. in Annales des Sciences,* 12. 68. (1827); *Dec. Prodr.* 3. 95. (1828); *Lindl. Synops.* 61. (1829).

DIAGNOSIS. Polypetalous dicotyledons, with definite hypogynous stamens, concrete carpella, a 1-celled ovarium with placentæ at the base, no stipulæ, shrubby stems, comose seeds, and a 4- or 5-parted calyx.

ANOMALIES.

ESSENTIAL CHARACTER.— *Calyx* 4- or 5-parted, persistent, with an imbricated æstivation. *Petals* inserted into the base of the calyx, withering, with an imbricated æstivation. *Stamens* hypogynous, either equal to the petals in number, or twice as many, distinct or monadelphous. *Ovarium* superior; *style* very short; *stigmata* 3. *Capsule* 3-valved, 1-celled, many-seeded; *placentæ* 3, either at the base of the cavity, or along the middle of the valves. *Seeds* erect or ascending, comose; *albumen* none; *embryo* straight, with an inferior radicle.— *Shrubs* or *herbs*, with rod-like branches. *Leaves* alternate, resembling scales, entire. *Flowers* in close spikes or racemes.

AFFINITIES. According to Decandolle (*Prodr.* 3. 95.), who places the order among those with perigynous stamens, related to Portulaceæ (or Illecebreæ), on account of the resemblance between their flowers and those of Telephium; but they differ in their parietal exalbuminous comose seeds. Also allied to Lythrariæ and Onagrariæ, but differing from the former in the imbricated æstivation, the petals arising from the bottom of the calyx, and parietal seeds; and from the latter in their superior ovarium, and the imbricated æstivation of the calyx. Dr. Ehrenberg asserts the order to have hypogynous stamens (*Ann. des Sc.* 12. 77.), and this agrees with my own observations. The same botanist, in separating the Tamarix songarica of Willdenow from Tamariscineæ, and referring it to the vicinity of Reaumuria, establishes the affinity of Tamariscineæ to the order of Reaumurieæ. Its true station appears to me to be next Frankeniaceæ.

GEOGRAPHY. Exclusively confined to the northern hemisphere, and even to its eastern half, that is, to the old world, on which they extend as far as the Cape de Verds. They usually grow by the sea-side, but occasionally by the edges of rivers and torrents. The maximum of species and of individuals also is found in the basin of the Mediterranean. The order appears bounded on the south by the 8° or 9° parallel of N. lat., and on the north by that of 50° and 55° in Siberia, Germany, and England. *Ehrenb.*

PROPERTIES. The bark is slightly bitter, astringent, and probably tonic. T. gallica and africana are remarkable for the quantity of sulphate of soda which their ashes contain. *Dec.* Dr. Ehrenberg found that the Manna of Mount Sinai is produced by a variety of Tamarix gallica. This substance, being analysed by M. Mitscherlich, was found to contain no crystallisable Mannite, but to consist wholly of pure mucilaginous sugar. *Ann. des Sc.* l. c.

EXAMPLES. Tamarix, Myricaria.

CXLIII. ELATINEÆ. THE WATER-PEPPER TRIBE.

ELATINEÆ, *Cambessédes in Mem. Mus.* 18. 225. (1829).

DIAGNOSIS. Polypetalous dicotyledons, with definite hypogynous distinct stamens, concrete carpella, an ovarium of several cells with the placentæ in the axis, an imbricated calyx, symmetrical flowers, indefinite exalbuminous seeds with a straight embryo, capitate stigmas, a fruit with the valves alternate with the septa, and a persistent axis and herbaceous stems.

ANOMALIES.

ESSENTIAL CHARACTER.— *Sepals* 3-5, distinct, or slightly connate at the base. *Petals* hypogynous, alternate with the sepals. *Stamens* hypogynous, usually twice as numerous as the petals. *Ovarium* with from 3 to 5 hypogynous cells, an equal number of styles, and capitate stigmas. *Fruit* capsular, 3-5-celled, with the valves alternate with the septa, which usually adhere to a central axis, but in Merimea to the valves separating from the axis. *Seeds* numerous, with a straight *embryo*, whose *radicle* is turned to the hilum, and no *albumen.—Annuals*, found in marshy places. *Stems* fistulous, rooting. *Leaves* opposite, without stipulæ.

AFFINITIES. This little order has been recently established by M. Cambessédes, who distinguishes them from Caryophylleæ, with which a part of them had been confounded, by their capitate stigmata, by the dehiscence of their fruit, and by their want of albumen. They agree with Hypericineæ in many respects, even in the presence of receptacles of resinous secretions; but differ in having a persistent central axis in the fruit, definite stamens, and so forth.

GEOGRAPHY. Found in marshes in the four quarters of the globe. The Elatines are natives of Europe, Bergias of the Cape of Good Hope and the East Indies, and Merimea of South America.

PROPERTIES. Unknown.

EXAMPLES. Elatine, Bergia, Crypta, Merimea.

CXLIV. PORTULACEÆ. THE PURSLANE TRIBE.

PORTULACEÆ, *Juss. Gen.* 313. (1789) *in part; A. St. Hil. Mem. Plac. Cent.* 42. (1815); *Dec. Prodr.* 3. 351. (1828); *Lindl. Synops.* 62. (1829); *Dec. Mem. de la Soc. d'Hist. Nat. de Paris,* (*Aug.* 1827).

DIAGNOSIS. Polypetalous dicotyledons, with unsymmetrical perigynous stamens, concrete carpella, a 1-celled ovarium, herbaceous stems, stamens opposite the petals or twice as many, 2 sepals, and naked seeds with the embryo curved round the albumen.

ANOMALIES. Sepals 5 in Trianthema and Cypselea. Petals sometimes wanting.

ESSENTIAL CHARACTER.—*Sepals* 2, seldom 3 or 5, cohering by the base. *Petals* generally 5, occasionally 3, 4, or 6, either distinct or cohering in a short tube, sometimes wanting. *Stamens* inserted along with the petals irregularly into the base of the calyx, variable in number, all fertile, sometimes opposite the petals; *filaments* distinct; *anthers* versatile, with 2 cells, opening lengthwise. *Ovarium* superior, 1-celled; *style* single, or none; *stigmata* several, much divided. *Capsule* 1-celled, dehiscing either transversely or by 3 valves, occasionally 1-seeded and indehiscent. *Seeds* numerous, if the fruit is dehiscent; attached to a central placenta; *albumen* farinaceous; *embryo* curved round the circumference of the albumen, with a long radicle. — Succulent *shrubs* or *herbs*. *Leaves* alternate, seldom opposite, entire, without stipulæ, or sometimes with membranous ones on each side at the base. *Flowers* axillary or terminal, usually ephemeral, expanding only in bright sunshine.

AFFINITÍES. Related in every point of view to Caryophylleæ, from which they scarcely differ except in their perigynous stamens, which are opposite the petals when equal to them in number, and two sepals; the latter character is not, however, very constant. The presence of scarious stipulæ in several Portulaceæ, although perhaps an anomaly in the order, indicates their affinity with Illecebreæ, from which the monospermous genera of Portulaceæ are distinguished by the want of symmetry in their flowers, and by the stamens being opposite the petals instead of the sepals. So close is the relationship between these orders, that several of the genus Ginginsia in Portulaceæ have been referred to Pharnaceum in Caryophylleæ, and several Portulaceæ have been described by authors as belonging to genera of Illecebreæ. Decandolle remarks, that his Ginginsia brevicaulis resembles certain species of Androsace, and that Portulaceæ have been more than once compared to Primulaceæ (*Mem.* p. 14.); and the same author remarks, in another place (*Prodr.* 3. 351.), that the genera with indefinite stamens and hairy axillæ approach Cacteæ, while the apetalous genera tend towards apetalous Ficoideæ.

GEOGRAPHY. A fourth of the order inhabits the Cape of Good Hope, rather more than another fourth is found in South America, 1 only in Guinea, 2 in New Holland, 1 in Europe, and the remainder in various parts of the world. They are always found in dry parched places.

PROPERTIES. Insipidity, want of smell, and a dull green colour, are the usual qualities of this order, of which the only species of any known use are common Purslane and Claytonia perfoliata, which resemble each other in property.

EXAMPLES. Portulaca, Montia, Talinum.

CXLV. FOUQUIERACEÆ.

FOUQUIERACEÆ, *Dec. Prodr.* 3. 349. (1828)

DIAGNOSIS. Succulent polypetalous dicotyledons, with perigynous stamens, concrete carpella, a superior ovarium with several cells, and a terminal style, regular flowers, the petals of which cohere in a tube, indefinite ovula, and no disk.

ANOMALIES.

ESSENTIAL CHARACTER. — *Sepals* 5, imbricated, ovate, or roundish. *Petals* 5, combined in a long tube, arising from the bottom of the calyx or torus, regular. *Stamens* 10 or 12, arising from the same line as the petals, but distinct from them, exserted; *anthers* 2-celled. *Ovarium* superior, sessile; *style* filiform, trifid at the apex; *ovules* numerous. *Capsule* 3-cornered, 3-celled, 3-valved; *valves* bearing the dissepiments in the middle. *Seeds* in part abortive, compressed, winged, affixed to the axis; *embryo* straight, in the centre of thin fleshy *albumen*; *cotyledons* flat. — *Trees* or *shrubs*. *Leaves* entire, oblong, fleshy, clustered in the axilla of a spine or a cushion. *Flowers* scarlet, arranged in a terminal spike or panicle.

AFFINITIES. Separated from Portulaceæ by Decandolle, as he tells us

(*Mém. Portul.* 4.), for the following reasons: 1. because their petals cohere in a long tube of the same nature as that of gamopetalous Crassulaceæ; 2. because their capsule consists of three loculicidal cells, that is to say, which separate through the middle, forming three septiferous valves; and, 3. because their embryo is straight, with flat cotyledons, and stationed in the centre of fleshy albumen. They approach the monopetalous Crassulaceæ in the structure of their flower; and Turneraceæ and Loaseæ in the form of their fruit. *Dec.*

GEOGRAPHY. All Mexican.

PROPERTIES. Unknown.

EXAMPLES. Fouquiera, Bronnia.

CXLVI. GALACINEÆ.

GALACINÆ, *Don in Edinb. New Phil. Journal, Oct.* (1828).

DIAGNOSIS. Polypetalous dicotyledons, with perigynous definite stamens which are alternately sterile, concrete carpella, a superior ovarium of several cells, several sepals, and indefinite ovules.

ANOMALIES.

ESSENTIAL CHARACTER.—*Calyx* 4-6-parted, persistent. *Petals* equal in number to the segments of the calyx, into the base of which they are inserted. *Stamens* perigynous, twice or 4 times as many as the petals, alternately barren; monadelphous or distinct; *anthers* 2-celled or 1-celled. *Ovarium* 3- or 4-celled, superior, with numerous ovula attached to the axis; *stigma* sessile, 3-4-lobed. *Capsule* 3-4-celled, with 3 or 4 valves, bearing the septa in their middle. *Seeds* indefinite. . . . .— *Herbaceous* plants. *Leaves* radical, simple or lyrate, without stipulæ. *Flowers* in terminal racemes. *Pedicels* with a bractea at the base.

AFFINITIES. This obscure order has been lately defined by Mr. Don; but its affinities can scarcely be determined, until something is known of the seeds. According to this botanist, it should be placed near Philadelpheæ and Saxifrageæ; but, in the opinion of Adrien de Jussieu, it, or at least Francoa, is akin to Crassulaceæ. The latter considers the stamens perigynous, the former describes them as hypogynous.

GEOGRAPHY. Natives of the temperate parts of North and South America.

PROPERTIES. Unknown.

EXAMPLES. Galax, Francoa.

Obs. This order requires to be reconsidered.

CXLVII. CRASSULACEÆ. THE HOUSE-LEEK TRIBE.

SEMPERVIVÆ, *Juss. Gen.* 207. (1789).—SUCCULENTÆ, *Vent. Tabl.* 3. 271. (1799).— CRASSULÆ, *Juss. Dict. des Sc. Nat.* 11. 369. (1818).—CRASSULACEÆ, *Dec. Bull. Philom.* n. 49. p. 1. (1801); *Fl. Fr. ed.* 3. v. 4. p. 271. (1805); *Mémoire* (1828), *Prodr.* 3. 381. (1828); *Lindl. Synops.* 63. (1829).—SEDEÆ, *Spreng.*

DIAGNOSIS. Succulent polypetalous dicotyledons, with definite perigynous stamens, superior distinct ovaria surrounded at the base by hypogynous scales, indefinite albuminous seeds, sepals in a single row, and exstipulate leaves.

ANOMALIES. Penthorum is not succulent. This genus and Diamorpha have the ovaria concrete. Some are monopetalous, particularly the genus Cotyledon. Petals and stamens often almost hypogynous. Tillæa has definite ovules.

ESSENTIAL CHARACTER.—*Sepals* from 3 to 20, more or less united at the base. *Petals* inserted in the bottom of the calyx, either distinct or cohering in a monopetalous corolla. *Stamens* inserted with the petals, either equal to them in number and alternate with them, or twice as many, those opposite the petals being shortest, and arriving at perfection after the others ; *filaments* distinct, subulate ; *anthers* of 2 cells, bursting lengthwise. *Hypogynous scales* several, 1 at the base of each ovarium, sometimes obsolete. *Ovaria* of the same number as the petals, opposite to which they are placed around an imaginary axis, 1-celled, tapering into stigmata. *Fruit* consisting of several follicles, opening by the suture in their face. *Seeds* attached to the margins of the suture, variable in number ; *embryo* straight in the axis of the albumen, with the radicle pointing to the hilum.—Succulent *herbs* or *shrubs*. *Leaves* entire or pinnatifid ; *stipulæ* none. *Flowers* usually in cymes, sessile, often arranged unilaterally along the divisions of the cymes.

AFFINITIES. These are all remarkable for the succulent nature of their stems and leaves, in which they resemble Cacteæ, Portulaceæ, and certain genera of Euphorbiaceæ, Asclepiadeæ, and Asphodeleæ ; but this analogy goes no further. Their real affinity is probably with Saxifrageæ through Penthorum, and with Illecebreæ through Tillæa, as Decandolle has remarked. In both those orders the hypogynous scales of Crassulaceæ are wanting. Are not these bodies analogous to the scales out of which the stamens of Zygophylleæ spring ? If so, an unsuspected affinity exists between these orders. Decandolle observes (*Mémoire*, p. 5.) that there is no instance of a double flower in the order, although this might have been expected from their analogy in structure with Caryophylleæ. Sempervivum tectorum exhibits almost constantly the singular phenomenon of anthers bearing ovules instead of pollen.

GEOGRAPHY. It appears, from Decandolle's researches, that of the 272 species of which the order consists, 133 are found at the Cape of Good Hope, 2 in South America beyond the tropics, 2 in the same country within the tropics, none in the West Indies or the Mauritian Islands, 8 in Mexico, 7 in the United States, 12 in Siberia, 18 in the Levant, 52 in Europe, 18 in the Canaries, 1 in southern Africa beyond the limits of the Cape, 9 in Barbary, 3 in the East Indies, 4 in China and Japan, and 2 in New Holland. They are found in the driest situations, where not a blade of grass nor a particle of moss can grow, on naked rocks, old walls, sandy hot plains, alternately exposed to the heaviest dews of night and the fiercest rays of the noon-day sun. Soil is to them a something to keep them stationary, rather than a source of nutriment, which in these plants is conveyed by myriads of mouths, invisible to the naked eye, but covering all their surface, to the juicy beds of cellular tissue which lie beneath them.

PROPERTIES. Refrigerant and abstergent properties, mixed sometimes with a good deal of acridity, distinguish them. The fishermen of Madeira rub their nets with the fresh leaves of Sempervivum glutinosum, by which they are rendered as durable as if tanned, provided they are steeped in some alkaline liquor. Malic acid exists in Sempervivum tectorum combined with lime. *Turner*, 634.

EXAMPLES. Sempervivum, Crassula, Cotyledon.

CXLVIII. FICOIDEÆ.

Ficoideæ, *Juss. Gen.* 315. (1789); *Dict. Sc. Nat.* 16. 528. (1820); *Dec. Prodr.* 3. 415. (1828).

Diagnosis. Succulent polypetalous dicotyledons, with definite perigynous stamens, concrete carpella, an inferior ovarium of several cells, and indefinite seeds with the embryo lying on the outside of mealy albumen.

Anomalies. Tetragonia and Miltus have no petals, and definite seeds. Sesuvium and Aizoon have no petals.

Essential Character.— *Sepals* definite, usually 5, but varying from 4 to 8, more or less combined at the base, either cohering with the ovarium, or nearly distinct from it, equal or unequal, with a quincuncial or valvate æstivation. *Petals* indefinite, coloured, opening beneath bright sunshine, sometimes wanting, but in that case the inside of the calyx is coloured. *Stamens* arising from the calyx, definite or indefinite, distinct; *anthers* oblong, incumbent. *Ovarium* inferior, or nearly superior, many-celled; *stigmata* numerous, distinct. *Capsule* either surrounded by the fleshy calyx, or naked, many-celled, often 5-celled, opening in a stellate manner at the apex. *Seeds* definite, or more commonly indefinite, attached to the inner angle of the cells; *embryo* lying on the outside of mealy *albumen*, curved or spiral.— *Shrubby* or *herbaceous* plants. *Leaves* succulent, opposite, simple. *Flowers* usually terminal.

Affinities. The embryo curved round mealy albumen, along with the superior calyx, and distinctly perigynous stamens, characterises these among their neighbours, independently of their succulent habit. With Crassulaceæ, Chenopodeæ, and Caryophylleæ, they are more or less closely related. Reaumurieæ and Nitrariaceæ, combined with Ficoideæ by Decandolle, are families different in affinity.

Geography. The hottest sandy plains of the Cape of Good Hope nourish the largest part of this order. A few are found in the south of Europe, north of Africa, Chile, China, Peru, and the South Seas.

Properties. The succulent leaves of a few are eaten, as of Tetragonia expansa, Mesembryanthemum edule, and Sesuvium portulacastrum; others yield an abundance of soda. Mesembryanthemum nodiflorum is used in the manufacture of Maroquin leather.

Examples. Mesembryanthemum, Tetragonia.

CXLIX. NITRARIACEÆ.

Diagnosis. Polypetalous dicotyledons, with perigynous stamens, concrete carpella, a superior ovarium of several cells, a deeply-divided calyx, regular flowers, an inflexed valvular æstivation, a terminal single style, pendulous exalbuminous seeds, and a straight embryo.

Anomalies.

Essential Character.— *Calyx* inferior, 5-toothed, fleshy. *Corolla* of 5 petals, which arise from the calyx, with an inflexed valvular æstivation. *Stamens* 3 times the number of the petals, perigynous; *anthers* innate, with 2 oblique longitudinal lines of dehiscence. *Ovarium* superior, 3- or more celled, with a continuous fleshy style, at the apex of which are as many stigmatic lines as there are cells; *ovula* pendulous, by means of a long funiculus. *Fruit* drupaceous, opening by 3 or 6 valves. *Seeds* solitary, with no *albumen*, and a straight *embryo*, with the *radicle* next the *hilum*.— *Shrubs* with deciduous succulent alternate *leaves*, which are sometimes fascicled. *Flowers* in cymes, or solitary.

Affinities. I take Nitraria to be the type of an order related on the one hand to Ficoideæ, and on the other to Rhamneæ, agreeing with both in a multitude of characters, and with the latter in habit. Decandolle includes Nitraria and Reaumuria among his Ficoideæ spuriæ, at the same time expressing a doubt whether they belong either to that or the same order. To me it appears that the affinities of Reaumuria are greater with Hypericum, and I accordingly adopt Dr. Ehrenberg's proposed separation of that genus along with Hololachna, the Tamarix songarica of Pallas, into a little order to be called Reaumurieæ. The affinity of Nitraria with Ficoideæ is undoubtedly great, especially with Tetragonia; but its very different embryo, and the peculiar æstivation of the petals, which is much more like that of Rhamneæ, remove it from that order.

GEOGRAPHY. Natives of western Asia and the north of Africa. One species is described from New Holland.

PROPERTIES. Slightly saline. Otherwise unknown.

EXAMPLE. Nitraria.

CL. ILLECEBREÆ.

HERNIARIÆ, *Cat. Hort. Par.* (1777). — ILLECEBREÆ, *R. Brown Prodromus*, 413. (1810); *Lindl. Synops.* 60. (1829). — PARONYCHIEÆ, *Aug. St. Hil. Mém. Plac. lib.* p. 56. (1815); *Juss. Mém. Mus.* 1. 387. (1815); *Dec. Prodr.* 3. 365. (1828); *Mémoire sur les Paronych.* (1829).

DIAGNOSIS. Polypetalous dicotyledons, with perigynous stamens opposite the 5 sepals, minute petals, concrete carpella, a 1-celled ovarium, and leaves with scarious stipulæ.

ANOMALIES. Petals very often wanting. Stamens sometimes hypogynous.

ESSENTIAL CHARACTER. — *Sepals* 5, seldom 3 or 4, sometimes distinct, sometimes cohering more or less. *Petals* minute, inserted upon the calyx between the lobes, occasionally wanting. *Stamens* perigynous, exactly opposite the sepals, if equal to them in number, sometimes fewer by abortion; *filaments* distinct; *anthers* 2-celled. *Ovarium* superior; *styles* 2 or 3, either distinct or partially combined. *Fruit* small, dry, 1-celled, either indehiscent, or opening with 3 valves. *Seeds* either numerous, upon a free central placenta, or solitary and pendulous from a funiculus originating in the base of the cavity of the fruit; *albumen* farinaceous; *embryo* lying on one side of the albumen, curved more or less, with the radicle always pointing to the hilum; *cotyledons* small.—*Herbaceous* or *half-shrubby* branching plants, with opposite or alternate, often fascicled, sessile, entire *leaves*, and scarious *stipulæ*. *Flowers* minute, with scarious *bracteæ*.

AFFINITIES. Very near Portulaceæ, Amarantaceæ, and Caryophylleæ, from which they are distinguished with difficulty. By excluding Sclerantheæ, which I consider, with Mr. Brown, a distinct order, their scarious stipulæ will distinguish them from the two last; and there is scarcely any other character that will; for there are Caryophylleæ that have perigynous stamens, as Larbrea and Adenarium, and Illecebreæ which have hypogynous ones, as Polycarpæa, Stipulicida, and Ortegia. From Portulaceæ they are scarcely to be known with absolute certainty, except by the position of the stamens before the sepals instead of the petals. With Crassulaceæ, particularly Tillæa, they agree very much in habit, but their concrete carpella will always distinguish them. Decandolle comprehends in the order various plants which have not stipulæ; but as the latter organs seem to be an essential part of the character, I should exclude his Queriaceæ, and Minuartieæ, which will be found elsewhere. The remaining tribes will be: —

1. TELEPHIEÆ.

Calyx 5-parted. Petals and stamens 5, arising from the bottom of the calyx. Styles 3, distinct, or slightly cohering at the base. — Leaves alternate.

EXAMPLES. Telephium, Corrigiola.

2. ILLECEBREÆ VERÆ.

Calyx 5-parted. Petals 5, or none. Stamens from 2 to 5, arising from the calyx. Styles distinct, or partially cohering. Capsule indehiscent, 1-seeded; an umbilical cord arising from the bottom, and bearing a some-what pendulous seed upon the apex. — Herbs, rarely under-shrubs. Leaves acute, opposite.

EXAMPLES. Illecebrum, Herniaria, Gymnocarpum.

3. POLYCARPÆÆ.

Calyx 5-parted. Petals 5, or none. Stamens from 1 to 5, arising from the bottom of the calyx. Styles 2 or 3, either distinct down to the base, or combined. Capsule 1-celled, many-seeded. Seeds attached to a central placenta. — Herbs or under-shrubs. Leaves opposite.

EXAMPLES. Polycarpæa, Stipulicida.

4. POLLICHIEÆ.

Calyx 5-toothed, with an urceolate tube. Stamens 1 or 2, arising from the throat. Petals none. Stigma bifid. Utriculus valveless, 1-seeded. Bracteæ (and perhaps also the calyx) enlarged after flowering, fleshy, and resembling a berry. — A suffruticose herb. Leaves opposite, somewhat whorled.

EXAMPLE. Pollichia.

GEOGRAPHY. The south of Europe and the north of Africa are the great stations of the order, where the species grow in the most barren places, covering with a thick vegetation soil which is incapable of bearing any thing else. A few are found at the Cape of Good Hope; and North America, including Mexico, comprehends several.

PROPERTIES. A trace of astringency pervades the order, and is the only sensible property that it is known to possess.

CLI. AMARANTACEÆ. THE AMARANTH TRIBE.

AMARANTHI, *Juss. Gen.* 87. (1789).—AMARANTHACEÆ, *R. Brown Prodr.* 413. (1810); *Von Martius Monogr.* (1826); *Lindley's Synopsis*, 213. (1829).

DIAGNOSIS. Apetalous dicotyledons, with erect seeds, an embryo curved round mealy albumen, radicle next the hilum, hypogynous stamens, and scarious bracteolate calyxes.

ANOMALIES. Stamens sometimes perigynous.

ESSENTIAL CHARACTER. — *Calyx* 3- or 5-leaved, hypogynous, scarious, persistent, occasionally with 2 bracteolæ at the base. *Stamens* hypogynous, either 5, or some multiple of that number, either distinct or monadelphous, occasionally partly abortive; *anthers* either 2-celled or 1-celled. *Ovarium* single, superior, 1- or few-seeded; the *ovules* hanging from a free central funiculus; *style* 1 or none; *stigma* simple or compound. *Fruit* a membra-nous utricle. *Seeds* lentiform, pendulous; *testa* crustaceous; *albumen* central, farinaceous; *embryo* curved round the circumference; *radicle* next the hilum; *plumula* inconspicuous.— *Herbs* or *shrubs*. *Leaves* simple, opposite or alternate, without stipulæ. *Flowers* in heads or spikes, usually coloured, occasionally unisexual, generally hermaphrodite. *Pubescence* simple, the hairs divided by internal partitions.

AFFINITIES. Different as this order appears to be from Chenopodeæ in habit, especially if we compare such a genus as Gomphrena with Chenopodium itself, it is so difficult to define the differences that distinguish the two orders, that, beyond habit, nothing certain can be pointed out. Mr. Brown remarks (*Prodr.* 413.), that he has not been able to ascertain any absolute diagnosis to distinguish them by; for the hypogynous insertion attributed to their stamens is not only not constant in the order, but is also found in some Chenopodeæ. Dr. Von Martius, in a learned dissertation upon the order, describes Chenopodeæ as being apetalous, and Amarantaceæ as polypetalous, considering the bracteolæ of these latter as a calyx, and that which I call a calyx a corolla. But it seems to me that this view of their structure is not borne out by analogy, and that it is impossible to believe the floral envelopes of the two orders to be of a different nature. I am certainly unable to indicate any better mode of distinguishing them than has been pointed out by those that have gone before me; and at the same time I cannot hesitate to keep asunder orders which it is evident that nature has divided. Bartling combines these plants in a single class, along with Caryophylleæ, Phytolacceæ, Sclerantheæ, and Illecebreæ; and there is no doubt of the near affinity borne to each other by all these, as is pointed out by their habit and by the structure of their seeds.

GEOGRAPHY. These plants grow in crowds or singly, either in dry, stony, barren stations, or among thickets upon the borders of woods, or a few even in salt marshes. They are much more frequent within the tropics than beyond them, and are unknown in the coldest regions of the world. 53 are found in tropical Asia, 105 in tropical America, but 5 in extratropical Asia, and but 21 in extra-tropical America; 5 are natives of Europe, 28 of New Holland, and 9 of Africa and its islands. See *Von Martius Monogr.*

PROPERTIES. Many of the species are used as potherbs, on account of the wholesome mucilaginous qualities of the leaves. Amaranthus obtusifolius is said to be diuretic. Several are objects of interest with gardeners for the beauty of their colouring and the durability of their blossoms. Gomphrena officinalis and macrocephala have a prodigious reputation in Brazil, where they are called Para todo, Perpetua, and Raiz do Padre Salerma: as the first of these names imports, they are esteemed useful in all kinds of diseases, especially in cases of intermittent fevers, colics, and diarrhœa, and against the bite of serpents. *Plantes Usuelles*, nos. 31. and 32.

EXAMPLES. Amaranthus, Gomphrena, Celosia.

CLII. SCLERANTHEÆ.

SCLERANTHEÆ, *Link Enum.* 417. (1821); *Dec. Prodr.* 3. 377. (1828) *a § of* Illecebreæ, *Lindley's Synopsis*, 217. (1829).—QUERIACEÆ, *a § of* Illecebreæ, Dec. l. c. (1828). —? MINUARTIEÆ, *ibid.*

DIAGNOSIS. Apetalous dicotyledons, with a single seed attached to a cord arising from the base of the cell, an inferior tubular indurated calyx, perigynous stamens, and an embryo curved round mealy albumen, with the radicle next the hilum.

ANOMALIES.

ESSENTIAL CHARACTER. — *Flowers* hermaphrodite. *Calyx* 4- or 5-toothed, with an urceolate tube. *Stamens* from 1 to 10, inserted into the orifice of the tube. *Ovarium* simple, superior, 1-seeded. *Styles* 2 or 1, emarginate at the apex. *Fruit* a membranous utricle enclosed within the hardened calyx. *Seed* pendulous from the apex of a funiculus, which arises from the bottom of the cell; *embryo* cylindrical, curved round farinaceous albumen.—Small *herbs.* *Leaves* opposite, without stipules. *Flowers* axillary, sessile.

AFFINITIES. Referred by Decandolle to Illecebreæ, from which they differ in absence of petals and stipules, these plants appear to me to constitute a distinct order, more nearly related to Chenopodeæ, from which they chiefly differ in the indurated tube of the calyx, from the orifice of which the stamens proceed, and in the number of the latter exceeding that of the divisions of the calyx. The tribe of Minuartias is probably not distinguishable from Scleranthæ, notwithstanding the supposed presence of petals, which would perhaps be more properly called abortive stamens.

GEOGRAPHY. Natives of barren fields in Europe, Asia, and North America, and in sterile places in countries of the southern hemisphere beyond the tropics. A single species is described from Peru.

PROPERTIES. Uninteresting weeds, of no known use.

EXAMPLES. Mniarum, Scleranthus.

CLIII. CHENOPODEÆ. THE GOOSEFOOT TRIBE.

ATRIPLICES, *Juss. Gen.* 83. (1789).—CHENOPODEÆ, *Vent. Tabl.* 2. 253. (1799); *R. Brown Prodr.* 405. (1810); *Lindley's Synopsis*, 213. (1829).

DIAGNOSIS. Apetalous dicotyledons, with erect seeds, an embryo curved round mealy albumen, radicle next the hilum, perigynous stamens, and herbaceous ebracteate calyxes.

ANOMALIES. Stamens sometimes hypogynous.

ESSENTIAL CHARACTER.— *Calyx* deeply divided, sometimes tubular at the base, persistent, with an imbricated æstivation. *Stamens* inserted into the base of the calyx, opposite its segments, and equal to them in number, or fewer. *Ovarium* single, superior, or occasionally adhering to the tube of the calyx, with a single *ovulum* attached to the base of the cavity; *style* in 2 or 4 divisions, rarely simple; *stigmas* undivided. *Fruit* membranous, not valvular, sometimes baccate. *Embryo* curved round farinaceous albumen, or spiral, or doubled together without albumen; *radicle* next the hilum; *plumula* inconspicuous.— *Herbaceous* plants or *under-shrubs.* *Leaves* alternate without stipulæ, occasionally opposite. *Flowers* small, sometimes polygamous.

AFFINITIES. The difficulty of distinguishing these from Amarantaceæ has been discussed under the latter order. They are distinguished from Phytolacceæ, independently of the simplicity of the structure of their ovarium, by their stamens never exceeding the number of the segments of the calyx, to which they are opposite: in Phytolacceæ, if they are not more numerous than the segments of the calyx, they are alternate with them.

GEOGRAPHY. Weeds inhabiting waste places in all parts of the world, but, unlike Amarantaceæ, abounding least within the tropics, and most in extra-tropical regions. They are exceedingly common in all the northern parts of Europe and Asia.

PROPERTIES. Some of these are used as potherbs, as Basella, Spinage, Garden Orach (Atriplex hortensis), and Chard Beet; the roots of others form valuable articles of food, as Beet and Mangel Wurzel. Many of them possess an essential oil, which renders them tonic and antispasmodic; such are Chenopodium ambrosioides and botrys. Chenopodium quinoa is a common article of food in Peru. But the most important of their qualities is the production of soda, which is yielded in immense quantities by the Salsolas Salicornias, and others. The essential oil of Chenopodium anthelminticum, known in North America under the name of Worm-seed Oil, is powerfully anthelmintic. *Barton*, 2. 187. The seeds of Atriplex hortensis are said to be so unwholesome as to excite vomiting. M. Chevallier has remarked the singular fact, that Chenopodium vulvaria exhales pure ammonia during its whole existence. This is the only observation upon record of a gaseous exhalation of azote by vegetables; and the facility with which this principle is abandoned by ammonia may perhaps explain the presence of azotic products in the vegetable kingdom. *Ann. des Sc. Nat.* 1. 444.

EXAMPLES. Chenopodium, Blitum, Atriplex.

CLIV. PHYTOLACCEÆ. THE VIRGINIAN POKE TRIBE.

PHYTOLACCEÆ, *R. Brown in Congo*, 454. (1818.)

DIAGNOSIS. Apetalous dicotyledons, with definite erect ovula, an inferior many-leaved calyx, distinct perigynous stamens, a multilocular ovarium, an embryo rolled round mealy albumen, with the radicle next the hilum, and terminal stigmas.

ANOMALIES. Rivina has only 1 carpellum.

ESSENTIAL CHARACTER. — *Calyx* of 4 or 5 petaloid leaves. *Stamens* either indefinite, or, if equal to the number of the divisions of the calyx, alternate with them. *Ovarium* of from 1 to several cells, each containing 1 ascending *ovulum; styles* and *stigmas* equal in number to the cells. *Fruit* baccate or dry, entire or deeply lobed, 1- or many-celled. *Seeds* ascending, solitary, with a cylindrical *embryo* curved round mealy *albumen*, with the *radicle* next the *hilum.* — *Under-shrubs* or *herbaceous* plants. *Leaves* alternate, entire, without stipulæ, often with pellucid dots. *Flowers* racemose.

AFFINITIES. Nearly related to Chenopodeæ and Polygoneæ, from the first of which they are distinguished by their multilocular ovarium, and by their stamens exceeding the number of divisions of the calyx; a circumstance which never occurs in Chenopodeæ. From Polygoneæ they are known by the radicle being turned towards the hilum, and the want of stipulæ. Rivina, which has the albumen very much reduced in quantity, and a unilocular fruit, connects Phytolacceæ with Petiveriaceæ. Mr. Brown remarks (*Congo*, 455) that these two orders, widely as they differ in the structure of the ovarium, are connected by a species of Phytolacca related to P. abyssinica, in which the 5 cells are so deeply divided that they merely cohere by their inner angles; and also by Gisekia, which has 5 distinct ovaria. But I do not think that the existence of these gradations of structure in the ovarium neutralises the remarkable differences that still exist between these two orders in embryo and stipulæ.

GEOGRAPHY. Natives of either America, within or without the tropics, Africa, and India. None have been found wild in Europe; but Phytolacca decandra is naturalised in some of the southern parts.

PROPERTIES. A tincture of the ripe berries of Phytolacca decandra seems to have acquired a well-founded reputation as a remedy for chronic and syphilitic rheumatism, and for allaying syphiloid pains. By some it is said to be more valuable than Guaiacum. Its pulverised root is an emetic. *Barton*, 2. 220. And a spirit distilled from the berries is stated to have killed a dog in a few minutes, by its violent emetic effects. According to Decandolle, this plant is also a powerful purgative. The leaves are extremely acrid, but the young shoots, which lose this quality by boiling in water, are eaten in the United States as Asparagus.

EXAMPLES. Phytolacca, Rivina.

CLV. PETIVERIACEÆ.

PETIVERIEÆ, *Agardh Classes*, (1825). — PETIVERIACEÆ, *Link Handb.* 1. 392. (1829.)

DIAGNOSIS. Apetalous dicotyledons, with definite erect ovula, an inferior many-leaved calyx, distinct perigynous stamens, an exalbuminous embryo with spiral cotyledons, and the radicle next the hilum.

ANOMALIES.

ESSENTIAL CHARACTER. — *Calyx* of several distinct leaves. *Stamens* perigynous, either indefinite, or, if equal to the segments of the calyx, alternate with them. *Ovarium* superior, 1-celled; *styles* 3 or more; *stigma* lateral; *ovulum* erect. *Fruit* 1-celled, indehiscent, dry. *Seed* erect, without *albumen ; embryo* straight ; *cotyledons* convolute; *radicle* inferior. — *Under-shrubs* or *herbaceous* plants, with an alliaceous odour. *Leaves* alternate, entire, with distinct stipulæ, often with minute pellucid dots. *Flowers* racemose.

AFFINITIES. Obviously akin both to Phytolacceæ and Polygoneæ, with the former of which Mr. Brown combines them. They are, however, distinguished from Phytolacceæ by the presence of stipulæ, and by their straight embryo destitute of albumen, and spiral cotyledons. From Polygoneæ they are known by the same characters, and also by the radicle being turned towards the hilum, and the stipulæ not having the form of Ochreæ.

GEOGRAPHY. West Indian or tropical American plants ; for the Seguiera asiatica of Loureiro probably does not belong to the order.

PROPERTIES. Nothing is known of their qualities, except that Petiveria alliacea yields a strong smell of garlic.

EXAMPLES. Petiveria, Seguiera.

CLVI. POLYGONEÆ. THE BUCK-WHEAT TRIBE.

POLYGONEÆ, *Juss. Gen.* 82. (1789) ; *R. Brown Prodr.* 418. (1810) ; *Lindl. Synops.* 209. (1829.)

DIAGNOSIS. Apetalous dicotyledons, with definite erect ovula, ochreate stipulæ, and a radicle remote from the hilum.

ANOMALIES. Eriogonum has not ochreate stipulæ.

ESSENTIAL CHARACTER. — *Calyx* divided, inferior, imbricated in æstivation. *Stamens* definite, inserted in the bottom of the calyx ; *anthers* dehiscing lengthwise. *Ovarium* superior, with a single erect ovulum ; *styles* or *stigmas* several. *Nut* usually triangular, naked, or protected by the calyx. *Seed* with farinaceous albumen, rarely with scarcely any ; *embryo* inverted, generally on one side ; *plumula* inconspicuous ; *radicle* at the end remote from the hilum.—*Herbaceous* plants, rarely *shrubs.* *Leaves* alternate, their stipulæ cohering round the stem in the form of an ochrea ; when young, rolled backwards. *Flowers* occasionally unisexual, often in racemes.

AFFINITIES. Mr. Brown remarks, that "the erect ovulum with a superior radicle together afford the most important mark of distinction between Polygoneæ and Chenopodeæ, a character which obtains even in the genus Eriogonum, in which there is no petiolar sheath, and scarcely any albumen, the little that exists being fleshy." Generally speaking, however, the cohesion of the scarious stipulæ into a sheath, technically called an ochrea, or boot, is sufficient to distinguish Polygoneæ from all other plants. For their relation to Begoniaceæ, see that order.

GEOGRAPHY. There are few parts of the world that do not acknowledge the presence of plants of this order. In Europe, Africa, North America, and Asia, they fill the ditches, hedges, and waste grounds, in the form of Docks and Persicarias; the fields, mountains, and heaths, as Sorrels and trailing or twining Polygonums; in South America and the West Indies they take the form of Coccolobas or sea-side grapes; in the Levant, of Rhubarbs; and even in the desolate regions of the North Pole they are found in the shape of Oxyria.

PROPERTIES. Sorrel on the one hand, and Rhubarb on the other, may be taken as the representatives of the general qualities of this order. While the leaves and young shoots are acid and agreeable, the roots are universally nauseous and purgative. To these two qualities is to be superadded a third, that of astringency, which is found in a greater or less degree in the whole order, but which becomes in Coccoloba uvifera so powerful as to rival Gum Kino in its effects. Some of the Polygonums are extremely acrid, as the P. Hydropiper, which is said to blister the skin. There is a species of Polygonum, called Cataya in the language of the Brazilian Indians, an infusion of the ashes of which is used to purify and condense the juice of the sugar-cane. It has a very bitter peppery taste, and is employed on the Rio St. Francisco with advantage in the disease called O Largo, which is an enlargement of the colon, caused by debility. *Pr. Max. Trav.* 71. The stem of the Rheum has been supposed to contain a peculiar acid called the rheumic, but this is now known to be the oxalic. *Turner,* 641. Rumex acetosa contains pure oxalic acid. *Ibid.* 623. The principle in which the active property of Rhubarb exists is supposed to be a peculiar chemical substance called Rhubarbarin. *Ibid.* 701. Some information may be found upon the Rhubarbs of India in the *Trans. of the Med. and Phys. Soc. of Calcutta,* 3. 438. by Dr. Royle; but nothing certain had been collected by him with regard to the plant producing the true officinal substance. Many species of Polygonum are used in dyeing. The seeds of P. fagopyrum and tataricum are used as food, for the sake of their mealy albumen; those of P. aviculare are said to be powerfully emetic and purgative; but this is doubted by Meisner. *Mon.* 49. The seeds of Polygonum barbatum are used as medicine by Hindoo practitioners, to ease the pain of griping in the colic. *Ainslie,* 2. 2. The leaves of P. hispidum are said by Humboldt to be substituted, in South America, for Tobacco. *N. G. and Sp.* 2. 178.

EXAMPLES. Rheum, Rumex, Coccoloba.

CLVII. BEGONIACEÆ.

BEGONIACEÆ, *R. Brown in Congo,* 464. (1818); *Link Handb.* 1. 309. (1829); *Martius H. Reg. Mon.* (1829.)

DIAGNOSIS. Apetalous dicotyledons, with a 3-celled winged ovarium, indefinite ovules, an irregular imbricated calyx, and membranous stipulæ.
ANOMALIES.

ESSENTIAL CHARACTER. — *Flowers* unisexual. *Sepals* superior, coloured; in the males 4, 2 within the others and smaller; in the females 5, imbricated, two smaller than the rest. *Stamens* indefinite, distinct or combined into a solid column; *anthers* collected in a head, 2-celled, continuous with the filaments, clavate, the connectivum very thick, the cells minute, bursting longitudinally. *Ovarium* inferior, winged, 3-celled, with 3 double polyspermous placentæ in the axis; *stigmas* 3, 2-lobed, sessile, somewhat spiral. *Fruit* membranous, capsular, winged, 3-celled, with an indefinite number of minute seeds; bursting by slits at the base on each side of the wings. *Seeds* with a transparent thin *testa* marked by reticulations, which are oblong at the sides and contracted at either extremity; *embryo* very cellular, without *albumen*, with a blunt round *radicle* next the hilum. — *Herbaceous* plants or *under-shrubs*, with an acid juice. *Leaves* alternate, toothed oblique at the base. *Stipulæ* scarious. *Flowers* pink, in cymes.

AFFINITIES. It is not easy to fix with precision the relative position of this order : I formerly thought it related to Hydrangeæ, chiefly on account of the striking resemblance in the areolations of the seeds, and the irregularity of the flowers. It is probable, however, that more importance should be attributed to the acid juice and membranous large stipulæ, in which case Begoniaceæ are most nearly related to Polygoneæ, many of which have a coloured calyx and 3-cornered fruit: from which they differ in the structure of the fruit and seed. Link places them near Umbelliferæ; but I know not upon what grounds.

GEOGRAPHY. Common in the West Indies, South America, and the East Indies. Mr. Brown remarks, that no species has been found on the continent of Africa, though several have been found in Madagascar and the Isles of France and Bourbon, and 1 in the island of Johanna. *Congo*, 464.

PROPERTIES. The roots are astringent and slightly bitter. Those of 2 species are used in Peru with success in cases of a flux of blood, or in other visceral diseases in which astringents are employed. They are also said to be useful in cases of scurvy, and in certain fevers.

EXAMPLE. Begonia.

CLVIII. NYCTAGINEÆ. THE MARVEL OF PERU TRIBE.

NYCTAGINES, *Juss. Gen.* 90. (1789); *R. Brown Prodr.* 421. (1810.)

DIAGNOSIS. Apetalous dicotyledons, with definite ascending ovula, an inferior tubular (often coloured) calyx hardening at the base, hypogynous stamens, and embryo surrounding floury albumen.

ANOMALIES.

ESSENTIAL CHARACTER.— *Calyx* tubular, somewhat coloured, contracted in the middle; its limb entire or toothed, plaited in æstivation, becoming indurated at the base. *Stamens* definite, hypogynous; *anthers* 2-celled. *Ovary* superior, with a single erect ovulum; *style* 1; *stigma* 1. *Fruit* a thin utricle, enclosed within the enlarged persistent tube of the calyx. *Seed* without its proper integuments, its testa being coherent with the utricle; *embryo* with foliaceous cotyledons, wrapping round floury *albumen; radicle* inferior; *plumula* inconspicuous. — *Stem* either herbaceous, shrubby, or arborescent. *Leaves* opposite, and almost always unequal; sometimes alternate. *Flowers* axillary or terminal, clustered or solitary, having an involucrum which is either common or proper, in one piece or in several pieces, sometimes minute.

AFFINITIES. The tubular calyx, the limb of which is plaited in æstivation, and the base of which becomes hardened round the ovarium, so that it resembles a woody pericarp, will, if taken with the curved embryo and farinaceous albumen, at all times distinguish Nyctagineæ; add to which, the articulations are tumid, as in Geraniaceæ. Its nearest affinity is perhaps with Polygoneæ, from which it, however, differs so much that it need not be compared with them.

GEOGRAPHY. Natives of the warmer parts of the world in either hemisphere, scarcely extending far beyond the tropics, except in the case of the Abronias found in North-west America.

PROPERTIES. In consequence of the generally purgative quality of the roots of species of this family, one of them was supposed to have been the true jalap plant, which is, however, now known to be a mistake. The flowers of several species of Mirabilis are handsome, as are those also of some of the Abronias; but the greater part of the order is composed of obscure weeds. The genus Pisonia consists of trees or shrubby plants.

EXAMPLES. Mirabilis, Boerhaavia, Oxybaphus.

CLIX. SAURUREÆ.

Saurureæ, *Rich. Anal.* (1808) ; *Meyer de Houttuynia atque Saurureis*, (1827) ; *Martius Hort. Monac.* (1829.)

Diagnosis. Achlamydeous dicotyledons, with 4 carpella, ascending ovules, and embryo in a sac.

Anomalies.

Essential Character. — *Flowers* naked, seated upon a scale, hermaphrodite. *Stamens* 6, clavate, hypogynous, persistent ; *filaments* slender ; *anthers* continuous with the filament, cuneate, with a thick connectivum and 2 lateral lobes bursting longitudinally. *Ovaria* 4, each distinct, with 1 ascending ovulum and a sessile recurved *stigma*, or connate into a 3- or 4-celled pistillum, with a few ovula ascending from the edge of the projecting semi-dissepiments. *Fruit* either consisting of 4 fleshy indehiscent nuts, or a 3- or 4-celled capsule, opening at the apex and containing a few ascending seeds. *Seeds* with a membranous integument ; *embryo* minute, lying in a fleshy lenticular sac, which is seated on the outside of hard mealy *albumen* at the end most remote from the hilum.—*Herbaceous* plants, growing in marshy places, or floating in water. *Leaves* alternate, with *stipulæ*. *Hairs* jointed. *Flowers* growing in spikes.

Affinities. Very near Piperaceæ, with which they agree in habit, but from which they differ in the compound nature of their ovarium, and their numerous stamens. From repeated examination of the embryo of Saururus, I have no doubt whatever that the embryo has no kind of vascular connexion with the sac that contains it ; and hence I adopt the opinion of Mr. Brown, that this sac is in reality nothing but the remains of the amnios surrounding the embryo. For the opinions of Mirbel and Richard upon this subject, see the figures and remarks of the former in *Ann. Mus.* 16. 449., and of the latter in *Humboldt and Bonpl. N. Gen. et Sp.* 1. 3. ; the latter being unquestionably wrong in considering the sac a portion of the embryo. This order is one of those which tend to destroy the distinction between Monocotyledons and Dicotyledons. Its affinity with Fluviales is indicated by the floating habit and general appearance of Aponogeton, and with Typhineæ by its anthers ; but its foliage and stipulæ are those of Dicotyledons, and the structure of the seed and the position of the embryo in a fleshy sac demonstrate its vicinity to Piperaceæ.

Geography. Natives of North America, China, the north of India, and the Cape of Good Hope, growing in marshes or pools of water.

Properties. Unknown.

Examples. Saururus, Aponogeton.

CLX. CHLORANTHEÆ.

Chlorantheæ, *R. Brown in Bot. Mag.* 2190. (1821) ; *Lindl. Collect. Bot.* 17. (1821) ; *Meyer de Houttuynia atque Saurureis*, 51. (1827) ; *Blume Flora Javæ*, (1829.)

Diagnosis. Achlamydeous herbaceous dicotyledons, with a 1-celled ovarium, a pendulous ovulum, opposite leaves, spiked flowers, and an embryo not enclosed in a sac.

Anomalies. Saururus, Aponogeton.

Essential Character. — *Flowers* naked, spiked, hermaphrodite, or unisexual, with a supporting scale. *Stamens* lateral ; if more than 1, connate, definite ; *anthers* 1-celled, bursting longitudinally, each adnate to a fleshy connectivum, which coheres laterally in various degrees (2-celled, according to some) ; *filament* slightly adhering to the ovarium. *Ovarium* 1-celled ; *stigma* simple, sessile ; *ovule* pendulous. *Fruit* drupaceous, indehiscent. *Seed* pendulous ; *embryo* minute, placed at the apex of fleshy albumen, with the radicle inferior, and consequently remote from the hilum ; *cotyledons* divaricate. — *Herbaceous* plants or *under-shrubs*, with an aromatic taste. *Stems* jointed, tumid under the articulations. *Leaves* opposite, simple, with sheathing petioles and minute intervening *stipulæ*. *Flowers* in terminal spikes.

AFFINITIES. Nearly allied to Saurureæ and Piperaceæ, from both which they differ in the want of a sac to the embryo, and in the pendulous ovule, and opposite leaves with intermediate stipulæ. Their anthers consist of a fleshy mass, upon the face of which the cell lies that bears the pollen; whether these anthers are 1- or 2-celled, is a matter of doubt; one botanist considering those which have 2 cells to be double anthers, another understanding those with 1 cell to be half anthers. Dr. Blume describes a calyx as being sometimes present in a rudimentary state, adhering to the ovarium, and hence he suspects some affinity between these plants and Opercularineæ. But I am persuaded that no such rudiment exists ; it is not represented in Dr. Blume's figures.

GEOGRAPHY. Natives of the hot parts of India and South America, the West Indies, and Society Islands.

PROPERTIES. The whole plant of Chl. officinalis has an aromatic fragrant smell, which is gradually dissipated in drying ; but its roots retain a fragrant camphorated smell, and an aromatic, somewhat bitter, flavour. They are found to possess very nearly the properties of Aristolochia serpentaria, and in as high a degree. There seems to be no doubt that it is a stimulant of the highest order. See *Blume Fl. Jav.*

EXAMPLES. Chloranthus, Ascarina, Hedyosmum.

CLXI. LACISTEMEÆ.

LACISTEMEÆ, *Martius N. G. et Sp. Pl.* 1. 154. (1824.)

DIAGNOSIS. Apetalous dicotyledons, with indefinite ovules, a 1-celled ovarium with parietal placentæ, dehiscent fruit, amentaceous hermaphrodite flowers, and hypogynous unilateral stamens.

ANOMALIES.

ESSENTIAL CHARACTER. — *Calyx* in several narrow divisions, inferior, covered over by a dilated bractea. *Corolla* wanting. *Stamens* hypogynous, standing on one side of the ovarium, with a thick 2-lobed connectivum, at the apex of each of which lobes is placed a single cell of an anther, bursting transversely. *Ovarium* superior, seated in a fleshy disk, 1-celled, with several *ovula* attached to parietal placentæ ; *stigmas* 2 or 3, sessile or on a style. *Fruit* capsular, 1-celled, splitting into 2 or 3 valves, each of which bears a placenta in its middle. *Seed* usually, by abortion, solitary, suspended, with a fleshy arillus ; *integument* crustaceous ; *albumen* fleshy ; *embryo* inverted, with plane *cotyledons* and a superior straight cylindrical *radicle*. — *Small trees* or *shrubs*. *Leaves* simple, alternate, with stipulæ. *Flowers* disposed in clustered axillary amenta.

AFFINITIES. Dr. Von Martius, the founder of this order, which he divides from Urticeæ, speaks of it thus : " The peculiar character consists in the presence of a distinct perianthum, while the amentaceous inflorescence is an indication of an affinity with apetalous orders of a lower grade." The same botanist indicates their affinity with Chlorantheæ in the structure of the filament, and with Samydeæ in that of their fruit, " the monadelphous stamens of both which may be perhaps considered a higher kind of evolution of the fleshy disk in the bottom of the flower of Lacistema." In habit they are something like Piperaceæ, but more arborescent.

GEOGRAPHY. Natives of low places in woods in equinoctial America.

PROPERTIES. Unknown.

EXAMPLE. Lacistema.

CLXII. PIPERACEÆ. The Pepper Tribe.

PIPERACEÆ, *Rich. in Humb. Bonpl. et Kunth N. G. et Sp. Pl.* 1. 39. t. 3. (1815);
Meyer de Houttuynia atque Saurureis, (1827.)

DIAGNOSIS. Achlamydeous dicotyledons, with a 1-celled ovarium, erect ovules, and an embryo enclosed in a sac.

ANOMALIES.

ESSENTIAL CHARACTER. — *Flowers* naked, hermaphrodite, with a bractea on the outside. *Stamens* definite or indefinite, arranged on one side or all round the ovarium, to which they adhere more or less; *anthers* 1- or 2-celled, with or without a fleshy connectivum; *pollen* smooth. *Ovarium* superior, simple, 1-celled, containing a single erect *ovulum; stigma* sessile, simple, rather oblique. *Fruit* superior, somewhat fleshy, indehiscent, 1-celled, 1-seeded. *Seed* erect, with the embryo lying in a fleshy sac placed at that end of the seed which is opposite the hilum, on the outside of the albumen. — *Shrubs* or *herbaceous* plants. *Leaves* opposite, verticillate, or alternate in consequence of the abortion of one of the pair of leaves, without *stipulæ*. *Flowers* usually sessile, sometimes pedicellate, in spikes which are either terminal, or axillary, or opposite the leaves.

AFFINITIES. As we approach the Monocotyledonous division of vegetables, we find the distinction between them and Dicotyledons, as derived from their anatomical structure, becoming weaker and weaker; but at the same time it appears to me that sufficient distinctions are still visible between these two modes of growth. Of this Piperaceæ are an instance. According to Richard, they are Monocotyledonous; an opinion in which Blume concurs, after an examination of abundance of species in their native places of growth. See *Ann. des Sc.* 12. 222. But if the medullary rays constitute the great anatomical difference between these divisions of the vegetable kingdom (and I know of no other which is absolute), then Piperaceæ are surely Dicotyledonous, as is shewn by Meyer (*Dissertatio de Houttuynia,* 38), and as may be ascertained by any one who will look at an old stem of any Pepper; add to this, the veins of their leaves having a distinct articulation with the stem, and the 2-lobed embryo; and it seems to me impossible to doubt their being properly stationed among Dicotyledons. In this view they are closely related to Polygoneæ, Saurureæ, and Urticeæ, from all which, however, they are distinguished by obvious characters; and also to Chorantheæ, from which they differ in the point of attachment of the ovule, and in the distinct existence of the remains of the amnios in the form of a sac around the embryo. In the opinion of those who believe Piperaceæ to be Monocotyledons, their station is near Aroideæ, with which, indeed, they must be considered in any point of view to be closely connected.

GEOGRAPHY. Exclusively confined to the hottest parts of the world. They are extremely common in tropical America and the Indian archipelago, but, according to Mr. Brown, are very rare in equinoctial Africa. Only 3 species have been found on the west coast; several exist at the Cape of Good Hope. *Congo,* 464.

PROPERTIES. Common Pepper, so well known for its pungent, stimulant, aromatic quality, represents the ordinary property of the order, which is not confined to the fruit only, but which pervades all the parts in a greater or less degree. The Cubebs of the shops, remarkable for their extraordinary power of allaying inflammation in the urethra and in the mucous membrane of the intestinal canal, are the dried fruit of Piper cubeba. *Ainslie,* 1. 98. The chemical principle called Piperin has been found in Black Pepper. *Turner,* 700. Piper anisatum has a strong smell of Anise, and a decoction of its berries is used to wash ulcers. Betel, an acrid stimulating substance, much used for chewing by the Malays, is the produce of Piper Betel, and Siriboa. Finally, P. inebrians possesses narcotic properties, of which the South Sea islanders avail themselves for preparing an intoxicating beverage. *Dec.*

EXAMPLES. Piper, Peperomia.

CLXIII. PODOSTEMEÆ.

PODOSTEMEÆ, *Richard and Kunth in Humb. N. G. et Sp.* 1. 246. (1815); *Martius Nov. G. et Sp.* 1. 6. (1822.)

DIAGNOSIS. Achlamydeous herbaceous dicotyledons, with a 2-celled polyspermous capsule, and solitary flowers.

ANOMALIES.

ESSENTIAL CHARACTER.—*Flowers* naked, hermaphrodite, bursting through an irregularly lacerated spatha. *Stamens* hypogynous, varying from 2 to an indefinite number, either placed all round the ovarium or on one side of it, monadelphous, alternately sterile; *anthers* oblong, 2-celled, bursting longitudinally. *Ovarium* 2-celled, with numerous ovula attached to a fleshy central placenta; *styles* or *stigmas* 2 or 3, and sessile. *Fruit* slightly pedicellate, ribbed, capsular, opening by 2 valves, which fall off from the dissepiment, which is parallel with them. *Seeds* numerous, minute, their structure unknown, or, according to Von Martius, entirely simple. — *Herbaceous* branched floating plants. *Leaves* capillary, or linear, or lacerated irregularly, or minute and densely imbricated, decurrent on the stem, with which they are not articulated. *Flowers* axillary or terminal, inconspicuous.

AFFINITIES. Little is at present known of the real characters of this curious order. Only 2 of its genera, Mniopsis and Lacis, have been well described, and even these are still but imperfectly understood. Dr. Von Martius has the following remarks upon it: " It is very doubtful in what part of the natural series Podostemeæ should be arranged; for they are connected with so many other orders, in so various and complicated a manner, that it is highly probable that several genera, the affinities of which will be more apparent, still remain to be discovered. Nothing can be more singular than the mixture of different characters which they exhibit. Thus, the structure of their spathes, and the want of a true calyx and corolla, approximate them to Naiades (Fluviales) and Aroideæ, while the character of their stamens and fruit is very much that of Juncagineæ; the former of these, however, differ in their lower degree of organisation, and the latter in the presence of a more or less perfect perianthium, and in the composition of their capsule. Lemna, a genus closely allied to Aroideæ, seems to be more related to them in its spatha, hypogynous stamens, habit, and mode of life, but is distinguished by its less highly developed few-seeded fruit. Again, Mniopsis, in its ramification, in the form and position of its leaves, and in its stipulæ, and Lacis and Podostemum in the character of their spatha and the emersion of their pedicels at the time of flowering, call remarkably to mind the habit of Jungermanniæ; so that we should probably not be far from the truth, if we were to say that this order forms a transition from Naiades (Fluviales) to Juncagineæ, on the one hand touching upon Aroideæ, thus being, as it were, a sort of noble analogy of Hepaticæ among monocotyledons." *Nov. G. et Sp.* 1. 7. Upon this it is difficult to make any additional remarks, without being in possession of a more complete knowledge of their structure. I must, however, observe, that it appears to me clear that Podostemeæ are not monocotyledons, as Von Martius, Kunth, and Richard, suppose, but dicotyledons; for which I have to offer the following reasons: In the first place their habit is that of dicotyledons, and not of monocotyledons; Podostemon being very like a starved Pepper, and Hydrostachys having its flowers in spikes resembling those of Saururus. Tristicha has minute scale-like leaves, imbricated in 3 rows, like which there is nothing among mo-nocotyledons. To this may be added the binary division of the ovarium, which is analogous to that of many dicotyledons, but a very rare structure among monocotyledons. Finally, the vernation of the leaves of Mourera of Aublet (t. 233), and of Marathrum, which is perhaps not distinct, is entirely that of dicotyledons, rather than of monocotyledons. I incline to place the order in the neighbourhood of Piperaceæ, to which it probably approaches more nearly than to any plants hitherto discovered.

GEOGRAPHY. Natives of still waters and damp places in South America and the islands off the east coast of Africa; 1 species is found in North America.

PROPERTIES. Unknown.

EXAMPLES. Lacis, Podostemum, Hydrostachys.

CLXIV. CALLITRICHINEÆ.

CALLITRICHINEÆ, *Link Enum.* 1. 7. (1821); *Dec. Prodr.* 3. 71. (1828); *a sect. of* Halorageæ. *Lindl. Synops.* 242. (1829.)

DIAGNOSIS. Achlamydeous herbaceous dicotyledons, with a 4-celled ovarium, and solitary peltate seeds.

ANOMALIES.

ESSENTIAL CHARACTER. — *Flowers* usually unisexual, monœcious, naked, with 2 fistular coloured bracteæ. *Stamen* single; *filament* filiform, furrowed along the middle; *anther* reniform, 1-celled, 2-valved; the valves opening fore and aft. *Ovarium* solitary, 4-cornered, 4-celled; *ovules* solitary, peltate; *styles* 2, right and left, subulate; *stigmas* simple points. *Fruit* 4-celled, 4-seeded, indehiscent. *Seeds* peltate; *embryo* inverted in the axis of fleshy *albumen;* *radicle* very long, curved, superior; *cotyledons* very short. — Small aquatic *herbaceous* plants, with opposite, simple, entire *leaves. Flowers* axillary, solitary, very minute.

AFFINITIES. I have remarked in my *Synopsis,* that " the affinity of this order to other dicotyledons appears to be of precisely the same nature as that borne by Lemna to monocotyledons: they each exhibit the lowest degree of organisation known in their respective classes." Mr. Brown considers it allied to Halorageæ; an opinion in which I concur, without adopting Decan-dolle's explanation of the structure of the flowers; but at the same time I confess that this affinity is less strong than could be wished; is it not rather an anomalous form of a reduced Euphorbiacea, or is it related to Podostemeæ? All this is still a problem.

GEOGRAPHY. Natives of still waters in Europe and North America.

PROPERTIES. Unknown.

EXAMPLE. Callitriche.

CLXV. CERATOPHYLLEÆ.

CERATOPHYLLEÆ, *Dec. Prodr.* 3. 73. (1828); *Lindl. Synops.* 225. (1829.)

DIAGNOSIS. Apetalous dicotyledons, with definite pendulous ovula, solitary flowers, a 1-celled ovarium, and many-parted calyx.

ANOMALIES.

ESSENTIAL CHARACTER.— *Flowers* monœcious. *Calyx* inferior, many-parted. *Male:* *Stamens* from 12 to 20; *filaments* wanting; *anthers* 2-celled. *Female:* *Ovarium* superior, 1-celled; *ovule* solitary, pendulous; *stigma* filiform, oblique, sessile. *Nut* 1-celled, 1-seeded, indehiscent, terminated by the hardened stigma. *Seed* pendulous, solitary; *albumen* 0; *embryo* with 4 cotyledons, alternately smaller; *plumula* many-leaved; *radicle* superior. (*Dec.*) — Floating *herbs*, with multifid cellular *leaves.*

AFFINITIES. These are not at all made out. In consequence of the number of its cotyledons, Richard placed it near Coniferæ, with which it seems to have no kind of affinity. Decandolle urges its relation to Hippuris and Myriophyllum, among Halorageæ, from which it differs in its superior ovarium; and he inquires whether Naias, which according to some is dicotyledonous, does not belong to the same order. Can this family have any relation to Podostemeæ? Agardh places it among Fluviales.

GEOGRAPHY. Found in ditches in Europe.

PROPERTIES. Unknown.

EXAMPLE. Ceratophyllum.

2. MONOPETALOUS PLANTS.

THE character by which this division of Dicotyledons is distinguished from the last (p. 2.), is the cohesion of the edges of the petals into a tube; whence the name Monopetalous, the petals forming together a single floral envelope. Generally it is easy to recognise this character, and the orders thus distinguished are individually perfectly natural; but occasionally certain genera in Polypetalous orders have flowers with a Monopetalous corolla, as in Crassulaceæ ; these cases are, however, rare, and are to be considered exceptions to the rule. For the most part, in Monopetalous plants belonging to Polypetalous orders, the petals are readily separable from each other, which is not the case in genuine Monopetalæ; but this is not always so. Apetalous exceptions are exceedingly uncommon : Glaux, among Primulaceæ, is a rare instance of this.

Monopetalous orders approach those which are Polypetalous, Apetalous, or Achlamydeous, at many points besides such as are adverted to at p. 2, especially by Ilicineæ, which are nearly allied to Rhamneæ.

LIST OF THE ORDERS.

CLXVI. ILICINEÆ. THE HOLLY TRIBE.

ILICINEÆ, *Ad. Brongniart Mémoire sur les Rhamnées,* p. 16. (1826); *Lindl. Synops.* p. 73. (1829).—AQUIFOLIACEÆ, *Dec. Théorie,* ed. 1. 217. (1813) ; *a sect. of* Celastrineæ, *Ib. Prodr.* 2. 11. (1825) ; *Martius H. R. Mon.* (1829.)

DIAGNOSIS. Monopetalous dicotyledons, with a superior 2-6-celled ovarium, regular flowers, definite pendulous ovules, a 4-6-lobed corolla, with the stamens equal to the number of its lobes, and albuminous seeds.

ANOMALIES.   Flowers unisexual in Prinos and Nemopanthes.

ESSENTIAL CHARACTER. — *Sepals* 4 to 6, imbricated in æstivation. *Corolla* 4- or 5-parted, hypogynous, imbricated in æstivation. *Stamens* inserted into the corolla, alternate with its segments; *filaments* erect; *anthers* adnate. *Disk* none. *Ovarium* fleshy, superior, somewhat truncate, with from 2 to 6 cells; *ovula* solitary, pendulous from a cup-shaped funiculus; *stigma* subsessile, lobed. *Fruit* fleshy, indehiscent, with from 2 to 6 stones. *Seed* suspended, nearly sessile; *albumen* large, fleshy; *embryo* small, 2-lobed, lying next the hilum, with minute *cotyledons*, and a superior *radicle*. — *Trees* or *shrubs*. *Leaves* alternate or opposite, coriaceous. *Flowers* small, axillary, solitary or fascicled.

AFFINITIES.   Included in Rhamneæ by most botanists, but well distinguished by Ad. Brongniart, who remarks that the suggestion of M. de Jussieu in his *Genera Plantarum*, that Ilicineæ ought probably to be placed among Monopetalæ, near Sapoteæ or Ebenaceæ, will probably be adopted.   From Celastrineæ, with which they are combined in most modern works, they differ in the form of their calyx and corolla, in the disposition and insertion of their stamens, and especially in the structure of their ovarium and fruit.   In these respects they are found by M. Brongniart to agree so completely with Ebenaceæ, that that order does not, in fact, differ essentially from Ilicineæ, except in characters of a secondary order, such as the calyx and corolla less deeply divided, the stamens often double the number of the segments of the corolla, the style being sometimes divided, the cells of the ovarium usually containing 2 collateral ovula, and finally in the cells of the fruit not becoming bony, as in most Ilicineæ.   Von Martius places them near Polygaleæ.

GEOGRAPHY.   Found in various parts of the world, especially in the West Indies, South America, and the Cape of Good Hope.   Several are found in North America; but 1, the common Holly, in Europe.

PROPERTIES.   The bark and berries of Prinos verticillatus possess, in an eminent degree, the properties of vegetable, astringent, and tonic medicines, along with antiseptic powers which are highly spoken of by American practitioners.   *Barton*, 1. 208.   Prinos glaber and Ilex Paraguensis are used as tea: the latter yields the famous beverage called Maté in Brazil.   Myginda Gongonha is diuretic.   *Dec.*

EXAMPLES.   Ilex, Prinos.

CLXVII. STYRACEÆ.

STYRACEÆ, *Rich. Anal. du Fr.* (1808); *Von Martius N. Gen. et Sp. Pl.* 2. 148. (1826). — EBENACEÆ, a § of Styraceæ, *Dec. and Duby*, 320. (1828). — SYMPLOCINEÆ, *Don Prodr. Nep.* 144. (1825). — STYRACINÆ, *Rich. in Humb. N. G. et Sp.* 3. 256. (1818); *Synops.* 2. 315. (1823). — HALESIACEÆ, *Don in Jameson's Journ.* (*Dec.* 1828); *Link Handb.* 1. 667. (1829.)

DIAGNOSIS.   Monopetalous dicotyledons, with an inferior ovarium of several cells, definite ovula, and alternate leaves.

ANOMALIES.

ESSENTIAL CHARACTER. — *Calyx* inferior or superior, with 5 divisions, persistent. *Corolla* hypogynous, monopetalous, the number of its divisions frequently different from that of the calyx; with imbricated æstivation. *Stamens* definite or indefinite, arising from the tube of the corolla, of unequal length, cohering in various ways, but generally in a slight degree only; *anthers* innate, 2-celled, bursting inwardly. *Ovarium* superior, or adhering to the calyx, with from 3 to 5 cells; *ovules* definite, the upper persistent, the lower pendulous, or *vice versâ*; *style* simple; *stigma* somewhat capitate. *Fruit* drupaceous, surmounted by or enclosed in the calyx, with from 1 to 5 cells. *Seeds* ascending or suspended, solitary, with the *embryo* lying in the midst of the *albumen*; *radicle* long, directed towards the hilum ; *cotyledons* flat, foliaceous. —*Trees* or *shrubs*. *Leaves* alternate, without stipulæ, usually toothed, turning yellow in drying. *Flowers* axillary, either solitary or clustered, with scale-like bracteæ. The *hairs* often stellate.

AFFINITIES. The plants comprehended under this name require a careful examination and settlement. They have been at one time combined with Ebenaceæ, or divided into the two orders of Styraceæ and Symplocaceæ, from both which Halesiaceæ have been again separated by Don and Link. From Ericeæ they differ in habit, in the definite number of their seeds, and their inferior ovarium ; from Ebenaceæ in the latter character, in the perigynous insertion of the stamens, in the peculiar circumstance of part of the ovules being erect and part inverted, and in the style being simple. Von Martius considers Styraceæ as gamopetalous rather than monopetalous ; but what is the real difference in the meaning of these two words ? Mr. Don says that Halesiaceæ are a group widely different from Styraceæ. *Jameson's Journ.* 1828. *Dec.* The genus Symplocos is rather different in habit from Styrax and Halesia, turning yellow in drying. Jussieu refers Styrax to Meliaceæ, with which family the order has no doubt much affinity. Decandolle considers them nearly akin to Ternströmiaceæ. *Essai Médic.* 203.

GEOGRAPHY. Found in North and South America within and without the tropics, and in tropical Asia and China.

PROPERTIES. Some of the genus Symplocos are used in dyeing yellow ; others, as Alstonia theiformis, are employed as tea, on account of a slight astringency in their leaves. Storax and Benzoin, two fragrant gum-resins, composed of resin, benzoic acid, and a peculiar aromatic principle, are the produce of two species of Styrax.

EXAMPLES. Styrax, Halesia, Symplocos.

CLXVIII. BELVISIACEÆ.

BELVISIEÆ, *R. Brown in Linn. Trans.* 13. 222. (1820.)

DIAGNOSIS. Monopetalous dicotyledons, with an inferior ovarium, a plaited many-lobed corolla, alternate leaves, and indefinite ovula.

ANOMALIES. Unknown.

ESSENTIAL CHARACTER.—*Calyx* of 1 piece, persistent, with a divided limb. *Corolla ?* monopetalous, plaited (many-lobed or undivided, simple or double), deciduous. *Stamens* either definite or indefinite, arising from the base of the corolla. *Ovarium* inferior ; *style* 1 ; *stigma* lobed or angular. *Fruit* berried, many-seeded. — *Shrubs.* *Leaves* alternate, entire, without stipulæ. *Flowers* axillary or lateral, solitary. *R. Br.*

AFFINITIES. Little is known of this obscure family, except that it is not referable to any order at present established. In fixing it near Styraceæ, it can only be said to resemble that order as much as any other.

GEOGRAPHY. African shrubs or trees.

PROPERTIES. Unknown.

EXAMPLE. Belvisia.

CLXIX. SAPOTEÆ. THE SAPPODILLA TRIBE.

SAPOTÆ, *Juss. Gen.* 151. (1789). — SAPOTEÆ, *R. Brown Prodr.* 528. (1810.)

DIAGNOSIS. Monopetalous dicotyledons, with a superior several-celled ovarium, regular flowers, definite erect ovules, an imbricated corolla, with seeds having a bony seed-coat and a large scar occupying the whole of one of their sides.

ANOMALIES.

ESSENTIAL CHARACTER. — *Flowers* hermaphrodite. *Calyx* divided, regular, persistent. *Corolla* monopetalous, hypogynous, regular, deciduous, its segments usually equal in number to those of the calyx, seldom twice or thrice as many. *Stamens* arising from the corolla, definite, distinct, the *fertile* ones equal in number to the segments of the calyx, and opposite those segments of the corolla which alternate with the latter, seldom more. *Anthers* usually turned outwards ; the *sterile* stamens as numerous as the fertile ones, with which they alternate, sometimes absent. *Ovarium* 1, with several cells, in each of which is 1 erect ovulum. *Style* 1. *Stigma* undivided, occasionally lobed. *Fruit* baccate with several 1-seeded cells, or by abortion with only 1. *Seeds* nut-like, sometimes cohering into a several-celled putamen. *Testa* bony, shining, its inner face opaque and softer than the rest. *Embryo* erect, large, white, usually enclosed in fleshy albumen. *Cotyledons*, when albumen is present, foliaceous ; when absent, fleshy and sometimes connate. *Radicle* short, straight, or a little curved, turned towards the hilum. *Plumula* inconspicuous. — *Trees* or *shrubs*, chiefly natives of the tropics, and abounding in milky juice. *Leaves* alternate, without stipulæ, entire, coriaceous. *Inflorescence* axillary.

AFFINITIES. This order is certainly near Ebenaceæ, with which it agrees in habit, arborescent stem, alternate entire leaves, and axillary inflorescence ; and moreover in its monopetalous regular hypogynous corolla, the absence of a hypogynous disk, an ovarium with several cells, and definite ovules and stamens. They, however, differ in several points. Sapoteæ have usually a milky juice, and therefore their wood is among the softer kinds ; their flowers are always hermaphrodite, the segments of the calyx and corolla are often placed in a double row ; their stamens are always in a single row, the fertile ones rarely more numerous than the segments of the calyx, and opposite the divisions of the corolla ; their style is undivided ; the cells of the ovarium are always 1-seeded, with erect ovules ; the testa is thick and bony ; the embryo is large with respect to the fleshy albumen, which is sometimes deficient ; the radicle is very short, and inferior. In Ebenaceæ there is no milk, and the wood is very hard ; the flowers are usually unisexual, the segments of the calyx and corolla are almost always in a single row ; the stamens are usually doubled, and either twice or four times as numerous as the segments of the corolla, or, if equal to them, alternate with them ; the style is generally divided, the cells of the ovarium sometimes 2-seeded, the ovules always pendulous, the testa thin and soft, the embryo middle-sized or small in respect to the cartilaginous albumen, which is always present ; the radicle is of middling length, or very long and superior. *R. Brown Prodr.* 529. It is worth remarking, that the woody shell of the seed of Sapoteæ is certainly testa, and not putamen, as is proved by the presence of the micropyle upon it.

GEOGRAPHY. Chiefly natives of the tropics of India, Africa, and America ; a few are found in the southern parts of North America, and at the Cape of Good Hope.

PROPERTIES. The fruit of many is esteemed in their native countries as an article of the dessert : such are the Sappodilla Plum, the Star Apple, the Medlar of Surinam, the Mimusops Elengi, and others ; they are described as having generally a sweet taste, with a little acidity. The seeds of Achras Sapota are aperient and diuretic ; those of some others are filled with a concrete oil, which is used for domestic purposes. A kind of thick oil, like butter, is obtained from the fruit of Bassia butyracea, the Mahva or Madhuca Tree. The flowers of the same tree are employed extensively in the distillation of a kind of arrack. *Ed. P. J.* 12. 192. The juice of the bark of Bassia longifolia is prescribed by the Indian doctors in rheumatic affections. *Ainslie*, 2. 100. The Butter Tree of Mungo Park was also a species of Bassia. The bark of 4 species of Achras is so astrin-

182

gent and febrifugal as to have been substituted for quinquina. The Cow Tree of Humboldt has been sometimes supposed to be referable to this order; but there seems no reason now to doubt its belonging to Artocarpeæ. The Tingi da Praya of Brazil, with which the Indians destroy fish, is the Jacquinia obovata. The branches are bruised and thrown into the water. It must not be confounded with another fish poison, called Tingi only, which is a species of Paullinia. *Pr. Max. Trav.* 166.

EXAMPLES. Achras, Mimusops.

CLXX. ERICEÆ. THE HEATH TRIBE.

ERICÆ, *Juss. Gen.* 159. (1789). —ERICEÆ, *R. Brown Prodr.* 557. (1810); *Lindl. Synops.* 172. (1829). —RHODODENDRA, *Juss. Gen.* 158. (1789). —ERICINEÆ, *Desv. Journ. Bot.* 28. (1813). — RHODORACEÆ *and* ERICACEÆ, *Dec. Fl. Fr.* 3. 671. *and* 675. (1815.)

DIAGNOSIS. Monopetalous shrubby dicotyledons, with regular flowers, a superior many-seeded ovarium, a single style, 2-celled dry anthers with appendages, apterous seeds, and embryo in the axis of albumen.

ANOMALIES. Azalea, Rhododendron, &c., have an irregular corolla, but their stamens are symmetrical. The petals of Ledum scarcely cohere. In Arctostaphylos the seeds are definite. There is a species of Erica with broad winged seeds, according to Mr. Brown.

ESSENTIAL CHARACTER. — *Calyx* 4- or 5-cleft, nearly equal, inferior, persistent. *Corolla* hypogynous, monopetalous, 4- or 5-cleft, occasionally separable into 4 or 5 pieces, regular or irregular, often withering, with an imbricated æstivation. *Stamens* definite, equal in number to the segments of the corolla, or twice as many, hypogynous, or inserted into the base of the corolla ; *anthers* 2-celled, the cells hard and dry, separate either at the apex or base, where they are furnished with some kind of appendage, and dehiscing by a pore or cleft. *Ovarium* surrounded at the base by a disk, or secreting scales, many-celled, many-seeded ; *style* 1, straight; *stigma* 1, undivided or toothed. *Fruit* capsular, many-celled, with central placentæ ; *dehiscence* various. *Seeds* indefinite, minute ; *testa* firmly adhering to the nucleus ; *embryo* cylindrical, in the axis of fleshy albumen ; *radicle* opposite the hilum. — *Shrubs* or *under-shrubs. Leaves* evergreen, rigid, entire, whorled, or opposite, without stipulæ. *Inflorescence* variable, the pedicels generally bracteate.

AFFINITIES. Formerly separated into two by Jussieu, who distinguished Ericeæ and Rhodoraceæ by the dehiscence of their capsule; a character which is not now esteemed of ordinal importance, and which is consequently abandoned. They differ from Vaccinieæ and Campanulaceæ in their superior ovarium, from Epacrideæ in the structure of their anthers, from Pyrolaceæ in the structure of their seeds and in habit, and from all the orders of which Scrophularineæ and Gentianeæ may be considered the representatives, in the number of cells of the ovarium agreeing with the lobes of the calyx and corolla.

GEOGRAPHY. Most abundant at the Cape of Good Hope, where immense tracts are covered with them; common in Europe and North and South America, both within and without the tropics; less common in northern Asia and India, and almost unknown in Australasia, where their place is supplied by Epacrideæ.

PROPERTIES. Their general qualities are, to be astringent and diuretic ; Azalea procumbens, Rhododendron ferrugineum and chrysanthemum, and Ledum palustre, being examples of the former, and Arctostaphylos Uva Ursi of the latter. This, Decandolle observes, has been confounded with Vaccinium Vitis Idea by some practitioners, but most improperly, the chemical composition of the two plants being extremely different. See *Essai Méd.* 194. An infusion of the leaves of Uva Ursi has been employed with success in cases of gonorrhœa of long standing. *Ibid.* The berries of the succulent-fruited kinds are usually grateful, and sometimes used as food. Gaultheria procumbens and Shallon, Arctostaphylos alpina, and Brossæa coccinea, are examples of this. In the island of Corsica an agreeable wine is said to be prepared from the berries of Arbutus Unedo. *Ed. P. J.* 2. 199. Gaultheria procumbens possesses stimulating and anodyne properties. In North America an infusion of it is used as tea. *Barton,* 1. 178. An infusion of the berries in brandy is taken in small quantities, in the same way as common bitters. *Ibid.* The fruit of Arbutus Unedo, taken in too great quantity, is said to be narcotic, and a similar quality no doubt exists in several other plants of the order; Ledum palustre renders beer heady, when used in the manufacture of that beverage; Rhododendron ponticum and maximum, Kalmia latifolia, and some others, are well known to be venomous. The honey which poisoned some of the soldiers in the retreat of the ten thousand through Pontus was gathered by bees from the flowers of Azalea pontica. The shoots of Andromeda ovalifolia poison goats in Nipal. *Don Prodr.* 149. It is stated by Dr. Horsfield that a very volatile heating oil, with a peculiar odour, used by the Javanese in rheumatic affections, is obtained from a species of Andromeda. *Ainslie,* 2. 107.

EXAMPLES. Erica, Andromeda, Ledum, Rhododendron, Azalea.

CLXXI. EPACRIDEÆ.

EPACRIDEÆ, *R. Brown Prodr.* 535. (1810); *Link Handb.* 1. 601. (1829), a § *of* Ericeæ.

DIAGNOSIS. Monopetalous dicotyledons, with regular flowers, a superior several-celled ovarium, an imbricated corolla, a single style, and dry 1-celled anthers.

ANOMALIES. Monotoca has but 1 cell in the ovarium.

ESSENTIAL CHARACTER. — *Calyx* 5-parted (very seldom 4-parted), often coloured, persistent. *Corolla* hypogynous, monopetalous, either deciduous or withering, sometimes capable of being separated into 5 pieces, its limb with 5 (rarely 4) equal divisions, sometimes, in consequence of the cohesion of the segments, bursting transversely; the æstivation valvular or imbricated. *Stamens* equal in number to the segments of the corolla, and alternate with them; very seldom fewer in number. *Filaments* arising from the corolla, or hypogynous. *Anthers* simple, with a single receptacle of pollen, which forms a complete partition sometimes having a border; undivided, opening longitudinally. *Pollen* either nearly round or formed of 3 connate grains. *Ovarium* sessile, usually surrounded at the base with 5 distinct or connate scales; with several, rarely a single, cell; *ovules* solitary or indefinite; *style* 1; *stigma* simple, or occasionally toothed. *Fruit* drupaceous, baccate, or capsular. *Seeds* with albumen. *Embryo* taper, straight, in the axis, more than half as long as the albumen. — *Shrubs* or *small trees*, their hair, when present, being simple. *Leaves* alternate, very rarely opposite, entire or occasionally serrated, usually stalked; their bases sometimes dilated, cucullate, overlapping each other and half sheathing the stem. *Flowers* white or purple, seldom blue, either in spikes or terminal racemes, or solitary and axillary; the calyx or pedicels with 2 or several bracteæ, which are usually of the same texture as the calyx.

AFFINITIES. This order differs from Ericeæ solely in the structure of the anther; but that organ being one of the principal features of Ericeæ, any material deviation from it acquires a peculiar degree of consequence. In Ericeæ the anther consists of 2 cells, usually furnished with peculiar appendages; in Epacrideæ it is simply 1-celled, with no appendages whatever. The order is remarkable for containing species with both definite and indefinite seeds.

GEOGRAPHY. All natives of Australasia or Polynesia, where they abound as Heaths at the Cape of Good Hope. It is remarkable that only 1 or 2 of the Heath tribe are found in the countries occupied by Epacrideæ.

PROPERTIES. The fruit of Lissanthe sapida, called the Australian cranberry, is eatable. Chiefly remarkable for the great beauty of the flowers of many species.

EXAMPLES. Epacris, Styphelia, Leucopogon, Sprengelia.

CLXXII. VACCINIEÆ. THE BILBERRY TRIBE.

VACCINIEÆ, *Dec. Théor. Elém.* 216. (1813); *Dec. and Duby*, 315. (1818); *Lindl. Synops.* 134. (1829.)

DIAGNOSIS. Monopetalous dicotyledons, with an inferior ovarium, a regular corolla, succulent fruit, indefinite ovules, alternate leaves, and calcarate anthers.

ANOMALIES.

ESSENTIAL CHARACTER.— *Calyx* superior, entire, or with from 4 to 6 lobes. *Corolla* monopetalous, lobed as often as the calyx. *Stamens* distinct, double the number of the lobes of the corolla, inserted into an epigynous disk; *anthers* with 2 horns and 2 cells. *Ovarium* inferior, 4- or 5-celled, many-seeded; *style* simple; *stigma* simple. *Berry* crowned by the persistent limb of the calyx, succulent, 4- or 5-celled, many-seeded. *Seeds* minute; *embryo* straight, in the axis of a fleshy albumen; *cotyledons* very short; *radicle* long, inferior. — *Shrubs,* with alternate coriaceous leaves.

AFFINITIES. Formerly combined with Ericeæ, from which it differs in its inferior ovarium and succulent fruit. It is confounded by Achille Richard with Escallonieæ, which are essentially distinguished by their flowers being polypetalous and the anthers bursting lengthwise. Myrtaceæ are obviously separated by being polypetalous, by the leaves being opposite and marked with transparent dots, &c.

GEOGRAPHY. Natives of North America, where they are found in great abundance as far as high northern latitudes; sparingly in Europe; and not uncommonly on high land in the Sandwich Islands.

PROPERTIES. Much the same as those of Ericeæ; their bark and leaves are astringent, slightly tonic, and stimulating. The berries of many are eaten, under the names of Cranberry, Bilberry, Whortleberry, &c. All the species are choice subjects of the gardener's care.

EXAMPLES. Vaccinium, Oxycoccus.

CLXXIII. PYROLACEÆ. THE WINTER GREEN TRIBE.

PYROLEÆ, *Lindl. Coll. Bot.* t. 5. (1821); *Synops.* 175. (1829). — MONOTROPEÆ, *Nutt. Gen.* 1. 272. (1818); *Dec. and Duby*, 319. (1828.)

DIAGNOSIS. Monopetalous dicotyledons, with regular flowers, a superior many-seeded ovarium, a single declinate style, 2-celled dry anthers with appendages, winged seeds, and a minute inverted embryo in fleshy albumen.

ANOMALIES. The style is not always declinate. There is a shrubby species of Pyrola.

ESSENTIAL CHARACTER. — *Calyx* 5-leaved, persistent, inferior. *Corolla* monopetalous, hypogynous, regular, deciduous, 4- or 5-toothed, with an imbricated æstivation. *Stamens* hypogynous, twice as numerous as the divisions of the corolla; *anthers* 2-celled, opening longitudinally, and furnished with appendages at the base. *Ovarium* superior, 4- or 5-celled, many-seeded, with a hypogynous disk; *style* 1, straight or declinate; *stigma* simple. *Fruit* capsular, 4- or 5-celled, dehiscent, with central placentæ. *Seeds* indefinite, minute, winged; *embryo* minute, inverted, at the extremity of a fleshy albumen. — *Herbaceous* plants, rarely *under-shrubs,* sometimes parasitical and leafless. *Stems* round, covered with scales ; in the frutescent species leafy. *Leaves* either wanting or simple, entire or toothed. *Flowers* in terminal racemes, rarely solitary.

AFFINITIES. However different the tribes of Ericeæ and Orobancheæ may seem, they are completely connected by this, which, with the regular corolla, having a slight tendency to irregularity in its declinate style, the 5 cells, and hypogynous dry spurred anthers of the former, combine the habit and peculiar structure of seed of the latter. They are known from Ericeæ by their winged seeds, minute embryo, often declinate style, and herbaceous often leafless habit. The latter character will not, however, alone point out the order; nor is it even universal in particular genera; for Pyrola itself, which has usually round bright green leaves, contains a species destitute of leaves, and having the habit of Pterospora.

GEOGRAPHY. Natives of Europe, North America, and the northern parts of Asia, in fir woods, or in similar situations.

PROPERTIES. Chimaphila umbellata is a most active diuretic; it is also found to possess valuable tonic properties. The leaves, applied to the skin, act as slight vesicatories. It is remarkable enough that C. maculata, a very closely allied species, should be asserted by American practitioners to be wholly inert. See *Barton,* 1. 28.

EXAMPLES. Pyrola, Chimaphila, Monotropa, Pterospora, Schweinitzia.

CLXXIV. CAMPANULACEÆ. THE CAMPANULA TRIBE.

CAMPANULÆ, *Juss. Gen.* 163. (1789) *in part.* — CAMPANULACEÆ, *R. Brown Prodr.* 559. (1810) ; *Lindl. Synops.* 135. (1829). — CAMPANULEÆ, *Alph. Dec. Monogr.* (1830.)

DIAGNOSIS. Monopetalous milky dicotyledons, with an inferior ovarium, a regular corolla, capsular fruit, indefinite ovules, alternate leaves, and round pollen.

ANOMALIES.

ESSENTIAL CHARACTER. — *Calyx* superior, usually 5-lobed (3-8), persistent. *Corolla* monopetalous, inserted into the top of the calyx, usually 5-lobed (3-8), withering on the fruit, regular. *Æstivation* valvate. *Stamens* inserted into the calyx alternately with the lobes of the corolla, to which they are equal in number. *Anthers* 2-celled, distinct. *Pollen* spherical. *Ovarium* inferior, with 2 or more polyspermous cells opposite the stamens, or alternate with them; *style* simple, covered with collecting hairs ; *stigma* naked, simple, or with as many lobes as there are cells. *Fruit* dry, crowned by the withered calyx and corolla, dehiscing by lateral irregular apertures or by valves at the apex, always loculicidal. *Seeds* numerous, attached to a placenta in the axis ; *embryo* straight, in the axis of fleshy albumen ; *radicle* inferior. — *Herbaceous* plants or *under-shrubs,* yielding a white milk. *Leaves* almost always alternate, simple, or deeply divided, without stipulæ. *Flowers* single, in racemes, spikes, or panicles, or in heads, usually blue or white, very rarely yellow.

AFFINITIES. While this work was going through the press, an excellent Monograph of the present order reached me from M. Alphonse Decandolle. I gladly avail myself of the valuable remarks of this skilful botanist in explaining the affinities of Campanulaceæ. He considers that they differ from Lobeliaceæ chiefly in their regular corolla, their stamens being almost always distinct, their pollen spherical (not oval), their stigmas generally long and velvety externally, in the abundance of collecting hairs on the style, and finally in their capsule usually opening laterally. " It is not only in the form," he proceeds, " but also in the number of the parts, that the flower of Campanulaceæ is more regular than that of Lobeliaceæ. Thus, in several Campanulas the cells of the ovarium are equal in number to the stamens and the divisions of the corolla and calyx, which points out the natural symmetry of the flower. In the Lobelias abortion is more frequent. In both groups the innermost organs are abortive more frequently than the outermost. Thus, the number of cells is often smaller (never greater) than that of the stamens; the number of stamens is sometimes smaller (but never larger) than that of the lobes of the corolla; and the same is true of the lobes of the corolla with respect to the calyx. Finally, Lobeliaceæ have sometimes a corolla of a fine bright red, a colour unknown among Campanulas; nine-tenths of the species of the latter have blue flowers; and those in which the colour varies, and into which a little red enters (as Canarina), are far from having the brilliancy of Lobelia cardinalis for instance. After Lobeliaceæ, the natural groups with which Campanulaceæ have the most relation are, no doubt, Goodenoviæ and Stylidieæ, which formed part of the Campanulaceæ of M. de Jussieu. The regular corolla of Campanulaceæ distinguishes them, at first sight, from both those groups, as well as from Lobeliaceæ. Besides, Campanulas have not the fringed indusium which terminates the style of Goodenoviæ, and surrounds their stigma. Although this organisation approaches that of Lobeliaceæ, and so Campanulaceæ, it is not less true that it affords an important mark of distinction, and that it is connected with essential differences in the mode of fecundation. Mr. Brown has also remarked, that the corolla of Goodenoviæ is sometimes polypetalous, which it never is in Campanulaceæ or Lobeliaceæ; that the æstivation of their corolla is induplicate, not valvate; that its principal veins are lateral, or alternate with the lobes, as in Compositæ; that in the species of Goodenoviæ with dehiscent fruit, the dehiscence is usually septicidal, while in the two other groups it is always loculicidal; finally, that Goodenoviæ have not the milky juice that characterises Campanulaceæ and Lobeliaceæ." Notwithstanding their polyspermous fruit and different inflorescence, these approach very closely to Compositæ; their milky juice is the same as that of Cichoraceæ; their species have, in many cases, the flowers crowded in heads; their stigma is similar to that of many Compositæ; they have the same collecting hairs on the style, in both cases intended to clear out the pollen from the cells of the anthers; and, finally, their habit is very like.

GEOGRAPHY. Chiefly natives of the north of Asia, Europe, and North America, and scarcely known in the hot regions of the world. In the meadows, fields, and forests of the countries they inhabit, they constitute the most striking ornament. Some curious species are found in the Canaries, St. Helena, and Juan Fernandez. M. Alphonse Decandolle remarks, that " it is within the 36° and 47° N. lat. that in our hemisphere the greatest number of species is found; the chain of the Alps, Italy, Greece, Caucasus, the Altai range, are their true country. In whatever direction we leave these limits, the number of species rapidly decreases. In the southern hemisphere, the Cape of Good Hope (lat. 34° S.) is another centre of habitation, containing not fewer than 63 species. This locality has a climate so different from that of our mountains, that it may be easily imagined that the species capable of living there differ materially from those of our own hemisphere: in fact, they belong to other genera." Of 300 species, only 19 are found within the tropics.

PROPERTIES. The milky juice is rather acrid, but nevertheless the roots and young shoots of some, particularly of Campanula Rapunculus, or Rampion, of Phyteuma spicata, of Canarina Campanula, &c., are an occasional article of food. The chief value of the order, however, is its beauty.

EXAMPLES. Campanula, Wahlenbergia.

CLXXV. LOBELIACEÆ.

CAMPANULACEÆ, § 2. *R. Brown Prodr.* 562. (1810). — LOBELIACEÆ, *Juss. Ann. Mus.* 18. 1. (1811); *Dec. and Duby*, 310. (1828); *Lindl. Synops.* 137. (1829.)

DIAGNOSIS. Monopetalous milky dicotyledons, with an inferior ovarium, an irregular corolla, syngenesious stamens, indefinite ovula, alternate leaves, and oval pollen.

ANOMALIES. Clintonia has a triangular 1-celled ovarium, with 2 parietal placentæ. Some have 5 petals. One species of Lobelia is diœcious.

ESSENTIAL CHARACTER. — *Calyx* superior, 5-lobed, or entire. *Corolla* monopetalous, irregular, inserted in the calyx, 5-lobed, or deeply 5-cleft. *Stamens* 5, inserted into the calyx alternately with the lobes of the corolla; *anthers* cohering; *pollen* oval. *Ovarium* inferior, with from 1 to 3 cells; *ovula* very numerous, attached either to the axis or the lining; *style* simple; *stigma* surrounded by a cup-like fringe. *Fruit* capsular, 1- or more-celled, many-seeded, dehiscing at the apex. *Seeds* attached either to the lining or the axis of the pericarpium; *embryo* straight, in the axis of fleshy albumen; *radicle* pointing to the hilum. — *Herbaceous* plants or *shrubs*. *Leaves* alternate, without stipulæ. *Flowers* axillary or terminal.

AFFINITIES. Yet more nearly related to Compositæ even than Campanulaceæ, especially in their cohering anthers and in the irregularity of their corolla, which consists in its being split, so that the segments cohere towards one side just like the 5 segments that make up the ligulate floret of a Composita. The stigma is surrounded by hairs, which are probably analogous to the indusium of Goodenoviæ, to which order Lobeliaceæ approach closely. Of course they participate in any and all the affinities of Campanulaceæ. M. Alphonse Decandolle criticises, with much justice, the character assigned to Lobeliaceæ in my *Synopsis* of the British Flora, particularly in regard to the *cup or* fringe assigned to their stigma: this was a misprint for cup-like. He is also, perhaps, right in considering Jasione more properly a Campanulaceous than a Lobeliaceous plant. The genus, however, seems to me to stand upon the limit between the two orders.

GEOGRAPHY. Unlike Campanulaceæ, these seem to prefer countries within or upon the border of the tropics to such as have a colder character. We find them abounding in the West Indies, Brazil, the Cape of Good Hope, and the Sandwich Islands; they are not uncommon in Chile, and New Holland.

PROPERTIES. All dangerous or suspicious, in consequence of the excessive acridity of their milk. Lobelia tupa yields a dangerous poison in Chile. The most active article of the North American Materia Medica is said to be the Lobelia inflata: it is possessed of an emetic, sudorific, and powerful expectorant effect, especially the first. When given with a view to empty the stomach, it operates vehemently and speedily; producing, however, great relaxation, debility, and perspiration, and even death, if given in over-doses. *Barton*, 1. 189. The anti-syphilitic virtues ascribed to Lobelia syphilitica are supposed to have resided in its diuretic property; they are, however, generally discredited altogether. *Ibid.* 2. 211. Lobelia longiflora, a native of some of the West India Islands, is one of the most venomous of plants. The Spanish Americans call it Rebenta Cavallos, because it proves fatal to horses that eat it, swelling them until they burst. Taken internally, it acts as a violent cathartic, the effects of which no remedy can assuage, and which end in death. The leaves are an active vesicatory. Lobelia cardinalis is an acrid plant which is reckoned an anthelmintic. *Ibid.* 2. 180.

EXAMPLES. Lobelia, Isotoma.

CLXXVI. GOODENOVIÆ.

CAMPANULÆ, *Juss. Gen.* 163. (1789) *in part.* — GOODENOVIÆ, *R. Brown Prodr.* 573. (1810).

DIAGNOSIS. Monopetalous dicotyledons, with a 2-4-celled inferior ovarium, an indusiate stigma, and indefinite seeds.

ANOMALIES. This order offers the singular anomaly of genera having, at the same time, an inferior calyx and a superior corolla; a circumstance which, it has been well observed by Mr. Brown, points out the real origin of both organs.

ESSENTIAL CHARACTER. — *Calyx* usually superior, rarely inferior, equal or unequal, in from 3 to 5 divisions. *Corolla* always more or less superior, monopetalous, more or less irregular, withering; its *tube* split at the back, and sometimes capable of being separated into 5 pieces, when the calyx only coheres with the base of the ovarium; its *limb* 5-parted, with 1 or 2 lips, the edges of the segments being thinner than the middle, and folded inwards in æstivation. *Stamens* 5, distinct, alternate with the segments of the corolla; *anthers* distinct or cohering, 2-celled, bursting longitudinally. *Pollen* simple or compound. *Ovarium* 2-celled, rarely 4-celled, with indefinite ovules, having sometimes a gland at its base between the 2 anterior filaments; *style* 1, simple, very rarely divided; *stigma* fleshy, undivided, or 2-lobed, surrounded by a membranous cup. *Fruit* a 2- or 4-celled capsule with many seeds, attached to the axis of the dissepiment, which is usually parallel with the valves, rarely opposite to them. *Seeds* usually with a thickened testa, which is sometimes nut-like; *albumen* fleshy, enclosing an erect *embryo*; *cotyledons* foliaceous; *plumula* inconspicuous. — *Herbaceous* plants, rarely *shrubs*, without milk, with simple or glandular hairs, if any are present. *Leaves* scattered, often lobed, without stipulæ. *Inflorescence* terminal, variable. *Flowers* distinct, never capitate, usually yellow, or blue, or pink.

AFFINITIES. The strict relation of these to Campanulaceæ and Lobeliaceæ cannot be doubted, from which they differ in the æstivation of the flower, and in the peculiar indusium of the stigma, a trace of which is to be found in Lobeliaceæ, and which exists in a remarkable degree in Brunoniaceæ. Scævoleæ differ only in their definite seeds. Upon the nature of the indusium of the stigma Mr. Brown makes the following observations.

" Is this remarkable covering of the stigma in these families merely a process of the apex of the style? or is it a part of distinct origin, though intimately cohering with the pistillum? On the latter supposition, may it not be considered as analogous to the glandular disk surrounding or crowning the ovarium in many other families? And, in adopting the hypothesis I have formerly advanced respecting the nature of this disk in certain families,—namely, that it is composed of a series of modified stamina,—has not the part in question a considerable resemblance, in apparent origin and division, to the stamina of the nearly-related family Stylideæ? To render this supposition somewhat less paradoxical, let the comparison be made especially between the indusium of Brunonia and the imperfect antheræ in the female flowers of Forstera. Lastly, connected with this view, it becomes of importance to ascertain whether the stamina in Stylideæ are opposite to the segments of calyx or of corolla. The latter disposition would be in favour of the hypothesis. This, however, is a point which will not be very easily determined, the stamina being lateral. In the mean time, the existence and division of the corona faucis in Stylidium render it not altogether improbable that they are opposite to the segments of the corolla." *R. Brown in Linn. Trans.* 12. 134. I am rather inclined to consider the indusium analogous to the collecting hairs of Campanulaceæ. In these they occupy the surface of the greater part of the style; in Lobelia they are arranged in a whorl, forming a cup-like fringe; and in Goodenoviæ the hairs, being still whorled, are consolidated into a uniform substance by their mutual cohesion.

GEOGRAPHY. Natives of New Holland, and other islands of the South Pacific Ocean.
PROPERTIES. Unknown.
EXAMPLES. Goodenia, Velleia, Leschenaultia.

CLXXVII. STYLIDIEÆ.

STYLIDEÆ, *R. Brown Prodr.* 565. (1810).

DIAGNOSIS. Monopetalous gynandrous dicotyledons.
ANOMALIES.

ESSENTIAL CHARACTER.— *Calyx* superior, with from 2 to 6 divisions, bilabiate or regular, persistent. *Corolla* monopetalous, falling off late; its limb irregular, rarely regular, with from 5 to 6 divisions, imbricated in æstivation. *Stamens* 2; *filaments* connate with the style into a longitudinal column; *anthers* twin, sometimes simple, lying over the stigma; *pollen* globose, simple, sometimes angular. *Ovarium* 2-celled, many-seeded, sometimes 1-celled, in consequence of the contraction of the dissepiment, often surmounted with a single gland in front, or two opposite ones; *style* 1; *stigma* entire or bifid. *Capsule* with 2 valves and 2 cells, the dissepiment between which being sometimes either contracted or separable from the inflexed margins of the valves, the capsule becomes as it were 1-celled. *Seeds* small, erect, sometimes stalked, attached to the axis of the dissepiment; *embryo* minute, enclosed within a fleshy, somewhat oily albumen.—*Herbaceous* plants or *under-shrubs*, without milk, having a stem or scape, their hair, where they have any, simple, acute, or headed with a gland. *Leaves* scattered, sometimes whorled, entire, their margins naked or ciliated, the radical ones clustered in the species with scapes. *Flowers* in spikes, racemes, or corymbs, or solitary; terminal, rarely axillary, the pedicels usually with three bracteæ.

AFFINITIES. Nearly allied both to Campanulaceæ and Goodenoviæ, from both which they are distinguished by their gynandrous stamens, and from the latter by the want of an indusium to the stigma. The structure of the sexual organs is highly curious; the stamens and style are closely combined in a solid irritable column, at the top of which is a cavity, including the stigma, and bounded by the anthers. A singular blunder was committed by Labillardiére, who mistook the epigynous gland for the stigma; and another by L. C. Richard, who considered the labellum to be the female organ.
GEOGRAPHY. Chiefly found in New Holland. Species have been discovered both in Ceylon and the South Sea Islands.
PROPERTIES. Unknown.
EXAMPLES. Stylidium, Forstera.

CLXXVIII. SCÆVOLEÆ.

GOODENOVIÆ, § Scævoleæ, *R. Brown Prodr.* 582. (1810).

DIAGNOSIS. Monopetalous dicotyledons, with a 1-4-celled inferior ovarium, an indusiate stigma, and definite erect seeds.
ANOMALIES. A Molucca species of Scævola exists, with opposite leaves. *R. Br.*

ESSENTIAL CHARACTER.— *Calyx* superior, equal or unequal, in 5 divisions, sometimes obsolete. *Corolla* superior, monopetalous, more or less irregular, withering, or deciduous; its *tube* split at the back; its *limb* 5-parted, with 1 or 2 lips, the edges of the segments being thinner than the middle, and folded inwards in æstivation. *Stamens* 5, distinct, alternate with the segments of the corolla; *anthers* distinct or cohering, 2-celled, bursting longitudinally; *pollen* simple. *Ovarium* 1- 2- or 4-celled, with 1, seldom 2, erect ovula in each cell; *style* 1, simple; *stigma* fleshy, surrounded by a membranous cup. *Fruit* inferior, indehiscent, drupaceous, or nut-like. *Seeds* with a thickened testa; *albumen* fleshy, enclosing an erect embryo; *cotyledons* foliaceous; *plumula* inconspicuous.— *Herbaceous* plants or *shrubs*, without milk, with simple or stellate hairs, if any are present. *Leaves* scattered, undivided, without stipulæ. *Inflorescence* axillary or terminal. *Flowers* distinct, never capitate, white, blue, or yellowish.

AFFINITIES. Combined, on account of their indusiate stigmas, by Mr. Brown, with Goodenovia and Brunoniaceæ, from the former of which they differ in habit, indehiscent fruit, and definite seeds; from the latter, in their inferior ovarium and habit.

GEOGRAPHY. Natives of the South Seas and the islands of the Indian archipelago. The species are abundant in New Holland.

PROPERTIES. Unknown.

EXAMPLES. Scævola, Diaspasis, Dampiera.

CLXXIX. BRUNONIACEÆ.

GOODENOVIÆ, § 2. *R. Brown Prodr.* 589. (1810).

DIAGNOSIS. Monopetalous dicotyledons, with regular flowers, a superior entire ovarium, a single erect ovulum, capitate flowers, and a stigma with an indusium.

ANOMALIES.

ESSENTIAL CHARACTER.— *Calyx* inferior, in 5 divisions, with 4 bracteæ at the base. *Corolla* monopetalous, almost regular, 5-parted, inferior, withering. *Stamens* definite, hypogynous, alternate with the segments of the corolla; *anthers* collateral, slightly cohering. *Ovarium* 1-celled, with a single erect ovulum; *style* single; *stigma* enclosed in a 2-valved cup. *Fruit* a membranous utricle enclosed within the indurated tube of the calyx. *Seed* solitary, erect, without *albumen*; *embryo* with plano-convex fleshy *cotyledons*, and a minute inferior *radicle.*— *Herbaceous* plants, without stems, and simple glandless hairs. *Leaves* radical, entire, with no stipulæ. *Flowers* collected in heads, surrounded by enlarged bracteæ, blue.

AFFINITIES. Placed by Mr. Brown as a section of Goodenoviæ, from which they, in my judgment, differ essentially in their superior 1-celled ovarium and capitate flowers, thus approaching some species of Dipsaceæ, from which they differ in the want of an involucellum, their erect ovulum, superior ovarium, and peculiar stigma. With reference to this, Mr. Brown says: " Brunonia agrees with Goodenoviæ in the remarkable indusium of the stigma, in the structure and connexion of the antheræ, in the seed being erect, and essentially in the æstivation of corolla. It differs from them in having both calyx and corolla distinct from the ovarium, in the disposition of vessels in the corolla, in the filaments being jointed at top, in the seed being without albumen, and in its remarkable inflorescence, compatible, indeed, with the nature of the irregularity in the corolla of Goodenoviæ, but which can hardly co-exist with that characterising Lobeliaceæ. With Compositæ it agrees essentially in inflorescence, in the æstivation of corolla, in the remarkable joint or change of texture in the apex of its filaments, and in the structure of the ovarium and seed. It differs from them in having ovarium liberum or superum, in the want of a glandular disk, in the immediately hypogynous insertion of the filaments, in the indusion of the stigma, and in the vascular structure of the corolla, whose tube has five nerves only, and these continued through the axes of the laciniæ, either terminating simply (as is at least frequently the case in Brunonia sericea), or (as in B. australis) dividing at top into two recurrent branches, forming lateral

nerves, at first sight resembling those of Compositæ, but which hardly reach
to the base of the laciniæ. It is a curious circumstance that Brunonia
should so completely differ from Compositæ in the disposition of vessels
of the corolla, while both orders agree in the no less remarkable structure
of the jointed filament; a character which had been observed in a very few
Compositæ only, before the publication of M. Cassini's second Dissertation,
where it is proved to be nearly universal in the order. In the opposite
parietes of the ovarium of Brunonia two nerves or vascular cords are observable, which are continued into the style, where they become approximated and parallel. This structure, so nearly resembling that of Compositæ, seems to strengthen the analogical argument in favour of the hypothesis
advanced in the present paper, of the compound nature of the pistillum in
that order, and of its type in phænogamous plants generally; Brunonia
having an obvious and near affinity to Goodenoviæ, in the greater part of
whose genera the ovarium has actually two cells with one or an indefinite
number of ovula in each; while in a few genera of the same order, as
Dampiera, Diaspasis, and certain species of Scævola, it is equally reduced
to one cell and a single ovulum." *R. Brown in Linn. Trans.* 12. 132. The
habit of this order is very much that of Globularineæ.

GEOGRAPHY. Natives of New Holland.
PROPERTIES. Unknown.
EXAMPLE. Brunonia.

CLXXX. PAPAYACEÆ. THE PAPAW TRIBE.

PAPAYÆ, *Agardh Classes.* (1824). — CARICEÆ, *Turpin in Atl. du Dict. des Sc. Nat.* (?) — PAPAYACEÆ, *Von Martius H. R. M.* (1829.)

DIAGNOSIS. Monopetalous dicotyledons, with regular unisexual flowers,
and a superior 1-celled ovarium with 5 parietal placentæ.
ANOMALIES.

ESSENTIAL CHARACTER.— *Flowers* unisexual. *Calyx* inferior, minute, 5-toothed.
Corolla monopetalous; in the *male* tubular, with 5 lobes and 10 stamens, all arising from
the same line, and of which those that are opposite the lobes are sessile, the others on short
filaments; *anthers* adnate, 2-celled, bursting longitudinally; in the *female* divided nearly
to the base into 5 segments. *Ovarium* superior, 1-celled, with 5 parietal polyspermous
placentæ; *stigma* sessile, 5-lobed, lacerated. *Fruit* succulent, indehiscent, 1-celled, with 5
polyspermous parietal placentæ. *Seeds* enveloped in a loose mucous coat with a brittle
pitted testa; *embryo* in the axis of fleshy *albumen*, with flat cotyledons and a taper radicle
turned towards the hilum. — *Trees* without branches, yielding an acrid milky juice. *Leaves*
alternate, lobed, on long taper petioles. *Flowers* in axillary racemes.

AFFINITIES. It was the opinion of Jussieu that the genus upon which
this order is founded held a sort of middle station between Urticeæ and
Cucurbitaceæ. Auguste St. Hilaire has, however, well remarked upon this
subject, that the only relation that it has with Urticeæ consists in the separation of sexes, its milky juice, its habit, which is like that of some species of
Ficus, its foliage, which is not very different from that of Cecropia, and the
position of its stigmas: and to these he wisely attaches very little importance.
Its fruit brings it near Cucurbitaceæ; but its true place is probably in the
vicinity of Passifloreæ, with which it altogether agrees in the appearance of
its testa, in its unilocular fruit with parietal polyspermous placentæ, and in
its dichlamydeous flowers; differing, however, widely in its habit and monopetalous flowers.

GEOGRAPHY. Natives of South America; unknown, except as objects of cultivation, beyond that continent.

PROPERTIES. The fruit of the Papaw is eaten, when cooked, and is esteemed by some persons; but it appears to have little to recommend it. Its great peculiarities are, that the juice of the unripe fruit is a most powerful and efficient vermifuge, the powder of the seed even answers the same purpose; and that a principal constituent of this juice is fibrine, a principle otherwise supposed peculiar to the animal kingdom and to fungi. The tree has, moreover, the singular property of rendering the toughest animal substances tender, by causing a separation of the muscular fibre; its very vapour even does this; newly-killed meat suspended among the leaves, and even old hogs and old poultry, becoming tender in a few hours, when fed on the leaves and fruit. See an excellent account of the Papaw by Dr. Hooker, in the *Bot. Mag.* 2898.

EXAMPLE. Carica.

CLXXXI. CUCURBITACEÆ. THE GOURD TRIBE.

CUCURBITACEÆ, *Juss. Gen.* 393. (1789); *Aug. St. Hil. in Mem. Mus.* 9. 190–221. (1823); *Dec. Prodr.* 3. 297. (1828); *Lindl. Synops.* 319. (1829).—NANDHIROBEÆ, *Aug. de St. Hil.* l. c. (1823); *Turpin Dict. des Sc. Atlas.* (?)

DIAGNOSIS. Monopetalous dicotyledons, with an inferior ovarium, parietal placentæ, succulent fruit, a regular corolla, and no albumen.

ANOMALIES. The ripe fruit is divided into 3 or 4 cells in some Momordicas, and is occasionally dry, opening by valves at the apex.

ESSENTIAL CHARACTER.—*Flowers* usually unisexual, sometimes hermaphrodite. *Calyx* 5-toothed, sometimes obsolete. *Corolla* 5-parted, scarcely distinguishable from the calyx, very cellular, with strongly marked reticulated veins, sometimes fringed. *Stamens* 5, either distinct, or cohering in 3 parcels; *anthers* 2-celled, very long and sinuous. *Ovarium* inferior, 1-celled, with 3 parietal placentæ; *style* short; *stigmas* very thick, velvety or fringed. *Fruit* fleshy, more or less succulent, crowned by the scar of the calyx, 1-celled, with 3 parietal placentæ. *Seeds* flat, ovate, enveloped in an arillus, which is either juicy, or dry and membranous; *testa* coriaceous, often thick at the margin; *embryo* flat, with no *albumen ; cotyledons* foliaceous, veined; *radicle* next the hilum.—*Roots* annual or perennial, fibrous or tuberous. *Stem* succulent, climbing by means of tendrils formed by abortive leaves (stipulæ, *St. Hil.*). *Leaves* palmated, or with palmate ribs, very succulent, covered with numerous asperities. *Flowers* white, red, or yellow.

AFFINITIES. Placed by Auguste de St. Hilaire and Decandolle between Myrtaceæ, to which they appear to me to have little affinity, and Passifloreæ, to which they are so closely allied, that they scarcely differ, except in their monopetalous corolla, sinuous stamens, unisexual flowers, and exalbuminous seeds, the habit of both being exactly the same. By the former of these two writers a very particular account of the structure of the order has been given in the *Mémoires du Muséum*. He adopts the opinion of Jussieu, that the apparent corolla of these plants is really a calyx, considering the apparent calyx to be merely certain external appendages. This view I cannot follow, any more than the notion of Passifloreæ being apetalous : however ingenious the reasoning may be upon which such theories are founded, they appear to me to be overstrained, and, entirely at variance with both analogy and actual structure. In discussing the affinities of the order, which he does much at length, he remarks, that Carica (now the type of the order Papayaceæ) should be excluded; that the tendrils of Cucurbitaceæ are transformed stipulæ, but scarcely analogous to the stipulæ of Passifloreæ; that there is an affinity between the order and Campanulaceæ, manifested in the perigynous insertion of the stamens, the inferior ovarium, the single style with several stigmas, the quinary division of the flower connected with the ternary division of the fruit, and, finally, some analogy in the nature of the floral envelopes. He, however, chiefly insists upon their affinity with Onagrariæ, with which, including Combretaceæ, they agree in their definite perigynous stamens, single style, exalbuminous seeds, fleshy fruit, and occasionally in the unisexual flowers and climbing stem, being connected in the latter point of view with Onagrariæ through Gronovia, a climbing genus of that order. He also points out the further connexion that exists between Cucurbitaceæ and Onagrariæ through Loaseæ, which, with an undoubted affinity to the latter, have all the habit of the former. With regard to the supposed affinity of Cucurbitaceæ to Myrtaceæ, this is founded upon the characters of a small group, called NANDHIROBEÆ, consisting of plants having the habit of Cucurbitaceæ, but some resemblance in the form of their fruit to that of Lecythideæ, which, as is well known, border closely upon Myrtaceæ: but beyond this resemblance in the fruit, which appears to be altogether a structure of analogy rather than of affinity, I find nothing to confirm the approachment. Indeed, I agree with Decandolle in estimating Nandhirobeæ no higher than a mere section of Cucurbitaceæ.

GEOGRAPHY. Natives of hot countries in both hemispheres, chiefly within the tropics; a few are found to the north in Europe and North America, and several are natives of the Cape of Good Hope. India appears to be their favourite station.

PROPERTIES. One of the most useful orders in the vegetable kingdom, comprehending the Melon, the Cucumber, the Choco, and the various species of Gourd, all useful as the food of man. A bitter laxative quality perhaps pervades all these, which, in the Colocynth gourd, is so concentrated as to become an active purgative principle. The Colocynth of the shops is prepared from the pulp of Cucumis Colocynthis: it is of so drastic and irritating a nature as to be classed by Orfila among his poisons; but, according to Thunberg, the gourd is rendered perfectly mild at the Cape of Good Hope, by being properly pickled. *Ainslie,* 1. 85. The bitter resinous matter in which the active principles of Colocynth are supposed to exist, is called by chemists Colocynthin. A waxy substance is secreted by the surface of the fruit of Benincasa cerifera. It is produced in the most abundance at the time of its ripening. *Delile Descript.* The leaf of Feuillea cordifolia is asserted by M. Drapiez to be a powerful antidote against vegetable poisons. *Ed. P. J.* 4. 221. The fruit of Trichosanthes palmata, pounded small and intimately blended with warm cocoa-nut oil, is considered a valuable application in India for cleaning and healing the offensive sores which sometimes take place inside of the ears. It is also supposed to be a useful remedy, poured up the nostrils, in cases of ozæna. *Ainslie,* 2. 85. The root of Bryonia possesses powerful purgative properties, but is said to be capable of becoming wholesome food if properly cooked. The perennial roots of all the order appear to contain similar bitter drastic virtues, especially that of the Momordica Elaterium, or Spirting Cucumber. An extremely active poisonous principle, called Elatine, has also been found in the placenta of this plant. It exists in such extremely small quantity, that Dr. Clutterbuck only obtained 6 grains from 40 fruit. *Ed. P. J.* 3. 307. An ingenious explanation of the cause of the singular ejection of the seeds of this plant will be found in Dutrochet *Nouvelles Recherches sur l'Exosmose.* The root of Bryonia rostrata is prescribed in India internally, in electuary, in cases of piles. It is also used as a demulcent, in the form of powder. That of Bryonia cordifolia is considered cooling, and to possess

virtues in complaints requiring expectorants. *Ainslie,* 2. 21. The root of
Bryonia epigæa was once supposed to be the famous Colombo root, to which
it approaches very nearly in quality. The tender shoots and leaves of Bryonia
scabra are aperient, having been previously roasted. *Ibid.* 2. 212. The seeds
of all the species are sweet and oily, and capable of forming very readily an
emulsion; those of Joliffia africana, an African plant, are as large as chest-
nuts, and said to be as excellent as almonds, having a very agreeable flavour;
when pressed they yield an abundance of oil, equal to that of the finest
Olives. Decandolle remarks, that the seeds of this family never participate
in the property of the pulp that surrounds them.

EXAMPLES. Cucumis, Bryonia, Cucurbita, Luffa.

CLXXXII. PLANTAGINEÆ. THE RIB-GRASS TRIBE.

PLANTAGINES, *Juss. Gen.* 89. (1789).—PLANTAGINEÆ, *R. Brown Prodr.* 423. (1810);
Lindl. Synops. 169. (1829).

DIAGNOSIS. Monopetalous tetrandrous dicotyledons, with a regular
corolla, a superior 2-4-celled ovarium, a simple filiform stigma, spiked
flowers, flaccid filaments, and a membranous pericarp dehiscing transversely.

ANOMALIES. In Littorella the flowers are solitary.

ESSENTIAL CHARACTER.—*Flowers* usually hermaphrodite, seldom unisexual. *Calyx*
4-parted, persistent. *Corolla* monopetalous, hypogynous, persistent, with a 4-parted limb.
Stamens 4, inserted into the corolla, alternately with its segments; *filaments* filiform,
flaccid, doubled inwards in æstivation; *anthers* versatile, 2-celled. *Ovarium* sessile, without a disk, 2-, very seldom 4-celled; *ovula* peltate or erect, solitary, twin, or indefinite;
style simple, capillary; *stigma* hispid, simple, rarely half bifid. *Capsule* membranous,
dehiscing transversely. *Seeds* sessile, peltate, or erect, solitary, twin, or indefinite; *testa*
mucilaginous; *embryo* in the axis of fleshy albumen; *radicle* inferior; *plumula* inconspicuous.—*Herbaceous* plants, usually stemless, occasionally with a stem; *hairs* simple, articulated. *Leaves* flat and ribbed, or taper and fleshy. *Flowers* in spikes, rarely solitary.

AFFINITIES. By Jussieu this is considered apetalous, the corolla being
called calyx, and the calyx bracteæ. But this appears so contrary to all
analogy, that it is impossible to adopt the opinion. The order seems to be
more near Plumbagineæ than any other, agreeing with them in habit, and
also in the general structure of the flower, but differing in having a 1-celled
ovarium, with a solitary ovulum, and several stigmas. Mr. Don (*Jameson's
Journal,* Jan. 1830, p. 166.) refers Glaux to Plantagineæ, " where it will
form the connecting link between that family and Primulaceæ."

GEOGRAPHY. Scattered over the whole world, in almost every quarter
of which they are found in one situation or another.

PROPERTIES. The herbage is slightly bitter and astringent, and they
have even been reckoned febrifuges. Their seeds are covered with mucus.
According to Decandolle, those of P. arenaria are exported in considerable
quantities from Nismes and Montpellier to the north of Europe, and are supposed to be consumed in the completion of the manufacture of muslins.
The seeds of Plantago Ispaghula are of a very cooling nature, and, like
those of Plantago Psyllium, form, with boiling water, a rich mucilage, which
is much used in India in catarrh, gonorrhœa, and nephritic affections.
Ainslie, 2. 116.

EXAMPLES. Plantago, Littorella.

CLXXXIII. PLUMBAGINEÆ. The Leadwort Tribe.

Plumbagines, *Juss. Gen.* 92. (1789).—Plumbagineæ, *R. Brown Prodr.* 425. (1810).

Diagnosis. Monopetalous dicotyledons, with regular flowers, a supe-rior 1-celled ovarium containing a single ovulum suspended from the apex·of an umbilical cord, and a naked stigma.
Anomalies.

Essential Character. — *Calyx* tubular, plaited, persistent. *Corolla* monopetalous or 5-petalous, regular. *Stamens* definite; in the monopetalous species hypogynous! in the polypetalous arising from the petals! *Ovarium* superior, single, 1-seeded; *ovulum* inverted, pendulous from the point of an umbilical cord, arising from the bottom of the cavity; *styles* 5! seldom 3 or 4; *stigmas* the same number. *Fruit* a nearly indehiscent utriculus. *Seed* inverted; *testa* simple; *embryo* straight; *radicle* superior.—*Herbaceous* plants or *under-shrubs*, variable in appearance. *Leaves* alternate or clustered, undivided, somewhat sheathing at the base. *Flowers* either loosely panicled, or contracted into heads, flowering irregularly.

Affinities. Distinguished from all other monopetalous orders by their plaited calyx and solitary ovulum, suspended from the apex of a cord which arises from the base of a 1-celled ovarium, with several stigmas. From Plan-tagineæ they are otherwise chiefly known by their inflorescence not being simply spiked, and their albumen not fleshy. The economy of the ovulum is highly curious; before fecundation it is suspended from the apex of a cord, or rather strap, which lies over the foramen or orifice through which the vivifying influence of the pollen has to be introduced; this foramen is presented to the summit of the cell immediately below the origin of the stigmas, but has no communication with that part of the cell, from contact with which it is further cut off by the overlying strap: but as soon as the pollen exercises its influence upon the stigmas, the strap slips aside from above the foramen, which is entered by an extension of the apex of the cell, and thus a direct communication is established between the pollen and the inside of the ovulum. This phenomenon is obscurely hinted at by several writers, but was first distinctly shewn me by Mr. Brown, and has lately been beautifully illustrated by Mirbel *Nouvelles Recherches sur l'Ovule*, tab. 4. Nyctagineæ are distinguished by their curved embryo, want of petals, and coloured calyx, the base of which hardens and contracts an adhesion with the pericarp, which is finally absorbed.

Geography. Many are inhabitants of the salt marshes and sea coasts of the temperate parts of the world, particularly of the basin of the Mediter-ranean and the southern provinces of the Russian empire; others grow from Greenland and the mountains of Europe, to the sterile volcanic regions of Cape Horn. A few are found within the tropics; of these Plumbago zeylanica extends from Ceylon to Port Jackson, and Ægialitis grows among the Mangroves of northern Australasia.

Properties. This order contains plants of very opposite qualities; part are tonic and astringent, and part acrid and caustic in the highest degree. The root of Statice caroliniana is one of the most powerful astrin-gents in the vegetable materia medica. *Bigelow*, 2. 55. The bruised fresh bark of the root of Plumbago zeylanica acts as a vesicatory, and is applied in India to buboes in their incipient state. *Ainslie*, 2. 77. Plumbago europæa is employed by beggars to raise ulcers upon their bodies to excite pity; and Plumbago scandens is remarkably acrid. Plumbago europæa is said by Duroques to have been used with considerable advantage in cases of cancer, for which purpose the ulcers were dressed twice daily with olive oil in which the leaves had been infused. *Ibid.* 2. 78. Plumbago scandens is called, on account of these properties, Herbe du Diable in St. Domingo. As garden plants, nearly the whole of the order is much prized for beauty, particularly the Statices, many of which are among the most lovely herbaceous plants we know.

EXAMPLES. Statice, Armeria, Taxanthema, Plumbago, Ægialitis, Vogelia, Theta.

CLXXXIV. DIPSACEÆ. THE SCABIOUS TRIBE.

DIPSACEÆ, *Juss. Gen.* 194. (1789); *Dec. et Duby Bot. Gall.* 255. (1828); *Lindl. Synops.* 139. (1829); *Coulter Mem. in Act. Genev.* 2. 13. (1823).

DIAGNOSIS. Monopetalous dicotyledons, with an inferior 1-celled ovarium, capitate flowers, distinct anthers, and albuminous pendulous seeds.

ANOMALIES. Ovarium sometimes partly superior.

ESSENTIAL CHARACTER.—*Calyx* superior, membranous, resembling pappus; surrounded by a scarious involucellum. *Corolla* monopetalous, tubular, inserted in the calyx; *limb* oblique, 4- or 5-lobed, with an imbricated aestivation. *Stamens* usually 4 or 5, alternate with the lobes of the corolla; *anthers* distinct. *Ovarium* inferior, 1-celled, with a single pendulous ovulum; *style* 1; *stigma* simple. *Fruit* dry, indehiscent, 1-celled, crowned by the pappus-like calyx; *embryo* straight, in the axis of fleshy albumen; *radicle* superior.—*Herbaceous* plants or *under-shrubs. Leaves* opposite or whorled. *Flowers* collected upon a common receptacle, and surrounded by a many-leaved *involucrum.*

AFFINITIES. The relation of this family is obviously in the first degree with Compositæ, from which it differs in its distinct stamens and its pendulous albuminous seeds; and next with Calycereæ, which have connate anthers and alternate leaves. But if we compare it with Caprifoliaceæ, different as it is in habit, we shall find very little beyond the capitate flowers and the presence of an involucellum to distinguish it absolutely. The same character of the capitate flowers, and the presence of albumen, forms the distinction between Dipsaceæ and Valerianeæ. What is called the involucellum is a curious organ, resembling an external calyx, and is to each particular flower of the head of Dipsaceæ what the partial involucrum of Umbelliferæ is to each partial umbel; and, accordingly, we ought to expect to find instances of more flowers than one being enclosed within this involucellum; and this is said by Coulter actually to take place in the genus Gundelia. This is, however, not the only peculiarity of the order. Mr. Brown has the following curious remarks.

"M. Auguste Saint Hilaire, in his excellent memoir on Primulaceæ, while he admits the correctness of M. Decandolle's account with respect to great part of Dipsaceæ, has at the same time well observed, that in several species of Scabiosa the ovarium is entirely united with the tube of the calyx. But neither of these authors has remarked the curious, and I believe peculiar, circumstance, of the base of the style cohering with the narrow apex of the tube of the calyx, even in those species of the order in which the dilated part of the tube is entirely distinct from the ovarium. This kind of partial cohesion between pistillum and calyx is directly opposite to what usually takes place, namely, the base of the ovarium being coherent, while its upper is distinct. It equally, however, determines the apparent origin or insertion of corolla and stamina, producing the unexpected combination of 'flos superus' with ovarium 'liberum.'" *Linn. Trans.* 12. 138.

GEOGRAPHY. Chiefly natives of the south of Europe, Barbary, the Levant, and the Cape of Good Hope; not affecting particular stations in any striking degree, except that they generally shun cold, and do not attain much elevation above the sea. *Coulter.*

PROPERTIES. Unimportant. The Teasel used by fullers in dressing cloth is the dried head of Dipsacus fullonum. Some of them are reputed febrifugal. Scabiosa succisa is said to yield a green dye, and also to be astringent enough to deserve the attention of tanners. *Gmel. Fl. Bad.* 1. 319.

EXAMPLES. Dipsacus, Scabiosa, Knautia.

CLXXXV. VALERIANEÆ. THE VALERIAN TRIBE.

VALERIANEÆ, *Dec. Fl. Fr.* ed. 3. v. 4. p. 232. (1815); *Dufr. Valer. Monogr.* 56. (1811); *Lindl. Synops.* 137. (1829).

DIAGNOSIS. Monopetalous dicotyledons, with an inferior 1-celled ovarium, distinct stamens, and exalbuminous pendulous seeds.

ANOMALIES.

ESSENTIAL CHARACTER.— *Calyx* superior; the *limb* either membranous, or resembling pappus. *Corolla* monopetalous, tubular, inserted into the top of the ovarium, with from 3 to 6 lobes, either regular or irregular, sometimes calcarate at the base. *Stamens* from 1 to 5, inserted into the tube of the corolla, and alternate with its lobes. *Ovarium* inferior, with 1 cell, and sometimes 2 other abortive ones; *ovulum* solitary, pendulous; *style* simple, *stigmas* from 1 to 3. *Fruit* dry, indehiscent, with 1 fertile cell and 2 empty ones. *Seed* solitary, pendulous; *embryo* straight, destitute of albumen; *radicle* superior.— *Herbs. Leaves* opposite, without stipulæ. *Flowers* corymbose, panicled, or in heads.

AFFINITIES. Distinguished from Dipsaceæ by their flowers not being in heads, by the want of albumen, by sensible properties, and the absence of an involucellum.

GEOGRAPHY. Natives of most temperate climates; sometimes at considerable elevations. They are abundant in the north of India, Europe, and South America, but uncommon in Africa and North America.

PROPERTIES. The roots of Valeriana officinalis, Phu, and celtica, are tonic, bitter, aromatic, antispasmodic, and vermifugal; they are even said to be febrifugal. The scent of these roots is not agreeable to a European; and yet those of some species are highly esteemed as perfumes. Eastern nations procure from the mountains of Austria the Valeriana celtica to aromatise their baths; the V. Jatamansi, or true Spikenard of the ancients, is valued in India, not only for its scent, but also as a remedy in hysteria and epilepsy. The young leaves of the species of Valerianella are eaten as salad, under the French name of Mâche, or the English one of Lamb's Lettuce. Red Valerian is also eaten in the same way in Sicily. *Dec.*

EXAMPLES. Valeriana, Valerianella, Patrinia.

CLXXXVI. COMPOSITÆ.

COMPOSITÆ, *Adans. Fam.* 2. 103. (1763); *Kunth in Humb. N. G. et Sp.* vol. 4. (1820); *Lindl. Synops.* 140. (1829).—SYNANTHEREÆ, *Rich. Anal.* (1808); *Cassini Dict. Sc. N.* 10. 131. (1818); *ibid.* 60. 563. (1830).—CORYMBIFERÆ, CYNAROCEPHALÆ, and CICHORACEÆ, *Juss. Gen.* (1789).

DIAGNOSIS. Monopetalous dicotyledons, with a 1-celled inferior ovarium, capitate flowers, syngenesious stamens, and erect ovula.

ANOMALIES.

ESSENTIAL CHARACTER.—*Calyx* superior, closely adhering to the ovarium, and un-distinguishable from it; its *limb* either wanting, or membranous, divided into bristles, paleæ, hairs, or feathers, and called *pappus*. *Corolla* monopetalous, superior, usually deci-duous, either ligulate or funnel-shaped; in the latter case, 4- or 5-toothed, with a valvate æstivation. *Stamens* equal in number to the teeth of the corolla, and alternate with them; the *anthers* cohering into a cylinder. *Ovarium* inferior, 1-celled, with a single erect ovulum; *style* simple; *stigmas* 2, either distinct or united. *Fruit* a small, indehiscent, dry pericarpium, crowned with the limb of the calyx. *Seed* solitary, erect; *embryo* with a taper, inferior radicle; *albumen* none.—*Herbaceous* plants or *shrubs*. *Leaves* alternate or opposite, without stipulæ, usually simple. *Flowers* (called *florets*) unisexual or hermaphro-dite, collected in dense *heads* upon a common *receptacle*, surrounded by an *involucrum*. *Bracteæ* either present or absent; when present, stationed at the base of the florets, and called *paleæ of the receptacle*.

AFFINITIES. One of the most natural and extensive families of the vegetable kingdom, at all times recognised by its syngenesious stamens and capitate flowers. Calycereæ and Dipsaceæ, neighbouring orders, are readily distinguished by their pendulous ovulum, and by the anthers being either wholly or partially distinct. In proportion to its strict natural limits, depending upon the uniformity of its characters, is the difficulty of sepa-rating it into sections or subordinate divisions, a measure absolutely neces-sary, on account of the vast number of species referable to the order. Jussieu has three; Corymbiferæ, the florets of which are flosculous in the middle, and ligulate at the circumference; Cichoraceæ, the florets of which are all ligulate; and Cynarocephalæ, all whose florets are flosculous; to which has since been added a tribe called *bilabiate*. Linnæus divided them according to the sexes of the florets of different parts of the same head. The former has been found unexceptionable, as far as it goes; the latter wholly unmanageable. Neither, however, have satisfied the views of modern botanists, who have divided the order into a considerable number of sections, to which each has given his own name; so that this order has become a perfect chaos to all who have not devoted years to its exclusive study. The most important of those who have undertaken to remodel Compositæ, are M. Cassini, who has written much upon them in the *Dictionnaire des Sciences Naturelles*, and elsewhere; M. Kunth, whose arrangement will be found in Humboldt's *Nova Genera et Species Plantarum*; Mr. Don, who has written several detached papers upon them; and Link, who has an arrangement of his own in his *Handbuch*, vol. 1. p. 685. The most pro-found writers upon their general structure are M. Cassini and Mr. Robert Brown, whose paper in the 12th volume of the *Transactions of the Linnean Society* is a masterpiece of careful investigation and acute reasoning, from which I extract the following remarks :—

" The whole of Compositæ agree in two remarkable points of structure of their corolla; which, taken together at least, materially assist in deter-mining the limits of the class. The first of these is its valvular æstivation; this, however, it has in common with several other families. The second I believe to be peculiar to the class, and hitherto unnoticed. It consists in the disposition of its fasciculi of vessels or nerves; these, which at their origin are generally equal in number to the divisions of the corolla, instead of being placed opposite to these divisions, and passing through their axes as in other plants, alternate with them; each of the vessels at the top of the tube dividing into two equal branches, running parallel to and near the margins of the corresponding laciniæ, within whose apices they unite. These, as they exist in the whole class, and are in great part of it the only vessels observable, may be called primary. In several genera, however, other vessels occur, alternating with the primary, and occupying the axes of the laciniæ; in some cases these secondary vessels being most distinctly visible in the laciniæ, and becoming gradually fainter as they descend the tube, might be regarded as recurrent, originating from the united apices of the primary branches; but in other cases, where they are equally distinct at the base of the tube, this supposition cannot be admitted. A monopetalous corolla not splitting at the base is necessarily connected with this structure, which seems also peculiarly well adapted to the dense inflorescence of Compositæ, the vessels of the corolla and stamina being united, and so disposed as to be least liable to suffer by pressure." *R. Brown Linn. Trans.* 12. 77.

GEOGRAPHY. All parts of the world abound in Compositæ, but in very different proportions. According to the calculations of Humboldt, they constitute $\frac{1}{7}$ of the phænogamous plants of France, $\frac{1}{8}$ of Germany, $\frac{1}{15}$ of Lapland, in North America $\frac{1}{4}$, within the tropics of America $\frac{1}{2}$; upon the authority of Mr. Brown, they only form $\frac{1}{16}$ of the Flora of the north of New Holland, and did not exceed $\frac{1}{23}$ in the collection of plants formed by Dr. Smith upon the western coast of Africa in Congo. *Congo*, 445. In Sicily they constitute rather more than $\frac{1}{2}$ (*Presl.*); the same proportion exists in the Balearic Islands (*Cambessédes*); but in Melville Island they are rather more than $\frac{1}{16}$ (*Brown*), a proportion nearly the same as that of the tropical parts of New Holland. It does not, therefore, appear that Compositæ, as an order, are subject to any very fixed ratio of increase or decrease corresponding with latitude. But much remains to be learned upon this subject. It is certain that Cichoraceæ are most abundant in cold regions, and Corymbiferæ in hot ones; and that while in the northern parts of the world Compositæ are universally herbaceous plants, they become gradually frutescent, or even arborescent, as we approach the equator; most of those of Chile are bushes, and the trees of St. Helena are chiefly Compositæ.

PROPERTIES. I shall extract the substance of Decandolle's excellent remarks upon the properties of this family, with some additions. See *Essai sur les Propriétés, &c.* 177.

They are best considered under the three principal heads of classification.

CORYMBIFERÆ.

There is a bitterness peculiar to all Compositæ, which in this section assumes a particular character, being combined with a resinous principle. If this latter exists in an inconsiderable quantity, and mixed with a bitter or astringent mucilage, we find tonic, stomachic, and febrifugal qualities, as in Tussilago Farfara, Camomile, Elecampane, Golden Rod, Matricaria Parthenium, the Stevia febrifuga of Mexico, and Eupatorium perfoliatum. The Inula Helenium, or Elecampane, has a root which is aromatic and slightly fœtid. It is said to be of little value as a stomachic; the French prepare from it a medicinal wine they call Vin d'Aulnée. *Ainslie*, 1. 120. Eupatorium perfoliatum is known in North America under the name of Boneset. It possesses very important tonic and diaphoretic properties; it is also slightly stimulant. See *Barton*, 2. 133. upon this subject. In proportion as this resinous principle increases, the stimulating properties are augmented. Some become anthelmintics, as Artemisia, Tansy, and Santolina; others emmenagogues, as Matricaria, Achillea and Artemisia. The seeds of Vernonia anthelmintica are accounted, in India, a very powerful anthelmintic. *Ainslie*, 2. 54. Artemisia chinensis and other species yield the Moxa of China, a substance which is used as a cautery, by burning it upon parts affected by gout and rheumatism. The leaves of A. maderaspatana are esteemed by the Indian doctors a valuable stomachic medicine; they are also sometimes used in antiseptic and anodyne fomentations. *Ibid.* 1. 482. Artemisia indica is considered in India a powerful deobstruent and antispasmodic. *Ibid.* 2. 194. Some are sudorifics, like Eupatorium, Achillea, Artemisia, and Calendula; others diuretic; and some possess both these qualities. A species of Conyza is highly esteemed in Mendoza as a diuretic. Erigeron philadelphicum and heterophyllum are both used in the United States as diuretics. They are commonly sold under the name of Scabions. *Barton,* 1. 234. The roots of several species of Liatris are active diuretics. *Ibid.* 2. 225. A decoction of the leaves and roots of Elephantopus scaber is given on the Malabar coast in cases of dysuria. *Ainslie,* 2. 17. A decoction of Cacalia sonchifolia is antifebrile. *Ibid.* 2. 213. The leaves of Cacalia alpina and sarracenica are recommended in coughs. *Ibid.* Many are sternutatories, as Ptarmica and Arnica; others excite salivation powerfully, as Spilanthus, Siegesbeckia orientalis, Anthemis pyrethrum, Coreopsis bidens, and Bidens tripartita: some are emetic. A decoction of Anthemis cotula is a strong and active bitter; in the dose of a teacupful it produces copious vomiting and sweating. *Barton,* 1. 169. Others are tonic and antispasmodic, such as Achillea, Camomile, Wormwood, Tansy, Eupatorium, &c. Many have been celebrated for their power of curing the bites of serpents, especially Eupatorium Ayapana, the leaves of which also form, in infusion, excellent diet drink; when fresh bruised, they are said to be a most useful application for cleaning the face of a foul ulcer. *Ainslie,* 2. 35. An infusion of another species is used by the Javanese in fevers. *Ibid.* A valuable antidote against the bite of serpents, Vijuco del guaco, much esteemed in Spanish America, is produced by Mikania guaco. *Humboldt Cinch. Forests,* p. 21. *Eng. ed.* But the power of this Mikania is denied in the most positive terms by Dr. Hancock (*Quarterly Journ. July* 1830, p. 334.), who suspects that the real Guaco antidote is some kind of Aristolochia. The peculiar and agreeable flavour of Tarragon (Artemisia dracunculus) is well known. A vinegar, not distinguishable in flavour from it, is prepared in the Alps from Achillea nana, as well as from several dwarf species of Artemisia. The seeds usually abound in a fixed oil, which, in some cases, has the reputation of being anthelmintic: it is extracted in abundance from Madia sativa, Verbesina sativa, and even Helianthus, the grains of which are made into cakes by the North American Indians. The genus Helianthus contains a species remarkable for its eatable, wholesome tubers (H. tuberosus, or Jerusalem Artichoke), while the roots of the Dahlia are extremely disagreeable. It is stated by M. Payen, that benzoic acid exists in the Dahlia. *Brewster,* 1. 376. A principle called Inulin is obtained from the roots of Inula Helenium. *Turner,* 700. The pith of the Sunflower has been stated by John to be a peculiar chemical principle, which he calls Medullin.

CINAROCEPHALÆ.

Characterised by intense bitterness, which depends upon the mixture of extractive with a gum which is sometimes yielded in great abundance. On this account some have been accounted stomachics, as Carduus benedictus; others slightly febrifugal, as Carduus marianus, Centaurea calcitrapa; the Artichoke and others sudorific and diaphoretic, as Carduus benedictus and Arctium Bardana. The modern Arabians consider the root of the Artichoke (Cynara scolymus) an aperient: they call the gum of it Kunkirzeed, and place it among their emetics. *Ainslie,* 1. 22. This bitterness is not, however, found in the unexpanded leaves or receptacles, on which account they are, in many cases, used as wholesome articles of food; as the leaves of the Cardoon, and the receptacle of the unexpanded flower of the Artichoke, the Carlina acanthifolia, and others. The flower of Echinops strigosus is used in Spain for tinder; the corollas of the Artichoke, the Cardoon, and of several thistles, are employed in the South of Europe for curdling milk; and those of Carthamus tinctorius yield a deep yellow dye, resembling Saffron. Their seeds are all oily and slightly bitter; some are purgative, as those of Carthamus; others diaphoretic, as Carduus benedictus; and, finally, some partake of all these qualities, as Arctium Bardana, whose seeds pass for diuretic, diaphoretic, and slightly purgative.

CICHORACEÆ.

These are very like Campanulaceæ in their medical and chemical properties, as might have been expected from the close affinity they bear that order botanically. Their juice is usually milky, bitter, astringent, and narcotic, as is well known to be the case in Succory, Endive, and even the common Lettuce, but more especially in Lactuca virosa and sylvestris, both of which yield an extract resembling Opium in its qualities, but less likely to produce the inconvenient consequences that often attend upon the use of that drug. Before this narcotic bitter secretion is formed, many of the species are useful articles of food; the Succory and Endive, for instance, when blanched, and the roots of Scorzonera and Tragopogon, or Salsafy.

EXAMPLES. Leontodon, Bellis, Carduus.

Since the foregoing was set in type, the last volume of the *Dictionnaire des Sciences Naturelles* has reached me. In that work M. Cassini has at length given the differential characters of his tribes, and a complete Index of the places in which his observations are to be found. This will render the study of the genera and divisions of this very accurate and learned botanist more accessible than it has hitherto been. I do not extract the names of the tribes and their characters, as they would, in the first place, occupy more space than could be conveniently afforded, and, secondly, because they cannot be considered sufficiently settled.

CLXXXVII. CALYCEREÆ.

CALYCEREÆ, *R. Brown in Linn. Trans.* 12. 132. (1816); *Rich. in Mém. Mus.* 6. 76. (1820). — BOOPIDEÆ, *Cassini in Dict. des Sc.* 5. 26. *Supp.* (1817.)

DIAGNOSIS. Monopetalous dicotyledons, with an inferior 1-celled ovarium, capitate flowers, half syngenesious stamens, and pendulous ovula.

ANOMALIES.

ESSENTIAL CHARACTER. — *Calyx* superior, of 5 unequal pieces. *Corolla* regular, funnel-shaped, with a long slender tube and 5 segments, each of which has 3 principal veins; glandular spaces below the stamens and alternate with them. *Stamens* 5, monadelphous; *anthers* combined by their lower half in a cylinder. *Ovarium* inferior, 1-celled; *ovulum* solitary, pendulous; *style* simple, smooth; *stigma* capitate. *Fruit* an indehiscent pericarpium, crowned by the rigid spiny segments of the calyx. *Seed* solitary, pendulous, sessile; *embryo* in the axis of fleshy albumen; *radicle* superior. — *Herbaceous* plants. *Leaves* alternate, without stipulæ. *Flowers* collected in heads, which are either terminal or opposite the leaves, surrounded by an involucrum. *Florets* sessile, hermaphrodite, or neuter.

AFFINITIES. A very small and curious tribe, differing from Compositæ in nothing but their albumen, pendulous ovulum, and half distinct anthers, and from Dipsaceæ in their filaments being monadelphous and their anthers partly connate. They may therefore be considered to hold a middle station between these two families. Richard's monograph, in the work above quoted, is worthy of the high reputation of that distinguished botanist.

GEOGRAPHY. All natives of South America.

PROPERTIES. Unknown.

EXAMPLES. Acicarpha, Boopis, Calycera.

CLXXXVIII. GLOBULARINEÆ.

GLOBULARINEÆ, *Dec. Fl. Fr.* 3. 427. (1815); *Cambessédes in Ann. des Sciences,* 9. 15. (1826); *Link Handb.* 1. 675. (1829.)

DIAGNOSIS. Monopetalous dicotyledons, with irregular capitate flowers, and a superior 1-celled indehiscent fruit.

ANOMALIES.

ESSENTIAL CHARACTER. — *Calyx* persistent, 5-cleft, usually equal, sometimes 2-lipped. *Corolla* hypogynous, tubular, bilabiate, rarely 1-lipped, made up of 5 petals, *Stamens* 4, the uppermost being wanting, arising from the top of the tube of the corolla, somewhat didynamous; *anthers* reniform, bursting longitudinally, the 2 cells confluent into 1. *Ovarium* superior, 1-celled, with a single pendulous *ovulum; style* filiform, emarginate at the apex. *Fruit* small, indehiscent, pointed with the persistent style. *Albumen* fleshy; *embryo* straight, in its axis; *radicle* superior, about as long as the ovate *cotyledons.* — *Shrubs,* or small low *under-shrubs,* or perennial *herbs. Leaves* alternate, often fascicled, turning black in drying. *Flowers* collected in small heads, upon a convex paleaceous receptacle.

AFFINITIES. These were placed near Primulaceæ both by Jussieu and Decandolle; but their closest affinity is now known to be with Dipsaceæ, with which Globularineæ agree in a multitude of particulars, especially in habit, but differ in having a superior ovarium, and in so little besides, that it may be doubted whether, considering the peculiar nature of the cohesion of the calyx and ovarium of Dipsaceæ, they and Globularineæ are not the same family. They were united by Lamarck in the same order as Proteaceæ.

GEOGRAPHY. Natives of the hot and temperate parts of Europe; Dantzic is their most northern station.

PROPERTIES. Bitter, tonic, and purgative herbaceous plants.

EXAMPLE. Globularia.

CLXXXIX. STELLATÆ. THE MADDER TRIBE.

RUBIACEÆ, Sect. I. *Juss. Gen.* 196. (1789). — STELLATÆ, *Linn.; R. Brown in Congo,* (1818); *Lindl. Synops.* 128. (1829). — GALIEÆ, *Turp. in Atlas du Nouv. Dict. des Sc.* (?)

DIAGNOSIS. Monopetalous dicotyledons, with an inferior didymous fruit, solitary erect ovula, angular stems, and verticillate scabrous leaves without stipulæ.

ANOMALIES.

ESSENTIAL CHARACTER. — *Calyx* superior, 4- 5- or 6-lobed. *Corolla* monopetalous, rotate or tubular, regular, inserted into the calyx; the number of its divisions equal to those of the calyx. *Stamens* equal in number to the lobes of the corolla, and alternate with them. *Ovarium* simple, 2-celled; *ovules* solitary, erect; *style* simple; *stigmata* 2. *Fruit* a dry indehiscent pericarpium, with 2 cells and 2 seeds. *Seeds* erect, solitary; *embryo* straight in the axis of horny albumen; *radicle* inferior; *cotyledons* leafy. — *Herbaceous* plants, with whorled *leaves,* destitute of *stipulæ;* square *stems; roots* staining red; *flowers* minute.

AFFINITIES. There can be little doubt that the inconspicuous weeds of which this order is composed have as strong claims to be separated from Cinchonaceæ as that order from Apocyneæ or Caprifoliaceæ. It is true that no very positive characters are to be obtained from the fructification, but the want is abundantly supplied by the square stems and verticillate leaves without stipulæ, forming a kind of star, from which circumstance the name Stellatæ is derived. Properly speaking, the appellation Rubiaceæ

should be confined to this group, as it comprehends the genus Rubia; but
that name has been so generally applied to the larger mass now comprehended under the name of Cinchonaceæ, that I find it better to abolish the
name Rubiaceæ altogether.

GEOGRAPHY. Natives of the northern parts of the northern hemisphere,
where they are extremely common weeds.

PROPERTIES. First among them stands Madder, the root of Rubia
tinctoria, one of the most important dyes with which we are acquainted; a
quality in which many other species of Stellatæ participate in a greater or
less degree. The roots of Rubia Manjista yield the Madder of Bengal
(*Ainslie*, 1. 203.) The torrefied grains of Galium are said to be a good
substitute for coffee. The flowers of Galium verum are used to curdle milk.
An infusion of Asperula cynanchica has a little astringency, and has been
used as a gargle. Asperula odorata, or Woodruff, is remarkable for its
fragrance when dried; it passes for a diuretic. Rubia noxa is said to be
poisonous. *Ed. Phil. Journ.* 14. 207.

EXAMPLES. Galium, Rubia, Asperula, Sherardia, Crucianella.

CXC. CINCHONACEÆ. THE CINCHONA TRIBE.

RUBIACEÆ, *Juss. Gen.* 196. (1789) *for the most part; Ann. Mus.* 10. 313. (1807); *Mém.
Mus.* 6. 365. (1820); *Dict. des Sciences*, 46. 385. (1827). — OPERCULARINEÆ,
Juss. Ann. Mus. 4. 418. (1804.)

DIAGNOSIS. Monopetalous dicotyledons, with an inferior ovarium, and
opposite entire leaves, with intermediate stipulæ.

ANOMALIES. Opercularia has but 1 cell and 1 seed, and the number of
stamens is incongruous with the lobes of the corolla.

ESSENTIAL CHARACTER. — *Calyx* superior, simple, with a definite number of divisions or none, and connate bracteæ at its base. *Corolla* superior, tubular, regular, with
a definite number of divisions, which are valvate or imbricated in æstivation and equal to
the segments of the calyx. *Stamens* arising from the corolla, all on the same line, and
alternate with its segments; *pollen* elliptical. *Ovarium* inferior, surmounted by a disk,
usually 2-celled, occasionally with several cells; *ovula* numerous and attached to a central
placenta, or few and erect or ascending; *style* single, inserted, sometimes partly divided;
stigma usually simple, sometimes divided into a definite number of parts. *Fruit* inferior,
either splitting into 2 cocci, or indehiscent and dry or succulent, occasionally many-celled.
Seeds definite or indefinite; in the former case erect or ascending, in the latter attached to
a central axis; *embryo* small, oblong, surrounded by horny albumen; *cotyledons* thin; *radicle* longer, turned towards the hilum. — *Trees, shrubs,* or *herbs.* Leaves simple, quite
entire, opposite or verticillate, with interpetiolary stipules. *Flowers* arranged variously,
usually in panicles or corymbs.

AFFINITIES. This well-marked and strictly limited order is nearly allied
to Compositæ, from which its distinct stamens, bilocular or plurilocular ovarium, and inflorescence, distinguish it; and consequently it participates in all
the relationship of that extensive group. From Apocyneæ the æstivation of
the corolla, the presence of stipulæ, and the inferior ovarium, distinctly
divide it; yet, according to Mr. Brown, there exists a genus in equinoctial
Africa which has the interpetiolary stipules and seeds of Rubiaceæ, and the
superior ovarium of Apocyneæ, thus connecting these two orders. *Congo,*
448. The close proximity of Caprifoliaceæ has been adverted to in speaking
of that order. A tribe called Opercularineæ, referred here by Mr. Brown
(*Ibid.* 447) and others (*A. Rich. Elém. ed.* 4. 483), is remarkable for having
but 1 seed, and the number of stamens unequal to the lobes of the corolla,

and occupies an intermediate position between genuine Cinchonaceæ and Dipsaceæ. A good monograph is much wanted of this extensive order, a very large proportion of the species belonging to which remains still unpublished. I have been constrained to alter the name of Rubiaceæ, because the genus Rubia does not belong to the order, as I limit it.

Schlechtendahl and Chamisso divide the order thus:—

Linnæa, 3. 309. &c. (1828.)

§ 1. Anthospermeæ.
Fruit capsular, 2-celled, 2-seeded, usually splitting into 2 pieces, rarely indehiscent. Leaves somewhat whorled, with a simple stipula between the leaves.

Examples. Anthospermum, Ambraria, Galopina, Phyllis.

§ 2. Spermacoceæ.
Fruit capsular, 2- 3- or 4-celled; cells 1-seeded. Leaves opposite, connected by a bristly ciliated stipula. Flowers in regular cymes, branched bi- or trichotomously.

Examples. Spermacoce, Borreria, Mitracarpum, Psyllocarpus, Richardsonia, Diodia, Staelia.

§ 3. Psychotriaceæ.
Ovarium generally with 2 cells, each containing 1 ovulum. Fruit drupaceous or berried. — Shrubs, usually with opposite leaves.

Examples. Declieuxia, Psychotria, Ixora, Coffea, Chiococca, Machaonia, Palicurea, Tetramerium.

§ 4. Cephaelideæ.
Flowers in capitate fascicles. Berry 2-seeded.

Examples. Cephaelis, Geophila.

§ 5. Coccocypseleæ.
Flowers in capitate fascicles. Berry 2-celled, many-seeded.

Examples. Coccocypselum, Burchellia.

§ 6. Cephalantheæ.
Flowers in round heads. Fruit variable.

Examples. Cephalanthus, Nauclea, Morinda.

§ 7. Hedyotideæ.
Capsule 2-celled, with a loculicidal dehiscence (indehiscent in Dentella). Cells many-seeded.

Examples. Dentella, Hedyotis, Gerontogea, Kohautia, Kadua, Xanthophytum, Metabolos, Rondeletia, Sipanea.

§ 8. Manettieæ.
Capsule 2-celled, with a septicidal dehiscence. Cells many-seeded. Stamens 4.

Example. Manettia.

§ 9. Cinchoneæ.
Capsule 2-celled, with a septicidal dehiscence. Cells many-seeded. Stamens 5, or more.

Examples. Cinchona, Buena, Exostemma, Augusta.

§ 10. Guettardeæ.
Drupe either with 1 stone and many seeds, or with several 1-seeded stones.

Examples. Guettarda, Chomelia, Burneya.

§ 11. Hameliaceæ.
Berry many-celled; cells many-seeded.

Examples. Hamelia, Sabicea, Axanthes, Gonzalagunia.

§ 12. GARDENIACEÆ.

Æstivation contorted.

EXAMPLES. Gardenia, Hillia.

This last section is intermediate between Cinchonaceæ and Strychnaceæ.

GEOGRAPHY. Almost exclusively found in the hotter parts of the world, especially within the tropics, where they are said to constitute about 1-29th of the whole number of flowering plants. In America the most northern species is Pinchneya pubens, a shrub inhabiting the southern states of North America ; the most southern is Nerteria depressa, a small herb found in the Straits of Magellan. The order is represented in northern regions by Stellatæ.

PROPERTIES. Powerful febrifugal or emetic properties are the grand features of this order, the most efficient products of which, in these two respects, are Quinquina and Ipecacuanha. The febrifugal properties depend upon the presence of a bitter, tonic, astringent principle, which exists in great abundance in the bark ; those of Cinchona are known to depend upon the presence of two alkalies, called cinchonia and quina, both of which are combined with kinic acid ; two principles which, though very analogous, are distinctly different, standing in the same relation to each other as potassa and soda. *Turner*, 648. Dr. Sertürner has obtained some other vegeto-alkalies from Cinchona, one of which he calls chinioidia. *Brande*, 12. 417. *N. S.* But the existence of this is denied by MM. Henry and Delondre. *Ibid. July* 1830, p. 422. A detailed account of the qualities, synonymes, and commercial names of the species of Cinchona is given in Mr. Lambert's *Illustration of the Genera Cinchona*, 4to. London, 1821. In the same work is a translation of Baron Humboldt's account of the Cinchona forests of South America. Three species of Cinchona, the C. ferruginea, Vellozii, and Remijiana, are found in Brazil, where they are used for the same purposes as the Peruvian bark, to which, however, they are altogether inferior. *Pl. Usuelles*, no. 2. The bark of French Guiana, possessing properties analogous to those of Cinchona, is obtained from Portlandia hexandra, the Coutarea speciosa of Aublet. *Humb. Cinch. For.* 43. *Eng. ed.* The Quinquina Piton and Quinquina des Antilles are produced by species of the genus Exostema, and are remarkable for possessing properties similar to those of true Quinquina, but without any trace of either cinchonine or quinine. *Pl. Usuelles*, no. 3. A kind of fever bark is obtained at Sierra Leone from Rondeletia febrifuga. Besides these, a great number of other species possess barks more or less valuable : Pinckneya pubens is the fever bark of Carolina ; Macrocnemum corymbosum, Guettarda coccinea, Antirhea and Morinda Royoc, are all of the same description. A lightish brown, bitter, and powerfully astringent extract, called Gambeer, is obtained at Malacca by boiling the leaves of Nauclea Gambeer ; it is sometimes substituted for Gum Kino. *Ainslie*, 2. 106. A decoction of the leaves as well as root of Weberea tetrandra is prescribed in India in certain stages of flux, and the last is supposed to have anthelmintic qualities, though neither have much sensible taste or smell. The bark and young shoots are also used in dysentery. *Ibid.* 2. 63. Among the emetics, Ipecacuanha holds the first rank : it is the root of Cephaelis Ipecacuanha, a little creeping-rooted, half-herbaceous plant, found in damp shady forests in Brazil. Similar properties are found in the roots of other Cinchonaceæ of the same country, as in Richardsonia rosea and scabra, Spermacoce ferruginea and Poaya, &c. A peculiar alkaline principle called Emetia is found in Ipecacuanha, which contains 16 per cent of it. *Turner*, 653. The Raiz Preta, which is celebrated for its power in curing dropsy, and in destroying the dangerous consequences of bites of serpents, is said to be related to Ipecacuanha. *Ed. P. J.* 1. 218. Several species of Psychotria, as emetica and herbacea, are substitutes for Ipecacuanha. The spurious barks called Quinquina Piton are capable of exciting vomiting. The powdered fruit of Gardenia dumetorum is a powerful emetic. An infusion of the bark of the root is administered to nauseate in bowel complaints. *Ainslie*, 2. 186. According to Roxburgh, the root bruised and thrown into ponds where there are fish intoxicates them as Cocculus indicus. *Ibid.* Psychotria noxa and Palicourea Marcgraavii, both called Erva de rata, are accounted poisonous in Brazil; but nothing very certain seems to be known of their properties. *Ed. P. J.* 14. 267. The leaves of Oldenlandia umbellata are considered by the native doctors of India as expectorant. *Ainslie*, 2. 101. Coffee is the roasted seeds of a plant of this order, Coffea arabica, and is supposed to owe its characters to a peculiar chemical principle called Caffein. *Turner*, 699. The part roasted is the albumen, which is of a hard horny consistence; and it is probable that the seed of all Cinchonaceæ or Stillatæ whose albumen is of the same texture would serve as a substitute. This would not be the case with those with fleshy albumen. The fruit of some species of Gardenia, Genipa, and of Vangueria, the Voa Vanga of Madagascar, are succulent and eatable.

EXAMPLES. See above.

CXCI. CAPRIFOLIACEÆ. THE HONEYSUCKLE TRIBE.

CAPRIFOLIA, *Juss. Gen.* 210. (1789) *in part.* — CAPRIFOLIACEÆ, *Dec. and Duby*, 244. (1828); *Lindl. Synops.* 131. (1829.)

DIAGNOSIS. Monopetalous dicotyledons, with an inferior many-celled ovarium, pendulous ovula, and opposite leaves without stipulæ.

ANOMALIES. Hedera, a doubtful citizen, is polypetalous. Hydrangea is both polypetalous and polyspermous.

ESSENTIAL CHARACTER. — *Calyx* superior, usually with 2 or more bracteæ at its base, entire or lobed. *Corolla* superior, monopetalous or polypetalous, rotate or tubular, regular or irregular. *Stamens* equal in number to the lobes of the corolla, and alternate with them. *Ovarium* with from 1 to 5 cells, 1 of which is often monospermous, the others polyspermous; in the former the ovulum is pendulous; *style* 1; *stigmas* 1 or 3. *Fruit* indehiscent, 1- or more celled, either dry, fleshy, or succulent, crowned by the persistent lobes of the calyx. *Seeds* either solitary and pendulous, or numerous and attached to the axis; *testa* often bony; *embryo* straight, in fleshy albumen; *radicle* superior. — *Shrubs* or *herbaceous* plants, with opposite *leaves*, destitute of *stipulæ*. *Flowers* usually corymbose, and often sweet-scented.

AFFINITIES. Whether this order comprehends the rudiments of four, namely, Hederaceæ, Hydrangeaceæ, Sambucineæ, and Lonicereæ (the true Caprifoliaceæ), or whether these are mere forms of one and the same order, it is not easy to say. They are usually combined; and yet the different habits of those sections, the separation of the petals in Hedera and Hydrangea, and some hints that have been thrown out by Mr. Brown, render it probable that there are weighty grounds for their disunion. In the mean while it is most advisable to retain the order in its present state until some skilful botanist shall have taken the subject up, especially as there can be no doubt that, whether distinct or the same, they are very nearly related to each other. Taking Lonicereæ, or the Honeysuckle tribe, for the type of the order, we find a striking affinity with Cinchonaceæ, in the monopetalous tubular corolla, definite stamens, inferior ovarium, and opposite leaves, an affinity which is confirmed by the corolla of the latter being occasionally regular or irregular. With Apocyneæ they will have, for the same reasons, an intimate alliance, differing chiefly in their qualities, in the non-connivence of their anthers, the æstivation of the corolla, and the structure of the ovarium. To Lorantheæ they also approach, but differ in the relation of the anthers to the lobes of the corolla, and in other points. But if we consider the tribe called Sambucineæ, our view of the affinities of the order will take a different turn, and we shall find an approach to an order the relationship of which would hardly have been suspected, viz. Saxifrageæ : this is established through the intervention of Hydrangea, a genus usually referred to Saxifrageæ, but which it appears more advisable to station by the side of Viburnum, from which it is undistinguishable in habit, and with which it accords in inflorescence and in the constant disposition of its flowers to become radiant, but which differs in being polypetalous and polyspermous. Besides these points of affinity, Caprifoliaceæ probably tend towards Umbelliferæ through Sambucineæ.

The following are the characters of the sections, if they be sections, of this order :—

1. LONICEREÆ. The Honeysuckle Tribe.

Lonicereæ, *Ach. Rich. Elém. de la Bot. ed.* 4. 484. (1828). — Caprifolieæ, *Dec. and Duby*, 244. (1828.)

Corolla tubular. Berry 2- to 4-celled, with 1 or many-seeded cells. Style 1. Leaves opposite.

True Caprifoliaceæ are said by Mr. Brown to be distinguished from the other genera hitherto associated with them, in the raphe being on the outer instead of inner side of the ovulum. *Brown in Wallich, Pl. As.* p. 15.

EXAMPLES. Caprifolium, Lonicera, Linnæa, Abelia, Triosteum, Diervilla, Schöpfia.

2. SAMBUCINEÆ. The Elder Tribe.

Sambucineæ, *A. Rich. Dict. Class.* 3. 173. (1823); *Dec. and Duby*, 244. (1828); *Link Handb.* 1. 662. (1829.)

Corolla rotate. Ovarium 3- or 4-celled, with solitary pendulous ovules. Styles 3 or 4. Flowers in cymes, the lateral ones often radiant. Leaves opposite.

These pass into Lonicereæ through Viburnum davuricum, which has the tubular corolla of a Lonicera, and into Hydrangeaceæ through the radiant-flowered species of Viburnum. With Hedera they are connected through Cornus.

EXAMPLES. Viburnum, Sambucus.

3. HEDERACEÆ. The Ivy Tribe.

Hederaceæ, *Ach. Rich. Bot. Med.* 2. 449. (1823); *Dec. and Duby*, 244. (1828.)

Corolla polypetalous. Disk epigynous. Style 1. Drupe or berry with 1-seeded cells. Leaves opposite or alternate.

Ach. Richard considers this a distinct order, on account of its polypetalous corolla and epigynous disk.

EXAMPLES. Hedera, Cornus.

4. HYDRANGEACEÆ. The Hydrangea Tribe.

Corolla polypetalous. Styles 2 to 5. Fruit succulent or capsular, 2- to 5-celled, many-seeded. Leaves opposite. Flowers in cymes, the lateral ones often radiant.

The characters of this tribe are so strongly marked as to justify its being established as an independent order; but the habit of the species is so entirely that of Viburnum, that I am not willing to separate them without absolute necessity. There is a remarkable resemblance between their seeds and those of Begonia.

EXAMPLES. Hydrangea, Adamia.

GEOGRAPHY. Natives of the northern parts of Europe, Asia, and America, passing downwards within the limits of the tropics; found very sparingly in northern Africa, and almost unknown in the southern hemisphere.

PROPERTIES. The fragrance and beauty of plants of the Honeysuckle tribe have been the theme of many a poet's song; but independently of such recommendations, they possess properties of considerable interest. Their bark is generally astringent; that of Lonicera corymbosa is used for dyeing black in Chile. The flowers of the Elder are fragrant and sudorific, its leaves fœtid, emetic, and a drastic purgative; qualities which are also possessed by the Honeysuckle itself, and the fruit of the Ivy. The fruit of the Viburnum is destitute of these properties, but has, instead, an austere astringent pulp, which becomes eatable after fermentation, and is made into a sort of cake by the North American Indians. Cornus mascula, or the Cornel tree, yields a fruit which is sometimes eaten, but which does not deserve much praise. The bark of Cornus florida and Cornus sericea is stated by Barton to be worthy of ranking among the best tonics of North America; nothing having been found in the United States that so effectually answers the purpose of the Peruvian bark in the management of intermittent fevers. *Barton*, 1. 51. It is a remarkable fact, that the young branches of Cornus florida, stripped of their bark and rubbed with their ends against the teeth, render them extremely white. *Ibid.* From the bark of the more fibrous roots the Indians obtain a good scarlet colour. *Ibid.* 1. 120. Triosteum perfoliatum is a mild cathartic; in large doses it produces vomiting. Its dried and roasted berries have been used as a substitute for Coffee. *Ibid.* 1. 63.

EXAMPLES. See above.

CXCII. LORANTHEÆ.

LORANTHEÆ, *Juss. and Rich. Ann. Mus.* 12. 292. (1808); *Dec. and Duby,* 246. (1828); *Lindl. Synops.* 133. (1829.)

DIAGNOSIS. Monopetalous dicotyledons, with an inferior 1-celled ovarium, a single pendulous ovulum, a naked stigma, and stamens opposite the lobes of the corolla.

ANOMALIES. Sometimes polypetalous.

ESSENTIAL CHARACTER.— *Calyx* superior, with 2 bracteæ at the base. *Corolla* with 4 or 8 petals, more or less united at the base. *Stamens* equal in number to the petals, and opposite to them. *Ovarium* 1-celled; *ovulum* pendulous; *style* 1 or none; *stigma* simple. *Fruit* succulent, 1-celled. *Seed* solitary, pendulous; *testa* membranous; *embryo* cylindrical, longer than the fleshy *albumen;* *radicle* naked, clavate, superior.—*Parasitical* half-shrubby plants. *Leaves* opposite, sometimes alternate, veinless, fleshy, without stipulæ. *Flowers* often monœcious, axillary or terminal, solitary, corymbose, or spiked.

AFFINITIES. Very near Caprifoliaceæ, from which they are readily known not only by their universally parasitical habit, but also by their stamens being opposite the lobes of the corolla, and not alternate with them. Viscum seems to bear about the same relation to Loranthus that Cornus does to Lonicereæ. Mr. Don has expressed an opinion that a connexion is established between this order and Araliaceæ, by means of Aucuba (*Jamtson's Journal,* Jan. 1830, p. 168); but this does not seem clearly made out. Mr. Brown (*Flinders,* 549) suggests their relation to Proteaceæ. The anther of Viscum is remarkable for having its substance broken up into a number of hollow cavities containing pollen, and not divided regularly into 2 lobes, each of which has a cavity containing pollen, and a longitudinal line of dehiscence.

A good figure of this will be found in the *Ann. du Muséum*, vol. 12. t. 27. fig. E. The germination of Viscum is exceedingly remarkable. It has afforded a subject for some curious experiments upon the nature of the vital energies of vegetables. See *Dutrochet sur la Motilité*, 114.

Geography. Judging from the collections of systematic botanists, it would appear that the tropics of America contain a greater number of species than all the rest of the world; but we now know, from the extensive researches of Dr. Wallich, that the Flora of India contains at least as large a proportion: the order would therefore seem to be equally dispersed through the equinoctial regions of both Asia and America; but on the continent of Africa to be much more rare, only 2 having been yet described from equinoctial Africa, and 5 or 6 from the Cape of Good Hope. Two are named from the South Seas, and 1 from New Holland; but this number requires, no doubt, to be largely increased.

Properties. The bark is usually astringent, as in the Mistletoe of the Oak. The berries contain a viscid matter like birdlime, which is insoluble in water and alcohol. The most remarkable quality that they possess, however, is the power of rooting in the wood of other plants, at whose expense they live. The habits of the common Mistletoe give an idea of those of all, except that in the genus Loranthus the corolla is tubular and usually richly coloured with scarlet.

Examples. Loranthus, Viscum.

CXCIII. POTALIACEÆ.

Potalieæ, *Martius N. G. et Sp.* 2. 91. *and* 133. (1828.)

Diagnosis. Monopetalous dicotyledons, with a superior simple ovarium, regular flowers, peltate sessile seeds, and a corolla with contorted convolute segments which are unequal to the number of lobes of the calyx.

Anomalies.

Essential Character. — *Calyx* inferior, with 4, 5, or 6 partitions. *Corolla* regular, with from 5 to 10 divisions, which are therefore not symmetrical with the segments of the calyx; the æstivation contorted, convolute. *Stamens* arising from the corolla, all upon the same line; *pollen* simple, elliptical. *Ovarium* superior; *style* continuous; *stigma* simple. *Fruit* succulent, with from 2 to 4 cells, and central placentæ. *Seeds* numerous, peltate; *testa* double; *embryo* supposed by Von Martius to be heterotropous (that is, to have its radicle not turned towards the hilum), lying in cartilaginous *albumen.* — *Trees* or *shrubs*, quite smooth. *Leaves* opposite, entire, united by interpetiolar sheathing stipulæ. *Flowers* terminal, with bracteæ, in panicles or corymbs.

Affinities. According to Von Martius, this lies between Loganieæ and Apocyneæ. Its chief characteristics are the inequality of the segments of the calyx and corolla and the stamens, and a 4-lobed placenta, which produces in Fagræa obovata, according to Dr. Wallich, a 4-celled berry. With that part of Apocyneæ to which Strychnos belongs they very nearly agree, differing principally in the above-mentioned character, the æstivation of the calyx, and the embryo not being foliaceous, agreeing in their peltate seeds and corneous albumen.

Geography. Natives of the tropics of Africa, America, and India.

Properties. An infusion of the leaves of Potalia resinifera is slightly mucilaginous and astringent, and is used in Brazil as a lotion for inflamed

P

eyes. *Von Martius*, 2. 90. Potalia amara is bitter like the Gentians, and acrid and emetic like Apocyneæ. *Dec. Prodr. Méd.* 217.

EXAMPLES. Potalia, Fagræa, Anthocleista.

CXCIV. LOGANIACEÆ.

LOGANIEÆ, *R. Brown in Flinders*, (1814); *Von Martius N. Gen. et Sp. Pl.* 2. 133. (1828.)

DIAGNOSIS. Monopetalous dicotyledons, with regular flowers, a superior 2-celled ovarium, a convolute corolla, and opposite leaves with interpetiolar stipules.

ANOMALIES. Stipulæ absent in some Loganias.

ESSENTIAL CHARACTER. — *Calyx* inferior, 5-parted. *Corolla* regular or irregular, with convolute æstivation. *Stamens* arising from the corolla, all placed upon the same line, 5 or 1, therefore not always symmetrical with the divisions of the corolla; *pollen* with 3 bands. *Ovarium* superior, 2-celled; *style* continuous; *stigma* simple. *Fruit* either capsular and 2-celled with placentæ finally becoming loose; or drupaceous, with 1- or 2-seeded stones. *Seeds* peltate, with a finely reticulated integument, sometimes winged; *albumen* fleshy or cartilaginous; *embryo* with the radicle turned towards the hilum. — *Shrubs*, *herbaceous* plants, or *trees*. *Leaves* opposite, entire, usually with stipulæ which are combined in the form of interpetiolary sheaths. *Flowers* racemose, corymbose, or solitary.

AFFINITIES. It is not clear, from the remarks upon Logania by Mr. Brown in his *Prodromus*, whether he intended to establish this order or not. He states that he has placed Logania at the end of Gentianeæ, on account of some affinity between it and Exacum and Mitrasacme, and also because it does not answer ill to the artificial character of that order; adding that it, however, might have a still closer connexion with Apocyneæ and with Usteria among Rubiaceæ (Cinchonaceæ). He further points out the close relation of Geniostoma to Logania, and concludes by inquiring whether those 2 genera do not, with Anasser, Fragræa, and Usteria, form an order intermediate between Apocyneæ and Rubiaceæ. This view has been adopted by Von Martius, with the exception of Fagræa, which he places among his Potalieæ; he founds the distinction of the order upon the want of symmetry between the parts of the calyx, corolla, and stamens, upon the æstivation of the corolla being convolute, not contorted, and in the presence of stipulæ combined in interpetiolary sheaths.

GEOGRAPHY. Found in tropical India and Africa, and in the temperate parts of New Holland.

PROPERTIES. Unknown.

EXAMPLES. Logania, Gærtneria, Pagamea.

CXCV. ASCLEPIADEÆ.

APOCYNEÆ, *Juss. Gen.* 143. (1789) *in part*; *Dec. and Duby Bot. Gall.* 323. (1828). — ASCLEPIADEÆ, *R. Brown in Wern. Trans.* 1. 12. (1809); *Prodr.* 458. (1810.)

DIAGNOSIS. Monopetalous dicotyledons, with a superior double ovarium, the apex of which is connected by a common tabular dilated stigma, regular flowers, waxy pollen, and contorted corolla.

ANOMALIES. Periploca and some others have granular pollen. Corolla valvate in Leptadenia.

ESSENTIAL CHARACTER.—*Calyx* 5-divided, persistent. *Corolla* monopetalous, hypogynous, 5-lobed, regular, with imbricated, very seldom valvular, æstivation, deciduous. *Stamens* 5, inserted into the base of the corolla, alternate with the segments of the limb. *Filaments* usually connate. *Anthers* 2-celled, sometimes almost 4-celled in consequence of their dissepiments being nearly complete. *Pollen* at the period of the dehiscence of the anther cohering in masses, either equal to the number of the cells, or occasionally cohering in pairs and sticking to 5 processes of the stigma either by twos, or fours, or singly. *Ovaria* 2. *Styles* 2, closely approaching each other, often very short. *Stigma* common to both styles, dilated, 5-cornered, with corpusculiferous angles. *Follicles* 2, 1 of which is sometimes abortive. *Placenta* attached to the suture, finally separating. *Seeds* numerous, imbricated, pendulous, almost always comose at the hilum. *Albumen* thin. *Embryo* straight. *Cotyledons* foliaceous. *Radicle* superior. *Plumula* inconspicuous. — *Shrubs*, or occasionally *herbaceous* plants, almost always milky, and often twining. *Leaves* entire, opposite, sometimes alternate or whorled, having ciliæ between their petioles in lieu of stipulæ. *Flowers* somewhat umbelled, fascicled, or racemose, proceeding from between the petioles. *R. Br.*

AFFINITIES. So closely are these plants allied to Apocyneæ, that the affinities of the one are precisely the same as those of the other; I shall therefore, in this place, speak of the difference between those two orders, and of the peculiarities of that more immediately under consideration. Mr. Brown, who distinguishes them, admits (*Flinders*, 564) that they differ solely in the peculiar character of their sexual apparatus; but this is of so unusual a kind in Asclepiadeæ, that it justifies a deviation from the general rule, that orders cannot be established upon solitary characters. In Apocyneæ the stamens are distinct, the pollen powdery (that is to say, in the ordinary state), the stigma capitate and thickened, but not particularly dilated, and all these parts distinct the one from the other. But in Asclepiadeæ the whole of the sexual apparatus is consolidated into a single body, the centre of which is occupied by a broad disk-like stigma, and the grains of pollen cohere in the shape of waxy bodies attached finally to the 5 corners of this stigma, to which they adhere by the intervention of peculiar glands. For a long time this structure was misunderstood; but Mr. Brown, in a dissertation in the Transactions of the Wernerian Society, placed its true nature beyond doubt. I subjoin the explanation given by this celebrated botanist, who thus describes the flower of Asclepias syriaca : —

" The flower-bud of this plant I first examined, while the unexpanded corolla was yet green and considerably shorter than the calyx. At this period the gland-like bodies which afterwards occupy the angles of the stamen were absolutely invisible; the furrows of its angles were extremely slight, and, like the body of the stigma, green; the antheræ, however, were distinctly formed, easily separable from the stigma, and their cells, which were absolutely shut, were filled with a turbid fluid, the parts of which did not so cohere as to separate in a mass; of the cuculli, which in the expanded flower are so remarkable, and constitute the essential character of the genus, there was no appearance.

" In the next stage submitted to examination, where the corolla nearly equalled the calyx in length, the gland-like bodies of the stigma were become visible, and consisted of 2 nearly filiform, light brown, parallel, contiguous, and membranaceous substances, secreted by the sides of the furrow, which was now somewhat deeper. Instead of the filiform processes, a gelatinous matter occupied an obliquely descending depression proceeding from towards the base of each side of the angular furrow.

" In a somewhat more advanced stage, the membranes which afterwards become glands of the stigma were found to be linear, closely approximated, and to adhere at their upper extremity. At the same time the gelatinous substance in the oblique depression had acquired a nearly membranaceous texture and a light brown colour; and on separating the gland from its furrow, which was then practicable, this membrane followed it. At this period, too, the contents of each cell of the anthera had acquired a certain degree of solidity, a determinate form, and were separable from the cell in one mass; the cuculli were also observable, but still very small and green, nearly scutelliform, having a central papilla, the rudiment of the future horn-like process. Immediately previous to the bursting of the cells of the antheræ, which takes place a little before the expansion of the corolla, the cuculli are completely formed, and between each, a pair of minute, light green, fleshy teeth are observable, the single teeth of each pair being divided from each other by the descending alæ of the antheræ. The glands of the stigma have acquired a form between elliptical and rhomboidal, a cartilaginous texture, and a brownish black colour; they are easily separable from the secreting furrow, and on their under surface there is no appearance of a suture, or any indication of their having originally consisted of two distinct parts: along with them separate also the descending processes, which are compressed, membranous, and light brown; their extremity, which is still unconnected, being more gelatinous, but not perceptibly thickened. The pollen has acquired the yellow colour, and the degree of consistence which it afterwards retains. On the bursting of the cells, the gelatinous extremity of each descending process becomes firmly united with the upper attenuated end of the corresponding mass of pollen. The parts are then in that condition in which they have been commonly examined, and are exhibited in the figures of Jacquin, who, having seen them only in this state, naturally considered these plants as truly gynandrous, regarding the masses of pollen as the antheræ, originating in the glands of the stigma, and merely immersed in the open cells of the genuine antheræ, which he calls antheriferous sacs; an opinion in which he has been followed by Rottbœll, Kœlreuter, Cavanilles, Smith, and Desfontaines. The conclusion to be drawn from the observations now detailed is sufficiently obvious; but it is necessary to remark, that these observations do not entirely apply to all the plants which I have referred to the Asclepiadeæ; some of them, especially Periploca, having a granular pollen, applied in a very different manner to the glands of the stigma: they all, however, agree in having pollen coalescing into masses, which are fixed or applied to processes of the stigma, in a determinate manner; and this is, in fact, the essential character of the order. Dr. Smith, in the second edition of his valuable *Introduction to Botany*, has noticed my opinion on this subject; but, probably from an indistinctness in the communication, which took place in conversation, has stated it in a manner somewhat different from what I intended to convey to him; for, according to his statement, the pollen is *projected* on the stigma. The term projection, however, seems to imply some degree of impetus, and at the same time presents the idea of something indeterminate respecting the part to which the body so projected may be applied. But nothing can be more constant than the manner in which the pollen is attached to the processes of the stigma in each species."

This order is one of those which contain indifferently what are called succulent plants and such as are in the usual state of other plants; this excessive development of the cellular tissue of the stem, and reduction of that of the leaves, is in its greatest degree in Stapelia and Ceropegia; it is diminished in Dischidia, the succulence of which is confined to the leaves; and it almost disappears in Hoya, the stem of which is in the usual state, but the leaves between fleshy and leathery.

GEOGRAPHY. Africa must be considered as the great field of Asclepiadeæ, especially its southern point, where vast numbers of the succulent species occupy the dry and sterile places of that remarkable country. In tropical India and New Holland, and in all the equinoctial parts of America, they all abound. Two genera only are found in northern latitudes, one of which, Asclepias, abounds in species, and is confined apparently to the eastern side of North America; the other, Cynanchum, is remarkable for extending from 59° north latitude to 32° south latitude.

PROPERTIES. The roots are generally acrid and stimulating, whence some of them act as emetics, as Cyanchum tomentosum and Periploca emetica; others are diaphoretic and sudorific, as the purgative Asclepias decumbens, which has the singular property of exciting general perspiration without increasing in any perceptible degree the heat of the body; it is constantly used in Virginia against pleurisy. *Dec.* Their milk is usually acrid and bitter, and is always to be suspected, although it probably participates in a slight degree only in the poisonous qualities of that of Apocyneæ, if we can judge from the use of some species as articles of food. Asclepias lactifera is said to yield so sweet and copious a milk, that the Indians use it for aliment; and Pergularia edulis, Periploca esculenta, Asclepias aphylla and stipitacea, are all reported to be eatable. *Dec.* The Cow Plant of Ceylon, or Kiriàghuna plant, Gymnema lactiferum, yields a milk of which the Cingalese make use for food; its leaves are also used when boiled. But very little is known about the real qualities of such plants. The root and tender stalks of Asclepias volubilis L. sicken and excite expectoration. *Ainslie,* 2. 155. Asclepias tuberosa, or Butterfly weed, is a popular remedy in the United States for a variety of disorders; its properties seem to be those of a mild cathartic, and of a certain diaphoretic attended with no inconsiderable expectorant effect. *Barton,* 1. 244. The root of Diplolepis vomitoria has a bitterish and somewhat nauseous taste. The Indian doctors prize it for its expectorant and diaphoretic qualities. It possesses virtues somewhat similar to those of Ipecacuanha, and has been found an extremely useful medicine in dysenteric complaints. *Ainslie,* 2. 84. A decoction of Asclepias curassavica is said to be efficacious in gleets and fluor albus. *Lunan,* 1. 64. The root and bark, and especially the inspissated milk, of Calotropis gigantea, the Akund, Yercum, or Mudar plant of India, is a powerful alterative and purgative; it is especially in cases of leprosy, elephantiasis, intestinal worms, and venereal affections, that it has been found important. A variety of cases are mentioned in books upon Indian medicine; and there seems no doubt that this will form one of the most important of all the articles of the Materia Medica. See, for information upon this point, *Ainslie's Materia Medica,* 1. 486.; *Trans. of the Med. Chir. Soc.* vol. 10.; *Edinb. Med. Chir. Trans.* 1. 414.

EXAMPLES. Asclepias, Cynanchum, Stapelia, Pergularia, Gomphocarpus, Caralluma.

CXCVI. APOCYNEÆ.

APOCYNEÆ, *Juss. Gen.* 143. (1789) *in part; R. Brown Prodr.* 465. (1810); *Lindl. Synops.* 176. (1829). — CONTORTÆ, *Linn.* — STRYCHNEÆ, *Dec. Théorie, ed.* 1. 217. (1813). — VINCEÆ, *Dec. and Duby Bot. Gall.* 324. (1828), *a* § *of* Apocyneæ. STRYCHNACEÆ, *Blume Bijdr.* 1018. (1826); *Link Handb.* 1. 439. (1829.)

DIAGNOSIS. Monopetalous dicotyledons, with a superior double ovarium, the apex of which is connected by a common simple stigma, regular flowers, powdery pollen, and a contorted corolla.

ANOMALIES. Corolla valvate in Gardneria. Leaves subalternate in succulent species.

ESSENTIAL CHARACTER.— *Calyx* divided in 5, persistent. *Corolla* monopetalous, hypogynous, regular, 5-lobed, with contorted æstivation, deciduous. *Stamens* 5, arising from the corolla, with whose segments they are alternate. *Filaments* distinct. *Anthers* 2-celled, opening lengthwise. *Pollen* granular, globose, or 3-lobed, immediately applied to the stigma. *Ovaria* 2, or 1 2-celled, polyspermous. *Styles* 2 or 1. *Stigma* 1. *Fruit* a follicle, capsule, or drupe, or berry, double or single. *Seeds* with fleshy or cartilaginous *albumen ; testa* simple ; *embryo* foliaceous ; *plumula* inconspicuous ; *radicle* turned towards the hilum. — *Trees* or *shrubs*, usually milky. *Leaves* opposite, sometimes whorled, seldom scattered, quite entire, often having ciliæ or glands upon the petioles, but with no stipulæ. *Inflorescence* tending to corymbose.

AFFINITIES. These are strongest with Asclepiadeæ, in which they have already been discussed ; otherwise they lie between Cinchonaceæ and Gentianeæ. From Cinchonaceæ they are distinguished by their superior ovarium, contorted flowers, and absence of stipulæ ; in room of which are, however, sometimes produced certain ciliæ, or other appendages of the petiole, which the inexperienced observer may mistake for stipulæ. The same characters divide them from Gentianeæ : and I think the combination of these peculiarities is sufficient to destroy all doubt about the limits of any of these orders. From Potalieæ and Loganieæ they are distinguished almost entirely by the perfect symmetry of the calyx, corolla, and stamens, and the want of true stipulæ.

I agree with Von Martius, Brown, and other botanists, who consider Strychneæ a mere section of Apocyneæ, rather than a distinct order : it differs chiefly in its peltate naked seeds and simple succulent fruit. In consequence of its ciliated petioles, I am unwilling to refer Gardneria to Loganieæ.

Plumieria is the most succulent genus of the order.

GEOGRAPHY. Natives of nearly the same localities as Asclepiadeæ, with the exception that they are less abundant at the Cape of Good Hope.

PROPERTIES. Not very different from those of Asclepiadeæ, but perhaps rather more suspicious. The order contains species with the same purgative, the same acrid, the same febrifugal qualities. The bark of Cerbera Manghas is purgative ; that of Echites antidysenterica is astringent and febrifugal. The leaves of Nerium Oleander contain an abundance of gallic acid ; the Vahea of Madagascar and Urceola elastica a notable quantity of caoutchouc. The fruit of the succulent-fruited genera is emetic ; and yet that of Carissa edulis is eaten in Nubia. *Delile Cent.* 11. The bark of the root and the sweet-smelling leaves of Nerium odorum are considered by the native Indian doctors as powerful repellents, applied externally. The root, taken internally, acts as a poison. *Ainslie*, 2. 23. It would seem, from an examination by Mr. Arnott of flower-buds of a milk-tree called Hya-hya in Demerara, that this remarkable vegetable production belongs to this order. It is described by Mr. Smith, its European discoverer, to yield a copious stream of thick, rich, milky fluid, destitute of all acrimony, and only leaving a slight clamminess upon the lips. A tree which was felled on the banks of a small stream had completely whitened the water in an hour or two. Mr. Arnott calls it Tabernæmontana utilis. *Jameson's Journal, Ap.* 1830. The milk has been analysed by Dr. Christison, who finds it to consist of a small proportion of caoutchouc, and a large proportion of a substance possessing in some respects peculiar properties, which appear to place it intermediate between caoutchouc and the resins : it probably, therefore, has no nutritive qualities. *Ed. N. Ph. Journ. June* 1830, p. 34. The Cream fruit of Sierra Leone belongs here ; birdlime is obtained in Madagascar from the Voacanga ; and the caoutchouc of Sumatra is produced by the genus Urceola. *Brown in Congo,* 449. The root of Plumeria obtusa is used as a cathartic in Java. *Ainslie,* 2. 137. The Conessi Bark of the British

Materia Medica, the Palapatta of the Hindoos on the Malabar coast, is the produce of Wrightia antidysenterica: it is a valuable tonic and febrifuge. On the Coromandel side of India it seems chiefly to be given in dysenteric affections. The milky juice of the tree is used as a vulnerary. *Ibid.* 1. 88. The Wrightia tinctoria is extremely valuable as a dyer's plant, the blue colour it yields equalling Indigo. The Sarsaparilla of India is chiefly the root of Periploca indica: a decoction of it is prescribed by European practitioners in cutaneous diseases, scrofula, and venereal affections. *Ibid.* 1. 382. An infusion of the leaves of Allamanda cathartica is a valuable cathartic. *Ibid.* 2. 9. The leaves of Cynanchum Argel are used in Egypt for adulterating Senna. A powerful poison is yielded by the kernel of the Tanghin tree of Madagascar (Cerbera Tanghin), a single seed being sufficient to destroy twenty persons: see the *Botanical Magazine*, folio 2968, for an excellent account of this plant. The Strychnos colubrina is used in Java in intermittent fever, and as an anthelmintic. According to Horsfield, the Malays prepare from it an excellent bitter tincture. Virey says, in an over-dose it occasions tremors and vomiting. *Ainslie*, 2. 203. The St. Ignatius's bean (Strychnos St. Ignatii), called Papeeta in India, is prescribed by the native practitioners of India in cholera with success: it is mixed with Jehiree or Durreoaye Narriol (Cocos maldivica). If given in over-dose, vertigo and convulsions come on; but they are easily cured by lemonade drank largely. *Trans. M. and P. S. Calc.* 3. 432. The seeds of Strychnos Nux vomica are well known, under the latter name, for containing a dangerous narcotic property, which modern chemists have ascertained to depend upon the presence of a peculiar principle called strychnia. Small quantities of the extract have been given with uncertain success in cases of mania, gout, epilepsy, hysteria, and dysentery, and also in paraplegia and hemiplegia. *Ainslie*, 1. 321. This strychnia is one of the most violent poisons hitherto discovered: its energy is so great, that half a grain blown into the throat of a rabbit, occasioned death in the course of five minutes. Its operation is always accompanied with symptoms of locked jaw and other tetanic affections. *Turner*, 651. A peculiar acid, called by MM. Pelletier and Caventou the Igasuric acid, occurs in combination with strychnia in nux vomica and the St. Ignatius bean; but its existence, as different from all other known acids, is doubtful. *Ibid.* 641. It is remarkable, that one of the most valuable febrifuges of Brazil belongs to this order. The bark of the Strychnos Pseudo-quina is fully equal to Cinchona in curing intermittent fevers; it appears to possess some of the dangerous properties of nux vomica; but according to the analysis of Vauquelin, it contains no strychnia whatever. *Pl. Usuelles*, no. 1. The pulp of the fruit of S. pseudo-quina, and even of S. nux vomica, is eaten without inconvenience. *Ibid.* no. 1. M. Cailliaud found a species of Strychnos in Nubia, the fruit of which is sweet and not unwholesome; and M. Delile remarks, that the venomous species are always bitter. *Delile Cent.* 11.

EXAMPLES. Nerium, Wrightia, Apocynum, Tabernæmontana, Cerbera, Carissa, Gardneria.

CXCVII. GENTIANEÆ. The Gentian Tribe.

GENTIANEÆ, *Juss. Gen.* 141. (1789); *R. Brown Prodr.* 449. (1810); *Lindl. Synops.* 177. (1829); *Von Martius Nov. Gen. &c.* 2. 132. (1828.)

DIAGNOSIS. Monopetalous bitter dicotyledons, with regular flowers, a superior 1- or 2-celled ovarium, an imbricated withering corolla, indefinite seeds, capsular fruit, and opposite exstipulate entire leaves.

ANOMALIES. Menyanthes and Villarsia have alternate leaves.

ESSENTIAL CHARACTER. — *Calyx* monophyllous, divided, inferior, persistent. *Corolla* monopetalous, hypogynous, usually regular, withering or deciduous ; the limb divided, equal, its lobes of the same number as those of the calyx, generally 5, sometimes 4, 6, 8, or 10, with an imbricated twisted æstivation. *Stamens* inserted upon the corolla, all in the same line, equal in number to the segments, and alternate with them ; some of them occasionally abortive. *Pollen* 3-lobed or triple. *Ovarium* single, 1- or 2-celled, many-seeded. *Style* 1, continuous ; *stigmas* 1 or 2. *Capsule* or *berry* many-seeded, with 1 or 2 cells, generally 2-valved ; the margins of the valves turned inwards, and in the genera with 1 cell, bearing the seeds ; in the 2-celled genera inserted into a central placenta. *Seeds* small ; *testa* single ; *embryo* straight in the axis of soft fleshy albumen ; *radicle* next the hilum. — *Herbaceous* plants, seldom *shrubs*, generally smooth. *Leaves* opposite, entire, without stipulæ, sessile, or having their petioles confluent in a little sheath. *Flowers* terminal or axillary.

AFFINITIES. Very near Apocyneæ, from which they differ in their herbaceous habit, withering corolla, entire ovarium, imbricated, not contorted, æstivation, want of milk, and capsular fruit without naked seeds. Mr. Brown remarks, that this order is better known by its habit than by any particular character ; being, on the one hand, allied to Polemoniaceæ and Scrophularineæ, from the latter of which it is distinguished by its regular flowers, the stamens of which are equal to the lobes of the corolla, and from the former by the dehiscence of the capsule and the placentation of the seeds ; and, on the other hand, to certain Apocyneæ. From Scrophularineæ it is frequently difficult to distinguish this order, especially if the flowers are absent ; Loganieæ and Spigeliaceæ are also very closely allied. For remarks on the three last, see those orders respectively. Von Martius, however, points out some differences between Gentianeæ and Scrophularineæ, and their allies, which will further assist in distinguishing them. No Gentianeæ, except Tachia, have a hypogynous disk ; and the two carpellary leaves of which the fruit is formed are lateral, or right and left with respect to the common axis of the inflorescence, their placentæ being consequently anterior and posterior ; but in Scrophularineæ, Gesnereæ, Bignoniaceæ, Acanthaceæ, and their allies, a hypogynous disk is very common in the shape of a fleshy ring, or of glands, or teeth, and the two carpellary leaves are anterior and posterior, the dissepiment being consequently in the same transverse line as separates the upper from the lower lip. Menyanthes and Villarsia are probably the type of a small order distinguished by their alternate and sometimes compound toothed leaves, the characters of which are still to trace. Von Martius excludes them absolutely ; Mr. Brown places them at the end of the order, along with Anopterus, which seems to be distinct both from Gentianeæ and Menyanthes : it will be seen, further on, that their properties are absolutely the same as those of Gentianeæ.

GEOGRAPHY. A numerous order of herbaceous plants, extending over almost all parts of the world, from the regions of perpetual snow upon the summits of the mountains of Europe, to the hottest sands of South America and India. They, however, do not appear in the Flora of Melville Island ; but they form part of that of the Straits of Magellan.

PROPERTIES. The intense bitterness of the Gentian is a characteristic of the whole order ; it resides both in their stems and roots, and renders them tonic, stomachic, and febrifugal ; and it is very remarkable that there are no exceptions to these properties in the whole order, as it is now limited. The principal enumerated by Decandolle are, Gentiana lutea, employed in France and England ; G. rubra, substituted for it in Germany ; G. purpurea in Norway ; G. amarella, campestris, cruciata, Chlora perfoliata, G. peruviana, called Cachen in Peru, G. Chirita, the famous stomachic of the East Indies, and Coutoubea alba and purpurea. The root of Gentiana lutea, notwith-standing its bitterness, contains a considerable proportion of sugar: it is, on this account, sometimes manufactured into brandy, for which purpose it is exported from some parts of Switzerland. Menyanthes trifoliata and Vil-larsia nymphoides are bitter, tonic, and febrifugal; and the same has been remarked of Villarsia ovata. *Essai Méd.* 216. Sabbatia angularis is held in estimation in North America for its pure bitter, tonic, and stomachic vir-tues. *Barton*, 1. 259. The root of Frazera Walteri is a pure, powerful, and excellent bitter, destitute of aroma. It is accounted in North America not inferior to the Gentian or Columbo of their shops. In its recent state it is said to possess considerable emetic and cathartic powers. *Ibid.* 2. 109. The roots of Lisianthus pendulus are used by the Brazilians in decoction as a febrifuge: they are intensely bitter. Tachia guianensis exudes little yellow drops of pellucid resin from the axillæ of the leaves; its bitter root is used as a febrifuge. *Von Martius.*

EXAMPLES. Gentiana, Chironia, Sabbatia, Coutoubea.

CXCVIII. SPIGELIACEÆ. THE WORMSEED TRIBE.

SPIGELIACEÆ, *Martius N. G. et Sp.* 2. 132. (1828.)

DIAGNOSIS. Monopetalous dicotyledons, with regular flowers, a supe-rior 2-celled ovarium, several ovules, a valvate corolla, dry fruit, and opposite leaves.

ANOMALIES.

ESSENTIAL CHARACTER.— *Calyx* inferior, regularly 5-parted. *Corolla* regular, with 5 lobes, which have a valvate æstivation. *Stamens* 5, inserted into the corolla all in the same line; *pollen* 3-cornered, with globular angles. *Ovarium* superior, 2-celled; *style* articulated with it, inserted; *stigma* simple. *Fruit* capsular, 2-celled, 2-valved, the valves turned inwards at the margin and separating from the central placenta. *Seeds* several, small; *testa* single; *embryo* very minute, lying in copious fleshy *albumen*, with the radicle next the *hilum.*— *Herbaceous* plants or *under-shrubs. Leaves* opposite, entire, with sti-pulæ, or a tendency to produce them. *Flowers* arranged in 1-sided spikes. *Pubescence* simple or stellate.

AFFINITIES. This order was founded by Dr. Von Martius, from whose splendid work upon the Brazilian Flora I extract the following remarks: — " There are many reasons for separating Spigelia from Gentianeæ; and I am the more disposed to attend to those reasons, from seeing daily instances of the necessity of establishing new orders, to avoid weakening the characters of old ones. For example, Aquilarineæ, Datisceæ, Hamamelideæ, and other orders constructed upon a few species, are so many instances of this practice, by which the science is both embellished and strengthened by our most skilful botanists. With regard to Spigelia, if we retain it among Gentianeæ, I do not know how we are to distinguish that order with certainty from those in its neighbourhood; for this genus approaches Scrophularineæ in the divi-sion of the two valves of the fruit, and in the central, not parietal, origin of the placentæ; and Rubiaceæ in the insertion of the style into the ovarium, and the distension of the petiole into the form of a stipula. Scrophularineæ are, indeed, so nearly related to Gentianeæ, that the best botanists have admitted that there are scarcely any marks of distinction between them, besides the regular number of the stamens of the latter, and the simplicity of the valves of the capsule." (The position of the pericarpial leaves with relation to the axis of inflorescence, is now known to be a certain mark of distinction between Gentianeæ and Scrophularineæ.) " Some may possibly adduce the irregularity of the corolla of Scrophularineæ, and the origin of the placentæ from the mere inflexion of the valves of the capsule in Gentianeæ; but it must be remembered, that there are certain genera of Scrophularineæ, such as Limnophila, Xuaresia, Ourisia, and Veronica, the corolla of which is regular or nearly so; and that certain Gentianeæ, for instance Exacum and Schübleria, have central placentæ, which, although deriving their origin from the inflexion of the valves of the capsule, yet become loose and more or less distinct. Others may refer to the æstivation as another source of differences, it being in Gentianeæ, on account of the lateral and somewhat contorted twisting of the nearly equal segments, *contorted-convolutive*, and in Scrophularineæ, on account of the involution of the unequal segments towards the centre of the flower, merely *imbricated; b*ut these differences, on account of the different forms of the corolla in these extensive orders, are scarcely distinguishable, and are more available in theory than in practice. Besides, in Spigelia the æstivation is different from either, being valvate, with the margins of the segments often protruding into acute angles, and is more like that of Rubiaceæ (Cinchonaceæ). It must not be omitted, that while the seeds of Gentianeæ are uniformly indefinite, those of Spigelia are definite, or nearly so. Upon all these considerations, and to avoid confusing the distinctive characters of the orders, I have formed that of Spigeliaceæ, the distinction of which will depend upon the symmetry of the stamens, corolline and calycine segments, the division of the valves of the capsule, and the presence of stipulæ. In this last point they approach Rubiaceæ (Cinchonaceæ), as also in a tendency in their leaves to become whorled, their intruded style, and valvate æstivation; but differ in their superior ovarium, and the want of the glandular disk which covers the apex of the ovarium of Rubiaceæ (Cinchonaceæ); so establishing, along with other things, an affinity between that order and Compositæ and Umbelliferæ," &c. &c.

GEOGRAPHY. All American, chiefly natives of the southern hemisphere within the tropics.

PROPERTIES. Spigelia marilandica root is used in North America as a vermifuge: if administered in large doses, it acts powerfully as a cathartic. Its use is, however, attended occasionally with violent narcotic effects, such as dimness of sight, giddiness, dilated pupil, spasmodic motions in the muscles of the eyes, and even convulsions. *Barton,* 2. 80.

EXAMPLE. Spigelia.

CXCIX. CONVOLVULACEÆ. THE BINDWEED TRIBE.

CONVOLVULI, *Juss. Gen.* 133. (1789). — CONVOLVULACEÆ, *R. Brown Prodr.* 481. (1810); *Lindl. Synops.* 167. (1829). — CUSCUTINÆ, *a § of* Convolvulaceæ, *Link Handb.* 1. 594. (1829.)

DIAGNOSIS. Monopetalous dicotyledons, with a superior 2-4-celled ovarium, regular flowers, definite erect ovules, a plaited corolla, and shrivelled cotyledons.

ANOMALIES. Cuscuta is leafless and has no cotyledons.

ESSENTIAL CHARACTER. — *Calyx* persistent, in 5 divisions. *Corolla* monopetalous, hypogynous, regular, deciduous; the limb 5-lobed, generally plaited. *Stamens* 5, inserted into the base of the corolla, and alternate with its segments. *Ovarium* simple, with 2 or 4 cells, seldom with 1; sometimes in 2 or 4 divisions; few-seeded; the ovules definite and erect, when more than 1 collateral; *style* 1, usually divided at the top, sometimes down to the base.; *stigmas* obtuse or acute. *Disk* annular, hypogynous. *Capsule* with from 1 to 4 cells; the valves fitting, at their edges, to the angles of a loose dissepiment, bearing the seeds at its base; sometimes valveless, or dehiscing transversely. *Seeds* with a small quantity of mucilaginous albumen; *embryo* curved; *cotyledons* shrivelled; *radicle* inferior. — *Herbaceous* plants or *shrubs*, usually twining and milky, smooth, or with a simple pubescence. *Leaves* alternate, undivided, or lobed, seldom pinnatifid, with no stipulæ. *Inflorescence* axillary or terminal; *peduncles* 1- or many-flowered, the partial ones generally with 2 bracteæ.

AFFINITIES. The plaited corolla and climbing habit are the *primâ facie* marks of this order, which approaches Cordiaceæ in its shrivelled cotyledons, and through that tribe Boragineæ, with which Falkia agrees in the deeply-lobed ovarium. Nolana, to be found in Solaneæ, would seem to establish a relationship between Convolvulaceæ and that order also. Polemoniaceæ are known by their loculicidal dehiscence, which in Convolvulaceæ is always opposite the dissepiments. Hydroleæ are characterised by their indefinite seeds, and taper embryo lying in the midst of fleshy albumen.

GEOGRAPHY. Very abundant in all parts of the tropics, but rare in cold climates, where a few only are found : they twine round other shrubs, or creep among the weeds of the sea-shore.

PROPERTIES. Their roots abound in an acrid milky juice, which is strongly purgative; this quality depends upon a peculiar resin, which is the active principle of the Jalap, the Scammony, and the others whose roots possess similar qualities. Conv. Jalapa produces the real jalap, and C. Scammonia the scammony; besides which, C. Turpethum, C. Mechoacanus, sepium, arvensis, Soldanella, macrorhizus, maritimus, macrocarpus, and probably many others, may be used with nearly equal advantage. The root of Convolvulus panduratus is used in the United States as jalap; its operation is like that of rhubarb; it is supposed to be also diuretic. *Barton,* 1. 252. The roots of Conv. floridus and scoparius, and Ipomœa Quamoclit, are used as sternutatories; those of C. Batatas and edulis are useful articles of food : the former is the common sweet Potato of European gardens. The Cuscutas are remarkable for becoming parasitical after having originally germinated in the ground, from which they derive their nourishment until they fix themselves firmly upon the plant that is finally to maintain them.

EXAMPLES. Convolvulus, Evolvulus, Falkia.

CC. POLEMONIACEÆ. THE GREEK VALERIAN TRIBE.

POLEMONIA, *Juss. Gen.* 136. (1789). — POLEMONIDEÆ, *Dec. and Duby,* 329. (1828). — POLEMONIACEÆ, *Lindl. Synops.* 168. (1829). — COBÆACEÆ, *Don in Ed. Ph. Journ.* 10. 111. (1824); *Link Handb.* 1. 822. (1829.)

DIAGNOSIS. Monopetalous dicotyledons, with regular flowers, a superior 3-celled ovarium, peltate or ascending ovules, and a pentandrous 5-parted corolla, with imbricated æstivation.

ANOMALIES. Cobæa has a climbing habit.

ESSENTIAL CHARACTER. — *Calyx* inferior, monosepalous, 5-parted, persistent, sometimes irregular. *Corolla* regular, 5-lobed. *Stamens* 5, inserted into the middle of the tube of the corolla, and alternate with its segments. *Ovarium* superior, 3-celled, with a few or many ovula; *style* simple; *stigma* trifid; *ovules* ascending or peltate. *Capsule* 3-celled, 3-valved, few- or many-seeded, with a loculicidal or septicidal dehiscence; the valves separating from the axis. *Seeds* angular or oval, or winged, often enveloped in mucus, ascending; *embryo* straight in the axis of horny albumen; *radicle* inferior; *cotyledons* elliptical, foliaceous. — *Herbaceous* plants, with opposite, or occasionally alternate, compound, or simple *leaves*; *stem* occasionally climbing.

AFFINITIES. The ternary division of the ovarium connected with the pentandrous corolla and 5-lobed calyx bring this order near Convolvulaceæ, from which the habit, embryo, and corolla, distinguish it; from Gentianeæ, to which it also approaches, the 3-celled ovarium divides it. It is remarkable for the blue colour of the pollen, which is usually of that hue, whatever may be the colour of the corolla. In Collomia linearis I have noticed (in *Botanical Register*, folio 1166) that the dilatation of the mucous matter in which the seeds are enveloped, and which, when they are thrown into water forms around them like a cloud, depends upon the presence of an infinite multitude of exceedingly delicate and minute spiral vessels, lying coiled up, spire within spire, on the outside of the testa; when dry, these vessels are confined upon the surface of the seed by its mucus, without being able to manifest themselves; but the instant water is applied, the mucus dissolves and ceases to counteract the elasticity of the spiral vessels, which then dart forward at right angles with the testa, each carrying with it a sheath of mucus, in which it for a long time remains enveloped as if in a membranous case. I know of no parallel to this, except in Casuarina, in which the whole of the inside of the testa consists of minute spiral vessels.

GEOGRAPHY. Very abundant in both North and South America, in temperate latitudes, particularly on the north-west side. It is stated by Dr. Richardson, that the most northern limit in North America is 54°. *Edin. Phil. Journ.* 12. 209. In Europe and Asia they are much more uncommon. They are unknown in tropical countries.

PROPERTIES. None, or unknown.

EXAMPLES. Polemonium, Collomia, Ipomopsis, Cantua, Gilia.

N. B. Mr. Don distinguishes Cobæaceæ from this order; but the only differences of importance between the one and the other consist in the former having a septicidal dehiscence and climbing habit; characters, I fear, of too little moment to be admitted as ordinal distinctions. The characters of Cobæaceæ, as understood by Mr. Don, are these:—

Calyx leafy, 5-cleft, equal. *Corolla* inferior, campanulate, regular, 5-lobed, with an imbricate æstivation. *Stamens* 5, equal, arising from the base of the corolla; *anthers* 2-celled, compressed. *Ovarium* superior, 3-celled, surrounded with a fleshy secreting annular disk; *ovules* several, ascending; *style* simple; *stigma* trifid. *Fruit* capsular, 3-celled, 3-valved, with a septicidal dehiscence; *placenta* very large, 3-cornered, in the axis, its angles touching the line of dehiscence of the pericarpium. *Seeds* flat, winged, imbricated in a double row; their integument mucilaginous; *albumen* fleshy; *embryo* straight; *cotyledons* leafy; *radicle* (according to Don) inferior. — Climbing *shrubs*. *Leaves* alternate, pinnated, their petiole lengthened into a tendril. *Flowers* axillary, solitary.

CCI. HYDROLEACEÆ.

R. Brown Prodr. 482. (1810) *without a name; Id. in Congo* (1818).— HYDROLEACEÆ, *Kunth in Humb. N. G. et Sp.* 3. 125. (1818); *Synops.* 2. 234. (1823).—DIAPENSIACEÆ, *Link Handb.* 1. 595. (1829), *a § of* Convolvulaceæ.

DIAGNOSIS. Monopetalous dicotyledons, with a superior 2- or 3-celled ovarium, several styles, indefinite seeds, and a plaited or imbricated corolla.

ANOMALIES.

ESSENTIAL CHARACTER.—*Calyx* 5-parted, inferior, persistent, with imbricated æstivation. *Corolla* hypogynous, monopetalous, regular, not always agreeing with the calyx in the number of its divisions. *Stamens* arising from the corolla, regular, agreeing in number with the segments of the calyx; *anthers* deeply lobed at the base. *Ovarium* superior, surrounded by an annular disk, 2- or 3-celled; *styles* 2 or 3; *stigmas* thickened. *Fruit* capsular, enclosed in the calyx, 2- rarely 3-celled, splitting through the middle of the cells; *valves* therefore bearing the dissepiments in their middle; *placentæ* either single and fungous, or double and thin. *Seeds* indefinite, very small; *albumen* fleshy, in the axis of which lies a taper, straight *embryo*. — *Herbaceous* plants or *under-shrubs*, sometimes spiny. *Leaves* alternate, entire, or lobed, without stipulæ, often covered with glandular or stinging hairs. *Flowers* numerous, axillary and terminal.

AFFINITIES. Separated from Convolvulaceæ by Mr. Brown, on account of their indefinite seeds, and taper embryo with small flat cotyledons in the midst of fleshy albumen. To me they appear equally related to Boragineæ, with some of which Wigandia agrees in habit. Also related to Hydrophylleæ, the membranous plates lining the tube of the corolla of that order being, according to Von Martius (*N. G.* 2. 138), analogous to the dilated base of the filaments of Hydroleaceæ.

GEOGRAPHY. No particular geographical limits can be assigned to this order. Diapensia is found in Lapland, Wigandia in the Caraccas, Hydrolea in the West Indies, and Nama in both the East and West Indies.

PROPERTIES. Unknown, except that a bitter principle exists in Hydrolea.

EXAMPLES. Hydrolea, Nama, Sagonea, Wigandia, Diapensia.

CCII. EBENACEÆ. THE EBONY TRIBE.

GUAIACANÆ, *Juss. Gen.* 155. (1789) *part of the first sect.* — EBENACEÆ, *Vent. Tabl.* 443. (1799); *Brown Prodr.* 524. (1810). — EBENACEÆ, § Diospyreæ, *Dec. and Duby*, 320. (1829).

DIAGNOSIS. Monopetalous dicotyledons, with superior several-celled ovarium, regular (unisexual) flowers, definite pendulous collateral ovules, a 3-6-lobed corolla with the stamens some multiple of its lobes, and albuminous seeds.

ANOMALIES.

ESSENTIAL CHARACTER. — *Flowers* polygamous or diœcious, seldom hermaphrodite. *Calyx* in 3 or 6 divisions, nearly equal, persistent. *Corolla* monopetalous, hypogynous, regular, deciduous, somewhat coriaceous, usually pubescent externally, and smooth internally; its *limb* with 3 or 6 divisions, imbricated in æstivation. *Stamens* definite, either arising from the corolla, or hypogynous; twice as many as the segments of the corolla, sometimes 4 times as many, or the same number, and then alternate with them; *filaments* simple in the hermaphrodite species, generally doubled in the polygamous and diœcious ones, both their divisions bearing anthers, but the inner one generally smaller; *anthers* attached by their base, lanceolate, 2-celled, dehiscing lengthwise, sometimes bearded; *pollen* round, smooth. *Ovarium* sessile, without any disk, several-celled, the cells each having 1 or 2 ovules pendulous from their apex; *style* divided, seldom simple; *stigmas* bifid, or simple. *Fruit* fleshy, round or oval, by abortion often few-seeded, its pericarpium sometimes opening in a regular manner. *Seed* with a membranous testa of the same figure as the albumen, which is cartilaginous and white; *embryo* in the axis, or but little out of it, straight, white, generally more than half as long as the albumen; *cotyledons* foliaceous, somewhat veiny, lying close together, occasionally slightly separate; *radicle* taper, of middling length or long, turned towards the hilum; *plumula* inconspicuous. — *Trees* or *shrubs*, without milk, and a heavy wood. *Leaves* alternate, without stipulæ, obsoletely articulated with the stem, quite entire, coriaceous. *Inflorescence* axillary. *Peduncles* solitary, those of the males divided, of the females usually 1-flowered, with minute bracteæ. *R. Br.*

AFFINITIES. Very near Oleaceæ, with which they agree in the placentation of the seeds and other points of structure; distinguished by their alternate leaves, constantly axillary and usually unisexual flowers, the stamens of which are at least double the number of the lobes of the corolla. *R. Br.* They are also closely allied to Ilicineæ, from which they chiefly differ in the number of their stamens and their divided sexes. For their resemblance to Sapoteæ, see that order. Styraceæ were combined with them by Jussieu.

GEOGRAPHY. Chiefly Indian and tropical; a very few are found northwards as far as Switzerland in Europe, and the state of New York in North America.

PROPERTIES. Remarkable only for the hardness and blackness of the wood, and the eatable quality of the fruit. The former is well known under the name of Ebony and Ironwood; the latter are occasionally introduced from China as a dry sweetmeat. They are noted for their extreme acerbity before arriving at maturity. The bark of Diosp. virginiana is said to be a febrifuge.

EXAMPLES. Diospyrus, Maba, Ferreola.

CCIII. COLUMELLIACEÆ.

COLUMELLIEÆ, *Don in Edinb. New Phil. Journ.* (*Dec.* 1828).

DIAGNOSIS. Monopetalous diandrous dicotyledons, with an inferior 2-celled many-seeded ovarium, opposite leaves, and regular flowers.

ANOMALIES.

ESSENTIAL CHARACTER.—*Calyx* turbinate, superior, many-toothed. *Corolla* rotate, 5-8-parted, with a convolute æstivation. *Stamens* 2, inserted in the throat; *anthers* linear, either sinuous or straight, 1- or 2-celled. *Ovarium* inferior, 2-celled, with an indefinite number of ovules; *style* simple, declinate; *stigma* capitate. *Disk* perigynous. *Fruit* capsular, 2-celled, many-seeded, with a septicidal incomplete dehiscence. *Seeds* ascending; *testa* polished; *embryo* taper, erect, in the axis of fleshy albumen.—*Shrubs*, *trees*, or *herbaceous* plants. *Leaves* opposite, without stipulæ, entire. *Flowers* solitary, yellow.

AFFINITIES. Only known from the remarks of Mr. Don, from whom the foregoing has been abridged. He thinks them near Jasmineæ, with which they correspond " in the structure and æstivation of their corolla, in their bilocular ovarium, and erect (?) ovula; and they agree both with them and Syringa in the structure and dehiscence of their capsule. They differ, however, essentially from Jasmineæ, by having an adherent ovarium, by the presence of a perigynous disk, by the undivided stigma, and, lastly, by having an inferior capsule with polyspermous cells." Mr. Don further thinks they connect Jasmineæ with Oleaceæ.

GEOGRAPHY. Mexican and Peruvian plants.

PROPERTIES. Unknown.

EXAMPLES. Columellia, Menodora.

CCIV. JASMINEÆ. THE JASMINE TRIBE.

JASMINEÆ, *Juss. Gen. Plant.* 104. (1789) *in part; R. Brown Prodr.* 520. (1810).

DIAGNOSIS. Monopetalous dicotyledons, with regular flowers, a superior 2-celled ovarium with erect seeds, 2 stamens, and an imbricate corolla.

ANOMALIES.

ESSENTIAL CHARACTER.—*Calyx* divided or toothed, persistent. *Corolla* monopetalous, hypogynous, regular, hypocrateriform, with from 5 to 8 divisions, which lie laterally upon each other, being imbricated and twisted in æstivation. *Stamens* 2, arising from the corolla, enclosed within its tube. *Ovarium* destitute of a hypogynous disk, 2-celled, with 1-seeded cells, the ovules in which are erect; *style* 1; *stigma* 2-lobed. *Fruit* either a double *berry* or a *capsule* separable in two. *Seeds* either with no albumen, or very little; *embryo* straight; *radicle* inferior.—*Shrubs*, having usually twining stems. *Leaves* opposite, mostly compound, ternate or pinnate, with an odd one; sometimes simple, the petiole almost always having an articulation. *Flowers* opposite, in corymbs. *R. Br.*

AFFINITIES. Formerly combined with Oleaceæ, from which they are distinguished by Mr. Brown by their ovules being erect, their seed with no, or very little, albumen, in the æstivation of the corolla being imbricate, not valvate, and in the number of its divisions being 5 or more, and consequently not regularly a multiple of the stamens, instead of 4, which is a multiple of them. But Ach. Richard (*Ann. des Sc.* 350.) endeavours to shew that these differences are insufficient. He states, that the ovules of Jasmineæ are originally pendulous, as in Oleaceæ; but that they subsequently become erect in consequence of the growth of the ovarium, whose apex does not elongate, while its sides extend considerably during the growth of the fruit. He says, upon the authority of his father, that albumen does exist in Jasminum and Nyctanthes; a fact which had been previously mentioned by Mr. Brown in defining the orders, but to which that distinguished botanist attached no importance, because only a small quantity was found by him to exist, while it is very abundant in Oleaceæ; and he probably conceived, as I certainly do, that it is the difference of its quantity only which gives the albumen value as a mark of ordinal distinction. I confess it does not appear to me that these remarks lessen the propriety of dividing Jasmineæ and Oleaceæ, which are still known by abundantly sufficient characters. The affinity of Jasmineæ, otherwise, is with those monopetalous orders, in which the number of stamina is different from that of the divisions of the corolla, as Labiatæ, Scrophularineæ, Verbenaceæ, and the like, but particularly with the latter, which sometimes resemble them in their fruit, as Clerodendron. Mr. Brown stations them between Pedalineæ and Oleaceæ (*Prodr.*); Decandolle between Oleaceæ and Strychneæ (*Théorie*, ed. 2.); Don suggests their affinity to his order Columellieæ.

GEOGRAPHY. Chiefly inhabitants of tropical India, in all parts of which they abound. One Jasminum only is mentioned from South America, but there are at least 3 species of Bolivaria on that continent; a few are natives of Africa and the adjoining islands; New Holland contains several; and, finally, 2 extend into the southern climates of Europe.

PROPERTIES. Not very different from Oleaceæ in qualities, except that their oil is deliciously fragrant, and produced by the flowers, and not by the pericarp. The genuine essential oil of Jasmine of the shops is produced by Jasminum officinale and grandiflorum; but a similar perfume is also procured from Jasminum Sambac. The leaves of Jasminum undulatum are slightly bitter. The bitter root of Jasminum angustifolium, ground small and mixed with powdered Acorus Calamus root, is considered in India as a valuable external application in cases of ringworm. *Ainslie*, 2. 52. In India Proper the tube of the corolla of Nyctanthes arbor tristis is used as a dye. *Buchanan L. Tr.* 13. 484.

EXAMPLES. Jasminum, Nyctanthes, Bolivaria.

CCV. OLEACEÆ. The Olive Tribe.

Oleineæ, *Hoffmannsegg et Link Fl. Port.* (1806); *Brown Prodr.* 522. (1810); *Lindl. Synops.* 171. (1829).—Lilaceæ *Vent. Tabl.* 1. 306. (1799).

Diagnosis. Monopetalous dicotyledons, with regular flowers, a superior 2-celled ovarium with pendulous seeds, 2 stamens, and a valvate corolla.
Anomalies. Fraxinus is generally apetalous.

Essential Character.—*Flowers* hermaphrodite, sometimes diœcious. *Calyx* monophyllous, divided, persistent, inferior. *Corolla* hypogynous, monopetalous, 4-cleft, occasionally of 4 petals, connected in pairs by the intervention of the filaments, sometimes without petals; *æstivation* somewhat valvate. *Stamens* 2, alternate with the segments of the corolla or with the petals; *anthers* 2-celled, opening longitudinally. *Ovarium* simple, without any hypogynous disk, 2-celled; the *cells* 2-seeded; the *ovules* pendulous and collateral; *style* 1 or 0; *stigma* bifid or undivided. *Fruit* drupaceous, berried, or capsular, often by abortion 1-seeded. *Seeds* with dense, fleshy, abundant albumen; *embryo* about half its length, straight; *cotyledons* foliaceous, partly asunder; *radicle* superior; *plumula* inconspicuous.—*Trees* or *shrubs*. *Leaves* opposite, simple, sometimes pinnated. *Flowers* in terminal or axillary racemes or panicles; the *pedicels* opposite, with single bracteæ. *R. Br.*

Affinities. Very near Jasmineæ, with which they are combined by Ach. Richard; see the observations upon that order. To some, it, I believe, still appears expedient to separate the small tribe of Lilaceæ, the representative of which is the Lilac of the gardens; but I am not aware of there being any greater peculiarity in that plant than its capsular fruit, a character very rarely of importance in distinguishing orders. Decandolle suggests (*Essai Méd.* p. 204.) that the Ash is related to the Maple tribe. I also find in the same work the following very good observations upon this order:— " However heterogeneous the Olive tribe may appear as at present limited, it is remarkable that the species will all gráft upon each other; a fact which demonstrates the analogy of their juices and their fibres. Thus the Lilac will graft upon the Ash, the Chionanthus and the Fontanesia, and I have even succeeded in making the Persian Lilac live ten years on Phyllirea latifolia. The Olive will take on the Phyllirea, and even on the Ash: but we cannot graft the Jasmine on any plant of the Olive tribe; a circumstance which confirms the propriety of separating these two tribes."
Geography. Natives chiefly of temperate latitudes, inclining towards the tropics, but scarcely known beyond 65° N. lat. The Ash is extremely abundant in North America; the Phyllireas and Syringas are all European or Eastern plants. A few are found in New Holland and elsewhere within the tropics. One Ash is a native of Nipal.
Properties. This order offers almost the only instance of oil being contained in the pericarp; from which Olive oil is entirely expressed; in most other plants oil is yielded by the seed. The flowers are frequently slightly fragrant; those of Olea fragrans are employed in China for flavouring tea. The bark of the Olive, but especially of the Ash, is so bitter and astringent, that it has been not only highly celebrated as a febrifuge, but even compared with Quinquina (*Dec.*) for effect. The sweet gentle purgative, called Manna, is a concrete discharge from the bark of several species of Ash, but especially from Fraxinus rotundifolia. The sweetness of this substance is not due to the presence of sugar, but to a distinct principle, called Mannite, which differs from sugar in not fermenting with water and yeast. *Turner*, 682. A peculiar substance, called Olivile, is contained in the gum of Olea europæa. *Ibid.* 701.
Examples. Olea, Phyllirea, Ligustrum, Chionanthus, Fraxinus.

CCVI. MYRSINEÆ.

OPHIOSPERMA, *Vent. Jard. Cels.* 86. (1800).—MYRSINEÆ, *R. Brown Prodr.* 532. (1810.)

DIAGNOSIS. Monopetalous arborescent dicotyledons, with regular flowers, an entire superior 1-celled ovarium with a free central placenta, and indehiscent fleshy fruit.

ANOMALIES. Ægiceras has no albumen, and the cells of its anthers are cellular.

ESSENTIAL CHARACTER.—*Flowers* hermaphrodite or polygamous. *Calyx* 4- or 5-cleft, persistent. *Corolla* monopetalous, hypogynous, 4-5-cleft, equal. *Stamens* 4-5, opposite the segments of the corolla! into the bases of which they are inserted; *filaments* distinct, rarely connate, sometimes wanting, sometimes 5 sterile petaloid alternate ones; *anthers* attached by their emarginate base, with 2 cells, dehiscing longitudinally. *Ovarium* 1, with a single cell in a free central placenta, in the midst of which is immersed a definite or indefinite number of peltate *ovula*; *style* 1, often very short; *stigma* lobed or undivided. *Fruit* fleshy, mostly 1-seeded, sometimes 2-4-seeded. *Seeds* peltate, with a hollow hilum and a simple integument; *albumen* horny, of the same shape as the seed; *embryo* lying across the hilum, taper, usually curved; *cotyledons* short; *radicle*, if several seeds ripen, inferior. *Plumula* inconspicuous.— *Trees* or *shrubs*. *Leaves* alternate, undivided, serrated or entire, coriaceous, smooth; sometimes *under-shrubs*, with opposite or ternate leaves. *Inflorescence* in umbels, corymbs, or panicles, axillary, seldom terminal. *Flowers* small, white or red, often marked with sunken dots or glandular lines.

AFFINITIES. Scarcely different from Primulaceæ, except in their arborescent habit and fleshy fruit; the embryo always lies across the hilum, and the stamens are opposite the lobes of the corolla, as in that order; add to which, the connivence of the anthers in a cone, which is frequent in Primulaceæ, is common in Myrsineæ also. Mr. Brown remarks (l. c.), that the order is related to Sapoteæ through Jacquinia, and to Primulaceæ through Bladhia. The immersion of the ovules in a fleshy placenta is a peculiar character of this tribe.

GEOGRAPHY. Tropical plants without exception, and common both in India and America; but " no species has been met with in equinoctial Africa, though several exist both at the Cape of Good Hope and in the Canary Islands." *Brown Congo,* 465.

PROPERTIES. Almost unknown. Generally handsome shrubs, with fine evergreen leaves. Bread is said to be prepared from the pounded seeds of Theophrasta Jussiæi in St. Domingo, where it is called Le Petit Coco. *Hamilt. Prodr.* p. 27.

EXAMPLES. Ardisia, Embelia, Myrsine.

CCVII. PRIMULACEÆ. THE PRIMROSE TRIBE.

LYSIMACHIÆ, *Juss. Gen.* 95. (1789).—PRIMULACEÆ, *Vent. Tabl.* 2. 285. (1799); *R. Brown Prodr.* 427. (1810); *Lindl. Synops.* 182. (1820.)

DIAGNOSIS. Monopetalous herbaceous dicotyledons, with regular flowers, an entire superior 1-celled ovarium with a free central placenta, and capsular fruit.

ANOMALIES. Samolus has the ovarium half inferior, and 5 sterile stamens. Glaux is apetalous.

ESSENTIAL CHARACTER.—*Calyx* divided, 5-cleft, seldom 4-cleft, inferior, regular, persistent. *Corolla* monopetalous, hypogynous, regular; the limb 5-cleft, seldom 4-cleft. *Stamens* inserted upon the corolla, equal in number to its segments, and opposite them!

Ovarium 1-celled; *style* 1; *stigma* capitate. *Capsule* opening with valves; *placenta* central, distinct. *Seeds* numerous, peltate; *embryo* included within fleshy albumen, and lying across the hilum; *radicle* with no determinate direction.—*Herbaceous* plants. *Leaves* usually opposite, either whorled or scattered. *R. Br.*

AFFINITIES. Nearly allied to all the regular monopetalous orders with capsular superior fruit, especially to Solaneæ and Gentianeæ, from both which, and all others, they are readily known by the stamens being placed opposite the segments of the corolla, and not alternate with them. In this respect they agree with Myrsineæ, which differ principally in their fleshy fruit and arborescent habit. Another character of Primulaceæ is to have the embryo lying across the hilum within the albumen, so that the radicle is presented neither to the umbilicus nor to one extremity, but to one side. Trientalis differs a little in its somewhat succulent fruit. Glaux, an apetalous genus, is usually placed here; but, according to Mr. Don (*Jameson's Journal*, Jan. 1830, p. 166.), it should be referred to Plantagineæ, " where it will form the connecting link between that family and Primulaceæ."

GEOGRAPHY. Common in the northern and colder parts of the globe, growing in marshes, hedges, and groves, by fountains and rivulets, and even among the snow of cloud-capped mountains. The genus Douglasia was found by the traveller whose name it bears, blossoming while covered with snow, on the Rocky Mountains of America. They are uncommon within the tropics, where they usually occupy either the sea shore, or the summits of the most lofty hills.

PROPERTIES. As beautiful objects of culture, these rank among the most esteemed, both on account of their bright but modest-looking flowers, the earliest harbingers of spring, and also for the sake of their fragrance. Their sensible properties are feeble. The Cowslip is slightly narcotic, and the root of Cyclamen is famous for its acridity; yet this is the principal food of the wild boars of Sicily, whence its common name of Sowbread.

EXAMPLES. Primula, Dodecatheon, Androsace.

CCVIII. LENTIBULARIÆ.

LENTIBULARIÆ, *Richard in Flor. Paris*, p. 26. (1808).—UTRICULINÆ, *Hoffmannsegg et Link Fl. Port.*(1806).—LENTIBULARIÆ, *R. Brown, Prodr.* 429. (1810); *Lindl. Synops.* 186. (1829); *Link Handb.* 1. 511. (1829) *a sect. of* Personatæ.

DIAGNOSIS. Monopetalous dicotyledons, with irregular flowers, and a superior 1-celled ovarium, with a central free placenta.

ANOMALIES. Seed undivided in Utricularia.

ESSENTIAL CHARACTER.—*Calyx* divided, persistent, inferior. *Corolla* monopetalous, hypogynous, irregular, bilabiate, with a spur. *Stamens* 2, included within the corolla, and inserted into its base; *anthers* simple, sometimes contracted in the middle. *Ovarium* 1-celled; *style* 1, very short; *stigma* bilabiate. *Capsule* 1-celled, many-seeded, with a large central placenta. *Seeds* minute, without albumen; *embryo* sometimes undivided.—*Herbaceous* plants, living in water or marshes. *Leaves* radical, undivided; or compound, resembling roots, and bearing little vesicles. *Scapes* either with minute stipula-like scales, or naked; sometimes with whorled vesicles; generally undivided. *Flowers* single, or in spikes, or in many-flowered racemes; with a single bractea, rarely without bracteæ. *R. Br.*

AFFINITIES. The central free placenta and minute exalbuminous embryo are the principal points of distinction between these and Scrophularineæ, to which their habit nearly approximates them. They are known from Primulaceæ by their irregular flowers, exalbuminous embryo, and stamens.

GEOGRAPHY. Natives of marshes, or rivulets, or fountains, in all parts of the world, especially within the tropics.

PROPERTIES. Unknown.

EXAMPLES. Pinguicula, Utricularia.

CCIX. GESNEREÆ.

GESNERIEÆ, *Rich. et Juss. Ann. Mus.* 5. 428. (1804); *Kunth in Humb. N. G. et Sp.* 2. 392. (1817); *Lindley in Bot. Reg.* 1110. (1827).—GESNERIACEÆ, *Link Handb.* 1. 504. (1829) *a sect. of* Personatæ.—GESNEREÆ, *Von Martius Nov. Gen. Bras.* 3. 68. (1829.)

DIAGNOSIS. Monopetalous dicotyledons, with a half inferior ovarium, parietal projecting placentæ, a capitate stigma, irregular flowers, and an embryo in the axis of fleshy albumen.

ANOMALIES. Sarmienta is diandrous.

ESSENTIAL CHARACTER.— *Calyx* half superior, 5-parted, with a valvate æstivation. *Corolla* monopetalous, tubular, more or less irregular, 5-lobed, with an imbricate æstivation. *Stamens* didynamous; *anthers* cohering, 2-celled, innate, with a thick tumid connectivum; the rudiment of a fifth stamen is present. *Ovarium* half superior, 1-celled, with 2 fleshy 2-lobed parietal polyspermous placentæ; surrounded at its base by glands alternating with the stamens; *style* continuous with the ovarium; *stigma* capitate, concave. *Fruit* capsular or succulent, half superior, 1-celled, 2-valved, with loculicidal dehiscence and 2 opposite lateral placentæ, each consisting of 2 plates. *Seeds* very numerous, minute; *embryo* erect, in the axis of fleshy *albumen*; *testa* thin, with very close fine oblique veins.— *Herbaceous* plants or *under-shrubs*. *Leaves* opposite, rugose, without stipulæ. *Flowers* showy, in racemes, or panicles, rarely solitary.

AFFINITIES. Nearly allied to Bignoniaceæ through Eccremocarpus, from which they differ in their ovarium being 1-celled and partly inferior, in their apterous seeds, and in habit. Distinguished from Cyrtandraceæ only by their usually inferior 1-celled ovarium, with simple placentæ and albuminous seeds, the testa of which is twisted in a singular manner. From Scrophularineæ they are known by the same characters, with the exception of the albuminous seeds, in which respect they agree with that order. They also approach Orobancheæ, Acanthaceæ, and Pedalineæ, with all which they agree in the position of the pericarpial leaves being anterior and posterior with regard to the axis of inflorescence, and consequently the placentæ right and left.

GEOGRAPHY. Exclusively natives of the tropical parts of South America and of the West India Islands.

PROPERTIES. Generally beautiful herbaceous plants, bearing flowers, the prevailing colour of which is bright red, and having tuberous roots. The succulent fruits are mucilaginous, sweetish, and eatable. A dye is obtained from the calyxes and fruit of some of them for staining cotton, straw work, and domestic utensils.

EXAMPLES. Gesnera, Gloxinia, Hypocyrta, Alloplectus.

CCX. OROBANCHEÆ. THE BROOM-RAPE TRIBE.

OROBANCHEÆ, *Juss. Ann. Mus.* 12. 445. (1808); *Richard in Pers. Synops.* 2. 180. (1807); *Dec. and Duby Bot. Gall.* 348. (1828); *Lindl. Synops.* 193. (1829).—OROBAN-CHINÆ, *Link Handb.* 1. 506. (1829) *a sect. of* Personatæ.

DIAGNOSIS. Monopetalous, colourless, parasitical dicotyledons, with a superior 1-celled ovarium, irregular unsymmetrical flowers, and a minute embryo inverted in the apex of fleshy albumen.

ANOMALIES.

ESSENTIAL CHARACTER.—*Calyx* divided, persistent, inferior. *Corolla* monopetalous, hypogynous, irregular, persistent, with an imbricated æstivation. *Stamens* 4, didynamous. *Ovarium* superior, 1-celled, seated in a fleshy disk, with 2 or 4 parietal polyspermous placentæ; *style* 1; *stigma* 2-lobed. *Fruit* capsular, enclosed within the withered corolla, 1-celled, 2-valved, each valve bearing 1 or 2 placentæ in the middle. *Seeds* indefinite, very minute; *embryo* minute, inverted, at the apex of a fleshy albumen.—*Herbaceous* leafless plants, growing parasitically upon the roots of other species. *Stems* covered with brown or colourless scales.

AFFINITIES. Extremely near Gesnereæ in character, although very different in habit. They are distinguished by their seeds having a minute embryo lying in one end of fleshy albumen, and spherical pollen, while the embryo of Gesnereæ is cylindrical and erect, occupying the axis of the albumen, and the pollen elliptical, with a furrow on one side. In Gesnereæ the seeds are attached by rather long funiculi, while they are absolutely sessile in Orobancheæ. Moreover, there is a tendency in the latter to become pentandrous, or even hexandrous; but not only no such tendency exists in the former, but the reverse takes place, in the occasional increased sterility of the stamens. There is scarcely any trace of the glandular processes of the disk of Gesnereæ in Orobanche, or at least nothing more than a thin glandular coating to the base of the ovarium. See *Von Martius Nov. Gen. et Sp. Bras.* 3. 72. From Scrophularineæ they are known by their 1-celled ovarium and minute inverted embryo; from Melampyraceæ, by the former of these characters; and from all that have been mentioned, by their habit and parasitical mode of growth. In this respect they resemble Pyrolaceæ, from which they differ in their ovarium being composed of 2, not 5 carpella, and their irregular unsymmetrical flowers. According to the observations of M. Vaucher, of Geneva, the seeds of Orobanche ramosa will lie many years inert in the soil unless they come in contact with the roots of Hemp, the plant upon which the species grows parasitically, when they immediately sprout. See *Ferussac, Feb.* 1824, 136.

GEOGRAPHY. Not uncommon in Europe, particularly in the southern kingdoms, Barbary, middle and northern Asia, and North America; very rare in India.

PROPERTIES. The Orobanche virginiana is supposed to have formed, in conjunction with white oxide of arsenic, a famous cancer powder, which was known in North America under the name of " Martin's Cancer Powder." It is thought to participate in the powerful astringent properties of Orobanche major. *Barton*, 2. 38.

EXAMPLES. Orobanche, Lathræa, Phelypæa, Æginetia.

CCXI. SCROPHULARINEÆ. THE FIGWORT TRIBE.

SCROPHULARIÆ, *Juss. Gen.* 117. (1789).—SCROPHULARINEÆ, *R. Brown Prodr.* 433. (1810); *Lindl. Synops.* 187. (1829). — PEDICULARES, *Juss. Gen.* 99. (1789) *in part.*—PERSONATÆ, *Dec. Fl. Fr.* 3. 573. (1815).—ANTIRRHINEÆ, *Dec. and Duby*, 342. (1828) —HALLERIACEÆ, *Link Handb.* 1. 506. (1829) *a sect. of* Personatæ.—SCOPARIACEÆ, *Ib.* 822. *the same.*—ERINEÆ, *Ib.* 510. *the same.*

DIAGNOSIS. Monopetalous dicotyledons, with a superior 2-celled capsule, irregular unsymmetrical flowers, albuminous seeds, and an orthotropous embryo.

ANOMALIES. Scoparia has regular symmetrical flowers. Leaves sometimes alternate.

ESSENTIAL CHARACTER.—*Calyx* divided, persistent, inferior. *Corolla* monopetalous, hypogynous, usually irregular, deciduous, with an imbricated æstivation. *Stamens* 2, or 4, didynamous, very seldom equal. *Ovarium* superior, 2-celled, many-seeded; *style* 1, continuous; *stigma* 2-lobed. *Fruit* capsular, very seldom succulent, with from 2 to 4 valves, which are either entire or bifid; the dissepiment either double, arising from the incurved margins of the valves; or simple, and in that case, either parallel with, or opposite to, the valves. *Placentæ* central, either adhering to the dissepiment or separating from it. *Seeds* indefinite; *embryo* included within fleshy albumen; *radicle* turned towards the hilum (orthotropous).—*Herbaceous* plants, seldom *shrubs*, with opposite *leaves*. *Inflorescence* very variable.

AFFINITIES. The capsular monopetalous genera of Dicotyledons, with a superior ovarium, albuminous seeds, and irregular diandrous or didynamous stamens, were separated by Jussieu into two orders, which he called Scrophulariæ and Pediculares, distinguished from each other by the dehiscence of the former being septicidal, and of the latter loculicidal. Mr. Brown, in his *Prodromus*, pointed out the insufficiency of this character, which is often not even of generic value, and he combined the orders of Jussieu under the common name of Scrophularineæ. This opinion has been adopted by subsequent writers, with the exception of Decandolle, who, in Duby's *Botanicon Gallicon* (1828), adheres to the old division of Jussieu, their names being changed into Antirrhineæ and Rhinanthaceæ. Notwithstanding this almost universal assent to the identity of the two orders of Jussieu, some separations have been made upon different principles from those of that learned botanist. Thus Orobancheæ have been distinguished by himself; Gesnereæ by Nees Von Esenbeck; and Melampyraceæ by Richard. The two former are adopted by botanists without dissent; the latter has not been so generally received. In my *Synopsis* I admitted it, upon the ground of its definite ascending seeds and inverted embryo; but subsequent consideration has led me to think that by excluding from the character all consideration of the number and direction of the seeds, a tribe would be formed, agreeing in a peculiar habit, and in the radicle of the embryo not being presented to the hilum, to which the name of Rhinanthaceæ might conveniently be retained. Upon this view of the subject, Scrophularineæ will include no genus the embryo of which is not orthotropous, and in Rhinanthaceæ it must be antitropous or heterotropous. For the distinctions of Gesnereæ and Orobancheæ, see those orders respectively. Scrophularineæ agree with Rhinanthaceæ, Orobancheæ, Gesnereæ, Bignoniaceæ, Cyrtandraceæ, Verbenaceæ, Myoporineæ, Selagineæ, Pedalineæ, Acanthaceæ, and Solaneæ, in their ovarium being formed by the cohesion of two carpella, which stand fore and aft with respect to the axis of inflorescence; or, in other words, the back of one is presented to the upper lip of the corolla, that part in which the fifth stamen is abortive or rudimentary, and the back of the other to the middle lobe of the lower lip between the two anterior stamens; a curious arrangement, by attending to which no difficulty can be found in recognising Gentianeæ, which, when out of flower, are exceedingly like. Scrophularineæ differ from Bignoniaceæ and Pedalineæ in their habit and albuminous seeds; from Solaneæ in their diandrous or didynamous flowers, straight not curved embryo, and opposite not alternate leaves; from Verbenaceæ and Myoporineæ in their polyspermous fruit, which is usually dehiscent, or at least never drupaceous; from Selagineæ in the same characters and their opposite leaves; and from Acanthaceæ in their flowers not being surrounded by imbricating bracteæ, and in the presence of albumen. Verbascum and Celsia, two genera usually referred to Solaneæ, are by some botanists placed here; they, and Digitalis, which has alternate leaves, form connecting links between the two orders.

GEOGRAPHY. Found in abundance in all parts of the world, from the coldest regions in which the vegetation of flowering plants takes place, to the hottest places within the tropics. One species is found in Melville Island; in middle Europe they form about a 26th of the flowering plants, and in North America about a 36th. In all India, New Holland, and South America, they are common, and, finally, the sterile shores of Terra del Fuego are ornamented with several species.

PROPERTIES. Generally acrid, bitterish, suspected plants. The leaves and roots of Scrophularia aquatica, and perhaps nodosa, of Gratiola officinalis and peruviana, and of Calceolaria, act as purgatives, or even as emetics. In Digitalis, which is in many respects very near Solaneæ, this quality is so much increased, that its effects become highly dangerous. The powdered leaves, or an extract of them, produce vomiting, dejection, and vertigo, increase the secretion of the saliva and urine, lower the pulse, and even cause death. *Dec.* According to Vauquelin, the purgative quality of Gratiola depends upon the presence of a peculiar substance, analogous to resin, but differing in being soluble in hot water. The leaves of Mimulus guttatus are eatable as salad. The juice of the leaves of Torenia asiatica are considered, on the Malabar coast, a cure for gonorrhœa. *Ainslie,* 2. 122. An infusion of Scoparia dulcis is used by the Indians of Spanish America to cure agues. *Humboldt Cinch. Forests,* 22. *Eng. ed.*

Duvau, in an excellent memoir upon the general characters of Veronica, proposes the following sections of this order; see *Ann. des Sc.* vol. 8. p. 176. 1826.

VERONICEÆ.

EXAMPLES. Veronica, Sibthorpia, Disandra.

ERINACEÆ.

EXAMPLES. Manulea, Buchnera, Erinus.

SCROPHULARINEÆ.

EXAMPLES. Scrophularia, Antirrhinum, Mimulus, Gratiola, Chelone, Digitalis.

To these Link adds, as will be seen among the synonymes of the order, Halleriaceæ, containing the baccate genera, and Scopariaceæ, containing Scoparia alone.

CCXII. RHINANTHACEÆ. THE RATTLE TRIBE.

MELAMPYRACEÆ, *Rich. Anal. du Fruit.* (1808); *Lindl. Synops.* 194. (1829).— RHINANTHACEÆ, *Dec. Fl. Fr.* 3. 454. (1815); *Dec. and Duby Bot. Gall.* 351. (1828) *in part.*—PEDICULARES, *Juss. Gen.* 99. (1789) *in part; Duvau in Ann. des Sc. Nat.* 8. 180. (1826.)

DIAGNOSIS. Monopetalous dicotyledons, with a superior 2-celled capsule, irregular unsymmetrical flowers, crested bracteæ, albuminous seeds, and a heterotropous embryo.

ANOMALIES.

ESSENTIAL CHARACTER.— *Calyx* divided, persistent, unequal, inferior, foliaceous. *Corolla* monopetalous, hypogynous, deciduous, personate. *Stamens* 4, didynamous; *anthers* with acuminate lobes. *Ovarium* superior, 2-celled, 2-seeded; *style* 1; *stigma* obtuse. *Fruit* capsular, 2-celled, 2-valved, covered by the calyx. *Seeds* ascending; *embryo* minute, inverted (heterotropous) in fleshy *albumen.*— *Herbaceous* plants. *Leaves* opposite, without stipulæ. *Flowers* axillary, with coloured or crested floral leaves.

AFFINITIES. Distinguished from Scrophularineæ by the inverted or heterotropous embryo, the seeds being generally winged and few in number, often definite, and the bracteæ dilated and foliaceous: at least such is the only character which I can find for this group, which Duvau calls " tres-tranché et presqu'isolé." The habit is peculiar; Chelone is the genus among Scrophularineæ to which they most nearly approach. In my *Synopsis* I have followed Richard in distinguishing Melampyraceæ from Rhinanthaceæ, and placing the latter among Scrophularineæ; but I now entertain a different opinion: see Scrophularineæ. Duvau says he has observed that, in some species of Euphrasia, Bartsia, Rhinanthus, Melampyrum, and Pedicularis, the base of the corolla is persistent in the form of a collar; and he suggests the possibility of this character, which he has also remarked in Orobanche, being of importance.

GEOGRAPHY. Natives of Europe, Asia, and America, particularly in the more temperate parts; also of the Cape of Good Hope, South America, India, and New Holland.

PROPERTIES. Euphrasia officinalis is slightly bitter and aromatic, and was formerly employed in diseases of the eye, but is now disused. Cows are said to be fond of Melampyrum pratense; and Linnæus says the best and yellowest butter is made where it abounds. The Pedicularises are acrid, but are eaten by goats. Nearly all this tribe turn black in drying.

EXAMPLES. Rhinanthus, Pedicularis, Melampyrum.

CCXIII. SOLANEÆ. THE NIGHTSHADE TRIBE.

SOLANEÆ, *Juss. Gen.* 124. (1789); *R. Brown Prodr.* 443. (1810); *Lindl. Synops.* 180. (1829.)

DIAGNOSIS. Monopetalous dicotyledons, with regular flowers, a superior 2-celled ovarium, indefinite ovules, a plaited corolla, succulent fruit, and alternate leaves.

ANOMALIES. Verbascum has irregular flowers. The anthers of Solanum open by pores. Nolana has a deeply 5- or more-lobed ovarium. Nicotiana multivalvis has many cells in the capsule.

ESSENTIAL CHARACTER.—*Calyx* 5-parted, seldom 4-parted, persistent, inferior. *Corolla* monopetalous, hypogynous; the *limb* 5-cleft, seldom 4-cleft, regular, or somewhat unequal, deciduous; the *æstivation*, in the genuine genera of the order, plaited; in the spurious genera imbricated. *Stamens* inserted upon the corolla, as many as the segments of the limb, with which they are alternate, 1 sometimes being abortive; *anthers* bursting longitudinally, rarely by pores at the apex. *Ovarium* 2-celled, with 2 polyspermous placentæ; *style* continuous; *stigma* simple. *Pericarpium* with 2 or 4 cells, either a capsule with a double dissepiment parallel with the valves, or a berry, with the placentæ adhering to the dissepiment. *Seeds* numerous, sessile; *embryo* more or less curved, often out of the centre, lying in fleshy *albumen; radicle* next the hilum.—*Herbaceous* plants or *shrubs. Leaves* alternate, undivided, or lobed; the floral ones sometimes double, and placed near each other. *Inflorescence* variable, often out of the axillæ; the *pedicels* without bracteæ.

AFFINITIES. Mr. Brown remarks, that this order is chiefly known from Scrophularineæ by the curved or spiral embryo, the plaited æstivation of the corolla, and the flowers being usually regular, with the same number of stamens as lobes. Hence the genera with a corolla not plaited, and at the same time a straight embryo, should either be excluded, or placed in a separate section, along with such as have an imbricated corolla, a slightly curved embryo, and didynamous stamens. *Prodr.* 444. To this a third section should be added for Nolana, which has a deeply 5- or more-lobed ovarium, each lobe containing one or more cells, in each of which lies a single seed. Nolana paradoxa has a considerable number of little drupes crowded one above the other; so that this section would appear to differ from true Solaneæ nearly as Labiatæ from Verbenaceæ; but there is a similar tendency to an excessive multiplication of cells in Nicotiana multivalvis, a genuine plant of the order, in which an additional verticillus of pericarpial leaves is added to the outside of the two central ones, forming together a singular instance of a many-celled fruit. Through Nolana, Solaneæ approach Convolvulaceæ. The position of the placentæ and pericarpial leaves is the same in this order as in Scrophularineæ and their allies, from which its alternate leaves usually distinguish them. Verbascum and Celsia are very near Scrophularineæ, to which they are actually referred by Reichenbach; but they differ in their alternate leaves and pentandrous flowers.

GEOGRAPHY. Natives of most parts of the world without the arctic and antarctic circles, especially within the tropics, in which the mass of the order exists, in the form of the genera Solanum and Physalis. Verbascum is wholly extratropical.

PROPERTIES. At first sight this family would seem to offer a strong exception to the general uniformity of structure and property, containing as it does the deadly Nightshade and Henbane, and the wholesome Potato and Tomato; but a little inquiry will explain this apparent anomaly. The tubers of the Potato are well known to be perfectly wholesome when cooked, any narcotic property which they possess being wholly dissipated by heat. This is the case with other succulent underground stems in equally dangerous families, as the Cassava among Euphorbiaceæ; besides which, as Decandolle justly observes, — "Il ne faut pas perdre de vue que tous nos alimens renferment une petite dose d'un principe excitant, qui, s'il y était en plus grande quantité, pourrait être nuisible, mais qui y est nécessaire pour leur servir de condiment naturel." The leaves of all are narcotic and exciting, but in different degrees, from the Atropa Belladonna, which causes vertigo, convulsions, and vomiting; the well-known Tobacco, which will frequently produce the first and last of these symptoms; the Henbane and Stramonium, down to some of the Solanum tribe, the leaves of which are used as kitchen herbs. The juice of Datura Stramonium is used in the United States, in doses of from 20 to 30 grains, in cases of epilepsy, or of mania without fever. *Dec.* The Quina of Brazil is the produce of Solanum pseudoquina, and is so powerful a bitter and febrifuge, that the Brazilians scarcely believe that it is not the genuine Jesuits' Bark. It has been analysed by Vauquelin, who found that it contained $\frac{1}{50}$ of a bitter resinoid matter, slightly soluble in water, about $\frac{1}{12}$ of a vegetable bitter, and a number of other principles in minute quantities. *Plantes Usuelles*, 21. The juice of Atropa Belladonna is well known to produce a singular dilatation of the pupil of the eye. Duval found that the same property exists in Solanums of the Dulcamara tribe, but in a more feeble degree. It is in the fruit that the greatest diversity of character exists; Atropa Belladonna, Solanum nigrum, and others, are highly dangerous poisons; Stramonium, Henbane, some Cestrums, and Physalis, are narcotic; the fruit of Physalis Alkekengi is diuretic, for which quality it is employed by veterinary surgeons; that of Capsicum is pungent, or even acrid; some Physalis are subacid, and so wholesome as to be eaten with impunity; and, finally, the Egg plant, Solanum esculentum, and all the Tomato tribe of Solanum, yield fruits which are common articles of cookery. But it is stated that the poisonous species derive their properties from the presence of a pulpy matter which surrounds the seeds; and that the wholesome kinds are destitute of this pulp, their fruit consisting only of what botanists call the sarcocarp; that is to say, the centre of the rind, in a more or less succulent state. It must also be remembered, that if the fruit of the Egg-plant is eatable, it only becomes so after undergoing a particular process, by which all its bitter acrid matter is removed, and that the Tomato is always exposed to heat before it is eaten. The fruit of Solanum Jacquini is considered by the native practitioners of India as expectorant. The juice of that of Solanum bahamense is used in the West Indies in cases of sore throat. *Ainslie*, 2. 91. A decoction of the root of S. mammosum is bitter, and reckoned a valuable diuretic. *Ibid.* The roots of Physalis flexuosa are supposed by the Indian doctors to have deobstruent and diuretic qualities, and also to be alexipharmic. The leaves moistened with a little warm castor oil are a useful external application in cases of carbuncle. *Ibid.* 2. 15. The common Potato, in a state of putre-faction, is said to give out a most vivid light, sufficient to read by. This was particularly remarked by an officer on guard at Strasburgh, who thought the barracks were on fire, in consequence of the light thus emitted from a cellar full of potatos. *Ed. P. J.* 13. 376. It has been supposed that Potash may be advantageously obtained from the stalk of Potatos; but it appears, from the experiments of Dr. Macculloch and Sir John Hay, that the quantity they contain is so small as not to be worth the manufacture. *Ibid.* 2. 399. The deleterious principle of the Belladonna has been ascer-tained by Vauquelin to be a bitter nauseous matter, soluble in spirit of wine, forming an insoluble combination with tannin, and yielding ammonia when burnt. *Dec. Prodr.* 225. The active principle of Solanum Dulcamara is an alkali, called Solania, which is in that plant combined with malic acid. *Turner*, 654.

EXAMPLES. The sections above alluded to in this order are the following :—

§ 1. SOLANEÆ. The Genuine Nightshade Tribe.
Corolla with the limb usually plaited. Stamens equal to the number of the lobes of the corolla. Embryo curved much. *R. Br.*
Solanum, Physalis, Nicotiana, Datura, Lycium, Atropa.

§ 2. NOLANEÆ. The Nolana Tribe.
Nolaneæ. *Reichenb. Consp.* 125. (1829).
Corolla plaited. Stamens equal to the number of the lobes of the corolla. Ovarium divided into 5 or more lobes. Fruit drupaceous. Embryo much curved.
Nolana.

§ 3. VERBASCEÆ. The Mullein Tribe.
Corolla not plaited. Stamens 5 and unequal, or didynamous. Embryo slightly curved.
Verbascum, Celsia, Anthocercis.

N. B. Reichenbach refers the first and last to Scrophularineæ. (*Con-spectus*, p. 124.)

CCXIV. ACANTHACEÆ. THE JUSTICIA TRIBE.

ACANTHI, *Juss. Gen.* 102. (1789). — ACANTHACEÆ, *R. Brown Prodr.* 472. (1810); *Link Handb.* 1. 500. (1829) *a sect. of* Personatæ.

DIAGNOSIS. Monopetalous dicotyledons, with a superior 2-celled cap-sule, irregular unsymmetrical flowers, exalbuminous wingless seeds with hooked dissepiments, and imbricated flowers.

ANOMALIES. A singular depauperation of the calyx takes place in the genera Thunbergia, Mendozia, and Clistax, in which that organ is reduced sometimes to a mere obsolete ring, its place being supplied by bracteæ. Mendozia is also remarkable for its fruit being a 1-seeded drupe, with crum-pled chrysaloid cotyledons.

ESSENTIAL CHARACTER.—*Calyx* 4- or 5-divided, cleft or tubular, equal or unequal, occasionally multifid, or entire and obsolete, persistent. *Corolla* monopetalous, hypogynous, bearing the stamens, mostly irregular; the *limb* ringent or 2-lipped (the lower lip overlapping the upper in æstivation), occasionally 1-lipped, sometimes nearly equal, deciduous. *Stamens* mostly 2, both bearing anthers; sometimes 4, didynamous, the shorter ones being sometimes sterile; *anthers* either 2-celled, their cells being inserted equally or unequally, or 1-celled, opening lengthwise. *Ovarium* seated in a disk, 2-celled, the cells either 2- or many-seeded; *style* 1; *stigma* 2-lobed, rarely undivided. *Capsule* 2-celled, the cells 2- or many-seeded, by abortion sometimes becoming 1-seeded, bursting elastically with 2 valves. *Dissepiment* opposite the valves, separable into two pieces through the axis (the middle being sometimes open); these pieces attached to the valves, sometimes separating from them with elasticity; entire, or occasionally spontaneously separating in two, their inner edge bearing the seeds. *Seeds* roundish, hanging by subulate ascending processes of the dissepiment; *testa* loose; *albumen* none; *embryo* curved or straight; *cotyledons* large, roundish; *radicle* taper, descending, and at the same time centripetal, curved, or straight; *plumula* inconspicuous.— *Herbaceous* plants or *shrubs*, chiefly tropical; their hairs, if they have any, simple, occasionally capitate, very rarely stellate. *Leaves* opposite, rarely in fours, without stipulæ, simple, undivided, entire, or serrated; rarely sinuate, or having a tendency to become lobed. *Inflorescence* terminal, or axillary, in spikes, racemes, fascicles, or panicles; the flowers sometimes even solitary. *Flowers* usually opposite in the spikes, sometimes alternate, with 3 bracteæ, of which the lateral are now and then deficient; these bracteæ sometimes large and leafy, and enclosing a diminished calyx, which is occasionally obsolete. *R. Br.* chiefly.

AFFINITIES. In habit these approach Scrophularineæ, from which their want of albumen, elastically dehiscing fruit, and the hooked processes of the dissepiment, distinguish them; with Bignoniaceæ they agree so nearly in character, that they may be said to differ in nothing but their seeds not being winged, for the hooks are sometimes absent: generally, however, their flowers being intermixed with imbricated bracteæ, their many-leaved imbricated calyx, and their herbaceous habit, point them out sufficiently. To Pedalineæ they approach in character, but are at once known by their 2-celled ovarium and peculiar habit. Von Martius remarks (*Nov. Gen. et Sp.* 3. 27.), that the didynamy of Acanthaceæ is frequently different from that of Scrophularineæ in the posterior pair of stamens being the longest, and the anterior pair shortest.

GEOGRAPHY. Common in all tropical countries, and only found beyond them in very hot ones. In North America a few species extend to the northward as far as Pennsylvania: and in Europe two are found in the basin of the Mediterranean.

PROPERTIES. Scarcely known. Acanthus mollis is considered emollient; Justicia biflora is used in Egypt for poultices; J. Ecbolium is said to be diuretic. *Dec.* The flowers, leaves, and root of Justicia Adhatoda are supposed to possess antispasmodic qualities. They are bitterish and subaromatic. *Ainslie*, 2. 3. Justicia pectoralis, boiled in sugar, yields a sweet-scented syrup, which is considered in Jamaica a stomachic. *Swartz.* 1. 32. The leaves and tender stalks of Justicia Gendarussa have, when rubbed, a strong and not unpleasant smell, and are, after being roasted, prescribed in India in cases of chronic rheumatism attended with swelling in the joints. *Ainslie*, 2. 68. The basis of a famous French bitter tincture, called Drogue Amère, highly valued for its stomachic and tonic properties, is the Justicia paniculata, called Creyat in India. *Ibid.* 1. 96. The leaves of Ruellia strepens are subacrid. *Ibid.* 2. 153. Another species is reckoned a diuretic in Java. *Ibid.*

EXAMPLES. Justicia, Lepidagathis, Ruellia, Acanthus.

CCXV. PEDALINEÆ. The Oil-Seed Tribe.

Pedalinæ, *R. Brown Prodr.* 519. (1810); *Lindley in Botan. Register,* 9. 934. (1825).— Sesameæ, *Kunth Synops.* 2. 251. (1823).— Martyniaceæ, *Link Handb.* 1. 504. (1829) *a sect. of* Personatæ.

DIAGNOSIS. Monopetalous dicotyledons, with a superior 1-celled or spuriously 4- or 6-celled short woody dehiscent or indehiscent fruit, a woody variously-lobed placenta, irregular unsymmetrical flowers, and exalbuminous apterous definite seeds.

ANOMALIES. Sesamum has indefinite seeds.

ESSENTIAL CHARACTER.—*Calyx* divided into 5 nearly equal pieces. *Corolla* monopetalous, hypogynous, irregular; the throat ventricose, the limb bilabiate. *Stamens* didynamous, included within the tube, together with a rudiment of a fifth. *Ovarium* seated in a glandular disk, unilocular or bilocular, with several 1- or 2-seeded spurious cells, formed by the splitting of two placentas and the divergence of their lobes; *ovules* either erect, or pendulous, or horizontal; *style* 1; *stigma* divided. *Fruit* drupaceous, juiceless, with several cells formed as those of the ovarium. *Seeds* pendulous, with a papery testa; *albumen* none; *embryo* straight.—*Herbaceous* plants. *Leaves* opposite. *Flowers* axillary, each with two bracteæ.

AFFINITIES. These differ from 'Bignoniaceæ in their wingless seeds, which are usually definite, and in their woody parietal lobed placentæ, which spread and divide variously in the inside of the pericarpium, so as to produce an apparently 4- or 6-celled fruit out of a 1-celled ovarium. For an explanation of the manner in which this takes place, see the *Botan. Register*, fol. 934. From Cyrtandraceæ they are known by their large seeds, free from all appendage at either end, by their woody placentæ, and short fruit. Sesamum may be considered a transition from the one to the other.

GEOGRAPHY. Found only within the tropics of Africa, Asia, and America.

PROPERTIES. The leaves of Sesamum are emollient. Its seeds contain an abundance of a fixed oil, as tasteless as that of Olive oil, for which it might be substituted, and which is expressed in Egypt in great quantities. The fresh leaf of Pedalium murex, when agitated in water, renders it mucilaginous, in which state it is prescribed by Indian doctors in cases of dysuria and gonorrhœa.

EXAMPLES. Pedalium, Pretrea, Josephinia, Martynia, Sesamum.

CCXVI. CYRTANDRACEÆ.

Cyrtandraceæ, *Jack in Linn. Trans.* 14. 23. (*read* 1822, *in May*).—Didymocarpeæ, *Don in Edinb. Phil. Journ.* 7. 82. (1822, *July*); *Prodr. Fl. Nep.* 121. (1825); *Martius H. R. Mon.* (1829.)

DIAGNOSIS. Monopetalous dicotyledons, with a superior 1-celled or spuriously 2-celled fruit, irregular unsymmetrical flowers, exalbuminous apterous minute seeds, and membranous double placentæ.

ANOMALIES.

ESSENTIAL CHARACTER.—*Calyx* campanulate, 5-cleft or 5-leaved, equal. *Corolla* tubular, irregular, 5-lobed, somewhat 2-lipped, the lobes imbricated in æstivation. *Stamens* 4, didynamous, of which 2 are sometimes sterile; *anthers* 2-celled. *Ovarium* superior, elongated, surrounded by an annular disk, 1-celled, with 2 many-seeded placentæ, each of which consists of 2 diverging plates; *style* filiform; *stigma* 2-lobed, or consisting of 2 plates. *Fruit* capsular or succulent; the former siliquose and 2-valved, 1-celled, with double longitudinal placentæ, which often cohere, so as to give the appearance of two cells. *Seeds* very numerous, minute, suspended, naked, or with a coma; *albumen* none; *embryo* straight, taper, orthotropous.— *Terrestrial* or *parasitical* plants, usually *herbaceous* and stemless, occasionally caulescent, and sometimes shrubby. *Leaves* usually opposite, one of them being dwarfed, radical, crenate and rugose, or smooth. *Flowers* umbellate, often purple or pink.

Affinities. Very closely allied to Gesnereæ, Bignoniaceæ, and Pedalineæ. From the former they differ in nothing except their never having any tendency to produce an inferior ovarium, their deeply-lobed placentæ, their usually siliquose fruit, and the want of albumen; agreeing entirely with them in habit. From Bignoniaceæ they are distinguished by their herbaceous mode of growth, their minute apterous seeds, 1-celled ovarium, with 2 double parietal placentæ. From Pedalineæ they differ in nothing whatever, except their minute indefinite seeds, and the membranous, not woody, texture of the fruit and placentæ. Sesamum forms a transition from the one order to the other, which would, perhaps, be better combined. Mr. Don appears to me to have been mistaken in assigning an heterotropous embryo to this tribe; the embryo is certainly orthotropous in Streptocarpus Rexii, with which the other genera no doubt agree. Von Martius refers Ramonda hither.

Geography. They occupy nearly the same station in the Old World as Gesnereæ in the New, being almost entirely confined to the tropics, unless the Ramonda of the Pyrenees should be found a genuine plant of the order, as Von Martius supposes.

Properties. Unknown.

Examples. Cyrtandra, Didymocarpus, Chirita, Incarvillea.

CCXVII. BIGNONIACEÆ. The Trumpet-Flower Tribe.

Bignoniæ, § 2. *Juss. Gen.* 137. (1789).—Bignoniaceæ, *R. Brown Prodr.* 470. (1810); *Link Handb.* 1. 503. (1829) *a sect. of* Personatæ.

Diagnosis. Monopetalous dicotyledons, with a superior 2-celled capsule, a central placenta, irregular unsymmetrical flowers, and exalbuminous winged seeds.

Anomalies. Eccremocarpus has a 1-celled fruit with parietal placentæ. The fruit is sometimes spuriously 4-celled.

Essential Character.—*Calyx* divided or entire, sometimes spathaceous. *Corolla* monopetalous, hypogynous, usually irregular, 4-5-lobed. *Stamens* 5, unequal, always 1, sometimes 3, sterile; *anthers* 2-celled, formed normally. *Ovarium* seated in a disk, 2-celled, or spuriously 4-celled, polyspermous; *style* 1; *stigma* of 2 plates. *Capsule* 2-valved, 2-celled, often long and compressed, sometimes spuriously 4-celled. *Dissepiment* either parallel with the valves, or contrary to them, finally becoming separate, bearing the seeds at the commissure along with the valves. *Seeds* transverse, compressed, often winged; *albumen* 0; *embryo* straight, foliaceous; *radicle* centrifugal.— *Trees* or *shrubs*, often twining or climbing. *Leaves* opposite, very rarely alternate, compound or occasionally simple, without stipulæ. *Inflorescence* terminal, somewhat panicled.

Affinities. Distinguished from Scrophularineæ and their immediate allies by the want of albumen, from Acanthaceæ by their winged seeds, and from both by their arborescent habit. Eccremocarpus is, however, an exception to the latter character, and also differs in having an unilocular ovarium and fruit; in the latter respect approaching Cyrtandraceæ and Pedalineæ, from which, however, its winged seeds divide it. This wing to the seed is a beautiful membrane formed of transparent cellular tissue, which, in an Indian unpublished genus given me by Dr. Wallich, offers an instance of reticulated cellules, analogous to those of Maurandya Barclaiana. There do not appear to be any very certain limits between Bignoniaceæ, Cyrtandraceæ, and Pedalineæ, which might be reunited without much inconvenience. Eccremocarpus may be considered the link between the two former, and Sesamum that between the two latter.

GEOGRAPHY. The tropics of either hemisphere are their chief station, from which they extend northwards in North America as far as Pennsylvania, and southwards into the southern provinces of Chile. In Europe they are unknown.

PROPERTIES. Little known, except the great beauty of their flowers. Chica is a red feculent substance obtained by boiling the leaves of Bignonia Chica in water; the Chica is quickly precipitated by adding some pieces of the bark of an unknown tree, called Arayana. The Indians use it for painting their bodies red; it is also becoming an article of importance to dyers. *Brewster*, 2. 370. It approaches in nature the resins, but contains some peculiar properties: it gives an orange red to cotton. *Ed. P. J.* 12. 417. The tough shoots of Bignonia Cherere are woven into wicker-work; and several kinds of Bignonias form large trees in the forests of Brazil, where they are felled for the sake of their timber; that called Ipe-tabacco furnishes durable ship-timber; the Ipeuna, another species, the hardest wood in Brazil. *Pr. Max. Travels*, p. 68. Another, called the Pao d'arco, supplies one of the best kinds of woods used for bows by the Brazilian Indians, especially the Botocudos of the Rio Grande de Belmonte, and the Patachos of the Rio do Prado. *Ibid.* 238.

EXAMPLES. Bignonia, Jacaranda, Spathodea.

CCXVIII. MYOPORINEÆ.

MYOPORINÆ, *R. Brown Prodr.* 514. (1810.)

DIAGNOSIS. Monopetalous dicotyledons, with irregular unsymmetrical flowers, a superior 2- or 4-celled ovarium with definite pendulous ovules, indehiscent fruit, a superior radicle, and albuminous seeds.

ANOMALIES.

ESSENTIAL CHARACTER. — *Calyx* 5-parted, persistent. *Corolla* monopetalous, hypogynous, nearly equal or 2-lipped. *Stamens* 4, didynamous, with sometimes the rudiment of a fifth one, which occasionally bears pollen. *Ovarium* 2- or 4-celled, the cells 1- or 2-seeded, with pendulous ovules; *style* 1; *stigma* scarcely divided. *Fruit* a drupe, with a 2- or 4-celled putamen, the cells of which are 1- or 2-seeded. *Seeds* with albumen; *embryo* taper; *radicle* superior. — *Shrubs*, with scarcely any pubescence. *Leaves* simple, without stipulæ, alternate or opposite. *Flowers* axillary, without bracteæ. *R. Br.*

AFFINITIES. The principal characters in the fructification of this order, by which it is distinguished from Verbenaceæ, are the presence of albumen in the ripe seed, and the direction of the embryo, whose radicle always points towards the apex of the fruit. The first of these characters is, however, not absolute, and neither of them can be ascertained before the ripening of the seed. *R. Brown in Flinders*, 567.

GEOGRAPHY. This order, with the exception of Bontia, a genus of equinoctial America, and of the species of Myoporum, found in the Sandwich Islands, has hitherto been observed only in the southern hemisphere, and yet neither in South Africa nor in South America beyond the tropics. Its maximum is evidently in the principal parallel of Terra Australis, in every part of which it exists; in the more southern parts of New Holland,

and even in Van Diemen's Island, it is more frequent than within the tropics. *R. Brown in Flinders*, 567.

PROPERTIES. The bark of Avicennia tomentosa, the White Mangrove of Brazil, is in great use at Rio Janeiro for tanning. *Pr. Max. Trav.* 206.

EXAMPLES. Myoporum, Stenochilus, Pholidia, Eremophila.

CCXIX. SELAGINEÆ.

SELAGINEÆ, *Juss. Ann. Mus.* 7. 71. (1806); *Richard in Pers. Synops.* 2. 146. (1807); *Choisy Mémoire*, (1823.)

DIAGNOSIS. Monopetalous dicotyledons, with irregular unsymmetrical flowers, a superior 2-celled ovarium with definite erect ovules, indehiscent fruit, a superior radicle, albuminous seeds, and alternate leaves.

ANOMALIES.

ESSENTIAL CHARACTER.—*Calyx* tubular, persistent, with a definite number of teeth, or divisions, rarely consisting of two sepals. *Corolla* tubular, hypogynous, more or less irregular, with 5 lobes. *Stamens* 4, usually didynamous, arising from the top of the tube of the corolla, seldom 2; *anthers* usually adnate to the dilated top of the filament, rarely versatile. *Ovarium* superior, very minute; *style* 1, filiform. *Fruit* 2-celled, the cells either separable or inseparable, 1-seeded, membranous. *Seed* solitary, erect; *embryo* in the axis of fleshy albumen; *radicle* inferior.—*Herbaceous* plants, or small branched *shrubs*. *Leaves* alternate, usually sessile, toothed, or entire, often fascicled. *Flowers* sessile, spiked, with large bracteæ.

AFFINITIES. Distinguished from Verbenaceæ by the radicle being superior, instead of inferior, and the leaves alternate; from Myoporineæ by the seeds being erect, not pendulous, and the embryo consequently antitropous, not orthotropous. M. Choisy remarks, that " if, on the one hand, we examine Selagineæ, Verbenaceæ, and Myoporineæ, and, on the other, Dipsaceæ, Compositæ, and Calycereæ, we shall find a perfect symmetry between their respective characters; thus Dipsaceæ differ from Compositæ exactly as Selagineæ from Verbenaceæ, by the inverted embryo and the presence of albumen, and Calycereæ differ from Compositæ as Myoporineæ from Verbenaceæ, by their pendulous ovulum; therefore, as every one admits Dipsaceæ and Calycereæ, it seems natural to admit Selagineæ and Myoporineæ." *Mémoire*, p. 9. Related to Scrophularineæ through Erinus and Manulea, and to Acanthaceæ through Eranthemum. The essential character is taken from M. Choisy.

GEOGRAPHY. All found at the Cape of Good Hope.

PROPERTIES. Unknown.

EXAMPLES. Selago, Polycenia, Agathelpis, Hebenstreitia.

CCXX. VERBENACEÆ. THE VERVAIN TRIBE.

VITICES, *Juss. Gen.* 106. (1789).—VERBENACEÆ, *Juss. in Ann. Mus.* 7. 63. (1806); *R. Brown Prodr.* 510. (1810); *Lindl. Synops.* 195. (1829.)

DIAGNOSIS. Monopetalous dicotyledons, with a superior undivided ovarium, a terminal style, irregular unsymmetrical flowers, indehiscent 2- or 4-celled fruit, opposite leaves, and solitary seeds with an inferior radicle.

ANOMALIES.

ESSENTIAL CHARACTER.— *Calyx* tubular, persistent, inferior. *Corolla* hypogynous, monopetalous, tubular, deciduous, generally with an irregular limb. *Stamens* usually 4, didynamous, seldom equal, occasionally 2. *Ovarium* 2- or 4-celled; *ovules* erect or pendulous, solitary or twin; *style* 1; *stigma* bifid or undivided. *Fruit* drupaceous, or baccate. *Seeds* erect or pendulous; *albumen* none, or in very small quantity; *embryo* erect.— *Trees* or *shrubs*, sometimes *herbaceous* plants. *Leaves* generally opposite, simple or compound, without stipulæ. *Flowers* in opposite corymbs, or spiked alternately; sometimes in dense heads; very seldom axillary and solitary.

AFFINITIES. The difference between these plants and Labiatæ consists in the concrete carpella of the former, their terminal style, and the usual absence of reservoirs of oil from their leaves, as contrasted with the deeply 4-lobed ovarium and aromatic leaves of the latter. There are, however, particular species of Labiatæ which approach Verbenaceæ very closely, so that Mr. Brown has remarked (*Congo*, 451.), that it has been difficult to distinguish the two orders. Verbenaceæ differ from Myoporineæ and Selagineæ in the position of the radicle, which in the former points to the base, and in the two latter to the apex of the fruit. There are also other points of difference, which will be mentioned under those orders. Acanthaceæ and Scrophularineæ differ in not having 1-or 2- seeded indehiscent cells. Mr. Brown remarks, that although all the genera of Verbenaceæ have an embryo whose radicle points towards the base of the fruit, yet many of them have pendulous seeds, and consequently a radicle remote from the umbilicus. *Flinders*, 567. Aug. de St. Hilaire asserts, that all, except Avicennia, have a sessile erect ovulum arising from the base of each cell. *Pl. Usuelles*, 40. Mr. Brown, however, places Avicennia in Myoporineæ.

GEOGRAPHY. Rare in Europe, northern Asia, and North America; common in the tropics of both hemispheres, and in the temperate districts of South America. In the tropics they become shrubs, or even gigantic timber, but in colder latitudes they are mere herbs.

PROPERTIES. Not of much importance in a medicinal or economical point of view. Callicarpa lanata bark has a peculiar subaromatic and slightly bitterish taste, and is chewed by the Cingalese when they cannot obtain Betel leaves; the Malays reckon the plant diuretic. *Ainslie*, 2. 180. Stachytarpheta jamaicensis is a plant to which the Brazilians attach the same false notions of powerful action as Europeans formerly did to the common Vervain. Its leaves are sometimes used to adulterate Chinese Tea, and have even been sent to Europe under the name of Brazilian Tea. *Pl. Usuelles*, p. 39. M. Auguste St. Hilaire speaks in terms of high praise of the agreeable properties of the aromatic Lantana pseudo-thea, used in infusion as tea. It is highly esteemed in Brazil, where it is vulgarly called Capitaô do matto, or Cha de pedreste. *Ibid.* p. 70. The root of Premna integrifolia is cordial and stomachic in decoction. *Ainslie*, 2. 210. Silex exists in abundance in the wood of the Teak Tree (Tectona grandis), which belongs here. *Ed. P. J.* 3. 413. The properties formerly ascribed to the Vervain appear to have been imaginary.

EXAMPLES. Verbena, Vitex, Clerodendron, Callicarpa.

CCXXI. LABIATÆ. THE MINT TRIBE.

LABIATÆ, *Juss. Gen.* 110. (1789); *R. Brown Prodr.* 499. (1810); *Mirbel in Ann. Mus.* 15. 213. (1810); *Lindl. Synops.* 196. (1829); *Bentham in Bot. Reg.* (1829.)

DIAGNOSIS. Monopetalous dicotyledons, with a superior 4-lobed ovarium, and irregular unsymmetrical flowers.

ANOMALIES.

ESSENTIAL CHARACTER. — *Calyx* tubular, 5- or 10-toothed, inferior, persistent, the odd tooth being next the axis; regular or irregular. *Corolla* monopetalous, hypogynous, bilabiate; the upper lip undivided or bifid, overlapping the lower, which is larger and 3-lobed. *Stamens* 4, didynamous, inserted upon the corolla, alternately with the lobes of the lower lip, the 2 upper sometimes wanting; *anthers* 2-celled; sometimes apparently unilocular in consequence of the confluence of the cells at the apex; sometimes 1 cell altogether obsolete, or the 2 cells separated by a bifurcation of the connectivum. *Ovarium* deeply 4-lobed, seated in a fleshy hypogynous disk; the lobes each containing 1 erect ovulum; *style* 1, proceeding from the base of the lobes of the ovarium; *stigma* bifid, usually acute. *Fruit* 1 to 4 small nuts, enclosed within the persistent calyx. *Seeds* erect, with little or no albumen; *embryo* erect; *cotyledons* flat.—*Herbaceous* plants or *under-shrubs. Stem* 4-cornered, with opposite ramifications. *Leaves* opposite, divided or undivided, without stipulæ, replete with receptacles of aromatic oil. *Flowers* in opposite, nearly sessile, axillary cymes, resembling whorls; sometimes as if capitate.

AFFINITIES. The 4-lobed ovarium, with a solitary style arising from the base of the lobes, has no parallel among monopetalous orders, except in Boragineæ, to which Labiatæ must be considered as most closely allied. They differ in the latter having not only an irregular corolla, but not more than 2 or 4 stamens, while the lobes of the corolla are 5, and opposite leaves; circumstances in which Labiatæ resemble Scrophularineæ and the orders allied to it. From all such they are known, in the absence of fructification, by their square stem and the numerous reservoirs of oil in their leaves. For some good remarks upon the anatomy of the stem of Labiatæ, see Mirbel in the *Annales du Muséum*, vol. 15. p. 223. The æstivation of the corolla of this order, first well pointed out by Mr. Brown (*Prodr.* 500), is an important consideration in determining whether a flower is resupinate or not. Prostanthera is remarkable for the appendages to its antheræ, and for the remains of albumen existing in the ripe seeds of several of its species. *Brown in Flinders,* 566. An arrangement of the genera has been published by Mr. Bentham in the *Botanical Register*, folios 1282, 1289, 1292, and 1300; a very difficult task, on account of the extremely close relationship which exists between all the species of this natural family, but one which has been executed in a most skilful and satisfactory manner. According to Dr. Griesselich, the reservoirs of oil in the leaves of Labiatæ are not analogous to those of Oranges and other plants, but are little utricules having an open orifice; and hence he calls them pores. *Ferussac, Jan.* 1830, p. 96. 200 GEOGRAPHY. Natives of temperate regions in greater abundance than elsewhere, their maximum probably existing between the parallels of 40° and 50° N. latitude. They are found in abundance in hot, dry, exposed situations, in meadows, hedgerows, and groves; not commonly in marshes. In France they form 1-24th of the Flora; in Germany, 1-26th: in Lapland, 1-40th; the proportion is the same in the United States of North America, and within the tropics of the New World (*Humboldt*); in Sicily they are 1-21 of flowering plants (*Presl.*); in the Balearic islands, 1-19th. About 200 species are mentioned in Dr. Wallich's *Catalogue of the Indian Flora*, a large proportion of which is from the northern provinces. They were not found in Melville Island.

PROPERTIES. Their tonic, cordial, and stomachic qualities, due to the presence of an aromatic volatile oil and a bitter principle, are the universal feature of Labiatæ, which do not contain a single unwholesome or even suspicious species. On account of the bitter qualities, several are used as febrifuges, as the Ocymum febrifugum of Sierra Leone; and many as aromatics in our food, such as Savory, Mint, Marjoram, and Basil. Others are found useful in the preparation of slightly tonic beverages, such as Glechoma hederacea, Sage, Balm of Gilead, &c. When the volatile oil is in great abundance, as in Lavender and Thyme, an agreeable perfume is the result. Rosemary is the herb used in the manufacture of Hungary water.

The leaves of Ocymum album are considered by the natives of India stomachic, and their juice is prescribed in the catarrhs of children. *Ainslie*, 2. 92. The fresh juice of Anisochilus (Lavandula carnosa *L.*) mixed with powdered sugarcandy, is prescribed by the native practitioners of India in cynanche. *Ibid.* 2. 144. Tonic and stimulant properties have been ascribed to the Origanum Dictamnus. *Ibid.* 1. 112. It is asserted that the juice of the bruised leaves of Phlomis esculenta, drawn up the nose, is a specific against the bite of serpents ; but there is reason to doubt the truth of this statement, as the plant, which is a common weed in Bengal, possesses but a slight aromatic scent, and has scarcely any flavour. *Trans. M. and P. Soc. Calc.* 2. 405. Hedeoma pulegioides, the Pennyroyal of the North Americans, has a great popular reputation as an emmenagogue. *Barton*, 2. 168. Cunila mariana is beneficially employed in infusion in slight fevers and colds, with a view to excite perspiration. *Ibid.* 2. 175. The roots of Stachys palustris are described as an esculent by Mr. Joseph Houlton. The Panax Coloni of old botanists is the same thing. *Trans. Soc. Arts*, 46. 8. Perhaps the most singular quality of these plants is their containing an abundance of camphor, a substance which seems to exist in the whole tribe, and which is found so copiously in the oils of Sage and Lavender as to be capable of being advantageously extracted.

EXAMPLES. Lamium, Mentha, Stachys, Thymus.

CCXXII. BORAGINEÆ. THE BORAGE TRIBE.

BORAGINEÆ, *Juss. Gen.* 143. (1789) ; *R. Brown Prodr.* 492. (1810) ; *Lindl. Synops.* 163. 1829.)

DIAGNOSIS. Monopetalous dicotyledons, with regular flowers, a deeply lobed superior ovarium, and round stems.

ANOMALIES. Echium has rather irregular flowers ; Benthamia has 4 cotyledons.

ESSENTIAL CHARACTER.— *Calyx* persistent, with 4 or 5 divisions. *Corolla* hypogynous, monopetalous, generally regular, 5-cleft, sometimes 4-cleft, with an imbricate æstivation. *Stamens* inserted upon the petals, equal to the number of lobes of the corolla, and alternate with them. *Ovarium* 4-parted, 4-seeded ; *ovula* attached to the lowest point of the cavity (pendulous, *R. Br.*) ; *style* simple, arising from the base of the lobes of the ovarium ; *stigma* simple or bifid. *Nuts* 4, distinct. *Seed* separable from the pericarpium, destitute of albumen. *Embryo* with a superior *radicle* ; *cotyledons* parallel with the axis, plano-convex, sometimes 4 ! — *Herbaceous* plants or *shrubs*. *Stems* round. *Leaves* alternate, covered with asperities, consisting of hairs proceeding from an indurated enlarged base. *Flowers* in 1-sided spikes or racemes, or panicles, sometimes solitary and axillary.

AFFINITIES. Nearly allied to Labiatæ, from which they are essentially distinguished by the regularity of the corolla, the presence of 5 fertile stamens, the absence of resinous dots, the round (not square) figure of the stem, and the scabrous alternate leaves. On account of this last character, they are often called Asperifoliæ. From all other monopetalous orders they are known by the 4 deep lobes of the ovarium, called by Linnean botanists naked seeds. Hydrophylleæ, Heliotropiceæ, Cordiaceæ, and Ehretiaceæ, are all distinguished by their undivided ovarium, but, together with Boragineæ, are known by the quaternary structure of their ovarium and the quinary division of the corolla and stamens.

GEOGRAPHY. Natives principally of the temperate countries of the northern hemisphere ; extremely abundant in all the southern parts of Europe, the Levant, and middle Asia ; less frequent as we approach the arctic circle, and almost disappearing within the tropics. A few species only are found in such latitudes. In North America they are less abundant than in Europe. Pursh reckons but 22 species in the whole of his Flora; while the little island of Sicily alone contains 35, according to Presl.

PROPERTIES. Soft, mucilaginous, emollient properties, are the usual characteristics of this order; some are also said to contain nitre, a proof of which is shewn by their frequent decrepitation when thrown on the fire. Borago officinalis gives a coolness to beverage in which its leaves are steeped. Echium plantagineum, naturalised in Brazil, is used in that country for the same purposes as the Borago officinalis in Europe. *Pl. Usuelles*, 25. The roots of Anchusa tinctoria or Alkanet, Lithospermum tinctorium, Onosma echioides, Echium rubrum, and Anchusa virginica, contain a reddish brown substance used by dyers. This matter is thought to be a peculiar chemical principle approaching the resins.

EXAMPLES. Borago, Lycopsis, Anchusa.

CCXXIII. HELIOTROPICEÆ. THE HELIOTROPE TRIBE.

HELIOTROPICEÆ, *Martius N. G. et Sp.* 2. 75. *and* 138. (1828.)

DIAGNOSIS. Monopetalous dicotyledons, with regular flowers, a superior 4-celled ovarium with solitary pendulous ovules, 5 stamens, and exalbuminous seeds with plano-convex cotyledons.

ANOMALIES.

ESSENTIAL CHARACTER. — *Calyx* inferior, hypogynous, 5-parted, persistent. *Corolla* hypogynous, monopetalous, regular, with a 5-parted limb, the segments of which are imbricated in æstivation. *Stamens* arising from the tube of the corolla, and alternate with the segments; *anthers* innate; *pollen* globose. *Ovarium* entire, 4-celled, with 4 pendulous *ovula*; *style* terminal, simple; *stigma* simple. *Fruit* drupaceous, separable into 4 pieces, terminated by the persistent style. *Seeds* pendulous, solitary; *embryo* without albumen, with fleshy plano-convex *cotyledons* and a minute *radicle* curved downwards and turned towards the hilum. — Half *shrubby* and *herbaceous* plants, covered over with asperities. *Leaves* alternate, simple, without stipulæ. *Flowers* in terminal fascicles, cymes, or corymbs.

AFFINITIES. Distinguished from Boragineæ solely by having a style proceeding from the apex of an undivided ovarium of several cells, by the drupaceous fruit separating in pieces, and the absence of albumen.

GEOGRAPHY. Common in the hotter parts of South America, the East and West Indies, the north of Africa, and the Levant; a few are found in the south of Europe and the southern states of America, but none appear to dwell further north than the parallel of 45°.

PROPERTIES. Unknown, except that some of the species are remarkable for their fragrance. Most of them are insignificant weeds.

EXAMPLES. Heliotropium, Preslea.

CCXXIV. EHRETIACEÆ.

EHRETIACEÆ, *Martius N. G. et Sp.* 2. 136. (1828.)

DIAGNOSIS. Monopetalous dicotyledons, with regular flowers, a superior 2- or more-celled ovarium with suspended ovules, 5 lobes to the calyx, and albuminous seeds.

ANOMALIES.

ESSENTIAL CHARACTER.— *Calyx* inferior, 5-parted, imbricated in æstivation. *Corolla* monopetalous, tubular, with as many segments of its limb as the calyx, with an imbricated æstivation. *Stamens* alternate with the segments of the corolla, and equal to them in number, arising from the bottom of the tube; *anthers* innate; *pollen* minute, elliptical. *Ovarium* simple, seated in an annular disk, 2- or more celled; *style* terminal; *stigma* simple, 2-lobed; *ovules* suspended. *Fruit* drupaceous, with as many stems as there are true cells of the ovarium. *Seed* suspended, solitary; *testa* simple, thin; *embryo* in the midst of thin fleshy albumen; *radicle* superior; *cotyledons* plano-convex. — *Trees* or *shrubs*, with a harsh pubescence. *Leaves* simple, alternate, without stipulæ. *Flowers* corymbose.

AFFINITIES. Another branch of the old Boragineæ, distinguished by a terminal style proceeding from the apex of a perfectly concrete ovarium of 4 cells, a baccate fruit, and seeds furnished with thin fleshy albumen. Of these characters I conceive the former to be good, and the latter bad; and the order itself, which I adopt upon the authority of Dr. Von Martius, hardly tenable, differing from Heliotropiceæ chiefly in its succulent (not dry) separable fruit.

GEOGRAPHY. Tropical trees or shrubs, natives of either hemisphere.

PROPERTIES. The root of Ehretia buxifolia is reckoned in India one of those medicines which assist in altering and purifying the habit in cases of cachexia and venereal affections of long standing. *Ainslie*, 2. 81.

EXAMPLES. Ehretia, Tournefortia, Rhabdia, Beurreria?

CCXXV. CORDIACEÆ.

R. Brown Prodr. 492. (1810), *without a name; Martius N. G. et Sp.* 2. 138. (1828), *without a name.* — CORDIACEÆ, *Link Handb.* 1. 569. (1829). — ARGUZIÆ, *ib.*

DIAGNOSIS. Monopetalous dicotyledons, with regular flowers, a superior 4-celled ovarium with solitary pendulous ovules, 5 stamens, and exalbuminous seeds with plaited shrivelled cotyledons.

ANOMALIES.

ESSENTIAL CHARACTER.— *Calyx* inferior, 5-toothed. *Corolla* monopetalous, with the limb in 5 divisions. *Stamens* alternate with the segments of the corolla, out of which they arise; *anthers* versatile. *Ovarium* superior, 4. celled, with 1 pendulous ovulum in each cell; *style* continuous; *stigma* 4-cleft, with recurved segments. *Fruit* drupaceous, 4-celled; part of the cells frequently abortive. *Seed* pendulous from the apex of the cells by a long funiculus, upon which it is turned back; *embryo* inverted, with the *cotyledons* plaited longitudinally; *albumen* 0. — *Trees.* *Leaves* alternate, scabrous, without stipulæ, of a hard harsh texture. *Flowers* panicled, with minute bracteæ.

AFFINITIES. The plaited cotyledons and dichotomous style first induced the separation of this order from Boragineæ, with which it was formerly associated, chiefly, it is to be supposed, on account of the roughness of the leaves. Von Martius well remarks, that it is in fact much nearer Convolvulaceæ, from which it differs in its inverted embryo and drupaceous fruit. *Nov. Gen. l. c.*

GEOGRAPHY. Natives of the tropics of both hemispheres.

PROPERTIES. The flesh of their fruit is succulent, mucilaginous, and emollient, as is seen in the Sebesten Plums, the produce of Cordia Myxa and Sebestena.

EXAMPLES. Cordia, Geraschanthus, Cerdana, Varronia, Cordiopsis, Menais.

CCXXVI. HYDROPHYLLEÆ. THE WATERLEAF TRIBE.

R. Brown Prodr. 1. 492. (1810), *without a name.*—HYDROPHYLLEÆ, *Von Martius N. G. et Sp.* 2. 138. (1828); *Link Handb.* 1. 570. (1829), *a § of* Cordiaceæ.

DIAGNOSIS. Monopetalous dicotyledons, with regular flowers, a superior 1-celled ovarium with ovula attached to parietal or fungous stalked placentæ, and a naked stigma.

ANOMALIES.

ESSENTIAL CHARACTER.— *Calyx* with 5 or 10 divisions, inferior. *Corolla* monopetalous, regular, or nearly so, hypogynous, 5-lobed, with 2 lamellæ at the base of each lobe. *Stamens* alternate with the segments of the corolla, in æstivation inflexed ; *anthers* ovate, innate, 2-celled, bursting longitudinally. *Ovarium* simple, 1-celled, superior, with slight traces of a hypogynous disk ; *style* simple or divided, terminal ; *stigma* bifid ; *ovules* attached to 2 parietal or fungous stalked placentæ, either definite or indefinite. *Fruit* capsular, few- or many-seeded, invested with the permanent calyx. *Seeds* definite or indefinite ; *embryo* taper, lying towards the end of the *albumen*, which is abundant and somewhat cartilaginous ; its *radicle* superior, and next the hilum. — *Herbaceous* hispid plants. *Leaves* either opposite or alternate, but in the latter case lobed. *Peduncles* opposite the leaves.

AFFINITIES. Very near Boragineæ and the orders which have been recently separated from it, with which Hydrophylleæ agree in the roughness of their leaves and many other marks of obvious resemblance. They are, however, known by their undivided 1-celled ovarium, terminal style or styles, and ovula (if definite) attached to two stalked fungous placentæ, which arise from the base of the cell, having their ovula on their inner face, or (if indefinite) attached to parietal placentæ. They are further characterised by the presence of 2 scales or lamellæ at the base of each lobe of the corolla, the nature of which is unknown. The former mode of placentation is highly curious, and, as far as I know, unlike that of any other plants.

GEOGRAPHY. American herbaceous plants, found either in the north or among the most southern of the southern provinces; not known beyond that continent.

PROPERTIES. Unknown.

EXAMPLES. Hydrophyllum, Nemophila, Ellisia, Eutoca, Phacelia.

Tribe II. GYMNOSPERMÆ.

Synorhizæ, *Rich. Anal. du Fr. Eng. ed.* 81. (1819). — Phanerogames Gymno-spermes, *Ad. Brongniart Veget. Foss.* 88. (1828.)

These have nearly an equal relation to flowering and flowerless plants. With the former they agree in habit, in the presence of sexes, and in their vascular tissue being complete; with Ferns and Lycopodiums, among the latter, they also accord in habit, in the peculiar gyrate vernation of the leaves of Cycadeæ, in their spiral vessels being imperfectly formed, and in the sexes being less complete than in other flowering plants; the females wanting a pericarpial covering, and receiving impregnation directly through the foramen of the ovulum, without the intervention of style or stigma, and the males consisting of leaves imperfectly contracted into an anther bearing a number of pollen-cases upon their surface. So great is the resemblance between Lycopodiums and certain Coniferæ, that I know of no external character, except size, by which they can be distinguished; and it is, at least, as probable that some of those vegetables found in the ancient Flora of the world, which have been considered gigantic Lycopodiums, are Coniferæ, as that they are flowerless plants. Gymnospermæ are known from all other Vasculares by the vessels of their wood having large apparent perforations, to which nothing similar has yet been seen elsewhere. It is not, however, on this account to be understood that these differ in growth from other Exogenous plants; on the contrary, they are essentially the same, deviating in no respect from the plan upon which Exogenous plants increase, but having a kind of tissue peculiar to themselves.

LIST OF THE ORDERS.

227. Cycadeæ. | 228. Coniferæ.

CCXXVII. CYCADEÆ.

Cycadeæ, *Rich. in Pers. Synops.* 2. 630. (1807); *Brown Prodr.* 346. (1810); *Kunth in Humb. et Bonpl. Nov. Gen. et Sp.* 2. 1. (1817); *Synops.* 1. 349 (1822); *R. Brown in King's Voyage,* (1825); *Rich. Mémoire,* 195. (1826); *Ad. Brongniart in Ann. des Sc.* 16. 589. (1829.)

Diagnosis. Naked-seeded mucilaginous dicotyledons, with a round or cylindrical undivided trunk, and pinnated leaves having a gyrate verna-tion and parallel veins.

Anomalies.

Essential Character. — *Flowers* diœcious, terminal. *Males* monandrous, naked, collected in cones; each floret consisting of a single scale (or anther) bearing the pollen on its under surface in 2-valved cases which adhere in clusters of 2, 3, or 4. *Females* either collected in cones, or surrounding the central bud in the form of contracted leaves without pinnæ, bearing the ovula on their margins. *Ovula* solitary, naked, with no other pericar-pium than the scale or contracted leaf upon which they are seated. *Embryo* in the midst of fleshy or horny albumen ; the *radicle* next the apex of the seed, from which it hangs by a long funiculus with which it has an organic connexion. —*Trees*, with a simple cylindrical trunk, increasing by the development of a single terminal bud, and covered by the scaly bases of the leaves; the wood consisting of concentric circles, the cellular zones between which are exceedingly loose. *Leaves* pinnated, not articulated, having a gyrate vernation.

AFFINITIES. One of the botanists who originally noticed the plants that constitute this order referred them to the Fern tribe ; an opinion to which Linnæus, having first adopted the idea of Adamson that they were related to Palms, finally acceded. He was followed by other botanists, until, after some suggestions by Ventenat that the genera Cycas and Zamia ought to form a particular tribe, the present order was finally characterised by the late M. Richard in Persoon's *Synopsis*, in 1807, with the observation that it was intermediate between Ferns and Palms. The opinion of their affinity to Ferns seems to have been thus generally adopted in consequence of their striking resemblance in the mode of developing their leaves; but the supposed relation to Palms was suggested rather by a vague notion of some general resemblance, as, for instance, in their cylindrical trunks, than by any precise knowledge of the structure of Cycadeæ. It is only within a few years that a more accurate knowledge of their structure has determined the real nature of their affinities. In 1825, the publication of Mr. Brown's remarks upon the ovulum, in which he demonstrated the similarity of conformation between the flowers of Cycadeæ and Coniferæ, suggested new ideas of the affinities of both tribes; and the determination, in 1829, by M. Adolphe Brongniart, of the exact resemblance between these two tribes in the structure of the vessels of their wood, while it decided the near relation of Coniferæ and Cycadeæ, confirmed the proximity of the former to Ferns, and shewed the inaccuracy of the ideas formerly held of a close resemblance between the latter and Palms. As this is still a matter but ill understood in general, it may be useful to make some further remarks upon the subject.

It has been said that the dissimilarity between Cycadeæ and Coniferæ is such as to render it impossible to admit of their close approximation in any natural arrangement; and that the affinity of Cycadeæ being with Palms, the former must necessarily be widely apart from Firs. These views of the subject appear to have arisen either from an imperfect knowledge of the real vegetation of the stem of Cycadeæ, or from a too superficial consideration of such points as were really well known. The affinity of Cycadeæ and Palms does at first sight appear probable, in consequence of the large pinnated leaves and simple cylindrical stems of both tribes; but here I think the resemblance stops. Cycadeæ have a gyrate, Palms a convolute vernation; Cycadeæ are naked-seeded and bear their seeds on the margins of a contracted leaf, Palms have the ordinary inflorescence of flowering plants; Cycadeæ are dicotyledonous, Palms monocotyledonous; and finally, the internal structure of the trunk of Cycadeæ is essentially exogenous, as is now perfectly well known : the affinity of Cycadeæ is therefore not with Palms. With regard to the nature of the evidence by which their strict relation to the Pine tribe is to be established, it may be observed, that they are both dicotyledonous in seed, both have naked ovula constructed in a similar remarkable manner, and borne in both cases not upon a rachis, but upon the margin or face of metamorphosed leaves ; that they have the same peculiar form of inflorescence, the same kind of male flowers, the same constant separation of sexes; that the arrangement of the veins of their leaves is peculiar and identical; that there is a like imperfect formation of spiral vessels, a most important consideration ; and finally, that they both agree in having the vessels of their wood apparently perforated with numerous holes ; a character, as far as is yet known, exclusively confined to these two tribes. The difference between the cylindrical simple stem of Cycadeæ and the branched conical one of Coniferæ arises from the terminal bud only of the former developing, its axillary ones all being uniformly latent, unless called into life by some accidental circumstance, as in the case recorded in the *Horticultural Transactions*, 6. 501. ; while in Coniferæ a constant tendency to a rapid evolution of leaf-buds takes place in every axilla. With regard to their foliage, on which the difference of their aspect chiefly depends, I have already stated that the arrangement of their veins is the same; but the leaves of Coniferæ are minute and undivided, while those of Cycadeæ are very large and pinnated; in both they are simple, and in Coniferæ there is a tendency to a higher development in the scales of the cones, while in Cycadeæ there is a corresponding contraction firstly in Cycas itself, and especially in Zamia, in which the contraction takes place to exactly the same point as the evolution of Coniferæ.

GEOGRAPHY. Natives of the tropics of America and Asia; not found in equinoctial Africa, although they exist at the Cape of Good Hope and in Madagascar. *Brown Congo*, 464.

PROPERTIES. The only remarkable quality in the order is the production of a kind of Sago, by the soft centre of Cycas circinalis. They all abound in a mucilaginous nauseous juice.

EXAMPLES. Cycas, Zamia.

CCXXVIII. CONIFERÆ. THE FIR TRIBE.

CONIFERÆ, *Juss. Gen.* 411. (1789); *Mirbel Elémens*, 2. 906. (1815); *Brown in King's Voyage, Appendix*, (1825); *Rich. Monogr.* (1826); *Dec. and Duby*, 431. (1828); *Lindl. Synops.* 240. (1829.)

DIAGNOSIS. Naked-seeded, resinous, dicotyledonous trees, with a branched trunk, and simple leaves with parallel veins.

ANOMALIES.

ESSENTIAL CHARACTER.— *Flowers* monœcious or diœcious. *Males* monandrous or monadelphous; each floret consisting of a single *stamen*, or of a few united, collected, in a deciduous amentum, about a common rachis; *anthers* 2-lobed or many-lobed, bursting outwardly; often terminated by a crest, which is an unconverted portion of the scale out of which each stamen is formed; *pollen* large, usually compound. *Females* usually in cones, sometimes solitary. *Ovarium*, in the cones, spread open, and having the appearance of a flat scale destitute of style or stigma, and arising from the axilla of a membranous bractea; in the solitary flower apparently wanting. *Ovula* naked; in the cones in pairs on the face of the ovarium, having an inverted position, and consisting of 1 or 2 membranes open at the apex, and of a nucleus; in the solitary flower erect. *Fruit* consisting either of a solitary naked seed, or of a cone; the latter, formed of the scale-shaped ovaria, become enlarged and indurated, and occasionally of the bracteæ also, which are sometimes obliterated, and sometimes extend beyond the scales in the form of a lobed appendage. *Seeds* with a hard crustaceous integument. *Embryo* in the midst of fleshy oily albumen, with 2 or many opposite *cotyledons*; the *radicle* next the apex of the seed, and having an organic connexion with the albumen.— *Trees* or *shrubs*, with a branched trunk abounding in resin. *Leaves* linear, acerose or lanceolate, entire at the margins, or dilated and lobed, always having the veins parallel with each other; sometimes fascicled in consequence of the non-development of the branch to which they belong; when fascicled, the primordial leaf to which they are then axillary is membranous, and enwraps them like a sheath.

AFFINITIES. With the exception of Orchideæ, there is perhaps no natural order the structure of which has been so long and so universally misunderstood as Coniferæ. This has arisen from the exceedingly anomalous nature of their organisation, and from the investigations of botanists not having been conducted with that attention to logical precision which is now found to be absolutely indispensable. The description above given is that which I conceive proper to explain the views now taken upon the subject, in consequence of the discovery by Mr. Brown of the ovula of the whole order being naked; and it will probably be found to offer a more intelligible account of the fructification than is to be met with in even the most recent systematic works. It is not expedient to enter here upon an inquiry into the ideas that botanists have successively entertained upon this subject. Those who are desirous of informing themselves upon this point will find all they can desire in the Appendix to Captain King's *Voyage to New Holland*, and in Richard's *Mémoires sur les Conifères et les Cycadées*. It may, however, be useful to advert briefly to the principal theories which have met with advocates. These are, firstly, that the female flowers consist of a bilocular ovarium having a style in the form of an external scale, an opinion held by Jussieu, Smith, and Lambert; secondly, that they have a minute cohering perianthium, and an external additional envelope called the cupula: this view was taken by Schubert, Mirbel, and others; thirdly, that they have a monosepalous calyx cohering more or less with the ovarium, contracted and often tubular at the apex, with a lobed, or glandular, or minute entire limb, an erect ovarium, a single pendulous ovulum, no style, and a minute sessile stigma: this explanation is that of Richard, published in his memoir upon the subject in 1826. It appears, however, from the observations of Mr. Brown, that the female organ of Coniferæ is a naked ovulum, the integuments of which have been mistaken for floral envelopes, and the apex of whose nucleus has been considered a stigma. Of the accuracy of this view there is probably, at this time, little difference in opinion. These female organs, or naked ovula, are in the cone-bearing genera 2 in number, and they originate from the larger scales of the cone towards their base, have an inverted position, and occupy the same relative place in Coniferæ and in Zamia, a genus of Cycadeæ. Now, as there cannot be any doubt of the perfect analogy that exists between the scales of the cone of Zamia and the fruit-bearing leaves of Cycas, the former differing from the latter only in each being reduced to 2 ovula, and to an undivided state; so there can be no doubt of the equally exact analogy between the scales of Coniferæ and Zamia, and therefore the former would be called reduced leaves if the general character of the tribe was to produce a highly developed foliage; but as the foliage of Coniferæ is in a much more contracted state than the scales of their cones, the latter must be understood to be the leaves of Coniferæ in a more developed state than usual. That the scales of the cone really are metamorphosed leaves, is apparent not only from this reasoning, but from the following facts. They occupy the same position with respect to the bracteæ as the leaves do to their membranous sheaths; they surround the axis of growth as leaves do, and usually terminate it; but in some cases, as often in the Larch, the axis continues to elongate beyond them, and leaves them collected round it in the middle. In Araucaria they have absolutely the same structure as the ordinary leaves; and finally they sometimes assume the common appearance of leaves, as is represented in Richard's memoir, tab. 12., in the case of a monstrous Abies. The scales of the cones of Coniferæ and strobilaceous Cycadeæ are therefore to these orders, what carpellary leaves are to other plants With regard to the male flowers, it is obvious that in the Ginkgo, the Larch, the Cedar of Lebanon, the Spruce, and the like, each anther is formed of a partially converted scale, analogous to the indurated carpellary scale of the females; and therefore each amentum consists of a number of monandrous naked male flowers, collected about a common axis. Some botanists, however, consider each male catkin as a single monadelphous male flower, which is impossible. But in the Yew the male flowers consist of a peltate scale, around which are arranged several polliniferous cavities; while, in Araucaria, these cavities occupy one side only of an ordinary flat scale. In the former case it is probable that the stamens are really monadelphous; an hypothesis which appears to derive confirmation from Ephedra, in one species of which, E. altissima, they are solitary, while in the common species they are manifestly monadelphous. In Araucaria, and such genera as agree with it in structure, the anthers may be considered to consist of an uncertain number of lobes, and in this respect to recede from the usual structure of the male organs of plants: in Coniferæ, the anthers of which are normal, we have 2; in Ephedra, 4; in Juniperus, the like number; in Cunninghamia, but 3; in Agathis, 14; and in Araucaria, from 12 to 20. Mr. Brown remarks, what is certainly very remarkable, that in Cunninghamia the lobes of the anther agree in number, as well as insertion and direction, with the ovula! *King's Appendix*, 32. It would almost appear, from Mr. Brown's remarks upon Gnetum or Thoa, that he considers that singular genus related to Coniferæ. But, independently of its very different habit, I confess it does not seem to me certain that its ovula are naked, as Mr. Brown supposes: on the contrary, as the nucleus has three coatings, I should rather understand the external at least as analogous to a carpellum, if the two others are allowed to belong to the ovulum, which I think admits of some degree of doubt. Coniferæ occupy a position, as it were, intermediate between Cellulares and Vasculares, approximating almost equally to each, connected with the former through Lycopodiaceæ, and with the latter by the intervention of Myriceæ and Cupuliferæ, Salicineæ, and Betulineæ. With Lycopodiaceæ they agree in the general aspect of the leaves and stems of several species, and in the nearly total absence, or at least very imperfect formation, of spiral vessels; with all the latter in their amentaceous inflorescence, but especially with Myriceæ, which are both amentaceous and resinous. But their most immediate relation is undoubtedly with Cycadeæ, the following order, as is there explained. The aspect of Callitris is so much that of Equisetum and Casuarina, that it is difficult to doubt an affinity also existing between them.

GEOGRAPHY. Natives of various parts of the world, from the perpetual snows and inclement climate of arctic America, to the hottest regions of the Indian Archipelago. The principal part of the order is found in temperate climates; in Europe, Siberia, China, and the temperate parts of North America, the species are exceedingly abundant, and have an aspect very different from that of the southern hemisphere. In the former we have various species of Pines, the Larch, the Cedar, the Spruce, and the Juniper; the place of which is supplied in the latter by Araucarias, Podocarpuses, Dammars, and Dacrydiums.

PROPERTIES. No order can be named of more universal importance to mankind than this, whether we view it with reference to its timber or its secretions. Gigantic in size, rapid in growth, noble in aspect, robust in constitution, these trees form a considerable proportion of every wood or plantation in cultivated countries, and of every forest where nature remains in a savage state. Their timber, in commerce, is known under the names of Deal, Fir, Pine, and Cedar, and is principally the wood of the Spruce, the Larch, the Scotch Fir, the Weymouth Pine, and the Virginian Cedar; but others are of at least equal, if not greater, value: the Norfolk Island Pine is an immense tree, known to botanists as Araucaria excelsa; the Kawie Tree of New Zealand, or Dammara australis, attains the height of 200 feet, and yields a light compact wood, free from knots; the Dacrydium taxifolium, or Kakaterro, equals this in stature. *Ed. Ph. Journ.* 13. 378. But they are both surpassed by the stupendous Pines of north-west America, one of which, P. Lambertiana, is reported to attain the height of 230 feet, and the other, P. Douglasii (qu. Pinus taxifolia?), to equal or even to exceed it. The latter is probably the most valuable of the whole for its timber. Their secretions consist of various kinds of resin. Oil of turpentine, common and Burgundy pitch, are obtained from Pinus sylvestris; Hungarian balsam from Pinus Pumilio; Bourdeaux turpentine from P. Pinaster; Carpathian balsam from P. Pinea; Strasburg turpentine from Abies pectinata (P. Picea *L.*), our Silver Fir; Canadian balsam from Abies balsamea, or the Balm of Gilead Fir. The common Larch yields Venetian turpentine. Liquid storax is thought to be yielded by the Dammar Pine; and a substance called in India Dammar, or country resin, is procured from the same plant, or from a tree which Dr. Buchanan calls Chloroxylon Dupada. *Ainslie,* 1. 337. Sandarach, a whitish yellow, brittle, inflammable, resinous substance, with an acrid aromatic taste, is said by Dr. Thomson to exude from Juniperus communis; but upon the authority of Brongniart and Schousboe, it is the tears of Thuja articulata (or quadrivalvis). *Ibid.* 1. 379. The substance from which spruce beer is made is an extract of the branches of the Abies canadensis, or Hemlock Spruce; a similar preparation is obtained from the branches of Dacrydium in the South Seas. Great tanning powers exist in the bark of the Larch; as great, it is said, as in the Oak. *Ed. P. J.* 1. 319. The stimulating diuretic powers of the Savin, Juniperus Sabina, are well known, and are partaken of in some degree by the common Juniper, the berries of which are an ingredient in flavouring gin. The large seeds of many are eatable. The Stone Pine of Europe, the Pinus Cembra, the Ginkgo, the Pinus Lambertiana and Gerardiana, the Araucaria Dombeyi, and Podocarpus neriifolia, are all eatable when fresh. The succulent covering of the Yew fruit is fœtid, and said to be deleterious by Decandolle; we all know that its seeds, if eaten, are highly dangerous.

Examples. Pinus, Cunninghamia, Araucaria.

Sub-Class II. ENDOGENÆ, or MONOCOTYLEDONOUS PLANTS.

Monocotyledones, *Juss. Gen.* 21. (1789); *Desf. Mem. Inst.* 1. 478. (1796). — Endorhizeæ, *Rich. Anal.* (1808). — Monocotyledoneæ *or* Endogenæ, *Dec. Théorie*, 209. (1813). — Cryptocotyledoneæ *or* Graniferæ, *Agardh Aph.* 73. (1821.)

Essential Character. — *Trunk* usually cylindrical when a terminal bud only is developed, becoming conical and branched when several develope; consisting of cellular tissue, among which the vascular tissue is mixed in bundles, without any distinction of bark, wood, and pith, and destitute of medullary rays; increasing in diameter by the addition of new matter to the centre. *Leaves* frequently sheathing at the base, and not readily separating from the stem by an articulation, mostly alternate, with parallel simple veins, connected by smaller transverse ones. *Flowers* usually having a ternary division; the calyx and corolla either distinct, or undistinguishable in colour and size, or absent. *Embryo* with but 1 cotyledon; if with 2, then the accessory one is imperfect and alternate with the other; *radicle* usually enclosed within the substance of the embryo, through which it bursts when germinating.

Nothing can be more simple than the mode of distinguishing Monocotyledonous from Dicotyledonous plants, notwithstanding the difficulty of fixing upon any single character of separation. It is true that the structure of the stem is not sufficient, because it is frequently impossible, in annual plants, to ascertain if it be Exogenous or Endogenous; the parallel veins of the leaves of Monocotyledons are not always constant, because some genera have reticulated ones; the want of articulation between the stem and the leaves, although very prevalent in Monocotyledons, sometimes changes to perfect articulation, as in Orchideæ; the ternary division of the flower of Monocotyledons is often departed from, as in Aroideæ and the neighbouring orders; many Dicotyledons have also ternary floral envelopes; Monocotyledons have sometimes more than one cotyledon, as the common Wheat; finally, when the stem is capable of being strictly examined, a distinction between wood and pith occasionally exists, as in the common Rush and in the Bamboo; and the conical branched character of Dicotyledons is assumed in Grasses and Asphodeleæ. Hence it is by a combination of characters that the two great divisions are to be known, and not by any absolute single mark: for instance, in Grasses, in which the stem is, as an eminent botanist has justly remarked, less Endogenous than in almost any other Monocotyledons, the leaves, flowers, and seeds, well shew them to be at once of the latter structure; so in Juncus, in which pith is present, no other character is at variance with those of Monocotyledons; and again in Orchideæ, in which a complete disarticulation of the stem and leaves takes place, every other point of structure is that of Monocotyledons. Mr. Brown has remarked (*Congo*, 481), that the presence of albumen may be considered as the natural structure of this primary division; seeds without albumen occurring only in certain genera of the paradoxical Aroideæ, and in some other Monocotyledonous orders which are chiefly aquatic. It is a fact well deserving attention, that Monocotyledons differ from Dicotyledons in their geographical distribution as well as in structure; a remarkable proof of the hypothesis, that the forms of vegetation are controlled by peculiarities of climate, acting in an unknown manner. From the enquiries of Humboldt, it appears that Monocotyledons form, in equinoctial regions, about 1-6th of the flowering plants; in the temperate zone, between 36° and 52° latitude, 1-4th; and towards the polar circle, nearly 1-3d.

The most important substance that they produce is amylaceous matter, which exists in great quantity in some of them, which hence become of incalculable value as aliment for man: such are all the Corn tribe, Plantains, and some Palms, which contain it in their fruit; the Sago and other Palms, in which it occupies the trunk; and the eatable Aroideous plants, Orchises, Yams, &c., in which it is found in the root. Sugar, gluten, oil, and aromatic principles, are also frequently met with in Monocotyledons; but, as Humboldt well remarks, acids, bitters, resins, camphor, tannin, milk, or poisonous matter, are either wholly wanting or very uncommon. The latter chiefly exists in Aroideæ, some Amaryllideæ, and Melanthaceæ.

The orders of Monocotyledons are given in the state in which they now exist; but it must be confessed that the characters and limits of many of them are far from satisfactory. The whole of those which border upon Asphodeleæ require to be reconsidered by some botanist who is in possession of the means of examining them in great detail; their actual condition is, no doubt, attributable to the partial view that has hitherto been taken of them. Some one should do that for Asphodeleæ which the late M. Richard so admirably executed for Alismaceæ and their affinities.

Endogenous plants are conveniently divided into those in which the floral envelopes are verticillate (*Petaloideæ*), and those in which the flowers consist of imbricated bracteæ (*Glumaceæ*).

TRIBE I. PETALOIDEÆ.

These comprehend all Monocotyledons except Grasses and Sedges. They are known by their flowers being fully and normally developed; or, if there is no proper floral envelope, by the sexual apparatus being in that case naked, and not covered by imbricated bracteæ. Some of them have both the calyx and corolla equally formed, and coloured so as to be undistinguishable, unless by the manner in which those parts originate: these constitute the *Hexapetaloideous* form. Others have the calyx and corolla distinct, as in Dicotyledons, to which, in fact, they nearly approach in Butomeæ, which have a strong analogy with Nymphæaceæ, and in Alismaceæ, which cannot be considered widely apart from Ranunculaceæ: these are named *Tripetaloideous*. Lastly, there is a group of orders in which the floral envelopes have a manifest tendency to abortion, being always small, and of a herbaceous colour, if present; often altogether wanting; and frequently less than 6, the normal number of Monocotyledons: as many of them are arranged in a spadix, and as most of them have a distinct tendency to that kind of inflorescence, the form is called *Spadiceous*.

LIST OF THE ORDERS.

229. Alismaceæ.
230. Butomeæ.
231. Hydrocharideæ.
232. Commelineæ.
233. Xyrideæ.
234. Bromeliaceæ.
235. Hypoxideæ.
236. Burmanniæ.
237. Hæmodoraceæ.
238. Amaryllideæ.
239. Irideæ.

240. Orchideæ.
241. Scitamineæ.
242. Marantaceæ.
243. Musaceæ.
244. Junceæ.
245. Melanthaceæ.
246. Pontedereæ.
247. Asphodeleæ.
248. Gilliesieæ.
249. Smilaceæ.
250. Dioscoreæ.

251. Liliaceæ.
252. Palmæ.
253. Restiaceæ.
254. Pandaneæ.
255. Typhaceæ.
256. Aroideæ.
257. Balanophoreæ.
258. Fluviales.
259. Juncagineæ.
260. Pistiaceæ.

CCXXIX. ALISMACEÆ. The Water-Plantain Tribe.

ALISMACEÆ, *R. Brown Prodr.* 342. *in part* (1810) ; *Rich. in Mém. Mus.* 1. 365. (1815) ;
  Juss. Dict. Sc. Nat. 1. 217. (1822) ; *Lindl. Synops.* 253. (1829). — ALISMOIDEÆ,
  Dec. Fl. Fr. 3. 188. (1815.)

DIAGNOSIS. Tripetaloideous monocotyledons, with numerous, distinct,
superior carpella.
  ANOMALIES.

ESSENTIAL CHARACTER.— *Sepals* 3, herbaceous. *Petals* 3, petaloid. *Stamens* definite or indefinite. *Ovaries* superior, several, 1-celled ; *ovules* erect or ascending, solitary,
or 2 attached to the suture at a distance from each other. *Styles* and *stigmas* the same
number as the ovaries. *Fruit* dry, not opening, 1- or 2-seeded. *Seeds* without albumen ;
embryo shaped like a horse-shoe, undivided, with the same direction as the seed. — *Floating*
plants. *Leaves* with parallel veins.

AFFINITIES. This order is to Monocotyledons what Ranunculaceæ are
to Polypetalous Dicotyledons, and is in like manner recognised by its indefinite distinct carpella and hypogynous stamens ; from Butomeæ it is known
by the indefinite ovula of that order being scattered over the face of the cells.
Juncagineæ, sometimes referred to Alismaceæ, appear nearer Aroideæ, and
are distinguished by their depauperated floral envelopes, concrete carpella,
and straight embryo having a lateral slit for the emission of the plumula.
The plants belonging to Alismaceæ, Hydrocharideæ, Fluviales, Juncagineæ,
and Butomeæ, have all a disproportionately large radicle, whence the embryos of such were called by the late M. Richard, macropodal.
  GEOGRAPHY. Chiefly natives of the northern parts of the world. Several Sagittarias and Actinocarpus inhabit the tropics, the former of both
hemispheres.
  PROPERTIES. All aquatic plants with a lax tissue, and many with a
fleshy rhizoma, which is eatable ; such are Alisma and Sagittaria : a species
of the latter is cultivated for food in China. The herbage is acrid. Alisma
Plantago is one of the plants recommended in hydrophobia. *Agdh.*
  EXAMPLES. Sagittaria, Echinodorus, Alisma, Actinocarpus.

CCXXX. BUTOMEÆ. The Flowering Rush Tribe.

BUTOMEÆ, *Richard in Mém. Mus.* 1. 364. (1815) ; *Lindley's Synopsis,* 271. (1829) ; *Dec.*
  and Duby, 437. (1828) *a* § *of* Alismaceæ.

DIAGNOSIS. Tripetaloideous monocotyledons, with the placentæ covering the whole lining of the superior carpella.
  ANOMALIES. In Butomus the calyx is more coloured than usual.

ESSENTIAL CHARACTER.— *Sepals* 3, usually herbaceous. *Petals* 3, coloured, petaloid.
Stamens definite or indefinite, hypogynous. *Ovaries* superior, 3, 6, or more, either distinct or united into a single mass ; *stigmas* the same number as the ovaries, simple.
Follicles many-seeded, either distinct and rostrate, or united in a single mass. *Seeds*
minute, very numerous, attached to the whole of the inner surface of the fruit ; *albumen*
none ; *embryo* with the same direction as the seed.— *Aquatic* plants. *Leaves* very cellular, often yielding a milky juice, with parallel veins. *Flowers* in umbels, conspicuous,
purple, or yellow.

  AFFINITIES. Although an undoubted tripetaloideous order, yet Butomeæ stand between it and the hexapetaloideous ones, on account of the

coloured state of the calyx of Butomus itself. They are, however, readily known by the remarkable circumstance of the placenta extending over the whole lining of the fruit, which is formed either of separate or concrete carpella. In this respect there is an evident analogy with Nymphæaceæ, which Limnocharis resembles in the structure of its fruit. Butomeæ are most closely akin to Alismaceæ. M. Decandolle has a remark (*Syst.* 2. 42.), that no Endogenæ are lactescent; but Limnocharis yields milk in abundance. This genus offers a singular example of a large conspicuous open hole in the apex of its leaf, apparently destined by nature as an outlet for superfluous moisture, which is constantly distilling from it.

GEOGRAPHY. Natives of the marshes of Europe, and equinoctial America.

PROPERTIES. Butomus is acrid.

EXAMPLES. Butomus, Limnocharis, Hydrocleys.

CCXXXI. HYDROCHARIDEÆ. THE FROG-BIT TRIBE.

HYDROCHARIDES, *Juss. Gen.* 67. (1789).—HYDROCHARIDEÆ, *Dec. Fl. Fr.* 3. 265. (1815); *R. Brown Prodr.* 344. (1810); *Richard in Mém. Mus.* vol. 1. 365. (1815); *Agardh Aph.* 127. (1822); *Lindley's Synopsis,* 254. (1829).—VALLISNERIACEÆ and STRATIOTEÆ, *Link Handb.* 1. 281. (1829.)

DIAGNOSIS. Tripetaloideous monocotyledons, with an inferior ovarium, and exalbuminous antitropous embryo. Water plants.

ANOMALIES.

ESSENTIAL CHARACTER.—*Flowers* hermaphrodite or unisexual. *Sepals* 3, herbaceous. *Petals* 3, petaloid. *Stamens* definite or indefinite. *Ovary* single, inferior, 1- or many-celled; *stigmas* 3-6; *ovules* indefinite, often parietal. *Fruit* dry or succulent, indehiscent, with 1 or more cells. *Seeds* without albumen; *embryo* undivided, antitropous. —*Floating* plants. *Leaves* with parallel veins, sometimes spiny. *Flowers* spathaceous.

AFFINITIES. These water-plants are readily distinguished from all other monocotyledons by their tripetaloideous flowers, with an inferior ovarium; by this they are separated from Alismaceæ, with which they agree in habit and want of albumen, but from which they also differ, as Pomaceæ from Ranunculaceæ, in the carpella being definite, not indefinite. Commelineæ are at once recognised by their superior trilocular ovarium. Agardh refers here Trapa (see p. 58.); Linnæus placed Hydrocharideæ along with Palms! in his natural arrangement.

Link defines his Hydrocharideæ, Stratioteæ, and Vallisneriaceæ, thus:—

Hydrocharideæ. Aquatic herbs. Leaves with parallel veins connected with lateral ones; sheath separate. Calyx divided to the base. Corolla polypetalous. Pericarpium. Albumen none, unless the thickened part of the embryo. *Hydrocharis.*

Stratioteæ. Aquatic herbs. Leaves sheathing with parallel veins. Flowers spathaceous. Calyx tubular. Corolla polypetalous, inserted on the calyx. A berry. *Stratiotes.*

Vallisneriaceæ. Aquatic herbs. Diœcious, diclinous. *Males;* Flowers in a spadix, from which they finally separate. Corolla monopetalous. *Females;* Spathe 1-flowered. Peduncles spiral. Calyx 1-leafed. Corolla polypetalous. Capsules 1-celled, many-seeded. Seeds parietal. *Vallisneria.*

GEOGRAPHY. Natives of Europe, North America, and the East Indies One species is found in Egypt (Ottelia indica), and two Vallisnerias in New Holland.

PROPERTIES. Nothing known, unless that the fruit of Enhalus is eatable, and its fibres capable of being woven, according to Agardh (*Aph.* 128). The Janji of Hindostan, called Vallisneria alternifolia by Roxburgh, Hydrilla by Dr. Hamilton, is one of the plants used in India for supplying water mechanically to sugar in the process of refining it. *Brewster*, 1. 34.

EXAMPLES. Hydrocharis, Hydrilla, Blyxa, Limnobium, Boottia, Stratiotes.

CCXXXII. COMMELINEÆ. THE SPIDER-WORT TRIBE.

EPHEMEREÆ, *Batsch. Tab. Affin.* 125. (1802) *in part.*—COMMELINEÆ, *R. Brown Prodr.* 268. (1810); *Richard in Humb. Bonpl. N. Gen.* 1. 258. (1815); *Agardh Aph.* 168. (1823.)

DIAGNOSIS. Tripetaloideous monocotyledons, with a superior 3-locular capsule.

ANOMALIES.

ESSENTIAL CHARACTER.—*Sepals* 3, distinct from the petals, herbaceous. *Petals* coloured, sometimes cohering at the base. *Stamens* 6, or a smaller number, hypogynous, some of them either deformed or abortive. *Ovarium* 3-celled, with few-seeded cells; *style* 1; *stigma* 1. *Capsule* 2- or 3-celled, 2- or 3-valved, the valves bearing the dissepiments in the middle. *Seeds* often twin, inserted by their whole side on the inner angle of the cell, whence the hilum is linear; *embryo* pulley-shaped, antitropous, lying in a cavity of the albumen remote from the hilum; *albumen* densely fleshy.—*Herbaceous* plants. *Leaves* usually sheathing at the base.

AFFINITIES. Mr. Brown remarks upon this order (*Prodr.* 269.), that " it is very different from Junceæ both in habit and structure; it agrees better with Restiaceæ in the situation of the embryo and the sheathing leaves, although otherwise quite distinct; it has scarcely any affinity with Palms, except in its trochlear embryo, remote from the hilum, and indicated in both orders by an external papilla." Agardh adds, that they agree with Orchideæ in the structure of their seeds and stamens. I know not in what respect this resemblance is shewn. Xyrideæ are probably the most nearly allied to Commelineæ of any known plants.

GEOGRAPHY. Chiefly found in the East and West Indies, and Africa. A few are found in North America, but none in northern Asia or Europe.

PROPERTIES. Often mere weeds, sometimes beautifully-flowering plants; otherwise having no known properties.

EXAMPLES. Commelina, Aneilema, Tradescantia, Cartonema.

CCXXIII. XYRIDEÆ.

XYRIDEÆ, *Kunth in Humb. N. G. et Sp.* 1. 255. (1815) *a sect. of* Restiaceæ; *Agardh Aphorism.* 158. (1823); *Desvaux in Ann. des Sc.* 13. 49. (1828.)

DIAGNOSIS. Tripetaloideous monocotyledons, with superior concrete carpella, a 1-celled capsule with parietal placentæ, and capitate flowers.

ANOMALIES.

ESSENTIAL CHARACTER.—*Calyx* glumaceous, 3-leaved. *Corolla* petaloid, 3-petalled. *Fertile stamens* 3, inserted upon the claws of the petals; *anthers* turned outwards; *sterile stamens* alternate with the petals. *Ovarium* single; *style* trifid; *stigmas* obtuse, multifid or undivided. *Capsule* 1-celled, 3-valved, many-seeded, with parietal placentæ. *Seed* with the embryo on the outside of the albumen, and at the end most remote from the hilum.— *Herbaceous* plants with fibrous roots. *Leaves* radical, ensiform, with dilated equitant scarious bases. *Flowers* in terminal, naked, imbricated heads.

AFFINITIES. United with Restiaceæ by Mr. Brown and others, separated as a distinct order by Agardh and Desvaux, this appears to me to be essentially distinguished by the higher development of its floral envelopes, a character which I cannot but regard as more important than the mere accordance in the structure of the seed, in consequence of which chiefly it has been retained in Restiaceæ. Those who have distinguished this order have referred to it several genera which by no means enter into the idea I have of the limits that should be prescribed to it, particularly Aphyllanthes, which is surely a Juncea. Mr. Brown remarks, that the anomalous genus Philydrum, and even Burmannia, are related to Xyris; and that these plants agree in some respects with Orchideæ in the structure of the seed and stamen (*Prodr.* 264). To me it seems that the relation of Xyrideæ is very great with Commelineæ.

GEOGRAPHY. All natives of the hotter parts of the world, chiefly in the tropics of America, Asia, and Africa. Two or three species of Xyris are found in the southern states of North America.

PROPERTIES. The leaves and root of Xyris indica are employed against itch and leprosy. *Agardh.*

EXAMPLES. Xyris, Abolboda.

CCXXXIV. BROMELIACEÆ. THE PINE-APPLE TRIBE.

BROMELIÆ, *Juss. Gen.* 49. (1789); *Dict. Sc. Nat.* 5. 347. (1817).—BROMELIACEÆ, *Lindl. in Bot. Reg.* fol. 1068. (1827); *Dec. and Duby*, 472. (1828.)

DIAGNOSIS. Tripetaloideous hexandrous monocotyledons, with an inferior ovarium, and an albuminous embryo.

ANOMALIES. Some, as Tillandsia, have a superior ovarium.

ESSENTIAL CHARACTER.— *Calyx* 3-parted or tubular, persistent, more or less cohering with the ovarium. *Petals* 3, coloured, withering or deciduous, equal or unequal. *Stamens* 6, inserted into the base of the calyx and corolla. *Ovarium* 3-celled, many-seeded; *style* single; *stigma* 3-lobed, often twisted. *Fruit* capsular or succulent, 3-celled, many-seeded. *Seeds* numerous; *embryo* taper, recurved, lying in the base of mealy albumen.—*Stemless* or *short-stemmed* plants, with rigid channelled *leaves*, often covered with cuticular scales, and spiny at the edge or point. *Fruit* sometimes eatable.

AFFINITIES. Stratiotes among Hydrocharideæ has so much the foliage of this order as to render it probable, taking the fructification also into account, that the nearest affinity of the Pine Apple tribe is with the former. It is distinguished from other tripetaloideous orders, when its ovarium is inferior, by its albuminous seeds and hexandrous flowers, while, in those cases in which the ovarium is superior, it is recognised by its polyspermous trilocular fruit; Commelineæ and Xyrideæ, with which alone it can be confounded, differing in this respect. The habit of Bromeliaceæ is peculiar; they are hard dry-leaved plants, generally with a mealy surface, and having a calyx the rigidity of which is strongly contrasted with the delicate texture of the petals. The habit of Agave is that of Aloe in Asphodeleæ, to which Bromeliaceæ approach: it was probably this consideration which induced M. Desfontaines to place Pitcairnia with the latter order.

GEOGRAPHY. All, without exception, natives of the continent or islands of America, whence they have migrated eastward in such numbers, as to have established themselves as part of the present Flora of the west coast of Africa, and some parts of the East Indies.

PROPERTIES. The most remarkable is the Pine Apple, or Ananas, which is well known for the sweetness and fine aromatic flavour of its fruit. No other species is of the same interest. They are all capable of existing in a dry hot air without contact with the earth; on which account they are favourites in South American gardens, where they are suspended in the dwellings, or hung to the balustrades of the balconies; situations in which they flower abundantly, filling the air with their fragrance. The wild Agave of Mexico yields a copious juice when tapped, which is fermented into a wine called Pulque, from which a spirit, known under the name of Vino Mercal, is obtained. Ropes are made in Brazil of a species of Bromelia, called Grawatha. *Pr. Max. Trav.*

EXAMPLES. Ananassa, Billbergia, Bromelia, Pitcairnia.

CCXXXV. HYPOXIDEÆ.

HYPOXIDEÆ, *R. Brown in Flinders*, (1814); *Agardh Aph.* 164. (1823) *a sect. of* Asphodeleæ.

DIAGNOSIS. Hexapetaloideous monocotyledons, with an inferior ovarium, a regular 6-parted perianthium with equitant sepals, rostellate seeds with a hard black coat.

ANOMALIES.

ESSENTIAL CHARACTER.—*Perianthium* superior, regular, 6-parted, with an equitant æstivation. *Stamens* 6, inserted into the base of the segments. *Ovarium* inferior, 3-celled, many-seeded; *style* single; *stigma* 3-lobed. *Capsule* indehiscent, sometimes succulent and many-seeded. *Seeds* with a black brittle integument, and a lateral rostelliform hilum; *embryo* in the axis of fleshy albumen, its *radicle* having no certain direction. —*Herbaceous* stemless, or nearly stemless plants with plaited *leaves*, and yellow or white *flowers*.

AFFINITIES. First placed by Mr. Brown at the end of Asphodeleæ, and afterwards separated as a distinct order, characterised by having, along with the fruit of Asphodeleæ, a superior perianthium and rostellate seeds. Agardh retains them in Asphodeleæ. The rigidity and harshness of their leaves is very unlike any thing among genuine plants of that tribe.

GEOGRAPHY. Natives of the Cape of Good Hope, New Holland, the East Indies, and North America.

PROPERTIES. Unknown.

EXAMPLES. Hypoxis, Curculigo.

CCXXXVI. BURMANNIÆ.

BURMANNIÆ, *Spreng. Syst.* 1. 123. (1825); *Reichenb. Conspect.* 60. (1828) *a sect. of* Amaryllideæ.

DIAGNOSIS. Hexapetaloideous triandrous monocotyledons, with an inferior winged ovarium, and minute indefinite seeds.

ANOMALIES.

ESSENTIAL CHARACTER.—*Flowers* hermaphrodite. *Perianthium* tubular, superior, coloured, membranous, with 6 teeth, the 3 inner of which (petals) are minute, the 3 outer larger, and having a wing or keel at the back. *Stamens* 3, inserted in the tube opposite the petals; *anthers* sessile, 2-celled, opening transversely, with a fleshy connectivum; sometimes 3 sterile stamens, alternate with them. *Ovarium* inferior, 3-celled, many-seeded, with the dissepiments alternate with the wings of the perianthium; *style* single; *stigma* 3-lobed. *Capsule* covered by the withered perianthium, 3-celled, 3-valved, bursting irregularly. *Seeds* very numerous and minute, striated; *embryo* . . . .—*Herbaceous* plants, with tufted radical acute *leaves*, a slender nearly naked stem, and terminal flowers, sessile upon a 2- or 3-branched rachis, or solitary.

AFFINITIES. The single genus upon which this is founded,—for Sonerila, referred here by Sprengel and Reichenbach, is not even monocotyledonous! (it belongs to Melastomaceæ),—was placed by Jussieu in Bromeliaceæ; Mr. Brown stationed it as a doubtful genus at the end of Junceæ, with the remark, that it is extremely distinct both in flower, fruit, and inflorescence, and not really allied to any other known plant, but more nearly related to Xyris and Philydrum than to either Bromelia or Hypoxis. Von Martius, who has beautifully illustrated the Brazilian species, refers them to Hydrocharideæ. To me it seems that they are, upon the whole, nearest Hæmodoraceæ, with which they agree in their tubular perianthium, in having the stamens reduced to three and opposite the petals, a much enlarged connectivum, the ovarium inferior, and some resemblance in foliage and habit. It is, however, certain that there is no known monocotyledonous order to which these really approach very closely. See *Irideæ.*

GEOGRAPHY. Natives of the tropics of Asia, Africa, and America. The plants called Tripterella by North American botanists are found as far to the north as Virginia.

PROPERTIES. Unknown.

EXAMPLES. Burmannia (Tripterella Mich.), Maburnia.

CCXXXVII. HÆMODORACEÆ. THE BLOOD-ROOT TRIBE.

HÆMODORACEÆ, *R. Brown Prodr.* 299. (1810); *Agardh Aphor.* 170. (1823); *Von Martius N. Gen. et Sp. Pl. Braz.* 1. 13. (1824); *Ach. Rich. Nouv. Elém.* 436. (1828.)

DIAGNOSIS. Hexapetaloideous monocotyledons, with an inferior ovarium, a (woolly) tubular perianthium, the sepals of which are not equitant, and farinaceous albumen.

ANOMALIES. Wachendorfia has a superior ovarium. Some Barbacenias are tripetaloideous. Vellozia has equitant sepals and petals.

ESSENTIAL CHARACTER.— *Calyx* and *corolla* confounded, petaloid, superior, rarely inferior. *Stamens* arising from the sepals and petals, either 3 and opposite the petals, or 6, or more numerous, and polyadelphous; *anthers* bursting inwardly. *Ovarium* with the cells 1- 2- or many-seeded; *style* simple; *stigma* undivided. *Fruit* capsular, valvular, seldom indehiscent, somewhat nucamentaceous. *Seeds* either definite and peltate, or indefinite; *testa* papery; *embryo* minute, orthotropous, in farinaceous albumen.—*Leaves* equitant, or arranged spirally or alternately, usually linear or linear-lanceolate, rarely acerose. *Flowers* often showy, the petals and sepals being highly developed.

AFFINITIES. The principal distinction between these and Amaryllideæ consists in their perianthium not having the regular equitant position of sepals and petals which is found in the latter, in their peculiar Iris-like or Bromelia-like habit, in the regularity of their flowers, which have frequently a woolly or papillose outer surface, and, finally, in the embryo being placed in *mealy* albumen. From Irideæ they are divided by the number of their stamens, by their anthers turning inwards, or, if their stamens are reduced to three by those organs then being opposite the petals, by their simple stigma, and by the texture of their albumen. From Bromeliaceæ, to which they approach by Barbacenia and Vellozia, they are known by being generally hexapetaloideous, not tripetaloideous. According to Mr. Don, the genera Vellozia, Barbacenia, and Xerophyta, probably constitute an intermediate group between the Hypoxideæ and Bromeliaceæ (*Jameson's Journal, Jan.* 1830, p. 166). Mr. Don finds the seeds of Barbacenia purpurea to be " compressed, cuneiform, and truncate at the apex, and narrowed towards the base, which is furnished with a protuberance arising from the elongation of the testa and umbilical cords. The testa is coriaceous, and marked outwardly with numerous shallow furrows." In this order, as well as in Gethyllis among Amaryllideæ, there are polyandrous species; a remarkable anomaly in monocotyledons, which rarely exceed the number 6 in their stamens. The Vellozias are singular in the tribe for their arborescent dichotomous trunks and tufted leaves.

GEOGRAPHY. Found in North America sparingly, abundantly at the Cape of Good Hope and in high land in Brazil, and 12 are described chiefly from the more temperate parts of New Holland.

PROPERTIES. M. Decandolle remarks, that the red colour found in the roots of Dilatris tinctoria in North America, where it is used for dyeing, prevails in Hæmodorum and Wachendorfia, and deserves to be studied in the rest of the order.

EXAMPLES. Hæmodorum, Conostylis, Dilatris, Lanaria.

CCXXXVIII. AMARYLLIDEÆ. The Narcissus Tribe.

NARCISSI, the second section, *Juss. Gen.* 54. (1789). — AMARYLLIDEÆ, *R. Brown Prodr.* 296. (1810); *Herbert Appendix to the Bot. Mag.* (1821); *Dec. and Duby,* 454. (1828); *Lindl. Synops.* 264. (1829).—NARCISSEÆ, *Agardh Aph.* 173. (1823.)

DIAGNOSIS. Hexapetaloideous bulbous hexandrous monocotyledons, with an inferior ovarium, a 6-parted perianthium with equitant sepals, and flat spongy seeds.

ANOMALIES. Gethyllis is polyandrous. Clivia and Doryanthes have fascicled roots.

ESSENTIAL CHARACTER.—*Calyx* and *corolla* confounded, superior, regular, coloured, the former overlapping the latter. *Stamens* 6, arising from the sepals and petals, sometimes cohering by their dilated bases into a kind of cup; sometimes an additional series of barren stamens is present, often forming a cup which surmounts the tube of the perianthium; *anthers* bursting inwardly. *Ovarium* 3-celled, the cells many-seeded, or sometimes 1- or 2-seeded; *style* 1; *stigma* 3-lobed. *Fruit* either a 3-celled, 3-valved *capsule*, with loculicidal dehiscence, or a 1-3-seeded berry. *Seeds* with either a thin and membranous, or thick and fleshy testa; *albumen* fleshy; *embryo* nearly straight, with its radicle turned towards the hilum.—Generally *bulbous*, sometimes *fibrous*-rooted. *Leaves* ensiform, with parallel veins. *Flowers* usually with spathaceous bracteæ.

AFFINITIES. The only orders with which this need be compared are Asphodeleæ and Liliaceæ, from which it is known by its inferior ovarium; Irideæ, which are distinguished by being triandrous, with the anthers turned outwards; and Hæmodoraceæ, which see. No one has ever thought of dismembering it, since Mr. Brown founded it upon Jussieu's 2d section of Narcissi; and it can scarcely be said to comprehend an anomalous genus, unless Clivia and Doryanthes be so considered, on account of their fascicled roots, and Gethyllis, because of its being polyandrous. The latter deviation from the ordinary character of the order will probably be considered of less importance, if we bear in mind the polyandrous structure of some Hæmodoraceæ, and especially if, in the first place, the genuine Amaryllideous genera Phycella and Placea be attended to, the former of which has a tendency to produce additional stamens, and the latter having them in a highly developed petaloid state; and if, secondly, the corona of Narcissus itself is borne in mind, which is in fact an organ representing an extra number of stamens. I have elsewhere remarked (*Bot. Reg.* 1341.) that this is connected with a strong tendency in the whole order to form another set of male organs between the perianthium and those stamens that actually develope. Hence a curious instance is exhibited, to which several parallels may, however, be found in other families, of the force of developement being generally confined to a series of organs originating within those which should be formed according to the ordinary laws of structure. Of course, in all such orders a multiplication of the usual number of stamens is more to be expected than where this peculiar circumstance does not exist.

GEOGRAPHY. A very few only are found in the north of Europe and the same parallel; these are plants of the genera Narcissus and Galanthus. As we proceed south they increase. Pancratium appears on the shores of the Mediterranean; Crinums and Pancratiums abound in the West and East Indies; Hæmanthus is found for the first time with some of the latter on the Gold Coast; Amaryllides shew themselves in countless numbers in Brazil, and across the whole continent of South America; and, finally, at the Cape of Good Hope the maximum of the order is beheld in all the beauty of Hæmanthus, Crinum, Clivia, Cyrtanthus, and Brunsvigia. A few are found in New Holland, the most remarkable of which is Doryanthes.

PROPERTIES. One of the few monocotyledonous orders in which any poisonous properties are found. These are principally apparent in the viscid juice of the bulbs of Hæmanthus toxicarius, in which the Hottentots are said to dip their arrow-heads, and in some neighbouring species. The bulbs of Narcissus poeticus have for ages been known as emetic; and it has recently been shewn by M. Loiseleur Deslongchamps that a similar power exists in Narcissus Tazetta, odorus, and Pseudo-Narcissus, and Pancratium maritimum. The flowers of Narcissus Pseudo-Narcissus are also said to be emetic. Decandolle considers the principle found in Amaryllideæ analogous to that of the Squill (*Essai*, p. 290). Sternbergia lutea is purgative, Alströmeria salsilla diaphoretic and diuretic, Amaryllis ornata astringent. *Agardh Aph.* 178.

EXAMPLES. Amaryllis, Phycella, Nerine, Vallota, Calostemma.

CCXXXIX. IRIDEÆ. THE CORNFLAG TRIBE.

IRIDES, *Juss. Gen.* 57. (1789).—ENSATÆ, *Ker in Ann. of Botany,* 1. 219. (1805).— IRIDEÆ, *R. Brown Prodr.* 302. (1810); *Ker. Gen. Irid.* (1827); *Dec. and Duby,* 451. (1828); *Lindl. Synops.* 254. (1829.)

DIAGNOSIS. Hexapetaloideous triandrous dicotyledons, with an inferior ovarium, anthers turned outwards, and equitant leaves.

ANOMALIES. Crocus leaves are not equitant.

ESSENTIAL CHARACTER.—*Calyx* and *corolla* superior, confounded, their divisions either partially cohering, or entirely separate, sometimes irregular, the 3 petals being sometimes very short. *Stamens* 3, arising from the base of the sepals; *filaments* distinct or connate; *anthers* bursting externally lengthwise fixed by their base, 2-celled. *Ovarium* 3-celled, cells many-seeded; *style* 1; *stigmas* 5, often petaloid, sometimes 2-lipped. *Capsule* 3-celled, 3-valved, with a loculicidal dehiscence. *Seeds* attached to the inner angle of the cell, sometimes to a central column, becoming loose; *albumen* corneous, or densely fleshy; *embryo* enclosed within it.—*Herbaceous* plants, or very seldom *undershrubs*, usually smooth; the hairs, if there are any, simple. *Roots* tuberous or fibrous. *Leaves* equitant, distichous, except in Crocus. *Inflorescence* terminal, in spikes, corymbs, or panicles, or crowded. *Bracteæ* spathaceous, the partial ones often scarious; the *sepals* occasionally rather herbaceous.

AFFINITIES. They differ from Amaryllideæ essentially, in being triandrous, with the anthers turned outwards; from Orchideæ, to which they approach very nearly in some respects, in not being gynandrous, and in all their anthers being distinct; from Scitamineæ and Marantaceæ their three perfect stamens divide them, independently of the structure of the leaves, which are extremely different. The Iris represents the general structure of the order; but a departure from the form of perianthium found in that genus takes place in the Crocus, the flower of which is extremely like that of Gethyllis and Sternbergia among Amaryllideæ on the one hand, and of Colchicum among Melanthaceæ on the other; the latter is known by its superior triple ovarium. The dilated stigma found in Iris is characteristic of the whole order; in Crocus it is rolled up instead of being spread open. Mr. Brown observes, that Burmannia appears at first sight to agree with Irideæ, especially in its equitant leaves, coloured superior triandrous perianthium, and 3 dilated stigmas: it cannot, however, be united with them, on account of its fertile stamens being opposite the inner segments of the perianthium, and alternating with an equal number of sterile ones, on account of the transverse dehiscence of the anthers, and also the structure of the seeds. In Xyris some resemblance with this order is discoverable, especially in the disposition of the leaves, the triandrous flowers, and anthers turned outwards; but that genus is very distinct in its inferior perianthium, the outer segments of which are glumaceous, and the inner distinctly petaloid, in the ungues bearing their stamens at the apex, in their sterile alternate stamens, and especially in the structure of the seed. *Prodr.* 302.

GEOGRAPHY. Principally natives either of the Cape of Good Hope, or of the middle parts of North America and Europe. A few only are found within the tropics, and the order is generally far from abundant in South America, if compared with the numbers that exist at the Cape. The genera Marica and Moræa appear to occupy the same station in hot climates that Iris, a closely related genus, does in cooler latitudes.

PROPERTIES. More remarkable for their beautiful fugitive flowers than for their utility. The rhizoma of some of them is slightly stimulating, as the violet-scented Orris root, the produce of Iris Florentina. A few, such as Iris tuberosa are purgative; and Iris versicolor and verna are used as cathartics in the United States. The substance called Saffron is the dried stigmas of a Crocus; the colouring ingredient is a peculiar principle, to which the name of Polychroite has been given. It possesses the remarkable properties of being totally destroyed by the action of the solar rays, of colouring in small quantity a large body of water, and of forming blue and green tints when treated with sulphuric and nitric acid, or with sulphate of iron. *Dec.* According to Mr. Gray, the roasted seeds of Iris pseud-acorus very nearly approach Coffee in quality. *Suppl. Pharmac.* 237.

EXAMPLES. Iris, Moræa, Ixia, Gladiolus.

CCXL. ORCHIDEÆ. The Orchis Tribe.

Orchides, *Juss. Gen.* 64. (1789). — Orchideæ, *R. Brown Prodr.* 309. (1810); *Rich. in Mém. Mus.* 4. 23. (1818); *Lindl. Synops.* 256. (1829); *Id. Genera and Species of Orch.* (1830.)

Diagnosis. Gynandrous monocotyledons, with 3 parietal placentæ.
Anomalies. Apostasia, if belonging to the order, has a trilocular ovarium and distinct stamens.

Essential Character. — *Perianthium* superior, ringent. *Sepals* 3, usually coloured, of which the odd one is uppermost in consequence of a twisting of the ovarium. *Petals* 3, usually coloured, of which 2 are uppermost in consequence of the twisting of the ovarium, and 1, called the *lip*, undermost; this latter is frequently lobed, of a different form from the others, and very often spurred at the base. *Stamens* 3, united in a central column, the 2 lateral usually abortive, the central perfect, or the central abortive, and the 2 lateral perfect; *anther* either persistent or deciduous, 2- or 4- or 8-celled; *pollen* either powdery, or cohering in definite or indefinite waxy masses, either constantly adhering to a gland or becoming loose in their cells. *Ovarium* 1-celled, with 3 parietal placentæ; *style* forming part of the column of the stamens; *stigma* a viscid space in front of the column, communicating directly with the ovarium by a distinct open canal. *Impregnation* taking effect by absorption from the pollen masses through the gland into the stigmatic canal. *Capsule* inferior, bursting with 3 valves and 3 ribs, very rarely baccate. *Seeds* parietal, very numerous; *testa* loose, reticulated, contracted at each end, except in one or two genera; *albumen* none; *embryo* a solid, undivided, fleshy mass. — *Herbaceous* plants, either destitute of a stem, or forming a kind of above-ground tuber (pseudo-bulbus) by the cohesion of the bases of the leaves, or truly caulescent. *Roots* in the herbaceous species fleshy, divided or undivided, or fasciculate; in the caulescent species tortuous, and green and proceeding from the stem. *Leaves* simple, quite entire, often articulated with the stem. *Pubescence* rare; when present, sometimes glandular. *Flowers* in terminal or radical spikes, racemes, or panicles; sometimes solitary.

Affinities. It is not necessary to enter, in this place, into an historical inquiry as to the gradual alteration that has taken place in the views of botanists with regard to the structure of the sexual apparatus of these most curious of plants, or to explain what degree of error existed in the descriptions of those who mistook masses of pollen for anthers, or a column of stamens for a style: such errors could only have occurred at a period when the laws of organisation were totally unknown. They have been corrected, in a more or less perfect manner, by various writers; most completely by Mr. Brown in his *Prodromus*, published in 1810, and subsequently by the late most accurate and indefatigable Richard. But long before the publication of any rational explanation of the structure of the Orchis tribe, while botanists were in utter darkness upon the subject, it had been most fully investigated by a gentleman unrivalled for the perfection of his microscopical analyses, the beauty of his drawings, and the admirable skill with which he follows Nature in her most secret workings; and let me add, which is a still rarer quality, the generous disinterestedness with which he communicates to his friends the result of his patient and silent labours. I have sketches before me by Mr. Bauer, executed from 1794 to 1807, in which not only all that has been published since that period is shewn in the most distinct and satisfactory manner, but in which much more is represented than botanists are even now aware of. I hope to be the humble means of giving some of these extraordinary productions of the pencil to the world, in an illustration of the Genera and Species of Orchideous Plants, which is now in preparation. If the sexual apparatus of an Orchideous plant is examined, it will be found to consist of a fleshy body stationed opposite the labellum, bearing a solitary anther at its apex, and having in front a viscid cavity, upon the upper edge of which there is often a slight callosity. This cavity is the stigma, and the callosity is the point through which the fertilising matter of the pollen passes into the tissue communicating with the ovules. Hence such a plant would appear to be monandrous; it will be seen, however, in Scitamineæ and Marantaceæ, the only other monandrous orders of Monocotyledons, that, while only one perfect stamen is developed, two others exist in a rudimentary state; so that the ternary number prevalent in Monocotyledons is not departed from. So it is in Orchideæ: the column does not consist of a single filament cohering with a style, but of three filaments firmly grown together, the central of which is antheriferous, the lateral sterile, or, as in Cypripedium, the central sterile, the two lateral antheriferous. This is proved, in the former case, by the frequent presence of callosities, or processes in the place of the sterile stamens; by imperfectly-formed anthers occasionally appearing at the side of the perfect one; and, if any further evidence were wanted, by monsters, in which a regular structure is exchanged for the ordinary irregular one. Such an instance in Orchis latifolia is described by M. Achille Richard, in the *Mémoires de la Soc. d'Hist. Nat. of Paris*, in which the flowers were perfectly triandrous, with no trace of irregularity in any part of the floral envelopes.

Orchideæ are remarkable for the bizarre figure of their multiform flower, which sometimes represents an insect, sometimes a helmet with the visor up, and sometimes a grinning monkey: so various are these forms, so numerous their colours, and so complicated their combinations, that there is scarcely a common reptile or insect to which some of them have not been likened. They all, however, will be found to consist of three outer pieces belonging to the calyx, and three inner belonging to the corolla; and all departures from this number, six, depends upon the cohesion of contiguous parts, with the solitary exception of Monomeria, in which the lateral petals are entirely abortive. Sometimes two of the sepals cohere into one, as in Cypripedium, and then the calyx has the appearance of consisting of but two sepals; sometimes the lateral petals are connate with the column, as in Gongora and probably Lepanthes, and then the column appears furnished with two wings. In nearly the whole order the odd petal, called the labellum, arises from the base of the column, and is opposite it; but in the Cape genus Pterygodium, the lip sometimes grows from the apex of the column, and sometimes is stalked and turned completely over between the fork of the inverted anther, and thus seems to belong to the back of the column. Nor is the anther less subject to modification, although constant to its place: sometimes it stands erect, the line of dehiscence of its lobes being turned towards the labellum; sometimes it is turned upside down, so that its back regards the lip; often it is prone upon the apex of the column, where a niche is excavated for its reception. The pollen is not less curious: now we have it in separate grains, as in other plants, but cohering to a meshwork of cellular tissue, which is collected into a sort of central elastic strap; now the granules cohere in small angular indefinite masses, and the central elastic strap becomes more apparent, has a glandular extremity, which is often reclined in a peculiar pouch especially destined for its protection; again the pollen combines into larger masses, which are definite in number, and attached to another modification of the elastic strap; and finally a complete union of the pollen takes place, in solid waxy masses, without any distinct trace of this central elastic tissue. Such is a part of the singularities of Orchideous plants, and upon these the distinctions of their tribes and genera are naturally founded. Whoever studies them must bear in mind that their fructification is always reducible to 3 sepals, 3 petals, a column consisting of 3 stamens grown firmly to one another, and to a single style and stigma; and, with this in view, he will have no difficulty in understanding the organisation of even the most anomalous Cape species. For a long time it was supposed that no deviation from the general structure existed, and that we had not in Orchideæ any very decided link between that family and others; but the discovery of a remarkable Indian plant by Blume and Wallich, called Apostasia by the former botanist, which, with many of the peculiarities of Orchideæ, is triandrous with a regular corolla and 3-locular fruit, seems to shew that even in this tribe there are gradations which tend to destroy the value of the technical differences of botanists. It does not, however, appear to me certain that this genus, although referred to Orchideæ by Blume, is not really of a different tribe.

If the following diagram be compared with those employed to illustrate the distinctions of Marantaceæ and Scitamineæ, p. 269, the relation borne to those orders by Orchideæ will be distinctly seen. In the diagram the parts are arranged as they are in nature before the ovarium twists; that is, with the labellum next the axis, or uppermost, and the stamen undermost. Let C, C, C represent the outer series of floral envelopes or calyx, and PP, P, P the inner, or corolla, of which PP is the labellum ; then the position of the single fertile stamen will be at S, and of the sterile ones at *s, s*; that is to say, in the situation of the supernumerary petaloid stamens of Scitamineæ and Marantaceæ, while the second series of stamens, to which the fertile stamen of these orders belongs, is not developed in Orchideæ.

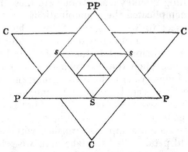

GEOGRAPHY. Found in almost all parts of the world, except upon the verge of the frozen zone, and in climates remarkable for dryness. In Europe, Asia, and North America, they are found growing every where, in groves, in marshes, and in meadows; in the drier parts of Africa they are either rare or unknown; at the Cape of Good Hope they abound in similar situations as in Europe; but in the hot damp parts of the West and East Indies, in Madagascar, and the neighbouring islands, in the damp and humid forests of Brazil, and on the lower mountains of Nipal, these Orchideous plants flourish in the greatest variety and profusion, no longer seeking their nutriment from the soil, but clinging to the trunks and limbs of trees, to stones and bare rocks, where they vegetate among ferns and other shade-loving plants, in countless thousands. Of the epiphytic class, one only is found so far north as South Carolina, growing upon the branches of the Magnolia, if we except the species from Japan, which, as I have elsewhere stated, appears to have a climate peculiar to itself, among countries in the same parallel of latitude. The number of species of this tribe is unknown, but probably is not less than 1500.

PROPERTIES. It often happens that those productions of nature which charm the eye with their beauty, and delight the senses with their perfume, have the least relation to the wants of mankind, while the most powerful virtues or most deadly poisons are hidden beneath a mean and insignificant exterior: thus Orchideæ, beyond their beauty, can scarcely be said to be of known utility, with a few exceptions. The nutritive substance called Salep is prepared from the subterraneous succulent roots of Orchis mascula and others: it consists almost entirely of a chemical principle called Bassorin. *Turner*, 699. The root of Bletia verecunda is said to be stomachic. *Lunan*. And some of the South American species, such as the Catasetums, Cyrtipodiums, &c., contain a viscid juice, which, being inspissated by boiling, becomes a kind of vegetable glue used for economical purposes in Brazil. The aromatic substance called Vanilla is the succulent fruit of a climbing West Indian plant of the order.

EXAMPLES. The following are the sections proposed in my *Orchidearum Sceletos* (1826).

§ I. Pollen simple, or consisting of granules in a lax state of cohesion.

Tribe 1. NEOTTIEÆ. Anther parallel with the stigma, and erect. (Goodyera, Spiranthes.)

Tribe 2. ARETHUSEÆ. Anther terminal, opercular. (Pogonia, Epipactis.)

§ II. Pollen cohering in granules, which finally become waxy, and are indefinite in number.

Tribe 3. GASTRODIEÆ. Anther terminal, opercular. (Gastrodia, Vanilla.)

Tribe 4. OPHRYDEÆ. Anther terminal, erect or inverted. Pollen masses with a caudicula. (Orchis, Ophrys.)

§ III. Pollen cohering in grains, which finally become waxy, and are definite in number.

Tribe 5. VANDEÆ. Pollen-masses attached to the stigma by a transparent caudicula and gland. (Oncidium, Brassia.)

Tribe 6. EPIDENDREÆ. Pollen-masses attached to the stigma by filiform, powdery, reflexed caudiculæ. (Bletia, Epidendrum.)

Tribe 7. MALAXIDEÆ. Pollen-masses loose, sometimes cohering at the apex by a viscid, or powdery, or granular matter. (Malaxis, Dendrobium.)

§ IV. Lateral anthers, fertile; the middle one sterile and petaloid.

Tribe 8. CYPRIPEDIEÆ. (Cypripedium.)

CCXLI. SCITAMINEÆ. THE GINGER TRIBE.

CANNÆ, *Juss. Gen.* 62. (1789), *in part.* — DRYMYRHIZEÆ, *Vent. Tabl.* (1799); *Dec. Ess. Med.* 281. (1816). — SCITAMINEÆ, *R. Brown Prodr.* 305. (1810); *Agardh Aph.* 182. (1823); *Rosc. Monogr.* — ZINGIBERACEÆ, *Rich. Anal. Fr.* (1808). — AMOMEÆ, *Juss. in Mirbel's Elém.* 854. (1815); *Ach. Rich. Nouv. Elém. ed.* 4. 438. (1828). — ALPINIACEÆ, *Link Handb.* 1. 228. (1829), *a* § *of* Scitamineæ.

DIAGNOSIS. Tripetaloideous monocotyledons, with a single 2-celled anther.

ANOMALIES. Hellenia abnormis has a unilocular monospermous ovarium.

ESSENTIAL CHARACTER. — *Calyx* superior, tubular, 3-lobed, short. *Corolla* tubular, irregular, with 6 segments in 2 whorls; the *outer* 3-parted, nearly equal, or with the odd segment, sometimes differently shaped; the *inner* (sterile stamens) 3-parted, with the intermediate segment (*labellum*) larger than the rest, and often 3-lobed, the lateral segments sometimes nearly abortive. *Stamens* 3, distinct, of which the 2 lateral are abortive, and the intermediate 1 fertile; this placed opposite the labellum, and arising from the base of the intermediate segment of the outer series of the corolla. *Filament* not petaloid, often extended beyond the anther in the shape of a lobed or entire appendage. *Anther* 2-celled, opening longitudinally, its lobes often embracing the upper part of the style. *Pollen* globose, smooth. *Ovarium* 3-celled, sometimes imperfectly so; *ovula* several, attached to a placenta in the axis; *style* filiform: *stigma* dilated.

hollow. *Fruit* usually capsular, 3-celled, many-seeded; occasionally berried (the dissepiments generally central, proceeding from the axis of the valves, at last usually separate from the latter, and of a different texture. *R. Br.*) *Seeds* roundish, or angular, with or without an arillus (*albumen* floury, its substance radiating, and deficient near the hilum. *R. Br.*); *embryo* enclosed within a peculiar membrane (*vitellus*, R. Br. Prodr.; *membrane of the amnios*, ibid. in King's Voyage, 21), with which it does not cohere. — Aromatic tropical *herbaceous* plants. *Rhizoma* creeping, often jointed. *Stem* formed of the cohering bases of the leaves, never branching. *Leaves* simple, sheathing, their *lamina* often separated from the sheath by a taper neck, and having a single midrib, from which very numerous, simple, crowded veins diverge at an acute angle. *Inflorescence* either a dense spike, or a raceme, or a sort of panicle, terminal or radical. *Flowers* arising from among spathaceous membranous bracteæ, in which they usually lie in pairs.

AFFINITIES. Formerly Scitamineæ and Marantaceæ were united in one tribe called Canneæ, and this is even still followed by some botanists : hence it is certain that they are at least more nearly related to each other than to any thing else, and that whatever is the affinity of the one will be that of the other. Taking the vegetation into account, these two tribes are exceedingly nearly allied to Musaceæ, in which is found the same kind of leaf, the veins of which are closely set, and diverge from the midrib to the margin, being connected by very weak and imperfect intermediate veins; the leaves have also the same distinct petiole, often with a thickened rounded space at the apex; Musaceæ are, however, pent- or hexandrous, with a calyx and corolla of the same texture. Irideæ are the next order with which Scitamineæ may be compared, agreeing in their superior flowers, which have sometimes an approach to the irregularity of Alpinia and the like, and also in the triple number of their stamens; but while these organs are all developed in Irideæ, two are abortive or deformed in both Scitamineæ and Marantaceæ. Bromeliaceæ have been identified with them of old, but their resemblance consists chiefly in the distinction of calyx and corolla, and their inferior ovarium. To Orchideæ they are related in consequence of the reduction of their three stamens to one by the abortion of two; but the cohesion of the stamens and style in the latter, and the want of any distinction between calyx and corolla, sufficiently separate them, besides which the series which produces the stamens in Orchideæ answers to the sterile stamens or inner limb of the corolla in Scitamineæ. For the differences between Scitamineæ and Marantaceæ, see the latter. There is a fine volume consecrated to plants of these two tribes by Mr. Roscoe, who first remodelled the genera and reduced them within certain limits. Between the embryo and the albumen is interposed a fleshy body enveloping the former : this has been called a process of the rostellum by Correa, a cotyledon by Smith, a vitellus by Gærtner and Brown, a central indurated portion of the albumen by Richard. It is now known to be the innermost integument of the ovulum, unabsorbed during the advance of this body to maturity.

Independently of the presence of this vitellus, the most remarkable part of the structure of Scitamineæ consists in the number of divisions of the floral envelopes, which consist of a tubular calyx, and of two more series instead of one. Mr. Brown, struck with this unusual deviation from the ordinary organisation of Monocotyledons, was disposed to consider the calyx an accessory part (*Prodr.* 305); but M. Lestiboudois' explanation appears more satisfactory. According to this botanist (as quoted in Ach. Richard's *Nouv. Elém.* 439), Scitamineæ are really hexandrous, like the nearly-related Musaceæ; but of their stamens the outer series is petaloid, and forms the inner limb of the corolla, and of the inner series of stamens the central one only developes, the lateral ones appearing in the form of rudimentary scales. This notion of M. Lestiboudois is confirmed by Marantaceæ, in which the inner stamens (even that which is antheriferous) become petaloid like the outer ; thus shewing that in these plants there is a strong and general tendency in the filaments to assume the state of petals.

GEOGRAPHY. All tropical, or nearly so. By far the greater number inhabit various parts of the East Indies; some are found in Africa, and a few in America. They form a part of the singular Flora of Japan.

PROPERTIES. Generally objects of great beauty, either on account of the high degree of developement of the floral envelopes, as in Hedychium coronarium and Alpinia nutans; or because of the rich and glowing colours of the bracteæ, as in Curcuma Roscoeana (*Wallich Plant. As. Rar.* vol. 1. tab. 9.) They are, however, principally valued for the sake of the aromatic stimulating properties of the roots or rhizoma, such as are found in Ginger (Zingiber officinalis), Galangale (Alpinia racemosa and Galanga), Zedoary (Curcuma Zedoaria and Zerumbet), and many other species of the latter genus. The warm and pungent roots of the greater and lesser Galangale are not only used by the Indian doctors in cases of dyspepsia, but are also considered useful in coughs, given in infusion. *Ainslie*, 1. 141. The seeds of many partake of the properties of the root. Cardamoms are the seeds of several plants of this order. On the eastern frontiers of Bengal the fruit of Amomum aromaticum is used ; the lesser Cardamom of Malabar is the Elettaria Cardamomum ; another sort is the produce of Amomum maximum ; and the greater Cardamoms are yielded by the Amomum Granum Paradisi. Others are known for their dyeing properties, such as Turmeric. This substance, obtained from Curcuma longa, is cordial and stomachic ; it is also considered by the native practitioners of India an excellent application in powder for cleaning foul ulcers. *Ibid.* 1. 455. The fruit of Globba uviformis is said to be eatable. Generally, in consequence of the presence of the aromatic oil that is so prevalent in the order, the roots or rhizomas, although abounding in fæcula, are not fit for the preparation of arrow-root; but an excellent kind is prepared in Travancore, in the East Indies, from Curcuma angustifolia. *Ibid.* 1. 19.

EXAMPLES. Amomum, Zingiber, Alpinia, Hellenia, Kæmpferia.

CCXLII. MARANTACEÆ. THE ARROW-ROOT TRIBE.

CANNÆ, *Juss. Gen.* 62. (1789) *in part.* — CANNEÆ, *R. Brown Prodr.* 1. 307. (1810); *Lindl. in Bot. Reg.* 932. (1825). — CANNEÆ *or* MARANTEÆ, *Brown in Flinders* (1814). — CANNACEÆ, *Agardh Aph.* 181. (1823); *Link Handb.* 1. 223. (1829), *a* § *of* Scitamineæ.

DIAGNOSIS. Tripetaloideous monocotyledons, with a single 1-celled anther, and a petaloid filament.

ANOMALIES. The ovarium of Thalia is monospermous.

ESSENTIAL CHARACTER. — *Calyx* superior, of 3 sepals, short. *Corolla* tubular, irregular, with the segments in 2 whorls ; the *outer* 3-parted, nearly equal ; the *inner* very irregular ; one of the lateral segments usually coloured, and formed differently from the rest ; sometimes by abortion fewer than 3. *Stamens* 3, petaloid, distinct, of which one of the laterals and the intermediate one are either barren or abortive, and the other lateral one fertile. *Filament* petaloid, either entire or 2-lobed, one of the lobes bearing the anther on its edge. *Anther* 1-celled, opening longitudinally. *Pollen* round (papillose in Canna coccinea, smooth in Calathea zebrina). *Ovarium* 3-celled ; *ovula* solitary and erect, or numerous and attached to the axis of each cell ; *style* petaloid or swollen ; *stigma* either the mere denuded apex of the style, or hollow, cucullate, and incurved. *Fruit* capsular, as in Scitamineæ. *Seeds* round, without arillus ; *albumen* hard, somewhat floury ; *embryo* straight, naked, its *radicle* lying against the hilum. — *Herbaceous* tropical plants, destitute of aroma. *Rhizoma* creeping, abounding in a nutritive fæcula. *Stem* often branching. *Leaves, inflorescence,* and *flowers*, as in Scitamineæ.

AFFINITIES. Under Scitamineæ, the relations of that order and the present to other monocotyledonous groups has been noticed. In this place the distinction between the two orders has to be explained. Mr. Brown was the first to propose the separation of them, in which he has not been followed generally; a circumstance which has possibly arisen from a belief that Marantaceæ differed from Scitamineæ only in the absence of aroma and vitellus, and in the imperfection of their anther. But, as I have formerly stated in the *Botanical Register,* folio 932, the distinction of the two orders depends upon a much more important consideration than either of these. In true Scitamineæ, as Mr. Brown has observed (*Prodr.* 305.), the stamen is always placed opposite the labellum or anterior division of the inner series of the corolla, and proceeds from the base of the posterior outer division; while the sterile stamens, when they exist, are stationed right and left of the labellum. But in Marantaceæ the fertile stamen is on one side of the labellum, occupying the place of one of the lateral sterile stamens of Scitamineæ. This peculiarity of arrangement indicates a higher degree of irregularity in Marantaceæ than in Scitamineæ, which also extends to the other parts of the flower. The suppression of parts takes place in the latter in a symmetrical manner; the two posterior divisions of the inner series of the perianthium, which are occasionally absent, corresponding with the abortion of the two anterior stamens. In Marantaceæ, on the contrary, the suppression of organs takes place with so much irregularity, that the relation which the various parts bear to each other is not always apparent: instead of the central stamen being perfect while the two lateral ones are abortive, as in Scitamineæ and most Orchideæ, or of the central stamen being abortive and the two lateral ones perfect, as in some Orchideæ, it is the central and one lateral one that are suppressed in Marantaceæ. In the perianthium of Canna only the most external within the calyx can properly be called corolla; the remainder of the segments being attempts to produce barren petaloid stamens analogous to what is called the inner limb of the corolla in Scitamineæ; and the characters upon which botanists found their specific distinctions depend upon the degree to which this developement of petaloid abortive stamens extends. When, for instance, they describe some as having an inner limb of 2 or of 3, or of 4 or of 5 segments, they should rather say 2, 3, 4, or 5 stamens are partially developed. For remarks upon the proof thus afforded of the affinity of Scitamineæ and Marantaceæ to Musaceæ, see the former order.

Perhaps it will be possible to put the relative structure of Scitamineæ and Marantaceæ in a clearer light by the following diagrams, in which the triangle C, C, C represents the calyx, the angles corresponding with the position of the sepals; the triangle P, P, P the corolla; R, *r, r* an outer series of petaloid stamens, of which *r, r* are rudimentary only; and S, *s, s* the inner series of stamens, of which S is the fertile and fully developed one.

SCITAMINEÆ.

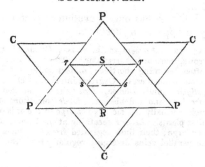

MARANTACEÆ.

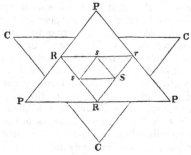

Agardh describes the albumen of Canna as a fungous elastic substance, formed of densely compact hyaline granules, white internally, gradually passing through yellow and brown into black, and more analogous to an internal membrane than to albumen, because it undergoes little change during germination. But the albumen is better understood now than in 1823. See Introduction, and *Outline of the First Principles of Botany*, par. 494, &c.

GEOGRAPHY. The greater part are found in tropical America and Africa; several are natives of India; some are known in a wild state beyond the tropics.

PROPERTIES. While the ginger tribe (Scitamineæ) are valued for their aromatic heating principle, the arrow-root tribe (Marantaceæ) is esteemed on account of the fæcula, which abounds in the rhizoma and root of both tribes, being destitute of that principle: on this account it is collected as a delicate article of food, both from Maranta arundinacea, Allouyia, and nobilis in the West Indies, and also from Maranta ramosissima in the East. The fleshy cormus of some Cannas is reported to be eaten in Peru. A tough fibre is obtained from Phrynium dichotomum; and the leaves of the South American Calatheas are worked into baskets, whence their name. The juice of Maranta arundinacea is said to be efficacious in poisoned wounds. *Agdh.*

EXAMPLES. Canna, Maranta, Calathea, Phrynium.

CCXLIII. MUSACEÆ. THE BANANA TRIBE.

MUSÆ, *Juss. Gen.* (1789). — MUSACEÆ, *Agardh Aph.* 180. (1823) ; *Ach. Rich. Nouv. Elém. ed.* 4. 436. (1828.)

DIAGNOSIS. Hexapetaloideous sub-hexandrous spathaceous monocotyledons, with an inferior ovarium, and leaves with veins diverging from the midrib to the margin.

ANOMALIES. Heliconia has only 1 ovulum in each cell. The lamina of the leaf occasionally disappears in Strelitzia.

ESSENTIAL CHARACTER.— *Flowers* spathaceous. *Perianthium* 6-parted, superior, petaloid, in 2 distinct rows, more or less irregular. *Stamens* 6, inserted upon the middle of the divisions, some often becoming abortive; *anthers* linear, turned inwards, 2-celled, often having a membranous petaloid crest. *Ovarium* inferior, 3-celled, many-seeded, rarely 3-seeded; *style* simple; *stigma* usually 3-lobed. *Fruit* either a 3-celled capsule with a loculicidal dehiscence, or succulent and indehiscent. *Seeds* sometimes surrounded by hairs, with an integument which is usually crustaceous; *embryo* in the axis of mealy albumen. — Stemless or nearly stemless plants. *Leaves* sheathing at the base, and forming a kind of spurious stem; often very large; their limb separated from the taper petiole by a round tumour, and having fine parallel veins diverging regularly from the midrib towards the margin.

AFFINITIES. These have been pointed out under Scitamineæ and Marantaceæ, with which the Banana tribe is strictly related. Agardh characterises it as gynandrous (*l. c.*), but it does not appear upon what principle. The flower of Musa is well described in the Appendix to the *Congo Expedition*, 471., in a note: that of Strelitzia is pentandrous and exceedingly irregular, and is admirably illustrated in Mr. Bauer's drawings, published some years since by Mr. Ker, under the title of *Strelitzia Depicta*. The hilum of the seed gives rise to a tuft of long hairs in Urania and Strelitzia.

GEOGRAPHY. Natives of the Cape of Good Hope, the islands of its south-east coast, and generally of the plains of the tropics, beyond which they do not naturally extend, unless in Japan, the climate of which seems to be much at variance with that of other countries in the same latitude.

PROPERTIES. Most valuable plants, both for the abundance of nutritiv food afforded by their fruit, and for the many domestic purposes to which the gigantic leaves of some species are applied. These are used for thatching Indian cottages, for a natural cloth from which the traveller may eat his food as a material for basket making, and finally they yield a most valuable flax (Musa textilis), from which some of the finest muslins of India are prepared. The stems are formed of the united petioles of the leaves, which are remarkable for the vast quantity of spiral vessels they contain: these exist in such numbers as to be capable of being pulled out by handfuls, and they are actually collected in the West Indies and sold as a kind of tinder. *Dec. Org.* 38. The number of threads in each convolution of these spiral vessels varies from 7 to 22. *Ibid.* 37. The young shoots of the Banana are eaten as a delicate vegetable. The root of Heliconia Psittacorum, and the seed of Urania speciosa, are said to be eatable. The juice of the fruit and the lymph of the stem of Musa are slightly astringent and diaphoretic. The juice of the fruit of Urania is used for dying. *Agdh.*

EXAMPLES. Musa, Heliconia, Strelitzia, Urania.

CCXLIV. JUNCEÆ. THE RUSH TRIBE.

JUNCI, *Juss. Gen.* (1789), *in part.*—JUNCEÆ, Dec. Fl. Fr. 3. 155. (1815); *R. Brown Prodr.* 257. (1810); *Dec. and Duby*, 474. (1828); *Lindl. Synops.* 273. (1829).— JUNCACEÆ, *Agardh Aphor.* 156. (1823), *in part.*

DIAGNOSIS. Hexapetaloideous herbaceous monocotyledons, with a superior ovarium, a half-glumaceous regular perianthium, a pale soft testa, a single style, capsular fruit, and an embryo next the hilum.

ANOMALIES. Flowers sometimes scarcely glumaceous.

ESSENTIAL CHARACTER.—*Flowers* hermaphrodite or unisexual. *Calyx* and *corolla* forming an inferior, 6-parted, more or less glumaceous *perianthium*. *Stamens* 6, inserted into the base of the segments; sometimes 3, and then opposite the calyx. *Anthers* 2-celled. *Ovarium* 1- or 3-celled, 1- or many-seeded, or 1-celled and 3-seeded. *Style* 1. *Stigmas* generally 3, sometimes only 1. *Fruit* capsular, with 3 valves, which have the dissepiment in their middle, sometimes destitute of valves, and 1-seeded by abortion. *Seeds* with a testa, which is neither black nor crustaceous; *albumen* firm, fleshy, or cartilaginous; *embryo* within it. *R. Br.* (1810.)—*Herbaceous* plants, with fascicled or fibrous roots. *Leaves* fistular, or flat and channelled with parallel veins. *Inflorescence* often more or less capitate. *Flowers* generally brown or green.

AFFINITIES. This order, in its most genuine state, may be said to stand between Petaloideous and Glumaceous Monocotyledons, agreeing with the former in the floral leaves having assumed the verticillate state necessary to constitute a perianthium, and with the latter in their texture. But while a glumaceous confounded calyx and corolla are the characteristic of one part of the order, another part, approaching Asphodeleæ, assumes a petaloid state; so that little is finally left to separate Junceæ from the latter, except the difference in the testa of their seed. Mr. Brown remarks that Junceæ are intermediate between Restiaceæ and Asphodeleæ, differing from the former in having an included embryo, a radicle usually centripetal, and the stamens, when there are only 3, opposite the sepals; from Asphodeleæ in the integument of the seed, in the texture of the perianthium, and in habit. *Prodr.* 258. Agardh combines Restiaceæ and Junceæ. *Aph.* 157.

From Palms they are distinguished, independently of their habit, by the texture of the perianthium, by the constant tendency to produce more than 1 ovulum in each cell, and by the embryo never being remote from the hilum. Juncus is an instance of a monocotyledonous plant having distinct pith. "Xerotes, in the structure and appearance of its flowers, and in the texture of albumen, has a considerable resemblance to Palms, but it wants the peculiar characters of the seed, and also the habit of that remarkable order. Flagellaria differs from Xerotes chiefly in its pericarpium, and in the form and relation of its embryo to the albumen, which is also of a different texture. In all these respects it approaches to Cyperaceæ, with some of whose genera it has even a certain resemblance in habit." *Brown in Flinders,* 578. From Melanthaceæ they are known by their concrete carpella, and anthers turned inwards.

GEOGRAPHY. Chiefly found in the colder parts of the world, some even in the coldest, two existing in the ungenial climate of Melville Island. Several, however, are known in the tropics. Eight are mentioned as inhabiting the tropical parts of New Holland alone. According to Humboldt (*Diss. Geogr.* 43), they constitute $\frac{1}{400}$ of the flowering plants in the equinoctial zone; in the temperate zone, $\frac{1}{90}$; in the frozen zone, $\frac{1}{25}$; in North America, $\frac{1}{150}$; in France, $\frac{1}{86}$. In Sicily, according to Presl, they do not form more than $\frac{1}{300}$.

PROPERTIES. Only employed for mechanical purposes, as the Rush and others for making the bottoms of chairs, &c.; the pith of the same for the wick of common candles. Juncus effusus is cultivated in Japan for making floor-mats. *Thunb.* The leaves of Flagellaria are said to be astringent and vulnerary.

EXAMPLES. Juncus, Luzula, Dasypogon.

CCXLV. MELANTHACEÆ. The Colchicum Tribe.

Melantheæ, *Batsch. Tab. Aff.* (1802). — Colchicaceæ, *Dec. Fl. Fr.* 3. 192. (1815);
Ess. Méd. 298. (1816). — Melanthaceæ, *R. Brown Prodr.* 272. (1810); *Lindl.
Synops.* 264. (1829); *Dec. and Duby*, 473. (1828). — Veratreæ, *Salisb. in Hort.
Trans.* 1. 328. (1812); *Agardh Aphor.* 166. (1823). — Merenderæ, *Mirb. accord-
ing to Decandolle.*

Diagnosis. Hexapetaloideous monocotyledons, with nearly separate
carpella, and anthers turned outwards.

Anomalies. Campynema has an inferior ovarium.

Essential Character. — *Perianthium* inferior, petaloid, in 6 pieces, or, in consequence of the cohesion of their claws, tubular; the pieces generally involute in æstivation.
Stamens 6; *anthers* mostly turned outwards. *Ovarium* 3-celled, many-seeded; *style* trifid
or 3-parted; *stigmas* undivided. *Capsule* generally divisible into 3 pieces; sometimes with
a loculicidal dehiscence. *Seeds* with a membranous testa; *albumen* dense, fleshy. *R. Br.*
—*Roots* fibrous, sometimes fascicled. *Rhizoma* sometimes fleshy. *Leaves* sheathing at the
base, with parallel veins. *Flowers* either arising from under the surface of the ground,
or arranged upon tall leafy stems in large panicles, or disposed in spikes or racemes upon a
naked scape.

Affinities. Mr. Brown, who restored this tribe, considers its station
to be between Asphodeleæ and Junceæ, from both which it is known by its
tripartible fruit, and anthers turned outwards. The genera differ very much
in habit, which renders it doubtful whether some further change in the order
will not be necessary. Their properties are more uniform than their appearance.

Geography. Frequent at the Cape of Good Hope, not uncommon in
Europe, Asia, and North America, and existing in the tropics of India and
New Holland, this order appears to be confined within no geographical
limits; it is, however, far more abundant in northern countries than elsewhere.

Properties. Poisonous in every species, but more especially in the
Colchicum and Veratrum. The cormus of the former is a well-known acrid
cathartic, narcotic, and diuretic; the latter is a nauseous dangerous emetic.
The medicinal properties of the root of Veratrum are, owing to a peculiar
alkaline principle, called Veratrin, which acts with singular energy on the
membrane of the nose, exciting violent sneezings, though taken in very minute
quantity. When taken internally in very small doses, it produces excessive
irritation of the mucous coat of the stomach and intestines; and a few grains
are found fatal to the lower animals. *Turner*, 652. Veratrum viride of
North America is an acrid, emetic, and powerful stimulant, followed by
sedative effects. *Bigelow*, 2. 125. Veratrin is found in the root of the
Colchicum. *Turner*, 652. Gloriosa superba is recorded to possess similar acrid powers. The root of Helonias dioica in infusion is anthelmintic,
but its tincture is bitter and tonic. *Dec.*

Examples. Colchicum, Melanthium, Uvularia, Bulbocodium, Tofieldia.

CCXLVI. PONTEDEREÆ.

Pontedereæ, *Kunth in Humb. et Bonpl. N. G.* 1. 211. (1815); *Agardh Aph.* 169.
(1823); *Hooker in Bot. Mag.* 2932. (1829). — Pontederiaceæ, *Ach. Rich.
Nouv. Elém. ed.* 4. 427. (1828.)

Diagnosis. Hexapetaloideous monocotyledons, with a superior ovarium and irregular perianthium, involute after flowering.

Anomalies.

ESSENTIAL CHARACTER. — *Perianthium* tubular, coloured, 6-parted, more or less irregular, with a circinate aestivation. *Stamens* 3 or 6, unequal, arising from the calyx. *Ovarium* superior, or rarely half inferior, 3-celled, many-seeded; *style* 1; *stigma* simple. *Capsule* 3-celled, 3-valved, with loculicidal dehiscence. *Seeds* indefinite, attached to a central axis; *hilum* small; *embryo* orthotropous, in the axis of somewhat mealy *albumen.*— *Aquatic* or *marsh*-plants. *Leaves* sheathing at the base, with parallel veins. *Flowers* either solitary, or in spikes or umbels, spathaceous, frequently blue.

AFFINITIES. These were referred to Commelineæ by Mr. Salisbury, and are considered nearly related to that order by M. Ach. Richard, who, however, separates them, suggesting their being referable to the great receptacle of miscellaneous monocotyledons called Liliaceæ. It is not improbable that the nearest relation of Pontedereæ is with Asphodeleæ, (to which Link actually refers Pontedera) and Butomeæ, from both which they are known by their irregular flowers rolling inwards after expansion, independently of more minute characters derived from the structure of the seeds and fruit. Dr. Hooker, who has given an excellent figure of Pontederia azurea, states that each fibre of the roots has a calyptrate covering at the extremity, similar to that found on the roots of the Duck-weed.

GEOGRAPHY. Water-plants found exclusively in North and South America, the East Indies, and tropical Africa.

PROPERTIES. Plants with neat deep green leaves and showy flowers; of no known use.

EXAMPLES. Pontederia, Heteranthera.

CCXLVII. ASPHODELEÆ. THE ASPHODEL TRIBE.

ASPARAGI and ASPHODELI, *of Juss. chiefly,* (1789).—ASPHODELEÆ, *R. Brown Prodr.* 275. (1810). *Dec. and Duby,* 463. (1828) *a section of* Liliaceæ; *Lindl. Synops.* 266. (1829).—ALLIACEÆ, ALOINÆ, HYACINTHINÆ, DRACÆNACEÆ, *Link Handb.* vol. 1. (1829), *all sections of* Liliaceæ.—ASPARAGINÆ, *Ib.* 272. (1829.)

DIAGNOSIS. Hexapetaloideous monocotyledons, with a superior ovarium, anthers turned inwards, a coloured perianthium, a 3-celled fruit, a hard black brittle testa, and an undivided style.

ANOMALIES. Tricoryne has three distinct carpella.

ESSENTIAL CHARACTER.—*Calyx* and *corolla* forming a 6-parted or 6-cleft, petaloid, regular *perianthium*. *Stamens* 6, inserted upon the perianthium, or hypogynous; the 3 opposite the sepals sometimes either unlike the rest or wanting. *Ovarium* superior, 3-celled, with 2- or many-seeded cells; *ovules* when 2, ascending; *style* 1; *stigma* entire, or with 3 short lobes. *Fruit* mostly a 3-celled, 3-valved *capsule*, with a loculicidal dehiscence; occasionally succulent, and sometimes 3-parted. *Seeds* with a testa, which is black, brittle, and crustaceous; *albumen* fleshy; *embryo* included.—*Herbaceous* plants, or occasionally *trees*, with bulbs, or fascicled roots. *Leaves* with parallel veins. *Peduncles* articulated in the middle. *Flowers* coloured.

AFFINITIES. There is really no other absolute distinction between these and Junceæ on the one hand, than their more petaloid perianthium and hard brittle testa; or Liliaceæ, on the other, than their smaller flowers and testa. They are, nevertheless, properly established as an independent order, occupying a higher place in the scale of developement than the Rush tribe, and a lower than that of Lilies. From Melanthaceæ they are known by their anthers not being turned outwards; from Smilaceæ, their simple undivided style, narrow leaves, erect habit, and hard brittle testa, are marks of separation; at least it seems that, unless the two tribes are to be so distinguished, they must be considered the same. By some they are actually united; by others different limits have been sought; but the baccate and capsular genera can by no means be collected into two groupes. Mr. Brown justly remarks (*Prodr.* 275), that there is very commonly in this tribe an articulation in the middle, or at the apex of the peduncle, which is scarcely found in any of the neighbouring tribes, except in some Aneilemas, among Commelineæ, and in Sanseviera, a genus usually referred to Asphodeleæ, but which Mr. Brown appears to consider belonging to some other tribe, without stating to what, perhaps to his Hemerocallideæ, which are understood here to be the same as Liliaceæ. The greatest confusion exists in authors as to the limits of the orders near Asphodeleæ, particularly in regard to those now mentioned.

GEOGRAPHY. Scattered widely over the world; but much more abundant in temperate climates than in the tropics, where they chiefly exist in an arborescent state. Aloes are mostly found in the southern parts of Africa. One species is a native of the West Indies, and two or three more of Arabia and the East. Dracænas, the most gigantic of the order, attain their largest size in the Canaries. A Dracæna Draco is described in the *Annales des Sciences*, 14. 140. as being between 70 and 75 feet high, 46½ feet in circumference at the base, and it was known to have been a very ancient tree in the year 1496. The northern Flora comprehends for the most part plants of the genera Scilla, Hyacinthus, Allium, and Ornithogalum. In the East Indies Asphodeleæ are rare; in New Holland they form a distinctly marked feature of the vegetation.

PROPERTIES. The tribe consists almost entirely of beautiful flowers, general favourites in gardens. A bitter stimulant principle, contained in a gummy viscid juice, prevails in all, differing in the species chiefly in regard to its quantity and degree of concentration. The bulb of the Scilla maritima is nauseous and acrid; it acts either as an emetic, purgative, or expectorant and diuretic, in proportion to the dose in which it is given. Its properties are said to be due to a peculiar principle, called by M. Vogel, Scillitin. The Onion, Garlic, Shallot, Chive, Rocambole, all species of Allium, agree in their stimulant, diuretic, and expectorant effects, differing in their degree of activity. According to Dr. A. T. Thomson, the virtues of the genus Allium depend on an acrid principle, soluble in water, alcohol, acids, and alkalies. *Conspectus*, p. 9. In consequence of the free phosphoric acid which the common Onion bulbs contain, they are supposed to be useful in calculous cases. *Ibid.* Aloes act in like manner as stimulants, to which they owe their remarkable cathartic powers. Soccotrine Aloes, so called from being produced in Zocotora, are obtained from Aloë spicata, *Linn.* An inferior sort, sold in the East Indian bazars, is supposed to be the produce of Aloë perfoliata. *Ainslie*, 1. 9. This is the Barbadoes Aloes, or Hepatic Aloes of the shops. The root of Dracæna terminalis is considered by the Javanese a valuable medicine in dysenteric affections. *Ibid.* 2. 20. The juice of Dracæna Draco is the Gum Dragon, a styptic substance, well known in medicine; it flows from the plants abundantly when cut. The bitter resinous root of Aletris farinosa is tonic and stomachic, in small doses; but a dose of 20 grains occasions much nausea, with a tendency to vomit. *Bigelow*, 3. 96. The bulbs of Scilla Lilio-Hyacinthus, and the roots of Anthericum bicolor, are both purgative, according to Decandolle, *Propr. Med.* 296. The juice of common Asparagus contains a peculiar principle, called Asparagin. *Turner*, 699.

EXAMPLES. No good sections have been yet formed; those of Link, quoted above, are not sufficiently well defined. The principal types of structure are, Scilla, Asphodelus, Hyacinthus, Puschkinia, Brodiæa, Aloë, Aletris, Asparagus.

CCXLVIII. GILLIESIEÆ.

GILLIESIEÆ, *Lindl. in Bot. Reg.* 992. (1826); *Hooker in Bot. Mag.* 2716. (1827.)

DIAGNOSIS. Hexapetaloideous monocotyledons, with a superior ovarium, and irregular petaloid involucella.
ANOMALIES.

ESSENTIAL CHARACTER.—*Flowers* hermaphrodite, surrounded by bracteæ, the outer of which are petaloid and herbaceous, the inner depauperated and coloured. *Perianthium* minute, either a single labelloid lobe, or an urceolate 6-toothed body. *Stamens* 6, either all fertile, or 3 sterile and nearly obliterated. *Ovarium* superior, 3-celled; *style* 1; *stigma* simple. *Capsule* 3-celled, 3-valved, with a loculicidal dehiscence, many-seeded. *Seeds* attached to the axis, by means of a broad hollow neck; *testa* black and brittle; *embryo* curved in the midst of fleshy albumen.—Small *herbaceous* plants, with tunicated bulbs. *Leaves* grass-like. *Flowers* umbellate, somewhat spathaceous, inconspicuous.

AFFINITIES. The distinctions of many of the natural orders among Hexapetaloideous Dicotyledons are so slight, as far as technical characters are capable of being employed, that the separation of this tribe from Asphodeleæ seems justifiable, even now that the structure of the seeds is known, and that they are found to be essentially those of Asphodeleæ, except in having a crustaceous neck that connects them with the placenta. The tribe was originally proposed in the *Botanical Register*, from which, as that work is in few hands, I make the following rather long extract.

" The whole structure of this most remarkable plant is so peculiar, that we scarcely know whether the definition and description of the parts of fructification above given will not be considered more paradoxical than just; and yet, if the analogies the various organs bear to those of other plants be carefully considered, their structure will scarcely admit of any other interpretation. With respect to the five petaloid leaves, which are here described as bracteæ, and which bear a considerable degree of resemblance to a perianthium, it may be observed, that this appearance is more apparent than real; they neither correspond in insertion nor in number with the segments of a monocotyledonous perianthium, nor do they bear the same relation to the parts contained as a perianthium should bear. The three outer are not inserted on the same line, but are distinctly imbricated at the base; and the two inner do not complete the second series, as would be required in a regular monocotyledonous perianthium.

" But if we were to admit, for a moment, the possibility of these bracteæ being segments of a perianthium, what explanation could be given of the setiform processes proceeding from their base, or of the central fleshy slipper-like body from within which the stamens proceed? The former bear no determinate relation to the other parts of the flower in their insertion; they are subject to much diversity of form and number, being sometimes *eight*, consisting of *two* unequal subulate bodies proceeding from the edges of each lateral segment, the outermost of the two being wider than the innermost, and being, moreover, not unfrequently a manifest process of the margin of the segment itself; sometimes having their number reduced to *four* by the suppression of the exterior processes of each lateral segment; and occasionally having the outer processes suppressed on one segment, and not suppressed on the other. In the many flowers which have been under examination, the processes, moreover, were always constituted of cellular tissue alone, without either tracheæ or tubular vessels. These circumstances being considered, it will scarcely be proposed, we presume, to identify them with abortive stamina. If they are, notwithstanding what has been advanced, determined to be the perianthium itself, what becomes of the outer segments, which had previously been referred to perianthium? for it would be difficult to trace any analogy between the structure of Gilliesia and of those genera in which a third series is added to the usual senary division of Monocotyledones. But none of the peculiarities adverted to are opposed to those bodies being referred to depauperated or reduced bracteæ.

" With respect to the central body from which the stamens proceed, this body, which might be conveniently disposed of by referring it to what Linnæan botanists call a nectarium, consists, as we have seen, of a fleshy slipper-like lobe, with or without two auricles at the base, and within which the cup of stamens is inserted. The relation it bears, as regards insertion, to the parts which have been already noticed, is very obscure; it is always opposite the solitary external bracteæ; but whether it is anterior with respect to the common axis of inflorescence, or posterior, has not at present been ascertained. The reasons which have been offered for the view here taken of the parts surrounding this body, make it obvious that it must be considered the perianthium. But of this more will be said hereafter. For the present it will be sufficient to remark, that it manifestly bears an intimate relation to the stamens, being obliterated in the same direction and degree as they are.

" In this view, then, the petaloid segments are considered perfect bracteæ, the subulate interior processes abortive bracteæ, and the fleshy central labelloid body the perianthium.

" However paradoxical this description of Gilliesia may appear, and however inconclusive the arguments adduced in support of the view we have taken of it may have hitherto been considered, they will probably be found more deserving of attention if compared with a nearly-allied plant discovered in Chile, by our friend John Miers, Esq., after whom it has been named. This singular genus forms part of a most valuable and remarkable collection of botanical drawings, which were made by Mr. Miers during his long residence in Chile, and which, it is to be hoped, will, at some future day, be laid before the public. Having been kindly permitted to make use of the drawing and manuscript description of the plant alluded to, we shall endeavour to explain the analogies and relation which exist between it and Gilliesia.

" In Miersia the bracteæ are six in number, of which two are interior and four exterior, a still more valid reason against their being segments of a perianthium. The subulate processes assume a more regular form, and a more constant mode of insertion, but still bear no very apparent relation to the bracteæ; and the fleshy labelloid central body is represented by an urceolate six-toothed cup, within the orifice of which six fertile stamens are included. In Miersia, therefore, the perianthium, which was in Gilliesia subject to a certain degree of imperfection, in which the stamens also participated, is in the usual regular form of many Monocotyledones, no irregularity occurring in the stamens. As there can be no doubt of the strict analogy which exists between Gilliesia and Miersia in their fructification, and as there can also be little doubt that the central body of the latter genus is perianthium, it will follow as a necessary consequence, that as the supernumerary appendages of that genus are external with respect to the perianthium, and therefore neither perianthium nor stamens, so also will the analogous appendages of Gilliesia not be perianthium. And the central body having been ascertained to be perianthium, all the parts which surround it will necessarily be bracteæ, or modifications of bracteæ.

" The natural affinity of these two genera is extremely obscure; and till some accurate information can be obtained of the structure of their seeds, it must be a subject of much uncertainty. Even with the requisite information upon that point, it is not probable that they will be found to bear any very close relation to the other monocotyledonous orders at present known. Their tunicated bulbs, spathaceous inflorescence, and general appearance, place them near Asphodeleæ, with some genera of which, especially Muscari and Puschkinia, Miersia at least agrees in the structure of perianthium ; but we are acquainted with no genus of Asphodeleæ to which the fructification of Gilliesieæ can be otherwise compared. If the one-flowered species of Schoenus, in which a single naked flower is surrounded by several imbricated squamæ, be admitted as a form of inflorescence analogous to that under consideration, it may perhaps be allowable to carry this comparison yet further, and to suggest an identity of origin and function between the depauperated bracteæ of Gilliesia and the hypogynous setæ of Seirpus and other Cyperaceæ. But on account of the presence of a perianthium, and of their polyspermous three-celled capsule, Gilliesieæ may perhaps be with most propriety referred to the neighbourhood of Restiaceæ, whose imbricated inflorescence does not offer any very powerful obstacle."

At this time the structure of the seeds was unknown: I have since been able to ascertain their nature, in consequence of a supply having been given me by Mr. Cruikshanks. The result of their examination, while it strengthens the opinion of their vicinity to Asphodeleæ, and weakens that of a relation to Restiaceæ, does not induce me to alter my view of them as constituting a small but distinct order.

GEOGRAPHY. Chilian bulbs.
PROPERTIES. Unknown.
EXAMPLES. Gilliesia, Miersia.

CCXLIX. SMILACEÆ. THE SMILAX TRIBE.

ASPARAGI, *Juss. Gen.* (1789) *in part.*—SMILACEÆ, *R. Brown Prodr.* 292. (1810); *Lindl. Synops.* 270. (1829).—TRILLIACEÆ, *Dec. Ess. Méd.* 294. (1816).—ASPARAGEÆ, *Dec. and Duby*, 458. (1828).—ASPARAGINEÆ, *Ach. Rich. Dict. Class.* 2. 20. (1822); *Nouv. Élém.* ed. 4. 430. (1828).—SMILACINÆ, *Link Handb.* 1. 275. (1829).—PARIDEÆ, *Ib.* 277. (1829).—CONVALLARIACEÆ, *Ib.* 184. (1829) *a sect. of* Liliaceæ.

DIAGNOSIS. Hexapetaloideous monocotyledons, with a superior ovarium, anthers turned inwards, a coloured perianthium, a 3-celled succulent fruit, a membranous testa, and a triple style.

ANOMALIES. Tamus has the ovarium inferior. The parts of the flower are quaternary in Paris.

ESSENTIAL CHARACTER. — *Flowers* hermaphrodite or diœcious. *Calyx* and *corolla* confounded, inferior, petaloid, 6-parted. *Stamens* 6, inserted into the perianthium near the base; seldom hypogynous. *Ovarium* 3-celled, the cells 1- or many-seeded; *style* usually trifid; *stigmas* 3. *Fruit* a roundish berry. *Seeds* with a membranous testa (not black or brittle); *albumen* between fleshy and cartilaginous; *embryo* usually distant from the hilum. *R. Br.*— *Herbaceous* plants or *under-shrubs*, often with a tendency to climb. *Leaves* sometimes with reticulated veins.

AFFINITIES. So nearly the same as Asphodeleæ, that some botanists unite them, others separate them upon different principles from those adopted here, and others strike certain genera off from both the one tribe and the other. The leaves of Smilaceæ are broader and shorter, with more of a dicotyledonous appearance than the ensate or grassy ones of Asphodeleæ, and the stem has a frequent tendency to twine. Even in Ruscus some trace of this is visible, in R. racemosus, although there is nothing to indicate it in R. hypophyllum and the like.

GEOGRAPHY. Found in small quantities in most parts of the world, especially in Asia and N. America.

PROPERTIES. Best known for the diuretic demulcent powers of Smilax Sarsaparilla, which also exist in other species of the same genus. Smilax aspera is a common substitute in the south of Europe. Smilax China has a large fleshy root, the decoction of which is supposed to have virtues equal to that of Sarsaparilla in improving the health after the use of Mercury. According to the Abbé Rochon, the Chinese often eat it instead of Rice, and it contributes to make them lusty. *Ainslie*, 1. 70. The root of Medeola virginica is stated to be diuretic, and to have some reputation as a hydragogue. *Barton*, 2. 147. The roots of Trillium are generally violently emetic, and their mawkish, rather nauseous, berries are at least suspicious. *Dec.*

EXAMPLES. Trillium, Paris, Medeola, Convallaria, Streptopus, Smilax, Drymophila, Ripogonum.

CCL. DIOSCOREÆ. THE YAM TRIBE.

DIOSCOREÆ, *R. Brown Prodr.* 294. (1810); *Agardh Aphor.* 169. (1823.) *Ach. Rich. Nouv. Elém.* 434. (1828);

DIAGNOSIS. Hexapetaloideous monocotyledons, with an inferior ovarium, unisexual flowers, and a minute herbaceous spreading regular perianthium.

ANOMALIES.

ESSENTIAL CHARACTER.—*Flowers* diœcious. *Calyx* and *corolla* confounded, superior. *Males: Stamens* 6, inserted into the base of the sepals and petals. *Females: Ovarium* 3-celled, with 1- or 2-seeded cells; *style* deeply trifid; *stigmas* undivided. *Fruit* leaf-like, compressed, with two of its cells sometimes abortive. *Seeds* flat, compressed; *embryo* small, near the hilum, lying in a large cavity of cartilaginous albumen.—Twining *shrubs. Leaves* alternate, occasionally opposite, usually with reticulated veins. *Flowers* small, spiked, with from 1 to 3 bracteæ each.

AFFINITIES. Undoubtedly the nearest approach among monocotyledons to the dicotyledonous structure; according to Mr. Brown approaching Smilaceæ in structure and habit, but separable from them by the threefold character of inferior ovarium, capsular fruit, and albumen having a large cavity. Tamus is, however, between the two tribes, agreeing with Smilaceæ in its baccate, with Dioscoreæ in its inferior fruit. *Prodr.* 294. The leaves are altogether those of dicotyledons; the stem, flower, and seeds, of monocotyledons.

GEOGRAPHY. Found exclusively in tropical countries of either hemisphere, if Tamus be excluded.

PROPERTIES. The yams, so important a food in all tropical countries, because of their large, fleshy, mucilaginous, sweetish tubers, are the only remarkable plants of the order.

EXAMPLES. Dioscorea, Rajania, Oncus, Æchma.

CCLI. LILIACEÆ. The Lily Tribe.

LILIA, *Juss. Gen.* 48. (1789).—NARCISSI, *the first sect. Ibid.* 54. (1789).—HEMERO-
CALLIDEÆ, *R. Brown Prodr.* 295. (1810). — LILIACEÆ, *Dec. Théor. Elém.*
1. 249. (1813); *Dec. and Duby*, 461. (1828) *in part; Lindl. Synops.* 266. (1829).—
TULIPACEÆ, *Dec. Ess. Méd.* 297. (1816); *Dec. and Duby*, 461 (1828); *Link
Handb.* 1. 177. (1829) *a sect. of* Liliaceæ.—CORONARIÆ, *Agardh Aphor.* 165.
(1823.)

DIAGNOSIS. Hexapetaloideous monocotyledons, with a superior ovarium, highly developed perianthium, anthers turned inwards, a trilocular
polyspermous capsule, and seeds with a soft spongy coat.

ANOMALIES.

ESSENTIAL CHARACTER.— *Calyx* and *corolla* confounded, coloured, regular, occasionally cohering in a tube. *Stamens* 6, inserted into the sepals and petals. *Ovary* superior, 3-celled, many-seeded; *style* 1; *stigma* simple, or 3-lobed. *Fruit* dry, capsular, 3-
celled, many-seeded, with a loculicidal dehiscence. *Seeds* flat, packed one upon another in
1 or 2 rows, with a spongy, dilated, often winged integument; *embryo* with the same
direction as the seed, in the axis of fleshy *albumen.*—*Bulbs* scaly, or *stems* arborescent.
Leaves with parallel veins, either lanceolate or cordate. *Flowers* large, usually with
bright colours, often solitary.

AFFINITIES. Distinguishable from Asphodeleæ by their higher degree
of developement, and by the texture of the coat of their seeds. Various
degrees of cohesion between their sepals and petals occur, so that we have
tubular perianths and revolute ones even in the same genus (Lilium). Hence
Mr. Brown's Hemerocallideæ, which he states differ from Liliaceæ in almost
nothing but their tubular perianth, cannot be retained. Decandolle refers
Erythronium to Asphodeleæ in the *Botanicon Gallicum;* in the *Flore Française* he placed it in Melanthaceæ; but it surely ought to be stationed
here.

GEOGRAPHY. The temperate parts of America, Europe, and Asia, are
the favourite resort of this tribe, which stretches towards equinoctial countries upon the mountains of Mexico in the form of Calochortus, and in New
Holland in the shape of Blandfordia.

PROPERTIES. Chiefly remarkable for their large richly coloured flowers.
The bulbs of Lilium pomponium are roasted and eaten in Kamtschatka,
where it is as commonly cultivated as the potato with us. *Gard. Mag.*
6. 322. The roots of Erythronium indicum are employed in India in cases
of strangury and fever in horses. *Ainslie,* 1. 403. Polianthes tuberosa,
or the Tuberose, is well known for its delicious fragrance. This plant emits
its scent most strongly after sunset, and has been observed in a sultry evening, after thunder, when the atmosphere was highly charged with electric
fluid, to dart small sparks, or scintillations of lucid flame, in great abundance
from such of its flowers as were fading. *Ed. P. J.* 3. 415.

EXAMPLES. Lilium, Fritillaria, Hemerocallis, Funkia.

CCLII. PALMÆ. The Palm Tribe.

PALMÆ, *Juss. Gen.* (1789); *R. Brown Prodr.* 266. (1810); *Von Martius Palm. Braz.*
(1824); *Id. Programma* (1824.)

DIAGNOSIS. Hexapetaloideous arborescent monocotyledons, with rigid
divided leaves, a superior 3-celled ovarium, and an embryo lying in cartilaginous or fleshy albumen at a distance from the hilum.

ANOMALIES.

ESSENTIAL CHARACTER.—*Flowers* hermaphrodite, or frequently polygamous. *Perianthium* 6-parted, in two series, persistent; the 3 outer segments often smaller, the inner sometimes deeply connate. *Stamens* inserted into the base of the perianthium, usually definite in number, opposite the segments of the perianthium, to which they are equal in number, seldom 3; sometimes, in a few polygamous genera, indefinite in number. *Ovary* 1, 3-celled, or deeply 3-lobed, the lobes or cells 1-seeded, with an erect ovulum, rarely 1-seeded. *Fruit* baccate or drupaceous, with fibrous flesh. *Albumen* cartilaginous, and either ruminate, or furnished with a central or ventral cavity; *embryo* lodged in a particular cavity of the albumen, usually at a distance from the hilum, dorsal and indicated by a little nipple, taper or pulley-shaped; *plumula* included, scarcely visible; the cotyledonous extremity becoming thickened in germination, and either filling up a pre-existing cavity, or one formed by the liquefaction of the albumen in the centre.—*Trunk* arborescent, simple, occasionally shrubby and branched, rough with the dilated half-sheathing bases of the leaves or their scars. *Leaves* clustered, terminal, very large, pinnate or flabelliform, plaited in vernation. *Spadix* terminal, often branched, enclosed in a 1- or many-valved spatha. *Flowers* small, with bracteolæ. *Fruit* occasionally very large. *R. Brown* (1810).

AFFINITIES. The race of plants to which the name of Palms has been assigned is, no doubt, the most interesting in the vegetable kingdom, if we consider the majestic aspect of their towering stems, crowned by a still more gigantic foliage; the character of grandeur which they impress upon the landscape of the countries they inhabit; their immense value to mankind, as affording food, and raiment, and numerous objects of economical importance; or, finally, the prodigious developement of those organs by which their race is to be propagated. A single spatha of the Date contains about 12,000 male flowers; Alfonsia amygdalina has been computed to have 207,000 in a spathe, or 600,000 upon a single individual; while every bunch of the Seje Palm of the Oronoco bears 8000 fruit. They are very uniform in the botanical characters by which they are distinguished, especially in their fleshy colourless 6-parted flowers, enclosed in spathes, their minute embryo lying in the midst of albumen remote from the hilum, and their arborescent stems with rigid, plaited or pinnated, inarticulated leaves, called fronds; but their aspect and habits are extremely various. To use the words of the most accomplished traveller of our own, or any age;— " While some (Kunthia montana, Aiphanes Praga, Oreodoxa frigida) have trunks as slender as the graceful reed, or longer than the longest cable, (Calamus Rudentum, 500 feet), others (Jubæa spectabilis and Cocos butyracea) are 3 and even 5 feet thick; while some grow collected in groups (Mauritia flexuosa, Chamærops humilis), others (Oreodoxa regia, Martinezia caryotæfolia) singly dart their slender trunks into the air; while some have a low caudex (Attalea amygdalina), others exhibit a towering stem 160-180 feet high (Ceroxylon andicola); and while one part flourishes in the low valleys of the tropics, or on the declivities of the lower mountains, to the elevation of 900 feet, another part consists of mountaineers bordering upon the limits of perpetual snow." To which may be added, that while many have a cylindrical undivided stem, the Doom Palm of Upper Egypt and the Hyphæne coriacea are remarkable for their dichotomous repeatedly-divided trunk. In botanical affinity they approach as nearly to Junceæ as to any order, but they can hardly be said to be closely allied to those at present known. The relation that was supposed to exist between them and Cycadeæ was inferred from inaccurate or imperfect considerations; and there is nothing in Pandaneæ that can approximate that order, except their dichotomous trunks. The Calamus genus, and the siliceous secretions of their leaves, indicate an affinity with Gramineæ, which would hardly be anticipated, if the grasses of our European meadows are compared with the Cocoa Nuts of the Indies, but which becomes more apparent when the Bamboo is placed by the side of the Cane.

GEOGRAPHY. Von Martius, the great illustrator of this noble family, speaks thus of their habits and geographical arrangement:—" Palms, the splendid offspring of Tellus and Phœbus, chiefly acknowledge as their native land those happy regions seated within the tropics, where the beams of the latter for ever shine. Inhabitants of either world, they hardly range beyond 35° in the southern, or 40° in the northern hemisphere. Particular species scarcely extend beyond their own peculiar and contracted limits, on which account there are few countries favourable for their production in which some local and peculiar species are not found; the few that are dispersed over many lands are chiefly Cocos nucifera, Acrocomia sclerocarpa, and Borassus flabelliformis. It is probable that the number of species thus scattered over the face of nature will be found to amount to 1000 or more. Of these not a few love the humid banks of rivulets and streams, others occupy the shores of the ocean, and some ascend into alpine regions; some collect into dense forests, others spring up singly or in clusters over the plains." *Progr.* 6. But if this statement be true as to the probable number of Palms, how little can be now known of their structure, seeing that not more than 175 are at this moment described, of which 119 are South American, 14 African, and 42 Indian. The testimony of Von Martius is, however, confirmed by Humboldt, who also asserts that there must be an incredible number still to discover in equinoctial regions, especially if we consider how little is yet known of Africa, Asia, New Holland, and America. He and Bonpland discovered a new species in almost every 50 miles of travelling, so narrow are the limits within which their range is confined. A different opinion appears to be entertained by Schouw, a respectable Danish writer upon botanical geography, whose views deserve to be quoted, although he is far from having had such personal means of judging as Humboldt and Von Martius. He seems to consider that we are acquainted already with the greater part of the Palms; for he says, " it appears from the reports of travellers that such Palm woods as those of South America are less frequent in other parts of the world. Africa and New Holland seem to be less favourable to this tribe, for on the Congo, Smith found only from 3 to 4 Palms. In Guinea we know merely of the same number; and of the other African Palms, 6 belong to the Isles of Bourbon and France. New Holland has, in the torrid zone, three species, while Forster's *Prodromus* of the Flora of the South Sea Islands contains four." The most northern limit of Palms is that of Chamærops palmetto in N. America, in lat. 34°-36°, and of Chamærops humilis in Europe, near Nice, in 43°-44° N. lat. They are found in the southern hemisphere as low as 38° in New Zealand. " It is remarkable that no species of Palm has been found in South Africa, nor was any observed by M. Leschenault on the west coast of' New Holland, even within the tropic." *Brown in Flinders,* 577. If Palms were not, as some say, among the earliest plants that clothed the face of the globe, none of their remains existing, mixed with the Ferns and Equisetums of the old coal formations, it is at least certain that their creation dates long before that of the present Flora of the globe. But it is probable that they really did exist at the most remote periods; for the Nöggerathia foliosa of Sternberg from the coal-fields of Bohemia seems really to have been a Palm; and M. Adolphe Brongniart refers two other fossils of the same epoch to this family. It is at least certain that they appeared immediately after the developement of Cycadeæ ceased in European latitudes, and that of Coniferæ took a more decided form; as we find unquestionable traces of them in those deposits above the plastic clay which Brongniart calls Marno-charbonneux.

PROPERTIES. Wine, oil, wax, flour, sugar, salt, says Humboldt, are the produce of this tribe; to which Von Martius adds, thread, utensils, weapons, food, and habitations. The most remarkable is the Cocoa Nut, of which an excelleut account will be found in the *Trans. of the Wernerian Society*, vol. 5. The root is sometimes masticated instead of the Areca Nut; of the small fibres baskets are made in Brazil. The hard case of the stem is converted into drums, and used in the construction of huts; the lower part is so hard as to take a beautiful polish, when it resembles agate; the reticulated substance at the base of the leaf is formed into cradles, and, as some say, into a coarse kind of cloth. The unexpanded terminal bud is a delicate article of food ; the leaves furnish thatch for dwellings, and materials for fences, buckets, and baskets ; they are used for writing on, and make excellent torches ; potash in abundance is yielded by their ashes; the midrib of the leaf serves for oars ; the juice of the flower and stems is replete with sugar, and is fermented into excellent wine, or distilled into a sort of spirit, called Arrack; or the sugar itself is separated under the name of Jagery. The value of the fruit for food, and the delicious beverage which it contains, are well known to all Europeans. The fibrous and uneatable rind is not less useful; it is not only used to polish furniture and to scour the floors of rooms, but is manufactured into a kind of cordage, called Coir rope, which is nearly equal in strength to hemp, and which Dr. Roxburgh designates as the very best of all materials for cables, on account of its great elasticity and strength. Finally, an excellent oil is obtained from the kernel by expression. The juice which flows from the wounded spathes of Palms, especially of Cocos nucifera, is known in India by the name of Toddy. Independently of the grateful qualities of this fluid as a beverage, it is found to be the simplest and easiest remedy that can be employed for removing constipation in persons of delicate habit, especially European females. *Ainslie*, 1. 451. Palm oil is chiefly obtained from Elais guineensis, and this tree is also said to yield the best kind of Palm-wine. The succulent rind of the Date is one of the most agreeable of fruits. Sago is yielded by the trunk of nearly all, except Areca Catechu, but especially of Sagus farinifera and Phœnix farinifera. The well known Betel Nut is the fruit of Areca Catechu, and remarkable for its narcotic or intoxicating power; from the same fruit is prepared a kind of spurious Catechu. *Ibid.* 1. 65. The Brazilian Indians, especially the Puris, Patachos, and Botocudos, manufacture their best bows from the wood of a species of Cocoa Nut, called the Airi, or Brejeuba. *Pr. Max. Trav.* 238. The Ceroxylon andicola, or Wax Palm of Humboldt, has its trunk covered by a coating of wax, which exudes from the spaces between the insertion of the leaves. It is, according to Vauquelin, a concrete inflammable substance, consisting of 1-3d wax and 2-3ds resin. It is a very remarkable fact, first noticed by Mr. Brown (*Congo*, 456.), that the plants of this order whose fruit affords oil belong to a tribe called by him Cocoinæ, which are particularly characterised by the originally trilocular putamen having its cells when fertile perforated opposite the seat of the embryo, and when abortive indicated by foramina cæca. The dark-coloured inodorous and insipid resin, called Dragon's Blood, is obtained in the eastern islands of the Indian Archipelago by wounding the Calamus Draco ; it is said to be of finer quality than that procured from Pterocarpus.

EXAMPLES. The following are Von Martius's sections of the tribe. (*Programma*, p. 7.)

1. SABALINÆ. Spathes numerous, incomplete. Ovarium 3-celled. Berry or drupe 1-3-seeded. (Chamædorea, Thrinax.)

2. CORYPHINÆ. Spathes numerous, incomplete. Pistils 3, cohering inwardly, 1 only usually ripening. Berry or drupe many-seeded. (Rhapis, Phœnix.)

3. LEPIDOCARYA. Spathés numerous, incomplete. Flowers in catkins. Ovarium 3-celled. Berry 1-seeded, with a tessellated rind. (Mauritia, Calamus.)

4. BORASSEÆ. Spathes many, incomplete. Flowers in catkins. Ovarium 3-celled. Berry or drupe 3-seeded. (Borassus, Hyphæne.)

5. ARECINÆ. Spatha none, or one or more, complete. Ovarium 3-
celled. Berry 1-seeded. (Leopoldinia, Areca, Wallichia.)

6. COCOINÆ. Spatha one, or several, complete. Ovarium 3-celled.
Drupe 1-3-seeded. (Cocos, Elate, Bactris.)

CCLIII. RESTIACEÆ.

RESTIACEÆ, *R. Brown Prodr.* 243. (1810); *Kunth in Humb. N. G. et Sp.* 1. 251. (1815);
Agardh Aph. 156. (1823) *a sect. of* Junceæ; *Ach. Rich. Nouv. Elém.* ed. 4. 424.
(1828); *Lindl. Synops.* 272. (1829).—CENTROLEPIDEÆ *and* ERIOCAULONEÆ,
Desvaux in Ann. des Sc. 13. 36. (1828).—ELEGIEÆ, *Beauv. in eod. loc.* (1828.)

DIAGNOSIS. Hexapetaloideous monocotyledons, with a superior ovarium, axile placentæ, capsular fruit, capitate glumaceous flowers, and an
embryo lying on the albumen at the end most remote from the hilum.

ANOMALIES. Willdenowia has drupaceous fruit.

ESSENTIAL CHARACTER.—*Perianthium* inferior, 2-6-parted, seldom wanting. *Stamens* definite, 1-6; when they are from 2 to 3 in number, and attached to a perianthium
of 4 or 6 divisions, they are then opposite the inner segments (*petals*); *anthers* usually
unilocular. *Ovarium* 1- or more celled, cells monospermous; *ovules* pendulous. *Fruit*
capsular, or nucamentaceous. *Seeds* inverted; *albumen* of the same figure as the seed;
embryo lenticular, on the outside of the albumen, at that end of the seed which is most
remote from the hilum.—*Herbaceous* plants or *under-shrubs*. *Leaves* simple, narrow, or
none. *Culms* naked, or more usually protected by sheaths, which are slit, and have equitant margins. *Flowers* generally aggregate, in spikes or heads, separated by bracteæ, and
most frequently unisexual. *R. Br.* (1810).

AFFINITIES. The principal character distinguishing this family from
Junceæ and Cyperaceæ consists in its lenticular embryo being placed at the
extremity of the seed opposite to the umbilicus. From Junceæ it also differs
in the order of suppression of its stamina, which, when reduced to 3, are
opposite to the inner laciniæ of the perianthium; and most of its genera
are distinguishable from both these orders, as well as from Commelineæ,
by their simple or unilocular anthers. *Brown in Flinders*, 579. To this
may be added, that its habit is rather that of Cyperaceæ, especially when
Xyrideæ are excluded. From all the orders with spadiceous characters, the
glumaceous nature of its perianthium, when it is present, distinguishes it.
If the perianthium is absent, it is then only to be known from Cyperaceæ
by the position of the embryo, and by the sheaths of its leaves being slit.
M. Desvaux separates from the genera with a perianthium those in which
the flowers are actually naked, under the name of Centrolepideæ: he further
adopts the supposed order of Eriocauloneæ of the late M. de Beauvois,
which seems to differ from Restiaceæ simply in having 1-seeded cells in the
capsule, and irregular flowers. The Elegieæ of M. de Beauvois were distinguished by nothing but their 2 or 3 styles. While I adopt the opinion of all
these being parts of the same natural order, I cannot doubt that the tripe-
taloid flower and polyspermous fruit of Xyris, characters indicating a far
superior degree of evolution, are sufficient to separate that genus as the
representative of a peculiar order; a measure which Mr. Brown appears
to have anticipated when he remarked (*Prodr.* 244.), that the genus Xyris,

although placed by him at the end of Restiaceæ, is certainly very different from the other genera, in the inner segments of the perianthium being petaloid, with the stamens proceeding from the top of their ungues, and in their numerous seeds.

GEOGRAPHY. All, with the exception of Eriocaulon, extra European; chiefly found in the woods and marshes of South America, and in New Holland and southern Africa.

PROPERTIES. None, except that the tough wiry stems of some species are manufactured into baskets and brooms. Willdenowia teres is employed for the latter purpose, and Restio tectorum for thatching.

EXAMPLES. Centrolepis, Restio, Thamnochortus, Tonina, Eriocaulon.

CCLIV. PANDANEÆ. THE SCREWPINE TRIBE.

PANDANEÆ, *R. Brown Prodr.* 340. (1810); *Decand. Propr. Méd.* 278. (1816); *Agardh Aph.* 133. (1822); *Gaudichaud in Ann. des Sc.* 3. 509. (1824).—? CYCLANTHEÆ, *Poiteau in Mem. Mus.* 9. 34. (1822.)

DIAGNOSIS. Spadiceous monocotyledons, with naked flowers, and fibrous drupes collected in parcels into many-celled pericarpia.

ANOMALIES. Phytelephas has pinnate leaves; but it is a doubtful plant of the order.

ESSENTIAL CHARACTER.—*Flowers* diœcious or polygamous, arranged on a wholly covered spadix. *Perianthium* wanting. *Males: Filaments* with single anthers; *anthers* 2-celled. *Females: Ovaria* usually collected in parcels, 1-celled; *stigmas* as many as the ovaries, sessile, adnate (*ovula* solitary, erect). *Fruit* either fibrous drupes, usually collected in parcels, each 1-seeded; or many-celled *berries*, with polyspermous cells. *Albumen* fleshy; *embryo* in its axis, erect; *plumula* inconspicuous.—*Stem* arborescent, usually sending down aerial roots, sometimes weak and decumbent. *Leaves* imbricated, in three rows, long, linear-lanceolate, amplexicaul, with their margins almost always spiny. *Floral leaves* smaller, often coloured. *R. Br.*

AFFINITIES. This is a tribe of plants having the aspect of gigantic Bromelias, bearing the flowers of a Sparganium; while there is no analogy with the former in structure beyond the general appearance of the foliage; the organisation of the fructification bears so near a resemblance to the latter as to have led to the combination of Pandaneæ and Typhaceæ by botanists of the first authority. But when we contrast the naked flowers, the compound highly-developed fruit, the spathaceous bracteæ, the entire embryo, and the arborescent habit of the former, with the half-glumaceous flowers, the simple fruit, the want of spathaceous bracteæ, the slit embryo, and the herbaceous sedgy habit of the latter, it is difficult to withhold our assent from the proposition to separate them. Mr. Brown justly remarks (*Prodr.* 341.), that these have no affinity with Palms beyond their arborescent stems. Freycinetia, the genus to which the character of polyspermous cells, minute seeds, and a pulpy pericarpium belongs, is described by M. Gaudichaud as having a very minute embryo lodged in the upper part of semitransparent albumen. It is possible that this is the station of the remarkable plants described by Poiteau as having an inflorescence which may be compared to two folded ribands rolled spirally round a cylinder! one full of stamens, the other full of ovules!! and called Cyclantheæ. M. Poitean has unfortunately omitted to give a sufficient explanation of the analogy between the structure of these plants and more regular forms of inflorescence, and his figures do not afford such information as could be wished for; but it may be conjectured that his ribands are connate bracteæ, subtending, alternately, naked male and female flowers. Pandaneæ are remarkable among arborescent monocotyledons for their constant tendency to branch, which is always effected in a dichotomous manner. Their leaves have also a uniform spiral arrangement round the axis, so as to give the stems a sort of corkscrew appearance before the traces of the leaves are worn away. The Chandelier Tree of Guinea and St. Thomas's derives its name (Pandanus Candelabrum) from this peculiar tendency to branching.

GEOGRAPHY. Abundant in the Mascaren Islands, especially the Isle of France, where, under the name of Vaquois, they are found covering the sandy plains. They have peculiar means given them by nature to subsist in such situations in the shape of strong aerial roots, which are protruded from the stem, and descend towards the earth, bearing on their tips a loose cup-like coating of cellular integument, which preserves their tender newly-formed absorbents from injury until they reach the soil, in which they quickly bury themselves, thus adding at the same time to the number of mouths by which food can be extracted from the unwilling earth, and acting as stays to prevent the stems from being blown about by the wind. They are common in the Indian Archipelago, and in most tropical islands of the Old World, but are rare in America. From this continent Cyclanthus and Phytelephas are the only genera of Pandaneæ, if they really belong to the order, that have been described. The former, called Tagua, resembles Palms in its fronds, which equal those of the Cocoa Nut in dimensions, in its torulose scaly stem, and, finally, in the remarkable structure and weight of its fruit. *Humb. de Distr. Géogr.* 198.

PROPERTIES. The seeds of Pandanus are eatable. The flowers of Pandanus odoratissimus are fragrant and eatable. The fruit of several is also an article of food. The leaves are used for thatching and cordage. The immature fruit is reputed emmenagogue. Buttons are turned from the hard albumen of Phytelephas, or the Tagua plant. *Humb.* l. c.

EXAMPLES. Pandanus, Freycinetia.

CCLV. TYPHACEÆ. THE BULRUSH TRIBE.

TYPHÆ, *Juss. Gen.* 25. (1789). — AROIDEÆ, § 3. *R. Brown Prodr.* 338. (1810). — TYPHINÆ, *Agardh Aph.* 139. (1823). — TYPHACEÆ, *Dec. and Duby,* 482. (1828); *Lindl. Synops.* 247. (1829). — TYPHOIDEÆ and SPARGANIOIDEÆ, *Link Handb.* 1. 132. 133. (1829), *both sections of* Cyperaceæ.

DIAGNOSIS. Spadiceous triandrous monocotyledons, with 3 half-glumaceous sepals, clavate anthers, long lax filaments, a solitary pendulous ovulum, and dry fruit.

ANOMALIES.

ESSENTIAL CHARACTER. — *Flowers* unisexual, arranged upon a naked spadix. *Sepals* 3, or more. *Petals* wanting. *Males: Stamens* 3 or 6, *anthers* wedge-shaped, attached by their base to long filaments. *Females: Ovary* single, superior, 1-celled; *ovulum* solitary, pendulous; *style* short; *stigmas* 1 or 2, simple, linear. *Fruit* dry, not opening, 1-celled, 1-seeded. *Embryo* in the centre of *albumen,* straight, taper, with a cleft in one side, in which the plumula lies; *radicle* next the hilum. — *Herbaceous* plants, growing in marshes or ditches. *Stems* without nodi. *Leaves* rigid, ensiform, with parallel veins. *Spadix* without a spathe.

AFFINITIES. Jussieu, following Adanson, distinguishes these from Aroideæ, with which Mr. Brown re-unites them, retaining them, however, in a separate section. They are generally regarded as a distinct tribe by most writers, and are surely sufficiently characterised by their 3-sepaled half-glumaceous calyx, long lax filaments, clavate anthers, solitary pendulous ovules, and peculiar habit. They are connected with Aroideæ by Acorus, which belongs to the latter. Agardh refers Typhaceæ to glumaceous Mono-cotyledons, on account of the analogy between the calyx of Typha and the hypogynous hairs of Eriophorum, a genus of Cyperaceæ. They are com-bined with Pandaneæ by M. Kunth, but appear to be sufficiently distin-guished by the slit in the side of their embryo, their simple fruit, pendulous ovulum, trisepalous calyx, and habit.

GEOGRAPHY. Found commonly in the ditches and marshes of the northern parts of the world, but uncommon in tropical countries; a species is found in St. Domingo, and another in New Holland. Two are described from equinoctial America.

PROPERTIES. Of little known use. The powdered flowers have been used as an application to ulcers. The pollen of Typha is inflammable, like that of Lycopodium, and is used as a substitute for it. M. Decandolle remarks that it is probable the facility of collecting this pollen is the real cause of its use, and that any other kind would do as well.

EXAMPLES. Typha, Sparganium.

CCLVI. AROIDEÆ. THE ARUM TRIBE.

AROIDEÆ, *Juss. Gen.* 23. (1789); *R. Brown Prodr.* 333. (1810); *Dec. and Duby,* 480. (1828); *Lindl. Synops.* 246. (1829).—ACORINÆ, *Link Handb.* 1. 144. (1829), *a § of* Junceæ.

DIAGNOSIS. Spadiceous monocotyledons, with simple, succulent, or capsular fruit, a developed spatha, and sub-sessile anthers.

ANOMALIES. Albumen sometimes absent. In Tacca the ovarium is inferior. Spatha absent or rudimentary in some.

ESSENTIAL CHARACTER. — *Flowers* unisexual, arranged upon a spadix, frequently naked. *Perianthium* either wanting, or consisting of 4 or 6 pieces. *Males: Stamens* definite or indefinite, hypogynous, very short; *anthers* 1- 2- or many-celled, ovate, turned outwards. *Females: Ovarium* superior, 1-celled, very seldom 3-celled, and many-seeded; *ovules* erect, or pendulous, or parietal; *stigma* sessile. *Fruit* succulent or dry, not opening. *Seeds* solitary or several; *embryo* in the axis of fleshy or mealy albumen, straight, taper, with a cleft in one side, in which the plumula lies; (*radicle* obtuse, usually next the hilum, occasionally at the opposite extremity. *R. Br.*) — *Herbaceous* plants, frequently with a fleshy *cormus,* or *shrubs ;* stemless or arborescent, or climbing by means of aerial roots. *Leaves* sheathing at the base, either with parallel or branching veins; sometimes com-pound ! often cordate. *Spadix* generally enclosed in a *spathe.*

AFFINITIES. The Arum tribe may be considered the centre of a system of organisation, of which the other orders of Spadiceæ are rays of unequal length. Taking its diagnosis as given above, we shall have it specially known by its highly developed spatha; Typhaceæ will be distinguished by their long anthers and want of spatha, Pandaneæ by their arborescent habit and drupaceous compound fruit, Fluviales and Juncagineæ by their want of spatha and return from the spadiceous form of inflorescence, and Pistiaceæ by their reduction to the simplest state in which flowering plants can exist. The whole of these tribes, taken together, are known by their general ten-dency to develope their flowers upon a spadix, by their want of floral enve-lopes, or by those parts not assuming the distinct forms of calyx and corolla, but existing only in the state of herbaceous scales. With the exception of Pandaneæ, they are all also known by their plumula lying within a cleft of the embryo; a structure found in no other monocotyledonous plants, except Grasses, in which the embryo is otherwise widely different. Mr. Brown has remarked that in Dracontium polyphyllum and fœtidum, in which there is no albumen, the plumula consists of imbricated scales, and that it is sometimes double or even triple. In the former of these plants the external scales, in germination, quickly wither away, when other internal and larger ones appear, and remain for some time round the base of the primordial leaf, before the developement of which no rootlets are emitted. *Prodr.* 334. A similar economy has been noticed by Du Petit Thouars, in his genus Ouvirandra. In Tacca it is probable that there are several germinating points upon the embryo, analogous to the double or triple plumula of Dracontium : hence embryos of such a kind may be said to be tubers found in the seed itself. Mr. Brown considers a relation to be established between Aroideæ and Aristolochiæ by means of Tacca, in which the ovarium is inferior. Agardh distinguishes Acoroideæ from Aroideæ by their capsular fruit.

GEOGRAPHY. Natives of all tropical countries abundantly, but of temperate climates rarely, not extending in Europe further north than 64° north latitude, in the form of Calla palustris, which inhabits the deep, muddy, frozen marshes of southern Lapland. In cold or temperate climates they are usually herbaceous, while in tropical countries they are often arborescent and of considerable size, frequently clinging to trees by means of their aerial roots, which they protrude in abundance. In America, according to Humboldt (*Distr. Géogr.* 196), their principal station is on the submontane region between 1200 and 3600 feet of elevation, where the climate is temperate and the rains abundant. In the Andes, Pothos pedatus and P. quinquenervius rise to the height of 8400 feet.

PROPERTIES. A principle of acridity generally pervades this tribe, and exists in so high a degree in some of them as to render them dangerous poisons. The most remarkable is the Dumb Cane, or Caladium Seguinum, a native of the West Indies and South America, growing to the height of a man : this plant has the power, when chewed, of swelling the tongue and destroying the power of speech. Dr. Hooker relates an account of a gardener, who "incautiously bit a piece of the Dumb Cane, when his tongue swelled to such a degree that he could not move it; he became utterly incapable of speaking, and was confined to the house for some days in the most excruciating torments." *Exot. Bot.* 1. The same excellent botanist adds, that it is said to impart an indelible stain to linen. P. Browne states, that its stalk is employed to bring sugar to a good grain when it is too viscid, and cannot be made to granulate properly by the application of lime alone; Arum ovatum is used for the same purpose. The leaves of Arum esculentum excite violent salivation and a burning sensation in the fauces, as I have myself experienced. The fresh leaves of Dracontium pertusum are employed by the Indians of Demerara as vesicatories or rubefiants in cases of dropsy. Milk in which the acrid root of Arum triphyllum has been boiled has been known to cure consumption. *Dec.* Notwithstanding this acridity, the flat under-ground stems, called roots, and the leaves of many Aroideæ, are harmless, and even nutritive when roasted or boiled, as, for instance, the roots of Arum esculentum, Colocasia, mucronatum, violaceum, and others, which, under the names of Cocoa root, Eddoes, and Yams, are common articles of food in hot countries. The roots (cormi) of the Arum maculatum are commonly eaten by the country people in the Isle of Portland ; they are macerated, steeped, and the powder obtained from them is sent to London for sale under the name of Portland Sago. *Enc. of Pl.* 800. Medicinally, the root in its recent state is stimulant, diaphoretic, and expectorant. The root and seeds of the Skunk Cabbage, Symplocarpus fœtida, are powerful antispasmodics; they are also expectorants, and useful in phthisical coughs. They have considerable reputation in North America as palliatives in paroxysms of asthma. *Barton*, 1. 130. The prepared root of Dracontium polyphyllum is supposed in India to possess antispasmodic virtues, and is considered a valuable remedy in asthma; it is also used in hemorrhoids. *Ainslie*, 2. 50. The root of the Labaria plant of Demerara, which is probably the same thing, is thought by the Indians to be an antidote to the bite of serpents. *Ed. N. Ph. Journ.*, *June* 1830, p. 169. The root of Acorus calamus is aromatic and stimulant. The seeds of Orontium aquaticum and Arum sagittifolium are acrid, but become eatable by roasting. The spadixes of some species have a fetid putrid smell; others, such as Arum cordifolium, Italicum, and maculatum, are said to disengage a sensible quantity of heat at the time when they are about to expand. Agardh considers that the acrid principle, which, notwithstanding its fugacity, has been lately obtained pure, is no doubt of great power as a stimulant. *Aph*. 133.

The following are the principal natural divisions of this order : —

I. Flowers unisexual. Perianthium wanting.

Aroideæ veræ, *Brown Prodr*. 335. (1810.)
EXAMPLES. Arum, Caladium.

II. Flowers hermaphrodite. Perianthium present.

Orontiaceæ, *Brown Prodr*. 337. (1810). — Acoroideæ, *Agardh Aph*. 133. (1822.)
EXAMPLES. Dracontium, Pothos, Gymnostachys, Acorus.

CCLVII. BALANOPHOREÆ.

BALANOPHOREÆ, *Rich. in Mém. Mus.* 8. 429. (1822). — CYNOMORIEÆ, *Agardh Aph*. 203. (1825), *a* § *of* Urticeæ.

DIAGNOSIS. Spadiceous monocotyledons, with an inferior ovarium and monœcious flowers.

ANOMALIES.

ESSENTIAL CHARACTER.—*Flowers* monœcious, collected in dense heads, which are roundish or oblong, usually bearing both male and female flowers, but occasionally having the sexes distinct; the receptacle covered with scales or setæ variable in form, here and there bearing also peltate thick scales; rarely naked. *Male flowers* pedicellate ; *calyx* deeply 3-parted, equal, spreading, with somewhat concave segments ; in Cynomorium there is a thick, truncate, obconical scale in room of a calyx. *Stamens* 1-3 (seldom more), epigynous, with both united filaments and anthers ; the latter 3 ; in Cynomorium 1 only, connate, 2-celled ; each cell being divided into 2 cavities, sometimes turned inwards, sometimes outwards, opening by a longitudinal slit. *Female flowers* : *Ovarium* inferior, 1-celled, 1-seeded, crowned by the limb of the calyx, which is either marginal and nearly inverted, or consisting of from 2 to 4 unequal leaflets ; *ovulum* pendulous. *Style* 1, seldom 2, filiform, tapering ; *stigma* simple, terminal, rather convex. *Fruit* a roundish caryopsis, crowned by the remains of the limb of the calyx. *Pericarpium* rather thick ; *albumen* globose, fleshy-cellular, whitish, very large. *Embryo* very minute in proportion to the albumen, roundish, whitish, enclosed in a superficial excavation, undivided. — Fungus-like plants, parasitical upon roots ; *roots* fleshy, horizontal, branched ; *stem* naked, or covered by imbricated scales. *Rich.*

AFFINITIES. This highly curious order has the same relation to Monocotyledons as Cytineæ to Dicotyledons. The late M. Richard is the only botanist who has written specially upon it, and to him we owe an excellent Monograph. He observes that the nearest affinity of the order is with Hydrocharideæ, while at the same time it must be admitted that its relation is by no means intimate. The habit of the two orders is very different, and the structure of their floral organs is essentially unlike. In Hydrocharideæ the ovarium has generally several cells, and each cell contains many seeds, while in Balanophoreæ the ovarium is constantly 1-celled with a single ovulum. The former have no albumen; in the latter it is abundant. The tribe of Arums, in its habit and characters, has in general a more essential affinity with Balanophoreæ than Hydrocharideæ; they have both the same arrangement of flowers in spikes, the seeds have in both a fleshy albumen, and the habit of their several genera is much the same. But in Aroideæ the ovarium is superior! He then points out the affinity borne to Cytinus; an affinity about which nothing certain can be said, in the absence of a knowledge of the structure of the seed of the latter. Agardh places these in Urticeæ, changing the name to Cynomorieæ.

GEOGRAPHY. A small tribe, consisting entirely of leafless plants, parasitical upon roots, found in the West Indies, South America, some of the South Sea Islands, the Mediterranean, and the Cape of Good Hope.

PROPERTIES. Cynomorium is known for its astringency. Nothing has been stated of the rest.

EXAMPLES. Langsdorffia, Helosis, Cynomorium, Balanophora, Sarcophyte or Ichthyosma.

CCLVIII. FLUVIALES.

NAIADES, *Juss. Gen.* 18. (1789) *in part.* — FLUVIALES, *Vent. Tabl.* 2. 80. (1799). — POTAMOPHILÆ, *Rich. Anal. Fr.* (1808). — POTAMEÆ, *Juss. Dict. Sc. Nat.* 43. 93. (1826); *Dec. and Duby*, 439. (1828). — NAIADEÆ, *Agardh Aph.* 125. (1822). — FLUVIALES, *Rich. Mém. Mus.* 1. 364. (1815); *Lindl. Synops.* 248. (1829). — HYDROGETONES, *Link Handb.* 1. 282. (1829). — NAIADEÆ, *Ib.* 1. 820. (1829.)

DIAGNOSIS. Caulescent floating exalbuminous monocotyledons, with a slit embryo, definite stamens, and dry superior fruit with pendulous seeds.

ANOMALIES. Caulinia and some others are said to have no spiral vessels.

ESSENTIAL CHARACTER.—*Flowers* hermaphrodite or unisexual. *Perianthium* of 2 or 4 pieces, often deciduous, rarely wanting. *Stamens* definite, hypogynous. *Ovarium* 1 or more, superior; *stigma* simple; *ovule* solitary, pendulous. *Fruit* dry, not opening, 1-celled, 1-seeded. *Seed* pendulous; *albumen* none; *embryo* antitropous, with a lateral cleft for the emission of the plumula.—*Water-plants.* *Leaves* very cellular, with parallel veins. *Flowers* inconspicuous, usually arranged in terminal *spikes.*

AFFINITIES. In this order we have the nearest approach, except in Pistiaceæ, to the division of flowerless plants. The perianthium is reduced to a few imperfect scales, the habit is almost that of Coniferæ, and there is in some of the genera either a total absence of spiral vessels, or that form of tissue exists in a very rudimentary state. Pollini asserts, according to Decandolle (*Org. Veg.* 40), that spiral vessels do exist in them; but Amici, on the other hand, maintains that there is no trace of them, at least in Caulinia. *Ann. des Sc.* 2. 42. The manifest affinity of Fluviales to Juncagineæ determines a relation on the part of the former to Aroideæ, which is confirmed by the tendency to produce a rudimentary spatha in some of them, and by their undoubted resemblance to Pistiaceæ, which may be understood as reduced Aroideæ. It is remarkable that Adanson was aware of this relationship between Aroideæ and Fluviales, to which, however, Jussieu, whose Naiades are a very heterogeneous assemblage, did not assent. They are generally

U

translucent cellular plants, destitute of stomata, having no epidermoidal layer, and perishing rapidly upon exposure to air. M. Amici has seen the sap circulate in the transparent joints of Caulinia fragilis, which he states is the unknown plant upon which Corti made observations relating to the same subject. See Amici in *Ann. des Sc.* 2. 42. Agardh refers to this order both Ceratophyllum and Sparganium.

GEOGRAPHY. Common in extra-tropical countries, but also found near the equator. Potamogetons are in every ditch and swamp as far north as Iceland.

PROPERTIES. Very unimportant. The root of Potamogeton natans is said to be eaten in Siberia, and that of Aponogeton distachyum by Hottentots. Zostera, or Sea wrack, is a common material for packing, and for stuffing cottagers' cushions.

EXAMPLES. Naias, Zostera, Caulinia, Cymodocea, Thalassia, Ruppia, Zannichellia, Potamogeton.

CCLIX. JUNCAGINEÆ.

JUNCAGINEÆ, *Rich. Anal. Fr.* (1808); *Mém. Mus.* 1. 364. (1815); *Lindl. Synops.* 252. (1829); *Dec. and Duby*, 438. (1828). *a sect. of* Alismaceæ.

DIAGNOSIS. Caulescent exalbuminous monocotyledons, with a slit embryo, 6 stamens, and dry superior fruit with erect seeds.

ANOMALIES. Lilæa has no perianthium.

ESSENTIAL CHARACTER. — *Sepals* and *petals* both herbaceous, rarely absent. *Stamens* 6. *Ovaries* 3 or 6, superior, cohering firmly; *ovules* 1 or 2, approximated at their base, erect. *Fruit* dry, 1- or 2-seeded. *Seeds* erect; *albumen* wanting; *embryo* having the same direction as the seed, with a lateral cleft for the emission of the plumule. — *Herbaceous* bog-plants. *Leaves* ensiform, with parallel veins. *Flowers* in spikes or racemes, inconspicuous.

AFFINITIES. The plumula lying within a cleft on one side of the embryo fixes these plants nearer Aroideæ than Alismaceæ, to which they are sometimes referred, principally on account of their want of albumen; and the depauperated state of their floral envelopes confirms the relationship. Juncagineæ are most nearly allied to Fluviales, which are readily distinguished by their floating habit and pendulous ovules. The genus Scheuchzeria is a transition from Juncagineæ to Junceæ.

GEOGRAPHY. Marshy places in most parts of the world may be expected to indicate traces of this order, which is found in Europe, Asia, and North America, the Cape of Good Hope, and equinoctial America.

PROPERTIES. Unknown. Triglochin has a salt taste.

EXAMPLES. Lilæa, Cathanthes, Triglochin, Scheuchzeria.

CCLX. PISTIACEÆ. The Duckweed Tribe.

Pistiaceæ, *Rich. in Humb. et Bonpl. N. G. et Sp.* 1. 81. (1815); *Lindl. in Hooker's Fl. Scot.* 2. 191. (1821); *Synops.* 251. (1829). — Lemnaceæ, *Dec. and Duby,* 532. (1828.)

Diagnosis. Floating monocotyledons, with solitary naked spathaceous flowers, and the stem and leaves confounded.

Anomalies.

Essential Character. — *Flowers* 2, naked, enclosed in a spatha. *Male: Stamens* definite. *Female : Ovarium* 1-celled, with 1 or more erect *ovules; style* short; *stigma* simple. *Fruit* membranous or capsular, not opening, 1- or more-seeded. *Seeds* with a fungous testa, and a thickened indurated foramen; *embryo* either in the axis of fleshy albumen, and having a lateral cleft for the emission of the *plumule,* or at the apex of the nucleus. — *Floating* plants, with very cellular, lenticular, or lobed *stems* and *leaves* confounded. *Flowers* appearing from the margin of the stems.

Affinities. These are plants of a still simpler organisation than Fluviales, like them apparently destitute of spiral vessels, and not producing any separate stem or leaves, but a body formed out of both, from within the substance of which proceeds a membranous spathe containing one naked male and one naked female flower; a stem and two flowers thus constituting the whole of the plant. But if an abstraction be made of the simplicity of this structure, and the organisation be considered as if it belonged to plants of a more highly developed character, it will be found that these are really nothing but Aroideæ, the spadix of which is reduced to two flowers of different sexes. But while the accuracy of this view of the nature of Pistiaceæ is not likely to be questioned, it must be borne in mind that this very reduction of parts is inconsistent with the notion of Aroideæ, properly so called; and hence the necessity of constituting a particular order. I find from an examination of seeds of Pistia, most kindly procured from India for me by Dr. Wallich, that the embryo is a minute body lying at the apex of the albumen; in Lemna it occupies the axis; in both there is a fungous testa, with a remarkable induration of the foramen of the secundine. The embryo of Pistia is very minute, and perhaps solid; but in Lemna there is the slit on one side for the emission of the plumula, just as in Aroideæ. In Dr. Hooker's *Botanical Miscellany,* part 2, is an account of the germination of Lemna, by Mr. Wilson of Warrington, which is worth consulting. Agardh refers Lemna to Urticeæ, and places Nepenthes here.

Geography. Lemna inhabits the ditches of the cooler parts of the world; Pistia the tropics.

Properties. Pistia Stratiotes grows in water-tanks in Jamaica, where, according to P. Browne, it is acrid, and in hot dry weather impregnates the water with its particles to such a degree as to give rise to the bloody flux. *Hist. of Jam.* 330. A decoction of the same plant is considered by the Hindoostanees as cooling and demulcent, and they prescribe it in cases of dysuria. The leaves are also made into a poultice for the piles. *Ainslie.*

Examples. Pistia, Lemna.

Tribe II. GLUMACEÆ.

These are distinctly characterised by the want of a true perianthium, in the room of which the floral envelopes are formed by imbricated bracteæ. The paleæ of Grasses approach the nature of a calyx; but as they do not originate from the same plane, they cannot, practically, be confounded with a calyx, however near such an organ they may, upon theoretical principles, be considered to approach. The same may be said of the hypogynous setæ of Cyperaceæ, which, although probably of the nature of a perianthium, exist in so rudimentary a state as not to form a real exception to the character of Glumaceæ. Restiaceæ and Palms connect Petaloideous Monocotyledons with Glumaceæ; the former by approaching Cyperaceæ, the latter Grasses.

LIST OF THE ORDERS.

261. Gramineæ.   262. Cyperaceæ.

CCLXI. GRAMINEÆ. The Grass Tribe.

Gramina, *Juss. Gen.* 28. (1789). — Gramineæ, *R. Brown Prodr.* 168. (1810); *Palisot de Beauv. Agrostog.* (1812); *Kunth in Mém. Mus.* 2. 62. (1815); *Id. in N. G. et Sp. Humb. et Bonpl.* 1. 84. (1815); *Turpin in Mém. Mus.* 5. 426. (1819); *Trinius Fundam. Agrostol.* (1820); *Agardh Aphor.* 143. (1823); *Kunth Synops.* 1. 163. (1823); *Dumortier Agrost. Belg.* (1823); *Trinius Diss. de Gram. Unifl. et Sesquif.* (1824); *De la Harpe in Ann. Sc.* 5. 335. 6. 21. (1825); *Raspail in Ann. des Sc.* 4. 271. 422. 5. 287. 433. 6. 224. 384. (1825), 7. 335. (1826); *Link Hortus Botanicus,* 1. (1827); *Lindl. Synops.* 293. (1829); *Nees v. Esenbeck Agrostog. Brasil.* (1829).)

Diagnosis. Glumaceous monocotyledons, with cylindrical stems, slit leaf-sheaths, and a lenticular embryo lying on the outside of the albumen, with a naked plumula.

Anomalies.

Essential Character. — *Flowers* usually hermaphrodite, sometimes monœcious or polygamous; consisting of imbricated bracteæ, of which the most exterior are called *glumes*, the interior immediately enclosing the stamens *paleæ*, and the innermost at the base of the ovarium *scales*. *Glumes* usually 2, alternate; sometimes single, most commonly unequal. *Paleæ* 2, alternate; the lower or exterior simple, the upper or interior composed of 2 united by their contiguous margins, and usually with 2 keels, together forming a kind of dislocated calyx. *Scales* 2 or 3, sometimes wanting; if 2, collateral, alternate with the paleæ, and next the lower of them; either distinct or united. *Stamens* hypogynous, 1, 2, 3, 4, 6, or more, 1 of which alternates with the 2 hypogynous scales, and is therefore next the lower palea; *anthers* versatile. *Ovarium* simple; *styles* 2, very rarely 1 or 3; *stigmas* feathery or hairy. *Pericarpium* usually undistinguishable from the seed, membranous. *Albumen* farinaceous; *embryo* lying on one side of the albumen at the base, lenticular, with a broad cotyledon and a developed plumula; and occasionally, but very rarely, with a second cotyledon on the outside of the plumula, and alternate with the usual cotyledon.—*Rhizoma* fibrous or bulbous. *Culms* cylindrical, fistular, closed at the joints, covered with a coat of silex. *Leaves* alternate, with a split sheath. *Flowers* in little spikes called *locustæ*, arranged in a spiked, racemed, or panicled manner.

Affinities. This family is one which offers more singularities in its organisation than any other among flowering plants, and is perhaps that of which the organisation is to this day least understood, although it is among the most common and the most completely known, and is one in which, formerly, botanists the least suspected anomalies of organisation to exist. They found calyx and corolla and nectaries here with the same facility as they found them in a Ranunculus; and yet it may be doubted whether such organs exist in any one genus of Grasses.

Before I advert to the affinities of this tribe, it is indispensable that the real nature of this organisation should be understood. I shall therefore, without occupying myself with the views of Linnæus and his school, first cite Mr. Robert Brown's account of their structure, and then proceed to offer some observations upon the views that other botanists have taken of the subject.

Mr. Brown's statement is this:—

" The natural or most common structure of Gramineæ is to have their sexual organs surrounded by two floral envelopes, each of which usually consists of two distinct valves; but both of these envelopes are, in many genera of the order, subject to various degrees of imperfection or even suppression of their parts. The outer envelope, or gluma of Jussieu, in most cases containing several flowers with distinct and often distant insertions on a common receptacle, can only be considered as analogous to the bracteæ or involucrum of other plants. The tendency to suppression in this envelope appears to be greater in the exterior or lower valve; so that a gluma consisting of one valve may, in all cases, be considered as deprived of its outer or inferior valve. In certain genera with a simple spike, as Lolium and Lepturus, this is clearly proved by the structure of the terminal flower or spicula, which retains the natural number of parts; and in other genera not admitting of this direct proof, the fact is established by a series of species shewing its gradual obliteration, as in those species of Panicum which connect that genus with Paspalum. On the other hand, in the inner envelope, or calyx of Jussieu, obliteration first takes place in the inner or upper valve; but this valve having, instead of one central nerve, two nerves equidistant from its axis, I consider it as composed of two confluent valves, analogous to what takes place in the calyx and corolla of many irregular flowers of other classes; and this confluence may be regarded as the first step towards its obliteration, which is complete in many species of Panicum, in Andropogon, Pappophorum, Alopecurus, Trichodium, and several other genera. With respect to the nature of this inner or proper envelope of Grasses, it may be observed, that the view of its structure now given, in reducing its parts to the usual ternary division of Monocotyledones, affords an additional argument for considering it as the real perianthium. This argument, however, is not conclusive, for a similar confluence takes place between the two inner lateral bracteæ of the greater part of Irideæ; and with these, in the relative insertion of its valves, the proper envelope of Grasses may be supposed much better to accord than with a genuine perianthium. If, therefore, this inner envelope of Grasses be regarded as consisting merely of bracteæ, the real perianthium of the order must be looked for in those minute scales, which, in the greater part of its genera are found immediately surrounding the sexual organs. These scales are, in most cases, only two in number, and placed collaterally within the inferior valve of the proper envelope. In their real insertion, however, they alternate with the valves of this envelope, as is obviously the case in Ehrharta and certain other genera; and their collateral approximation may be considered as a tendency to that confluence which uniformly exists in the parts composing the upper valve of the proper envelope, and which takes place also between these two squamæ themselves, in some genera, as Glyceria and Melica. In certain other genera, as Bambusa and Stipa, a third squamula exists, which is placed opposite to the axis of the upper valve of the proper envelope, or, to speak in conformity with the view already taken of the structure of this valve, opposite to the junction of its two component parts. With these squamæ the stamina in triandrous Grasses alternate, and they are consequently opposite to the parts of the proper envelope; that is, one stamen is opposed to the axis of its lower or outer valve, and the two others are placed opposite to the two nerves of the upper valve. Hence, if the inner envelope be considered as consisting of bracteæ, and the hypogynous squamæ as forming the perianthium, it seems to follow, from the relation these parts have to the axis of inflorescence, that the outer series of this perianthium is wanting, while its corresponding stamina exist, and that the whole or part of the inner series is produced while its corresponding stamina are generally wanting. This may, no doubt, actually be the case; but as it would be, at least, contrary to every analogy in Monocotyledonous plants, it becomes in a certain degree probable that the inner or proper envelope of Grasses, the calyx of Jussieu, notwithstanding the obliquity in the insertion of its valves, forms in reality the outer series of the true perianthium, whose inner series consists of the minute scales, never more than three in number, and in which an irregularity in some degree analogous to that of the outer series generally exists. It is necessary to be aware of the tendency to suppression existing, as it were, in opposite directions in the two floral envelopes of Grasses, to comprehend the real structure of many irregular genera of the order, and also to understand the limits of the two great tribes into which I have proposed to subdivide it. One of these tribes, which may be called Paniceæ, comprehends Ischæmum, Holcus, Andropogon, Anthistiria, Saccharum, Cenchrus, Isachne, Panicum, Paspalum, Reimaria, Anthenantia, Monachne, Lappago, and several other nearly related genera; and its essential character consists in having always a locusta of two flowers, of which the lower or outer is uniformly imperfect, being either male or neuter, and then not unfrequently reduced to a single valve. Ischæmum and Isachne are examples of this tribe in its most perfect form, from which Anthenantia, Paspalum, and Reimaria, most remarkably deviate, in consequence of the suppression of certain parts : thus Anthenantia (which is not correctly described by Palisot de Beauvois) differs from those species of Panicum that have the lower flower neuter and bivalvular, in being deprived of the outer valve of its gluma; Paspalum differs from Anthenantia in the want of the inner valve of its neuter flower, and from those species of Panicum whose outer flower is univalvular, in the want of the outer valve of its gluma; and Reimaria differs from Paspalum in being entirely deprived of its gluma. That this is the real structure of these genera may be proved by a series of species connecting them with each other, and Panicum with Paspalum. The second tribe, which may be called Poaceæ, is more numerous than Paniceæ, and comprehends the greater part of the European genera, as well as certain less extensive genera peculiar to the equinoctial countries; it extends also to the highest latitudes in which Phænogamous plants have been found; but its maximum appears to be in the temperate climates, considerably beyond the tropics. The locusta in this tribe may consist of 1, 2, or of many flowers; and the 2-flowered genera are distinguished from Paniceæ by the outer or lower flower being always perfect, the tendency to imperfection in the locusta existing in opposite directions in the two tribes. In conformity with this tendency in Poaceæ, the outer valve of the perianthium in the single-flowered genera is placed within that of the gluma, and in the many-flowered locusta the upper flowers are frequently imperfect. There are, however, some exceptions to this order of suppression, especially in Arundo Phragmites, Campulosus, and some other genera, in which the outer flower is also imperfect ; but as all of these have more than two flowers in their locusta, they are still readily distinguished from Paniceæ." *Brown in Flinders*, 580.

According to this view, in a locusta of several florets, the scales at its base, or glumes, are bracteæ, and each floret consists of a calyx formed of one sepal remote from the rachis, and two cohering by their margins and next the rachis ; the little hypogynous scales are the rudiments of two petals, and the stamens alternate with these in the normal manner. This may be rendered more clear by the following diagram,

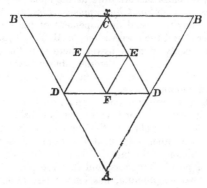

in which the triangle A B B represents the outer series, or paleæ, or calyx, A being the inferior valve, and B B the superior, formed of two sepals united by their contiguous margins at *x*. If the triangle C D D be understood to represent the next series, the position of the parts will be at the three angles ; and in reality the two scales that are usually developed do occupy the places D D ; while the third, whenever it is superadded, is stationed at C. The triangle E E F indicates by its angles the normal position of the first series of stamens, which are actually so situated, the stamen F which is opposite the sepal A alternating with the rudimentary petals D D.

The principal objection to this is, that the parts of the supposed calyx or paleæ are not inserted upon the same plane, or truly verticillate, and consequently do not answer exactly to what is required in a floral envelope; and it is on this account that M. Turpin rejects Mr. Brown's opinion, giving the paleæ the name of spathelle, and considering them bracteæ of a second order. But after all, this is a question of words rather than of facts ; for what are sepals but bracteæ of a second order? and what difficulty is there in identifying bracteæ having the near approach to a verticillate state, and the perfect symmetry of position that those of Grasses possess, with a kind of dislocated calyx ?

I know, however, from conversation with my friend M. Kunth, that he entertains a different view of the nature of the floral envelopes, considering the hypogynous scales to be analogous to the ligula, and the normal state of Grasses to be hexandrous; but as I unfortunately cannot discover the place in which he has explained this theory more fully, I refrain from dwelling upon it.

M. Raspail, in a memoir upon the structure of Gramineæ, hazards a strange theory, that the midrib of the bracteæ of Grasses is an axis of developement in cohesion with the bracteæ, and that when it separates, as in Phleum, Bromus, or Corynephorus, it is attempting to revert to the functions of ulterior developement, for which it is more especially destined. Among other things he states (*Ann. des Sc.* 4. 276. E.) that he should not be surprised one day to find some Grass in which the midrib of the lower palea actually became a new axis bearing other florets. I mention this for the sake of remarking that such a case is known, without however admitting that it is any confirmation of M. Raspail's views, which are at direct variance with the laws of vegetable developement, for reasons which are so obvious, as to render it altogether unnecessary to give them here. I have a monstrous Wheat, specimens of which I communicated in 1830 to M. Kunth and others, in which the midrib of the lower palea actually becomes saccate towards the apex, *bearing an imperfect floret, with stamens, ovarium, and hypogynous scales, in its cavity.* What we know of the tendency to special developement of buds in the margins of leaves, and, from Ferns and the observations of M. Turpin, in the whole substance of certain monocotyledonous leaves, there is nothing in this fact to excite surprise or to give rise to new theories; but it is worth mentioning as the only instance upon record of a flower-bud with sexual apparatus being developed under such circumstances.

The embryo is here described in conformity with the views that are most commonly taken of its nature; that is to say, it is considered to consist of a dilated lenticular cotyledon applied to the albumen on one side, and bearing a naked plumula on the other side, next the testa. It is proper, however, to remark, that the opinion of the late M. Richard, that the part commonly called cotyledon is a peculiar process, and that the plumula is a body contained within the apparent plumula, has been lately adopted by Professor Nees v. Esenbeck, in his *Agrostologia Brasiliensis*, but with some difference. Richard considered the cotyledon to be a part of the radicle, to which he gave the name of macropodal, in consequence of its great supposed enlargement in Grasses and some other families; Nees v. Esenbeck, on the contrary, seems to entertain the opinion that this cotyledon is a special organ, for which he retains Richard's name of hypoblastus, although he does not adopt the view that botanist took of its nature. But I think if we consider the improbability of any special organ being provided for Grasses, which is not found elsewhere, and if we consider how nearly alike are the embryos of Grasses and certain Aroideæ, in which the plumula lies within a cleft of the cotyledon, it is impossible to doubt the identity of the hypoblastus of Richard and Nees v. Esenbeck, and the cotyledons of other Monocotyledons. Indeed, the latter himself appears, in one place, to hesitate about the accuracy of distinguishing them, when he says (p. 9), " Tum vero hypoblastus pars quædam habenda est cotyledoni analoga, magisque ad interiora seminis quam ad externam corculi evolutionem spectans."

The structure of the stem of Grasses is so much at variance, apparently, with that of other Endogenous plants, as to have led Professor Agardh to remark, that it is the least monocotyledonous of all Monocotyledonous plants. It is probable, however, that its peculiarity does not depend so much upon any specific deviation from the ordinary laws of growth in Endogenæ, as upon a separation of the parts at an early period of their growth. The stem of a Grass, it must be remembered, exists in two different states, — that of the rhizoma, and of the culm: the rhizoma, which is the true trunk; and the culm, which may be considered ramifications of it. The rhizoma grows slowly, and differs in no respect from the stem of other Monocotyledons, as is evident in that of the Bamboo. The culm, on the contrary, which grows with great rapidity, is fistular, with a compact impervious diaphragm at each articulation; a fact which must be familiar to every one who has examined a straw or the joint of a Bamboo. In the beginning, when this culm was first developed, it was a solid body like the rhizoma, only infinitely smaller; but in consequence of the great rapidity of its developement, the cellular tissue forms more slowly than the woody vascular bundles which it connects, and in consequence a separation takes place between the latter and the former, except at the articulations, where, by the action of the leaves and their axillary buds, is formed a plexus of vessels, which grows as rapidly as the culm distends, and therefore never separates in the centre. Something analogous to this occurs in the flowering stem of the common Onion among Monocotyledons, and in Umbelliferæ among Dicotyledons.

The strict relation that exists between Palms and Grasses has been already adverted to in speaking of the former order: hence Nees considers Grasses to be a sort of Palms of a lower grade. In reality, the habit of the Calamus and Bambusa genera is nearly alike; the inflorescence of Grasses may be considered to be the same as that of Palms, the floral envelopes of the latter taken away, and only their bracteæ remaining; and, finally, their leaves are formed upon exactly the same plan, with this difference only, that those of Grasses are undivided. With Cyperaceæ, however, it is that Grasses are most properly to be compared: while a manifest tendency, at least to the degree of verticillation requisite to constitute a calyx, evidently takes place in the paleæ of Grasses, Cyperaceæ are destitute of all trace of such a tendency, unless the opposite connate glumes of the female flowers of Carex, or the hypogynous scales of certain Schœnus' and others, be considered an approach to the production of a perianthium. For this reason, Grasses are to be considered plants in a higher state of evolution than Cyperaceæ. Independently of this difference, the orders are readily known by the stems of Grasses being round, those of Cyperaceæ angular; the leaves of Grasses having a ligula at the apex of their sheath, which is split, while the sheath of Cyperaceæ is not split, and is destitute of this ligula; and, finally, the embryo of the two is at variance both in structure and position. With Asphodeleæ their relationship consists in nothing more than the tendency to branch which is observable in part of that order.

GEOGRAPHY. As nothing can be uninteresting which is connected with the habits of a tribe of such vast importance to man, I extract the following account of the geographical distribution of Grasses by Schouw, from Professor Jameson's *Philosophical Journal for April* 1825: —

" The family is very numerous: Persoon's *Synopsis* contains 812 species, 1-26th part of all the plants therein enumerated. In the system of Rœmer and Schultes there are 1800; and, since this work, were it brought to a conclusion, would probably contain 40,000 in all, it may be assumed that the Grasses form a 22d part. It is more than probable, however, that in future the Grasses will increase in a larger ratio than the other phænerogamic plants, and that perhaps the just proportion will be as 1 to 20, or as 1 to 16. Greater still will be their proportion to vegetation in general, when the number of individuals is taken into account; for, in this respect, the greater number, nay, perhaps the whole, of the other classes are inferior.

" With regard to *locality* in such a large family, very little can be advanced. Among the Grasses there are both land and water, but no marine, plants. They occur in every soil, in society with others, and alone; the last to such a degree as entirely to occupy considerable districts. Sand appears to be less favourable to this class; but even this has species nearly peculiar to itself.

" The *diffusion* of this family has almost no other limits than those of the whole vegetable kingdom. Grasses occur under the equator; and Agrostis algida was one of the few plants which Phipps met with on Spitzbergen.

On the mountains of the south of Europe, Poa disticha and other Grasses ascend almost to the snow-line; and, on the Andes, this is also the case with Poa malulensis and dactyloides, Deyeuxia rigida and Festuca dasyantha.

"The *distribution* is of greater importance. As to the chief groups and species, their distribution will not attain a real interest until we shall be in possession of a perfect natural classification; for in this respect we are still, in my opinion, far behind. The division of Beauvois appears to me too artificial, and in that of Brown the groups Paniceæ and Poaceæ are too large. The best, perhaps, is that of Kunth, according to which the Grasses are arranged under ten groups. In respect of latitude, the relation of the Grasses, in the system of Rœmer and Schultes, in the hot and temperate zone, is the following:—

GRASSES.	No. of Species.		Proportion of the Species to the whole of the Grasses.	
	Tor. Zone.	Temp. Zone.	Tor. Zone.	Temp. Zone.
Paniceæ..........	303	103	$\frac{1}{2}-\frac{1}{3}$	$\frac{1}{9}$
Stipaceæ..........	40	58	$\frac{1}{20}$	$\frac{1}{20}$
Agrostideæ........	58	220	$\frac{1}{14}$	$\frac{1}{5}$
Bromeæ	133	554	$\frac{1}{6}$	$\frac{1}{2}$
Chlorideæ	78	30	$\frac{1}{10}$	$\frac{1}{40}$
Hordeaceæ........	33	101	$\frac{1}{25}$	$\frac{1}{12}$
Saccharinæ.........	120	65	$\frac{1}{7}$	$\frac{1}{8}$
Oryzeæ	10	9	$\frac{1}{32}$	$\frac{1}{133}$
Olyreæ............	18	4	$\frac{1}{46}$	$\frac{1}{300}$
Bambusaceæ.......	6	3	$\frac{1}{137}$	$\frac{1}{400}$

"Hence it follows that not one of these groups belongs exclusively to either the one or the other zone, but that, on account of the proportionally greater number, the Paniceæ, Chlorideæ, Saccharinæ, Oryzeæ, Olyreæ, and Bambusaceæ, may be regarded as tropical, and Agrostideæ, Bromeæ, and Hordeaceæ, as extra-tropical forms; and that there is, consequently, a considerable contrast between the former of these two zones. On the contrary, the difference between the various continents and degrees of longitude is inconsiderable. Neither in the torrid nor temperate zones has any group in the continent a perceptible preponderance over another. The result also appears to be the same, on comparing the two hemispheres: we know, however, too little of the southern to state this precisely. In respect of elevation, the distribution, according to the degrees of latitude, is very similar; for, in the mountains of South America, the proportions of the larger groups are:—

	0–200 Toises.	200–1100 Toises.	1100–1600 Toises.	Above 1600 Toises.
Paniceæ.......	39	33	12	1
Agrostideæ.....	6	10	23	2
Bromeæ.......	7	7	37	8
Saccharinæ	16	20	20	2

" Between the genera the contrast is naturally greater, and manifests itself not only according to latitude, but also longitude. Thus, in the torrid zone, the genus Paspalus has a decided preponderance in the New World. Most of the genera, however, especially the larger, for example, Panicum, Andropogon, Chloris, are every where nearly equal, those that are peculiar being generally not at all numerous. The generic difference between North America and the temperate regions of the European continent is very small. In North America, however, a greater number of tropical forms appears. Between the two temperate zones also the distinction seems to be by no means considerable. Of 36 genera from the Cape, 30 occur in the temperate zone of the northern hemisphere, while, in other families, southern Africa has many peculiar to itself. In the extra-tropical part of New Holland the greater number of genera is found also in the north (about 2-3ds); and this appears to be still more the case in the southern parts of South America, as well as New Zealand. One of the most extensively distributed genera is Poa. It is found almost over the whole earth; and, although it reaches its maximum in the temperate, has also many species in the torrid zone.

" What has been said of the decided influence of the degrees of latitude on groups and genera, holds also of the *habitus* of vegetation in general. The greatest differences between tropical and extra-tropical Grasses appear to be the following :—

" 1. The tropical Grasses acquire a much greater height, and occasionally assume the appearance of trees. Some species of Bambusa are from 50 to 60 feet high.

" 2. The leaves of the tropical Grasses are broader, and approach more in form to those of the other families of plants. Of this the genus Paspalus affords many examples.

" 3. Separate sexes are more frequent in the tropical Grasses. Zea, Sorghum, Andropogon, Olyra, Anthistiria, Ischæmum, Ægilops, and many other genera, which only occur in the torrid zone, and are there found in perfection, are monœcious, or polygamous. Holcus is perhaps the only extra-tropical genus with separate sexes.

" 4. The flowers are softer, more downy, and elegant.

" 5. The extra-tropical Grasses, on the contrary, far surpass the tropical in respect of the number of individuals. That compact grassy turf, which, especially in the colder parts of the temperate zones, in spring and summer, composes the green meadows and pastures, is almost entirely wanting in the torrid zone. The Grasses there do not grow crowded together, but, like other plants, more dispersed. Even in the southern parts of Europe, the assimilation to the warmer regions, in this respect, is by no means inconsiderable. Arundo donax, by its height, reminds us of the Bamboo;

Saccharum Ravennæ, S. Teneriffæ, Imperata arundinacea, Lagurus ovatus, Lygeum spartum, and the species of Stipa, by their soft, downy, elegant flowers; and the species of Andropogon, Ægilops, &c. by separate sexes, exhibit tropical qualities. The Grasses are also less gregarious, and meadows seldomer occur, in the south than in the north of Europe.

"As to what relates to the distribution of individuals, the generality of species are social plants.

"Lastly,—Do we wish to know how this family is distributed, in respect of the number of species, and where they reach their maxima and minima? The following materials may supply, not indeed either a complete or faithful representation, because the Grasses are not treated of by botanists or travellers in general with the same care as the other families; but they will at least give some hints towards effecting that object. In Persoon's *Synopsis*, the Grasses of the torrid zone form 1-25th, and those of the temperate zone 1-22d of the whole vegetation; but when it is considered that the Grasses of the former have been less investigated than the European, the quotient would be nearly alike in both zones. In the systems of Römer and Schultes, tropical are to the European Grasses as 2 to 3; but this, from a probable conjecture, is also the proportion of all tropical and extra-tropical plants. In Persoon's *Synopsis* it is as 1 to 2; and since the publication of that work, the knowledge of tropical has been enlarged in a greater proportion than that of extra-tropical plants. Although, however, the quotients in the torrid and temperate zones may be nearly equal upon the whole, when taken in subdivisions there will be an inequality. In the warm regions of South America, the Grasses, under 200 toises elevation, form from 1-15th to 1-16th of the whole; in the West Indies 1-17th; on the river Essequibo, in Guyana, 1-12th to 1-15th; on the river Congo 1-12th to 1-13th; in Guyana 1-10th; (in the three last the local circumstances are peculiarly favourable for the Grasses); in the East Indies, according to Brown, 1-12th; in Arabia 1-15th; and in tropical New Holland 1-10th to 1-11th. Now, attending to the circumstance, that tropical are scarcely so well known as other phænerogamic plants, it is not improbable that the true quotient for the torrid zone is 1-10th to 1-12th. In the warmer parts of the temperate zone the Grasses appear to form a smaller proportion of the vegetation; for, in the extra-tropical parts of New Holland, they form from 1-24th to 1-25th, at the Cape 1-35th, in Greece 1-15th to 1-16th, in the Canary Islands 1-12th to 1-13th, in the Crimea and Caucasus 1-14th to 1-15th, in Naples 1-11th to 1-12th, in France 1-13th, and in Egypt (where, however, the circumstances are peculiarly favourable) 1-8th. Farther north the relative numbers seem to rise somewhat higher; in Germany 1-13th, in Great Britain 1-11th to 1-12th, in Denmark 1-10th to 1-11th, in Scandinavia 1-10th to 1-11th, in Kamchatka 1-7th to 1-8th, Lapland 1-10th, Iceland 1-8th to 1-9th, Greenland 1-8th to 1-9th, and in North America, according to Pursh, 1-14th to 1-15th. We may assume, perhaps, as a medium for the warmer parts of the temperate zone, 1-12th to 1-14th; for the colder, together with the polar regions, 1-8th to 1-10th. That almost in every Flora the quotient is considerably higher than in the works of Persoon, and of Römer and Schultes, affords another proof, that, in the rule, the distribution of the Grasses is more extensive that that of the other phænerogamic plants.

"In southern Europe the number of the Grasses seems to diminish according to the elevation, for in the Alpine Flora they are only 1-18th. Their distribution according to elevation does not, therefore, accord with that of the latitude; in South America the agreement is greater, for the relative numbers are, 0 to 200 toises, 1-15th to 1-16th; 200 to 1100 toises, 1-15th to 1-16th; 1100 to 1600 toises, 1-11th; above 1600 toises, 1-14th.

" A detailed representation of the distribution of the cultivated Gramina would certainly be very interesting. Here we must restrict ourselves to a short and general outline. We shall endeavour to specify those Gramina which are the prevailing ones in the large zones and continents, mentioning, in passing, those plants of other families which either supply the place of, or are associated with, the different kinds of grain, as the chief article of food. This distribution is determined, not merely by climate, but depends on the civilisation, industry, and traffic of the people, and often on historical events.

" Within the northern polar circle, agriculture is found only in a few places. In Siberia grain reaches at the utmost only to 60°, in the eastern parts scarcely above 55°, and in Kamchatka there is no agriculture even in the most southern parts (51°). The polar limit of agriculture on the north-west coast of America appears to be somewhat higher; for, in the more southern Russian possessions (57° to 58°), barley and rye come to maturity. On the east coast of America it is scarcely above 50° to 52°. Only in Europe, namely, in Lapland, does the polar limit reach an unusually high latitude (70°). Beyond this, dried fish, and here and there potatoes, supply the place of grain.

" The grains which extend farthest to the north in Europe are barley and oats. These, which in the milder climates are not used for bread, afford to the inhabitants of the northern parts of Norway and Sweden, of a part of Siberia and Scotland, their chief vegetable nourishment.

" Rye is the next which becomes associated with these. This is the pre-vailing grain in a great part of the northern temperate zone, namely, in the south of Sweden and Norway, Denmark, and in all the lands bordering on the Baltic; the north of Germany, and part of Siberia. In the latter, another very nutritious grain, buck-wheat, is very frequently cultivated. In the zone where rye prevails, wheat is also generally to be found; barley being here chiefly cultivated for the manufacture of beer, and oats supplying food for the horses.

" To these there follows a zone in Europe and western Asia, where rye disappears, and wheat almost exclusively furnishes bread. The middle, or the south of France, England, part of Scotland, a part of Germany, Hun-gary, the Crimea and Caucasus, as also the lands of middle Asia, where agriculture is followed, belong to this zone. Here the vine is also found; wine supplants the use of beer; and barley is consequently less raised.

" Next comes a district where wheat still abounds, but no longer exclu-sively furnishes bread, rice and maize becoming frequent. To this zone belong Portugal, Spain, part of France on the Mediterranean, Italy, and Greece; further, the countries of the East, Persia, northern India, Arabia, Egypt, Nubia, Barbary, and the Canary Islands; in these latter countries, however, the culture of maize or rice, towards the south, is always more con-siderable, and in some of them several kinds of Sorghum (Doura) and Poa Abyssinica come to be added. In both these regions of wheat, rye only occurs at a considerable elevation; oats, however, more seldom, and at last entirely disappear; barley affording food for horses and mules.

" In the eastern parts of the temperate zone of the Old Continent, in China and Japan, our northern kinds of grain are very unfrequent, and rice is found to predominate. The cause of this difference between the east and the west of the Old Continent appears to be in the manners and peculiarities of the people. In North America, wheat and rye grow as in Europe, but more sparingly. Maize is more reared in the Western than in the Old

Continent, and rice predominates in the southern provinces of the United States.

" In the torrid zone, maize predominates in America, rice in Asia, and both these grains in nearly equal quantity in Africa. The cause of this distribution is, without doubt, historical; for Asia is the native country of rice, and America of maize. In some situations, especially in the neighbourhood of the tropics, wheat is also met with, but always subordinate to these other kinds of grain. Besides rice and maize, there are, in the torrid zone, several kinds of grain, as well as other plants, which supply the inhabitants with food, either used along with them, or entirely occupying their place. Such are, in the New Continent, Yams (Dioscorea alata), the Manihot (Jatropha manihot), and the Batatas (Convolvulus batatas), the root of which, and the fruit of the Pisang (Banana, Musa), furnish universal articles of food. In the same zone, in Africa, Doura (Sorghum), Pisang, Manihot, Yams, and Arachis hypogæa. In the East Indies, and on the Indian Islands, Eleusine coracana, E. stricta, Panicum frumentaceum; several Palms and Cycadeæ, which produce the Sago; Pisang, Yams, Batatas, and the Bread-fruit (Artocarpus incisa). In the islands of the South Sea, grain of every kind disappears, its place being supplied by the Bread-fruit tree, the Pisang, and Tacca pinnatifida. In the tropical parts of New Holland there is no agriculture, the inhabitants living on the produce of the Sago, of various Palms, and some species of Arum.

" In the high lands of South America there is a distribution similar to that of the degrees of latitude. Maize, indeed, grows to the height of 7200 feet above the level of the sea, but only predominates between 3000 and 6000 of elevation. Below 3000 feet it is associated with the Pisang, and the above-mentioned vegetables; while, from 6000 to 9260 feet, the European grains abound; wheat in the lower regions, and rye and barley in the higher; along with which Chenopodium Quinoa, as a nutritious plant, must also be enumerated. Potatoes alone are cultivated from 9260 to 12,300 feet.

" To the south of the tropic of Capricorn, wherever agriculture is practised, considerable resemblance with the northern temperate zone may be observed. In the southern parts of Brazil, in Buenos Ayres, in Chile, at the Cape of Good Hope, and in the temperate zone of New Holland, wheat predominates; barley, however, and rye, make their appearance in the southernmost parts of these countries, and in Van Diemen's Land. In New Zealand the culture of wheat is said to have been tried with success; but the inhabitants avail themselves of the Acrostichum furcatum as the main article of sustenance.

" Hence it appears, that, in respect of the predominating kinds of grain, the earth may be divided into five grand divisions, or kingdoms. The kingdom of Rice, of Maize, Wheat, and Rye, and lastly of Barley and Oats. The first three are the most extensive; the Maize has the greatest range of temperature; but Rice may be said to support the greatest number of the human race."

PROPERTIES. The uses of this most important tribe of plants, for fodder, for food, and for clothing, require little illustration. The abundance of wholesome fæcula contained in all their seeds renders them peculiarly well adapted for the sustenance of man; and if the Corn tribe only, such as Wheat, Barley, Oats, Maize, Rice, and Guinea Corn, are the kinds commonly employed, it is because of the large size of their seeds compared with those of other Grasses, for none are unwholesome in their natural state, with the single exception of Lolium temulentum, a common weed in many parts of England, the effects of which are undoubtedly deleterious, although perhaps much exaggerated. In this respect an approach seems to be naturally made to the properties of half-putrid Wheat, which are known to be dangerous. The grain of Eleusine coracana is cultivated as corn, under the name of Natchenny, upon the Coromandel Coast. *Ainslie*, 1. 245. Independently of their nutritive fæcula, Grasses contain a large proportion of two other principles which deserve especial mention, viz. sugar and silex. The abundance of the former in the Sugar-cane is the cause of its extensive cultivation ; but a large quantity exists in many other Grasses, some of which, such as Holcus saccharatus, have actually been grown as substitutes for the Sugar-cane in Italy ; its presence in the nascent embryo of Barley is the cause of that grain being employed under the name of malt in the preparation of beer and of ardent spirits. Dr. Chisholm says, that the juice of the Sugar-cane is the best antidote to arsenic. *Ed. P. J.* 4. 221. That the cuticle of Grasses contains a large proportion of silex, is proved by its hardness, and by large masses of vitrified matter being found whenever a hay-stack or heap of corn is accidentally consumed by fire. In the joints of some Grasses a perfect siliceous deposit is found, particularly in a kind of jungle Grass mentioned in a letter from Dr. Moore to Dr. Kennedy of Edinburgh. *Ibid.* 2. 192. It is also said that Wheat-straw may be melted into a colourless glass with the blow-pipe, without any addition. Barley-straw melts into a glass of a topaz yellow colour. *Ibid.* 2. 194. The siliceous matter of the Bamboo is often secreted at the joints, where it forms the singular substance called tabasheer, of which see a very interestiug account in Dr. Brewster's *Journal*, 8. 268. It was found by Dr. Turner that the tabasheer of India consisted of silica containing a minute quantity of lime and vegetable matter. A coarse soft paper, of excellent quality, is manufactured in India from the tissue of the Bamboo. A cooling drink is prepared in India from the roots of Cynodon Dactylon. *Ainslie*, 2. 27. The fragrance of some Grasses, such as Anthoxanthum odoratum and Holcus odoratus, depends, according to Vögel, upon the presence of Benzoic acid. *Ed. P. J.* 14. 170. Sulphur exists, in combination with different bases, in Wheat, Barley, Rye, Oats, Maize, Millet, and Rice. *Ibid.* 172. The Arundo arenaria is an invaluable species for keeping together the blowing sands of the sea-coast, by its creeping suckers and tough entangled roots. It is employed in the Hebrides for many economical purposes, being made into ropes for various uses, mats for pack-saddles, bags, hats, &c. *Ibid.* 6. 155. Some of the Reeds of Brazil, called Taquarussa, grow from 30 to 40 feet high, with a diameter of six inches ; they form thorny impenetrable thickets, and are exceedingly grateful to hunters ; for, on cutting off such a reed below the joint, the stem of the younger shoots is found to be full of a cool pleasant liquid, which immediately quenches the most burning thirst. *Pr. Max. Trav.* 81. The roasted leaves of Andropogon Schænanthus are used in India, in infusion, as an excellent stomachic. An essential oil of a pleasant taste is extracted from the leaves in the Moluccas ; and the Javanese esteem the plant much as a mild aromatic and stimulant. *Ainslie*, 2. 58. This is the Grass oil of Nemaur, called in India Ivarancusa, and described in *Brewster's Journal*, 9. 333. Many others, such as Andropogon citratum and nardus, and Anthoxanthum odoratum, partake in the same qualities. The gluten of Wheat yields the two chemical principles called gliadine and zimome. *Ann. of Phil.* no. 89. p. 390. M. Decandolle truly remarks, that the dangerous effects of the ergot of Corn is no exception to the generally wholesome properties of the order, because in this the whole grain is in a state of disease. The ergot of Rye has been lately found to exercise a decidedly powerful stimulant effect upon the uterus, on which account it is now frequently and successfully employed by European practitioners in cases of difficult parturition. The ergot of Maize is, according to M. Roulin, very common in Colombia, and the use of it is attended with a shedding of the hair, and even the teeth, of both man and beast. Mules fed on it lose their hoofs, and fowls lay eggs without shell. Its action upon the uterus is as powerful as that of the Rye ergot, or perhaps more so. *Ann. des Sc.* 19. 279. The country name of the Maize thus affected is Maïs peladero. The best fodder Grasses of Europe are usually dwarf species, or at least such as do not rise more than 3 or 4 feet above the ground, and of these the larger kinds are apt to become hard and wiry; the most esteemed are Lolium perenne, Phleum, and Festuca pratense, Cynosurus cristatus, and various species of Poa and dwarf Festuca, to which should be added Anthoxanthum odoratum for its fragrance. But the fodder Grasses of Brazil are of a far more gigantic stature, and perfectly tender and delicate. We learn from Nees von Esenbeck, that the Caapim de Angola of Brazil, Panicum spectabile, grows 6 or 7 feet high; while other equally gigantic species constitute the field crops on the banks of the Amazons.

EXAMPLES. It is no easy matter to decide upon the arrangement of Grasses which is most likely to be eventually adopted, when we find such men as Brown, Kunth, Palisot, Link, and Trinius, advocating different methods; and it would be quite beyond my purpose to give all of them here. Upon the whole, the following, which is that employed by Nees v. Esenbeck in his excellent account of the Grasses of Brazil, has the best prospect of becoming established among botanists : —

1. Paniceæ, *Kunth.* (Panicum, Paspalus, Cenchrus.)
2. Olyreæ, *Kunth.* (Luziola, Pharus, Olyra.)
3. Saccharineæ, *Kunth.* (Saccharum, Andropogon, Anthistiria.)
4. Stipeæ, *Kunth.* (Stipa, Chætaria.)
5. Agrosteæ, *Kunth.* (Phalaris, Vilfa, Agrostis, Spartina.)
6. Chlorideæ, *Kunth.* (Pappophorum, Chloris, Eleusine.)
7. Hordeaceæ, *Kunth.* (Lolium, Triticum, Secale.)
8. Festucaceæ, *Kunth.*

§ 1. Avenaceæ, *Kunth.* (Avena.)
§ 2. Arundinaceæ, *Kunth.* (Arundo, Gynerium.)
§ 3. Festuceæ, *Kunth.* (Cynosurus, Bromus, Poa.)

9. Oryzeæ, *Kunth.* (Leersia, Oryza.)
10. Bambuseæ, *Kunth.*

§ 1. Triglossæ, *Link.* (Arundinaria.)
§ 2. Bambuseæ veræ, *Nees.* (Bambusa, Streptochæta.)

CCLXII. CYPERACEÆ. THE SEDGE TRIBE.

CYPEROIDEÆ, *Juss. Gen.* 26. (1789); *Link Hort. Botanic.* 1. (1827). — CYPERACEÆ, *R. Brown Prodr.* 212. (1810); *Lestiboudois Essai; Dec. and Duby,* 483. (1828); *Lindl. Synops.* 278. (1829.)

DIAGNOSIS. Glumaceous monocotyledons, with angular stems, entire leaf-sheaths, and an undivided embryo included within the albumen.

ANOMALIES. The glumes of Carex and Uncinia are united by their margins, so as to form an external covering to the pistillum.

ESSENTIAL CHARACTER. — *Flowers* hermaphrodite or unisexual, consisting of imbricated solitary bracteæ, very rarely enclosing other opposite bracteæ at right angles with the first, called *glumes. Perianthium* none, unless the glumes, when present, be so considered, or the hypogynous setæ. *Stamens* hypogynous, definite, 1, 2, 3, 4, 5, 6, 7, 10, 12 ; *anthers* fixed by their base, entire, 2-celled. *Ovary* 1-seeded, often surrounded by bristles called hypogynous setæ, probably constituting the rudiments of a perianthium; *ovulum* erect; *style* single, trifid, or bifid; *stigmas* undivided, occasionally bifid. *Nut* crustaceous or bony. *Albumen* of the same figure as the seed; *embryo* lenticular, undivided, enclosed within the base of the albumen; *plumula* inconspicuous.—*Roots* fibrous. *Stems* very often without joints, 3-cornered, or taper. *Leaves* with their sheaths entire. The lowermost bracteæ often sterile.

AFFINITIES. These so nearly resemble the last tribe in appearance, that the one may be readily mistaken for the other by incurious persons; they are, however, essentially distinguished by many important points of structure. In the first place, their stems are solid and angular, not round and fistular; there is no diaphragm at the articulations; their flowers are destitute of any other covering than that afforded them by a single bractea, in the axilla of which they grow, with the exception of Carex, Uncinia, and Diplacrum, in which 2 opposite glumes are added; and, finally, the seed has its embryo lying in one end of the albumen, within which its cotyledonar extremity is enclosed, and not on the outside, as in Grasses; a very important fact, which it is the more necessary to point out, as Mr. Brown describes it (*Prodr.* 212) as lenticular and placed on the outside of the albumen. The additional glumes above adverted to form what Linnean botanists call the nectary or arillus! Mr. Brown mentions a case where these glumes, which he calls a capsular perianthium, included stamens instead of a pistillum. According to Turpin, rudiments of them sometimes appear in different species of Mariscus. The close affinity of Cyperaceæ, on the one hand, to Grasses, is sufficiently apparent; on the other, they approach Junceæ and Restiaceæ, in the glumaceous state of the perianthium, and in general habit. They are at once known from Restiaceæ by the sheaths of the leaves not being slit. The species are extremely difficult to determine, and the distinctive characters of the genera are unsatisfactory.*

GEOGRAPHY. Found in marshes, ditches, and running streams, in meadows and on heaths, in groves and forests, on the blowing sands of the sea-shore, on the tops of mountains, from the arctic to the antarctic circle, wherever Phænogamous vegetation can exist. Humboldt remarks, that in Lapland Cyperaceæ are equal to Gramineæ, but that thence, from the temperate zone to the equator, in the northern hemisphere, the proportion of Cyperaceæ to Gramineæ very much diminishes. As we approach the line, the character of the order also changes: Carex, Scirpus, Schœnus, and their allies, cease to form the principal mass of the order, the room of which is usurped by Cyperus, Kyllinga, Mariscus, and the like, genera comparatively unknown in northern regions, or at least not forming any marked feature in the vegetation. A few species are common to very different parts of the world, as Scirpus triqueter and capitatus, and Fuirena umbellata, to New Holland and South America, and several Scirpuses to Europe and the southern hemisphere.

PROPERTIES. While Grasses are celebrated for their nutritive qualities, and for the abundance of fæcula and sugar they contain, Sedges are little less remarkable for the frequent absence of those principles: hence they are scarcely eaten by cattle. The roots of Carex arenaria, disticha, and hirta, have diaphoretic and demulcent properties, on which account they are called German Sarsaparilla. Those of Cyperuses are succulent, and filled with a nutritive and agreeable mucilage. In Cyperus longus a bitter principle is

* It is to be hoped that much light will be thrown upon the subject by M. Prescott, of St. Petersburgh, who has long been making these plants his especial study, and to whom all botanists who wish well to science ought to confide whatever materials they may be able to spare.

superadded, which gives its roots a tonic and stomachic quality. *Dec.* The
tubers of Cyperus rotundus are said by General Hardwicke to be administered successfully in cases of cholera by Hindoo practitioners, who call the
plant Mootha. Those of C. perennis, or Nagur-Mootha, are, when dried
and pulverised, used by Indian ladies for scouring and perfuming their hair.
Trans. M. and P. Soc. Calc. 2. 400. The root of Cyperus odoratus has a
warm aromatic taste, and is given in India, in infusion, as a stomachic.
Ainslie, 2. 58. Cyperus Hydra is said by Dr. Hamilton to be a pest to the
sugar-cane plantations of the West India Islands, overrunning them and
rendering them barren. The planters call it Nut Grass. *Prodr. Fl. Ind.*
p. 13. The root of Scleria lithosperma is supposed, upon the Malabar coast,
to have antinephritic virtues. *Ainslie*, 2. 121. The papyrus of the Egyptians
was obtained from a plant of this order, Cyperus Papyrus. Various Scirpuses and similar plants are applied to domestic purposes, such as making
the bottoms of chairs, the wicks of candles, the stuffing of cushions, &c.

EXAMPLES. M. Lestiboudois divides Cyperaceæ thus : —

§ 1. Scirpeæ. (Scirpus, Eriophorum.)
§ 2. Kobresieæ. (Elyna, Kobresia.)
§ 3. Cypereæ. (Cyperus, Kyllinga.)
§ 4. Chrysitriceæ. (Chorizandra, Chrysitrix.)

But this arrangement has little merit. M. Kunth uses the following : —

§ 1. True Cyperaceæ. (Cyperus, Kyllinga.)
§ 2. Scirpeæ. (Scirpus, Schœnus.)
§ 3. Sclerinæ. (Scleria.)
§ 4. Caricinæ. (Carex, Uncinia.)

Class II. CELLULARES, or FLOWERLESS PLANTS.

ACOTYLEDONES, *Juss. Gen.* 1. (1789). — EXEMBRYONATÆ *or* ARHIZÆ, *Rich. Anal. du Fr.* (1808). — CELLULARES, *Dec. Fl. Fr.* 1. 68. (1815); *Lindl. Synops.* p. 3. (1829). — ACOTYLEDONEÆ *and* PSEUDOCOTYLEDONEÆ, *Agardh Aph.* 72. (1821). — AGAMÆ, CRYPTOGAMOUS *or* ÆTHEOGAMOUS PLANTS *of authors; Ad. Brongniart in Dict. Class.* 5. 155. (1824).—NEMEA, *Fries. Syst. Orb. Veg.* 1. 30. (1825.)

ESSENTIAL CHARACTER.—Substance of the plant composed of cellular tissue chiefly, either in a spheroidal or elongated state ; spiral vessels wholly absent ; annular ducts present in some. *Cuticle* generally destitute of stomata. *Sexual organs*, and consequently *flowers*, absent. *Reproduction* taking place either by *sporules*, which are enclosed in particular cases called *thecæ*, or imbedded in the substance of the plant, or else by a mere dissolution of the utricles of cellular tissue ; *germination* occurring at no fixed point, but upon any part of the surface of the sporules.

Such are the characters by which this class of the vegetable kingdom is distinguished from the last ; characters of so marked a kind as to render it impossible to refer individuals of one to the other. The universal want of flowers, and of all kinds of sexual apparatus ; the total absence of spiral vessels, the place of which is only occasionally supplied by annular ducts ; and the non-existence of a true trunk (for the stipes of Ferns, composed only of the united bases of the leaves or fronds, is scarcely analogous to the trunk of Vascular plants) ; and, finally, the near approach in the most simple tribes, such as Arthrodieæ and Chaodineæ, to the nature of infusorial animalcules, are all facts, the accuracy of which is undisputed, and which have no parallel in flowering plants. It is true that sexual apparatus has been described by various authors in many of the tribes of Cellulares ; but it is equally certain, that if such a provision for propagation ever exists, which is extremely doubtful, it is in a most imperfect state, and by no means analogous to what we call the sexes in Vasculares ; and it is even conjectured that the simplest forms of Lichens, Fungi, and Algæ, are produced by a kind of equivocal generation, from a common form of matter having no inherent special tendency to control its mode of developement, but appearing as a Lichen, Alga, or Fungus, according to the peculiar conditions of soil and atmosphere under which it is called into action. Upon this subject more will be said, in speaking of those orders hereafter.

Flowerless may be said to approach Flowering plants by Ferns, which have a certain relation to Cycadeæ, by Lycopodineæ, which may be compared in many respects to Coniferæ, and by Equisetaceæ, which have a great external resemblance to Casuarina.

The subject of Cryptogamic botany is not less obscure than extensive ; it is usually, among botanists, an object of separate attention, especially in the lower tribes ; and I think I shall best consult the interest of readers of this work, by stating the opinions of those who have given the greatest attention to particular tribes, rather than by offering any thing novel myself. I trust, however, I may, without incurring the charge of presumption from those great cryptogamists whose lives have been devoted to the study of the subject, offer here and there a few remarks upon the analogy that exists

between the more anomalous forms of Cellulares and those of Flowering
Plants : I venture to do this with the more confidence, because the truth of
any opinions I may advance will have to be tried by the general laws of
vegetable organisation, and upon principles which do not depend upon an
extensive acquaintance with species.

We have seen that in Vascular plants the great divisions of Monocotyledons and Dicotyledons, or of Endogenous and Exogenous plants, have been
satisfactorily established. In Cellulares attempts have been made to establish parallel divisions, but, I fear, without much success ; these plants
appearing to be analogous rather to one of the two divisions of Vasculares,
than to comprehend within themselves groups of equally different organisation.

M. Decandolle refers Ferns and their immediate allies to Endogenous
plants, and separates the remainder into *Foliaceæ*, or plants with leafy expansions, and *Aphyllæ*, or those destitute of leaves : but to the first of these
there are grave objections ; the second nearly corresponds with the arrangement here adopted.

Agardh, in 1821, divided them thus : — ACOTYLEDONEÆ, or leafless
plants, with all the parts confluent, the colour not herbaceous, with no
sexes, and propagated by sporidia. (Sporidium est corculum nudum, radi-
culâ, cotyledone, et hilo destitutum. *Aph.* 71.) PSEUDOCOTYLEDONEÆ, or
leafy plants, the parts of which are sometimes confluent, the colour green,
with an attempt at producing sexes, and propagated by sporules enclosed
in capsules. (Spora est corculum perispermio (?) et membranâ simplici hilo
destitutâ inclusum, germinatione cotyledonidium (analogon cotyledoni folium)
explicans. *Ibid.* 71.) To Acotyledoneæ he refers only Fungi, Lichens, and
Algæ, and comprehends the remainder in Pseudocotyledoneæ. This arrangement is undoubtedly natural, but it is liable to objection, on the ground,
that although the two groups are distinct, yet it is extremely uncertain whether the characters assigned to each are founded upon accurate observation.
For instance, the distinction drawn between their modes of reproduction or
germination is altogether arbitrary. It is well known that Mosses and Confervæ are so similar when germinating, that young plants of the former have
been described as belonging to the latter tribe (see Mr. Drummond's paper
in the *Transactions of the Linnean Society*, 15. p. 20.); and yet one is said to
increase by sporules, and the other by sporidia. The confluence of all the
parts in Acotyledoneæ, and the separation of them in Pseudocotyledoneæ,
will not distinguish them ; witness Marchantia, Riccia, &c. in the latter, and
such species as Caulerpa hypnoides in the former. Colour is a still less
satisfactory difference : for example, what green have we in Mosses or
Ferns, or other Pseudocotyledoneæ, more intense than in Ulva and numerous Algæ among Acotyledoneæ ? As to a supposed tendency to developement of sexes in one and not in the other, this may possibly be the
case ; but it is no character of the two groups ; for what better proof have
we of any such tendency existing in Lycopodineæ or Hepaticæ, than in
Lichens.

Fries, in his *Plantæ Homonemeæ*, adopts these divisions, but assigns
them new names and characters. He calls the Acotyledoneæ of Agardh
Homonemea, and the Pseudocotyledoneæ he terms Heteronemea, with the
following characters : — HETERONEMEA. Germinating filaments, combining
in a heterogeneous body, with some analogy to the difference of sexes.
Tissue consisting of cellules regularly united. HOMONEMEA. Germinating
filaments, either distinct or combining in a homogeneous body, with no trace
of sexual differences. Tissue consisting of anomalous, somewhat filamentous
cellules.—I scarcely know whether to consider these definitions more satis-

factory than those of Agardh; perhaps they are : but their fault is evidently that of being too hypothetical, and of not distinctly deciding the position of Hepaticæ.

Struck, perhaps, with this objection, M. Adolphe Brongniart has more recently proposed a triple division of Cellular plants, in the following manner : — I. Neither vessels nor foliaceous appendages; no trace of sexual organs; sporules contained in indehiscent capsules, or bursting irregularly, with no kind of proper integument. These answer to the Acotyledones of Agardh and the Homonemea of Fries. II. No vessels, but foliaceous appendages; sexual organs doubtful; sporules contained in great numbers in capsules that burst regularly, and having a proper integument. Ex. Hepaticæ and Mosses. III. Vessels present, and foliaceous appendages; sexual organs certainly existing in some; sporules contained in polyspermous and dehiscent, or monospermous indehiscent capsules. Ex. Ferns and their allies, with Chara.—To the definitions of these, several objections might be taken, particularly to all that part which relates to the supposed presence of sexual organs; but the divisions themselves appear less exceptionable than any others that have been proposed. They are therefore adopted here, with such an alteration of their definitions as will render them less open to criticism. They are in conformity with the view that has been taken of the subject by Nees v. Esenbeck, in his and Ebermaier's excellent *Medical Botany*, which only reached me after the whole of the preceding matter had been written.

FLOWERLESS plants may be considered to exist in three principal forms: first, those in which a distinct vascular system exists; secondly, those in which no vascular system exists, but which have a central axis of developement; and thirdly, those which have neither a vascular system nor a central axis, but are mere homogeneous masses ramified irregularly. The two former have their reproductive bodies, or sporules, arranged in cases provided for their elaboration and ultimate dispersion ; in the latter the sporules lie in the substance of the plant, and can only be disseminated by its destruction. These may be called *Fern-like, Moss-like,* and *Leafless* Flowerless Plants.

Tribe I. FILICOIDEÆ, or FERN-LIKE PLANTS.

ENDOGENÆ CRYPTOGAMÆ, *Dec. Théor. Elém.* 249. (1829). — PSEUDOCOTYLEDONEÆ, Classes 2, 3, and 4; *Agardh Aph.* 103. (1822). — HETERONEMEA, *Fries Syst. Orb. Veg.* 33. (1825), *in part.* — ACOTYLEDONES, Class 3; *Ad. Brongn. in Dict. Class.* 5. 159. (1824). — CRYPTOGAMICÆ, 3d Circle, *T. F. L. Nees v. Esenbeck and Ebermaier Handb. der Med. Bot.* 1. 18. (1830.)

DIAGNOSIS. Flowerless plants, with a stem having a vascular system and distinct leaves; their sporules having a proper integument, and contained in distinct axillary or dorsal thecæ.

This differs from the third class of M. Brongniart in the exclusion of Characeæ, which are known to be destitute of a vascular system, and which more properly belong to the next section, connecting it with the third; as Marsileaceæ unite the first and second. Von Esenbeck and Ebermaier also exclude this family, referring it to the third or leafless tribe.

LIST OF THE ORDERS.

263 Equisetaceæ.
264 Filices.

265 Lycopodiaceæ.
266 Marsileaceæ.

CCLXIII. EQUISETACEÆ. The Horse-tail Tribe.

EQUISETACEÆ, *Dec. Fl. Fr.* 2. 580. (1815); *Agardh Aph.* 119. (1822); *Kaulfuss Enum. Filicum,* 1. (1824); *Greville Flora Edin.* xiii. (1824); *Adolphe Brongniart Hist. Veg. Foss.* 99. (1828.)

DIAGNOSIS. Flowerless plants, with their sporules surrounded by elastic clavate filaments, and enclosed in thecæ arising from the scales of terminal cones. Vernation straight.

ANOMALIES.

ESSENTIAL CHARACTER. — *Leafless* branched plants, with a striated fistular stem, beneath the cuticle of which silex is secreted; the *articulations* separable, and surrounded by a membranous toothed sheath. *Reproductive organs* consisting of 1-valved thecæ bursting longitudinally, and arranged upon cuneate scales, which are collected into strobiliform heads; *sporules* surrounded by minute granules, and having at their base 4 elastic clavate filaments, twisted spirally round them when dry, but expanding when moistened.

AFFINITIES. The very remarkable plants known by the vulgar name of horsetails, seem to have no very decided affinity to any existing tribes. With Ferns their relation is far from obvious, depending almost entirely upon the want of sexes, and the presence of annular ducts without spiral vessels. In the arrangement and appearance of their reproductive organs they have a striking resemblance to Zamia, and in general aspect to Casuarina. Their germination is that of Cellular plants, and approaches nearly to Mosses. Upon the whole, they must be considered an exceedingly anomalous tribe, approaching Coniferæ through Cycadeæ more closely than any thing else. The curious structure of their stem is well described by Ad. Brongniart in his *History of Fossil Vegetables,* as are, indeed, all the parts of their organisation: see Tables 11 and 12 of that work. This ingenious writer entertains the opinion that the green body, which is known to be the sporule, is a naked ovulum, and the 4 swollen filaments that surround it 4 grains of pollen united in pairs to the base of the ovulum. It is probable that the nearest approach to the structure of sexual organs does take place here, and that, considering the analogy between the thecæ of Equisetum and the lobes of the anther of Coniferæ, and the filaments of the former and the quaternary grains of pollen of Cycas, the parallel drawn by M. Brongniart is just; but it must, at the same time, I think, be admitted, that it is very doubtful whether, in this order, the parts are any thing more than representatives of the sexual apparatus, without the power of performing its functions.

The germination of the sporules has been explained, both by Agardh and Bischoff. The former (*Aphor.* 120) describes it thus: From 3 to 14 days after they are sown, they send down a filiform, hyaline, somewhat clavate, simple root, and protrude a confervoid, cylindrical, obtuse, articulated, torulose thread, either 2-lobed (in E. pratense) at the apex, or simple (in E. palustre). Some days after, several branches grow out and are agglutinated together, forming a body resembling a bundle of confervoid threads, each of which pushes out its own root. The account of Bischoff (*Nov. Act. Acad. N. Cur.* 14. t. 44.) is not materially different : he finds the confervoid threads or numerous processes of cellular developement go on growing and combining, until a considerable cellular mass is formed; then this mode of developement ceases, and a young bud is created, which springs up in the form of the stem of the Equisetum, at once completely organised, with its air-cells, its central cavity, and its sheaths, the first of which is formed before the elongation of the stem, out of the original cellular matter.

GEOGRAPHY. From the researches of M. A. Brongniart, it appears indisputable that plants very nearly the same as these in their organisation formed part, and a considerable part too, of the original vegetation of the globe; not, however, puny species, such as those of our days, with feeble stems, scarcely ever exceeding 3 or 4 feet in height, but gigantic vegetables, many yards long. If, indeed, certain striated fossils of the coal fields should be referable to this family, it will be found that some of them must have been vast trees. In our days they are found in ditches and rivers in most parts of the world, within and without the tropics; they have not, however, been yet seen in New Holland.

PROPERTIES. None of importance in a medicinal point of view; they are said to be slightly astringent and stimulating, and have even been recommended as diuretics and emmenagogues; they are, however, not now employed. In economical purposes they are found highly useful, for polishing furniture and household utensils; a property which is due to the presence of a great quantity of silex below their cuticle. According to the observations of Dr. John of Berlin, they contain full 13 per cent of siliceous earth. *Ed. P. J.* 2. 394. The ashes have been found by chemists to contain half their weight of silica. *Jameson's Journal, Jan.* 1830, p. 101. The quantity of silex contained beneath the cuticle of Equisetum hyemale is so great, that Mr. Sivright succeeded in removing the vegetable matter and retaining the form. *Grev. Fl. Edin.* 214. On subjecting a portion of the cuticle of Equisetum hyemale to the analysis of polarised light under a high magnifying power, Dr. Brewster detected a beautiful arrangement of the siliceous particles, which are distributed in two lines parallel to the axis of the stem, and extending over the whole surface. The greater number of the particles form simple straight lines, but the rest are grouped into oval forms connected together like the jewels of a necklace, by a chain of particles forming a sort of curvilinear quadrangle, these rows of oval combinations being arranged in pairs. Many of those particles which form the straight lines do not exceed the 500th of an inch in diameter. Dr. Brewster also observed the remarkable fact, that each particle has a regular axis of double refraction. In the straw and chaff of Wheat, Barley, Oats, and Rye, he noticed analogous phenomena ; but the particles were arranged in a different manner, and displayed figures of singular beauty. From these data the doctor concludes that the crystalline portions of silex and other earths, which are found in vegetable tissues, are not foreign substances of accidental occurrence, but are integral parts of the plant itself, and probably perform some important function in the process of vegetable life. *Grevill. Fl. Edinens.* 214.

EXAMPLE. Equisetum.

CCLXIV. FILICES. THE FERN TRIBE.

FILICES, *Juss. Gen.* 14. (1789); *Swartz Synops. Filicum* (1806); *Willd. Sp. Pl.* vol. 5. (1810); *R. Brown Prodr.* 145. (1810); *Agardh Aph.* 115. (1822); *Kaulfuss Enum.* (1824); *Spreng. Syst. Veg.* vol. 4. (1827); *Hooker and Greville Icones Filicum* (1827—1829.)

DIAGNOSIS. Flowerless plants, with their sporules either enclosed in thecæ arising from the back or margin of the leaves, or naked upon the back of deformed leaves. Vernation circinate.

ANOMALIES. In Ophioglosseæ the vernation is straight.

ESSENTIAL CHARACTER.— *Leafy* plants, producing a *rhizoma*, which creeps below or upon the surface of the earth, or rises into the air like the trunk of a tree ; this trunk consists of a hollow cylinder, of equal diameter at both ends, containing a loose cellular substance which often disappears ; it is coated by a hard, cellular, fibrous rind, which is much thicker next the root than at the apex, and is composed of the united bases of the leaves. *Leaves* (or *fronds*) coiled up in vernation, with annular ducts in the vascular tissue of their petiole, either simple or divided in various degrees, traversed by dichotomous veins of equal thickness, which are composed of elongated cellular tissue, with occasional ducts ; *cuticle* frequently with stomata. *Reproductive organs* consisting of *thecæ* or semitransparent cases arising from the veins upon the under surface of the leaves or from their margin. *Thecæ* either pedicellate, with the stalk passing round them in the form of an elastic ring, or sessile and destitute of such a ring ; either springing from beneath the cuticle, which they then force up in the form of a membrane (or *indusium*), or from the actual surface of the leaves. *Sporules* usually triangular, arranged without order within these thecæ. Sometimes the leaves are contracted about the thecæ, so as to assume the appearance of forming a part of the reproductive organs, and sometimes the place of theca is supplied by the depauperated lobes of the leaves.

AFFINITIES. These, which are by far the most gigantic of the Cellular class, sometimes having trunks 40 feet high, approach the nearest to the Vascular class by Cycadeæ, which may be considered to have much affinity with them, on account of the imperfect degree in which their vascular system is developed, their pinnate leaves with a gyrate vernation, and their naked ovules borne upon the margin of contracted leaves, as the thecæ of Ferns are upon the fronds of Osmunda. Their affinity with Equisetum, to which they were formerly joined, consists more in their want of flowers, and in the presence of annular ducts, than in any similarity of habit. Lycopodiaceæ are readily known by their axillary thecæ dehiscing by two regular valves. Marsileaceæ are so very different, that it is difficult to find points of comparison between them.

M. Bory de St. Vincent elevates Ferns to the rank of a class intermediate between Monocotyledons and Acotyledons ; but at the same time he attaches no importance to the descriptions of those writers who, having seen the germination of the sporules, have attempted to prove an identity between them and Monocotyledons in that respect. He justly observes, that the irregular unilateral scale which has been seen to sprout forth upon the first commencement of their growth is extremely different from the cotyledon of Monocotyledons, which pre-exists in the seed and never quits it, but swells during germination, and acts as a reservoir of nutriment for the young plantlet. He most properly regards it as an imperfectly developed primordial leaf.

The organ in Ferns which deserves the most particular attention is the theca, or case that contains the reproductive matter. By many it is named capsule; but as that kind of pericarpium is essentially connected with the power of conveying fertilisation from the male apparatus to the ovules, and implies the existence of a certain definite relation between the various parts that it contains, nothing of which kind is found in the theca of Ferns, it is not necessary to insist upon the impropriety of applying such a name to any sporule-case in Cellulares. Easy as it is to shew that the theca is not analogous to a capsule, it is far less so to demonstrate with what organs or modifications of organs it really has an analogy. I am not, indeed, aware that this had been attempted, all botanists seeming to consider it a special organ, until, in the *Outlines of the First Principles of Botany*, I ventured to hazard the following theory (par. 533): "The thecæ may be considered minute leaves, having the same gyrate mode of developement as the ordinary leaves of the tribe; their stalk the petiole, the annulus the midrib, and the theca itself the lamina, the edges of which are united." I was led to this opinion, first, by the persuasion that there was no special organ in Ferns to perform a function which in flowering plants is executed by modifications of leaves; and, secondly, by the examination of viviparous species. I need not here remark, that observation has shewn us that the leaves of Vasculares have the power of producing leaf-buds from their margin or any point of their surface; and the instance I have adduced in Grasses of a monstrous Wheat shews that they can produce flower-buds also. I found in Ferns, which are exceedingly subject to become viviparous, that the young plants often grow from the same places as the thecæ, or from the margin; and I was particularly struck with a viviparous Fern, of which a morsel was given me by Dr. Wallich, where the young plants form little clusters of leaves in the place of sori. Upon examining these young plants, I saw that the more perfect, though minute, fronds were preceded by still more minute primordial leaves or scales, the cellular tissue of which had nearly the same arrangement as the cellules of the theca; and I was most especially struck with the resemblance between the midrib of one of these scales and the annulus of a Polypodium. A view of the thecæ of various annulate Ferns produced a conviction of the truth of the theory I had formed, which I now submit with much deference to the consideration of the botanical world. It is, however, necessary that I should here add what is only implied in the little work from which the foregoing extract is taken, that this explanation applies only to the gyrate Ferns. With regard to those with striated thecæ, or with what is called a broad transverse ring, they may either be considered not to have the midrib of the young scale, out of which the theca is formed, so much developed; or the theca may be with still more probability considered a nucleus of cellular tissue, separating both from that which surrounds it and also from its internal substance, which latter assumes the form of sporules, in the same way as the internal tissue of an anther separates from the valves under the form of pollen. This conjecture is, I think, very much confirmed by the anatomical structure of those striated thecæ which consist of a cluster of sporule-like areolæ of cellular tissue at the base and apex, connected by extended cellules of the same description, as in Gleichenia, and is far from being weakened by such thecæ as those of Parkeria. In Ophioglosseæ another kind of provision is made for the production of sporules, which in those plants seem to have no theca whatever beyond the involute contracted segments of the frond which bears them. What are called the thecæ in Ophioglosseæ are improperly so termed, and are much more analogous to the involucrum of Marsilea.

GEOGRAPHY. The earliest Flora of the globe, that indicated by the fossil remains in the coal measures, was composed of Ferns, almost to the exclusion of other plants; and even in these islands, where the tribe now forms an inconspicuous feature in the vegetation, grasses, herbs, and trees, were represented by herbaceous and arborescent Ferns, and Fern-like plants. An approach to this enormous disproportion between Ferns and the rest of the Flora is even now exhibited in certain tropical islands, such as Jamaica, where they are 1-9th of the Phænogamous plants; New Guinea, where D'Urville found them as 28 to 122; New Ireland, where they were as 13 to 60; and in the Sandwich Islands, where they were as 40 to 160; and it is clear, from the collections of Dr. Wallich, that Ferns must form a most important feature in the Indian Archipelago. Upon continents, however, they are far less numerous: thus, in equinoctial America Humboldt does not estimate them higher than 1-36th; and in New Holland Mr. Brown finds them 1-37th. They decrease in proportion towards either pole: so that in France they are only 1-63d; in Portugal, 1-116th; in the Greek Archipelago, 1-227th; and in Egypt, 1-971st. Northwards of these countries their proportion again augments, so that they form 1-31st of the Phænogamous vegetation of Scotland; 1-35th in Sweden; 1-18th in Iceland; 1-10th in Greenland; and 1-7th at North Cape. (See a very good paper upon this subject by D'Urville, in the *Ann. des Sc. Nat.* 6. 51.; also *Brown's Appendix to the Congo Voyage*, 461.) Mr. Brown has observed (*Flinders*, 584), that it is remarkable, that although arborescent Ferns are found at the southern extremity of Van Diemen's Island, and even at Dusky Bay in New Zealand, in nearly 46° south latitude, yet they have in no case been found beyond the northern tropic.

PROPERTIES. The leaves generally contain a thick astringent mucilage, with a little aroma, on which account many are considered pectoral and lenitive, especially Adiantum pedatum and Capillus Veneris; but almost any others may be substituted for them. Capillaire is so called from being prepared from the Adiantum Capillus Veneris, a plant which is considered to be undoubtedly pectoral and slightly astringent; though its decoction, if strong, is, according to Dr. Ainslie, a certain emetic. The Peruvian Polypodium Calaguala, Acrostichum Huacsaro, and Polypodium crassifolium, are said to be possessed of important medicinal properties, especially the former; their effects are reported to be solvent, deobstruent, sudorific, and antirheumatic; antivenereal and febrifugal virtues are also ascribed to them. See the *Pharmacopœia Madritensis*, 1792, and Lambert's *Illustration of the Genus Cinchona*, 114. The leaves of Adiantum melanocaulon are believed to be tonic in India. *Ainslie*, 2. 215. The tubes of the pipes of the Brazilian negroes are manufactured from the stalk of Mertensia dichotoma, which they call Samanbaya. *Pr. Max. Trav.* 96. The bruised fronds of the fragrant Angiopteris evecta are employed in the Sandwich Islands to perfume the Cocoa-nut oil. Polypodium phymatodes is also used for the same purposes. *D'Urv.* The stem is, on the contrary, both bitter and astringent; whence that of many species, such as Aspidium Filix Mas, and Pteris aquilina, has been employed as an anthelmintic. They have also been given as emmenagogues and purgatives. Osmunda regalis has been employed successfully, in doses of 3 drachms, in the rickets. The rhizoma of Aspidium Filix Mas has been analysed, and found by M. Morin to contain, 1st, volatile oil; 2d, a fat matter composed of elaine and stearine; 3d, gallic and acetic acids; 4th, uncrystallisable sugar; 5th, tannin; 6th, soap; 7th, a gelatinous matter insoluble in water and in alcohol. It contains also the subcarbonate, sulphate, and hydrochlorate of potash, carbonate and phosphate of lime, alumine, silex, and oxyde of iron. *Brewster*, 2. 176. The roots of Nephrodium esculentum are eaten in Nipal, according to Dr. Buchanan. *Don Prodr*. 6. Those of Angiopteris evecta are used for food in the Sandwich Islands, under the name of Nehai. Diplazium esculentum, Cyathea medullaris, Pteris esculenta, and Gleichenia dichotoma, are also occasionally employed for food in different countries. Pteris aquilina and Aspidium Filix Mas have even been used in the manufacture of beer, and Aspidium fragrans as a substitute for tea. *Agdh.*

EXAMPLES. Ferns have been divided into several sections, of which the following are the most generally adopted : —

I. POLYPODIACEÆ.

Gyratæ, *Swartz Synopsis Filicum*, (1806). — Filices veræ, *Willd. Sp. Pl.* 5. 99. (1810). — Polypodiaceæ, *R. Brown Prodr*. 145. (1810); *Agardh Aph*. 116. (1822); *Kaulfuss Enumeratio*, 55. (1824); *Bory in Dict. Class*. 6. 586. (1824.)

Thecæ furnished with a vertical, usually incomplete, annulus; bursting irregularly and transversely. (Polypodium, Pteris, Adiantum.)

II. GLEICHENEÆ.

Schismatopterides, *Willd. l. c*. 69. (1810). — Gleicheneæ, *R. Br. l. c*. 160. (1810); *Kaulfuss l. c*. 36. (1824); *Bory, l. c*. (1824.)

Thecæ furnished with a transverse, occasionally oblique, annulus, nearly sessile, and bursting lengthwise internally. (Platyzoma, Gleichenia, Mertensia.)

III. OSMUNDACEÆ.

Osmundaceæ, *R. Br. l. c*. 161. (1810); *Agardh l. c*. 115. (1822); *Kaulfuss l. c*. 42. (1824); *Bory l. c*. (1824.)

Thecæ without any annulus, reticulated, striated with rays at the apex, bursting lengthwise, and usually externally. (Osmunda, Schizæa, Lygodium.)

IV. DANÆACEÆ.

Agyratæ, *Swartz Synops*. (1806). — Poropterides, *Willd. l. c*. 66. (1810). — Danæaceæ, *Agardh l. c*. 117. (1822). — Marattiaceæ, *Kaulf. l. c*. 31. (1824); *Bory l. c*. (1824.)

Thecæ sessile, without any ring, concrete into multilocular sub-immersed masses, opening at the apex. (Marattia, Danæa.)

V. OPHIOGLOSSEÆ.

Ophioglosseæ, *R. Br. l. c*. 163. (1810); *Agardh Aph*. 113. (1822); *Kaulfuss l. c*. 24. (1824); *Bory l. c*. (1824.)

Thecæ single, roundish coriaceous, opaque, without ring or cellular reticulation, half 2-valved. Vernation straight. (Ophioglossum, Botrychium.)

To which Dr. Hooker adds : —

PARKERIACEÆ.

Parkeriaceæ, *Hooker Exot. Fl*. t. 147. (1825); t. 231. (1827); *Hooker et Greville Icones Filicum*, t. 97. (1828.)

Thecæ scattered, sessile, marked with a broad, almost obsolete, very short annulus, which is sometimes distinct and nearly complete. Sporules large, 3-cornered, striated. (Parkeria, Ceratopteris.)

CCLXV. LYCOPODIACEÆ. The Club-Moss Tribe.

Lycopodineæ, *Swartz Synopsis Filicum* (1806); *R. Brown Prodr.* 164. (1810); *Agardh Aph.* 112. (1822); *Greville Flor. Edin.* xii. (1824).—Lycopodiaceæ, *Dec. Fl. Fr.* 2. 571. (1815); *Ad. Brongn. in Dict. Class.* 9. 561. (1826.)

Diagnosis. Flowerless plants, with the sporules enclosed in axillary thecæ, vernation circinate.

Anomalies.

Essential Character.—Often moss-like plants, with creeping stems and imbricated leaves, the axis abounding in annular ducts; or stemless plants, with erect subulate leaves, and a solid cormus. *Organs of reproduction* axillary sessile thecæ, either bursting by distinct valves, or indehiscent, and containing either minute powdery matter, or sporules, marked at the apex with three minute radiating elevated ridges upon their proper integument.

Affinities. Intermediate as it were between Ferns and Coniferæ on the one hand, and Ferns and Mosses on the other; related to the first of those tribes in the want of sexual apparatus, and in the abundance of annular ducts contained in their axis; to the second in the aspect of the stems of some of the larger kinds; and to the last in their whole appearance, Lycopodiaceæ are distinctly characterised by their organs of reproduction. These are generally considered to be of two kinds, both of which are axillary and sessile, and have from 1 to 3 regularly dehiscing valves, the one containing a powdery substance, the other bodies much larger in size, which have been seen to germinate. In conformity with the theory, that all plants have sexes, the advocates of that doctrine have found anthers in the former, and pistilla in the latter; but, as in other similar cases, this opinion is entirely conjectural, and founded upon no direct evidence: all that we really know is, that the larger bodies do germinate, and, if we are to credit Willdenow, the powdery particles grow also. He says he has seen them. I think it is hardly to be doubted that the latter are the abortive state of the former. According to Salisbury, in the *Linnean Transactions,* vol. 12. tab. 19. Lycopodium denticulatum emits two cotyledons upon germinating; but, supposing this observation, which requires confirmation, to be exact, it is much more probable that the two little scales so emitted are primordial leaves than analogous to cotyledons. The genus Isoetes is by some referred to Marsileaceæ, to which it forms a transition. I follow Decandolle and Brongniart in referring it here. M. Delile has published an account of the germination of Isoetes setacea, from which it appears that its sporules sprout upwards and downwards, forming an intermediate solid body, which ultimately becomes the stem, or cormus; but it is not stated whether the points from which the ascending and descending axes take their rise are uniform; as no analogy in structure is discoverable between these sporules and seeds, it is probable that they are not. M. Delile points out the great affinity that exists between Isoetes and Lycopodium, particularly in the relative position of the two kinds of reproductive matter. In Lycopodium, he says the pulverulent thecæ occupy the upper ends of the shoots, and the granular thecæ the lower parts; while, in Isoetes, the former are found in the centre, and the latter at the circumference. If this comparison is good, it will afford some evidence of the identity of nature of these thecæ, and that the pulverulent ones are at least not anthers, as has been supposed; for in Isoetes the pulverulent inner thecæ have the same organisation, *even to the presence of what has been called their stigma,* as the outer granular ones; so that, if Isoetes has sexes, it will offer the singular fact of its anther having a stigma.

GEOGRAPHY. It is the opinion of M. Ad. Brongniart, that in the earlier ages of the world these plants attained a gigantic size, equalled only by the timber-trees of our forests ; and it is certain that remains of what appear to have been species of this tribe are abundant in the coal measures, along with Ferns. At the present day they do not exceed the height of 2 or 3 feet in any instance, and are usually weak, prostrate plants, having the habit of Mosses. In geographical distribution they follow the same laws as Ferns, being most abundant in hot humid situations in the tropics, and especially in small islands. As they approach the north they become scarcer; but even in the climate of northern Europe, in Lapland itself, whole tracts are covered with Lycopodium alpinum and Selaginoides.

PROPERTIES. Lycopodium clavatum and Selago excite vomiting; the powder contained in the thecæ is highly inflammable, and is employed in the manufacture of fireworks. According to M. Vastring, they are likely to become of importance in dyeing. He asserts, that woollen cloths boiled with Lycopodiums, especially with L. clavatum, acquire the property of becoming blue when passed through a bath of Brazil wood. Lycopodium Phlegmaria is reputed an aphrodisiac.

EXAMPLES. Isoetes, Lycopodium, Psilotum, Tmesipteris.

CCLXVI. MARSILEACEÆ. THE PEPPERWORT TRIBE.

RHIZOCARPÆ, *Batsch. Tab. Aff.* (1802); *Agardh. Aph.* 111. (1822).—RHIZOSPERMÆ, *Roth. Dec. Fl. Fr.* 3. 577. (1815).— HYDROPTERIDES, *Willd. Sp. Pl.* 5. 534. (1810).—MARSILEACEÆ, *R. Brown Prodr.* 166. (1810); *Grev. Fl. Edinens.* xii. (1824); *Ad. Brongn. in Dict. Class.* 10. 196. (1826); *Dec. and Duby,* 542. (1828). —SALVINIEÆ, *Juss. in Mirb. Elémens,* 853. (1815.)

DIAGNOSIS. Flowerless plants, with their sporules enclosed in thecæ, contained within close involucra.

ANOMALIES.

ESSENTIAL CHARACTER.— Creeping or floating plants. *Leaves* either petiolate and divided (or petioles destitute of lamina), rolled up in vernation, or imbricated and sessile. *Reproductive organs* enclosed in leathery or membranous involucra, and of two kinds, the one consisting of membranous sacs, containing a body or bodies, which germinate, the other of similar sacs, containing loose granules.

AFFINITIES. It is probable that this tribe, as now constituted, comprehends two exceedingly different forms of organisation, of which one is represented by Marsilea and Pilularia, and the other by Azolla and Salvinia. I follow M. Adolphe Brongniart in this division, adopting from him many of the succeeding observations.

The tribe to which Pilularia and Marsilea belong consists of creeping plants, having the circinate vernation of Ferns, with their reproductive organs in indehiscent leathery cases, called involucra, springing either from the root, or from the petioles of the leaves. These involucra are separated internally by membranous partitions, and contain oval bodies of two kinds, one of which has been called anthers, and the other capsules.

Beautiful figures of Marsilea vestita and polycarpa have been published by Messrs. Hooker and Greville, at t. 159 and 160 of their noble *Icones Filicum.* From these it is clear that the involucrum of the genus consists of an involute frond, of the same degree of analogy to the true frond as a carpellary leaf to a true leaf. It further appears that the reproductive bodies arise from the veins of this involute frond, and are therefore analogous, as to position, to the sori of Ferns. What the nature of these bodies may be, is not so obvious. They are represented as being of two kinds; the first, called the capsule (?), being an oval stalked case, having two integuments, of which the outer is reticulated and hyaline, the inner oval, white, and opaque, with an apiculate tubercle at its base, and containing corpuscles of two kinds, the one angular and very minute, the other much larger and roundish; the second, much smaller bodies, called the anthers (?), being little sacs filled with yellowish roundish granules, and attached by fours to the stalk of the capsule.

The structure of Pilularia is of an analogous kind. The exact nature of the parts called anthers is unknown; from the name that has been given them, it has been supposed that they were similar to the male apparatus of flowering plants; but this view is altogether gratuitous, and has not been taken from any direct evidence. It seems more probable that they are abortive sacs, analogous to the larger bodies. With regard to the latter, M. A. Brongniart has the following passage:—" Experiments made upon the germination of Salvinia and Pilularia have long since shewn that in these plants the larger globules were true seeds; and analogy permitted us to entertain the same belief in regard to Marsilea and Azolla; but it remained to be proved that the other bodies were really male organs, the action of which is necessary to fertilise the seeds. This, Professor Savi, of Pisa, had appeared to have demonstrated. Salvinia grows abundantly near that city, and there was no difficulty in procuring fresh plants for the purpose of experiment. He put into different vessels, 1st, the seeds alone; 2d, the male globules alone; and, 3d, both mixed. In the first two vessels nothing appeared; in the 3d, the seeds rose to the surface of the water and fully developed. But M. G. L. Duverney has since published a dissertation upon this plant, in which he states that, having repeated the experiments of Savi, he has not obtained the same results, and that the seeds, when separated from the supposed male organs, developed perfectly." I am not acquainted with the particulars of these experiments, nor do I know with what degree of care the exact mode of germination in Salvinia has been observed; but it appears more consonant to the analogical structure of other plants, particularly of Ferns and Azolla, to consider the larger bodies, called seeds by these observers, as thecæ; in which I am the more confirmed, by finding it to be the view taken of their nature by Mr. Brown, and Drs. Hooker and Greville.

In Salvinia and Azolla the vegetation is that of Mosses, or of Jungermannia, and the organs of reproduction are quite different. The latter consist of two sorts of membranous bags, of which one contains bodies analogous to the larger bodies, or thecæ of Marsilea, and the other what have been considered male organs. These, in Salvinia, have been described by Brongniart as spherical grains, attached by long stalks to a central column, and much smaller than what he calls the seeds: their surface is reticulated in a similar manner, and they only burst by the action of water. In Azolla, M. Bauer represents, and Mr. Brown describes, them as from 6 to 9 in number, angular and inserted upon a central body, occupying the upper half of the involucrum, the lower being filled with a turbid fluid. If the real nature of these parts in Pilularia and Marsilea is involved in obscurity, that of the reproductive organs of Salvinia and Azolla is still more mysterious. Mr. Brown, who had good opportunities of studying Azolla in New Holland, with Mr. Ferdinand Bauer's acuteness and profound knowledge of structure to assist him, could arrive at no certain conclusion. The involute vernation of the leaves of some of these plants and their involucrum being formed out of the involute frond, as in Ophioglossum, indicate a close affinity to Ferns; but the habit of Azolla is rather that of some Hepaticæ. Marsileaceæ may be considered to occupy an intermediate position between these tribes. Authors have not stated whether ducts are to be found in Pilularia, Salvinia, and Azolla; they are present in abundance in Marsilea, where I have seen them; but they are so minute as to require to be magnified 200 times to be distinctly observed.

GEOGRAPHY. Of 20 species enumerated by writers, all are inhabitants of ditches or inundated places, in various part of the world. They do not appear to be affected by climate so much as by situation, whence they have been detected in various parts of Europe, Asia, Africa, and America; chiefly, however, in temperate latitudes.

PROPERTIES. Unknown.

EXAMPLES. § 1. MARSILEACEÆ, *Ad. Brongn. in Dict. Class.* 10. 196. (1826), Marsilea, Pilularia.

§ II. SALVINIEÆ, *Id.* l. c., Salvinia, Azolla.

TRIBE II. MUSCOIDEÆ, OR MOSS-LIKE PLANTS.

CELLULARES FOLIACEÆ, *Dec. Théor. Elém.* 249. (1819).—PSEUDOCOTYLEDONEÆ,
Class 1. *Agardh Aph.* 103. (1822).—HETERONEMEA, *Fries Syst. Orb. Veg.* 33.
(1825) *in part.*—ACOTYLEDONES, Class 2. *Ad. Brongniart in Dict. Class.* 5. 159.
(1824).—CRYPTOGAMICÆ, 2d Circle, *T. F. L. Nees v. Esenbeck and Ebermaier
Handb. der Med. Bot.* 1. 18. (1830.)

DIAGNOSIS. Flowerless plants, with a distinct stem having no vascular
system, but frequently furnished with leaves; their sporules having a proper
integument, and contained in distinct axillary, terminal, or superficial thecæ.

These are altogether intermediate between the first and third families,
and are distinguishable essentially by their having a distinct axis of growth
without any vascular system; they are connected with Marsileaceæ by
Jungermannia, and with Lichens by Riccia and Marchantia; to Algæ the
transition is by Characeæ, which have the evascular axis of Muscoideæ, with
the habit and propagating matter of Algæ. Von Esenbeck and Ebermaier
refer Characeæ to the next tribe, but their structure is scarcely reconcilable
with the character those authors give it, viz. "root, stem, and leaves, not
separately formed; all analogy with plants of a higher organisation is lost,
and the green matter, which is so characteristic of the vegetable kingdom,
scarcely makes its appearance," &c.

LIST OF THE ORDERS.

267. Musci. | 268. Hepaticæ. | 269. Characeæ.

CCLXVII. MUSCI. THE MOSS TRIBE.

MUSCI, *Juss. Gen.* 10. (1789); *Hedwig Descr. et Adumb.* (1787–1797); *Bridel Muscolog.
recentiorum* (1797–1803); *Hedw. Species Muscor. Frondos.* (1801); *Palisot Pro-
drome des 5 et 6 Fam. de l'Æthiogam.* (1805); *Bridel Suppl.* (1806–1819); *Weber
Tabul. Musc. Frondos.* (1813); *Dec. Fl. Fr.* 2. 438. (1815); *T. F. L. Nees de
Muscor. Propag.* (1818); *Hooker and Taylor Musc. Brit.* (1818); *Hooker Musci
Exotici* (1818–1820); *Agardh Aphor.* 105. (1822); *Greville and Arnott in Wern.
Trans.* 4. 109. &c. (1822); *Nees v. Esenbeck, Hornschuch, and Sturm, Bryolog.
Germ.* (1823); *Grev. Fl. Edin.* xiii. (1824); *Ad. Brongn. in Dict. Class.* 11. 248.
(1827); *Hooker Brit. Fl.* 1. 459. (1830.)

DIAGNOSIS. Flowerless plants, with the sporules contained in thecæ,
closed by an operculum.

ANOMALIES. In Andreæa the theca separates into 4 valves.

ESSENTIAL CHARACTER.—Erect or creeping, terrestrial or aquatic, cellular plants,
having a distinct axis of growth, destitute of a vascular system, and covered with minute,
imbricated, entire, or serrated leaves. *Reproductive organs* of two kinds, viz.; 1. *Axillary
bodies,* cylindrical or fusiform stalked sacs, containing a multitude of spherical or oval
particles, which are emitted upon the application of water; 2. *Thecæ,* hollow urn-like
cases seated upon a seta or stalk, covered by a membranous calyptra, closed by a lid or
operculum, within which are one or more rows of cellular rigid processes, called collectively the peristomium, and separately teeth, which are always some multiple of four,
and combined in various degrees; the centre of the theca is occupied by an axis or

columella, and the space between it and the sides of the theca is filled with sporules. *Sporules* in germination protruding confervoid filaments, which afterwards ramify, and form an axis of growth at the point of the ramifications.

AFFINITIES. These little plants, which form one of the most interesting departments of Cryptogamic Botany, are distinctly separated from all the other tribes by the peculiar structure of their reproductive organs, in which they resemble no others, except some Hepaticæ, which, however, approach them in this respect more in appearance than in reality. In their organs of vegetation they are strikingly similar to many Lycopodiums, which are always to be known by their vascular axis. The reproductive organs have been described above as of two kinds. Those which are called AXILLARY BODIES have been supposed to be anthers; with how little reason will be clear from the following extract from Dr. Greville and Mr. Arnott's excellent memoir, published in the 4th volume of the *Transactions of the Wernerian Society*, to which I refer those who are desirous of minute information upon the structure and history of Mosses.

" What the organs really are, in the plants under review, which the accurate Hedwig so well figured and described under the name of stamens, we leave to others to decide; but we cannot help entering our protest against those bodies called Stamina and Pistilla (the young thecæ) being regarded in a similar light with the same organs in more perfect plants. ' Though,' says Sprengel, ' I have formerly been a zealous advocate for Hedwig's *Theory of the Fructification of Mosses*, it has nevertheless appeared to me an insurmountable objection, that the supposed anther can again produce buds and strike roots, which is certainly the case with regard to the disks of Polytrichum commune, Bartramia fontana, Bryum palustre, undulatum, cuspidatum, punctatum, and with those of Tortula ruralis. In Bryum argenteum we see the buds containing the supposed anthers constantly drop off, strike root, and produce new plants; this I have observed myself times out of number. Still more in point is the experiment first made by David Meese, of sowing the stellulæ of Polytrichum commune, containing merely club-shaped bodies, when he found that plants came up, which in their turn produced fruit. Another excellent naturalist, Dr. Roth, has made similar observations with regard to Hypnum squarrosum and Bryum argenteum.' He afterwards adds, — ' It is more probable, therefore, that these supposed anthers are mere gemmæ, produced by the superabundance of the juices, and hence surrounded by succulent filaments.'"

It is not necessary to adopt the exact conclusion at which the learned botanist, whose opinions are thus quoted, arrived, to decide that these axillary bodies are not stamens. He has not expressed himself in regard to their nature very clearly, or perhaps he has been mistranslated; but this is of little consequence compared with the ascertained fact, that, be they what they may, they are not anthers. Nevertheless, in the face of this evidence, M. Adolphe Brongniart retains a belief in the sexuality of Mosses, and in the male functions of the axillary bodies; and he says, with justice, that it appears from Mr. Brown's mode of describing Mosses, that he entertains a similar opinion. It is to be hoped that these distinguished botanists will some day favour us with a statement of the evidence upon which their decision has been taken; for it is to be presumed that something beyond the conjectures advanced in the article Mousses in the *Dictionnaire Classique*, weighs down the positive testimony of those who have seen the germination of the powder in the axillary bodies. Whether or not they can be called gemmæ, will depend upon the sense in which that term is employed.

With regard to the theca there is now no difference of opinion, either as to its containing sporules, or as to the general nature of its organisation.

But I am not aware that any one has ever attempted to explain the analogy of its structure until I ventured to introduce the subject very briefly into my *Outline of the First Principles of Botany*. That perfect unity of design, which is visible in all parts of the vegetable creation, and the constant adherence to the construction of every organ of plants, except the stem out of modified leaves, seemed to be deviated from in the Cryptogamic class generally, and in Mosses in particular. An uninitiated person, reading the definition of a genus of Mosses, might suppose that it was in that tribe that the approach to the animal creation, of which so much has been said, takes place. Unacquainted with the exact meaning of the Latin words employed by Bryologists, he might understand by the peristomium a jaw, by the calyptra a nightcap, and by the struma a kind of goitre; and when he saw that teeth belonged to this jaw, he would naturally conclude that it was really a vegeto-animal of which he was reading. Struck with the evident absurdity of giving such names to parts of plants, without at the same time explaining their real nature, I ventured to call the attention of naturalists to the subject by the following paragraph in the little book above referred to.

" 539. The calyptra may be understood to be a convolute leaf; the operculum another; the peristomium one or more whorls of minute flat leaves; and the theca itself to be the excavated distended apex of the stalk, the cellular substance of which separates in the form of sporules."

It is now time to shew upon what evidence and reasoning this hypothesis may be sustained. Every one agrees in describing the calyptra as a membrane arising from between the leaves and the base of the young theca, and as enveloping the latter, but having no organic connexion with it: when the stalk of the theca lengthens, no corresponding extension of the parts of the calyptra takes place; so that it must be either ruptured at its apex (as in Jungermannia), or at the base; and in the latter case it would necessarily be carried up upon the tip of the theca, which it originally enveloped. Now, what can be more reasonable than that such an organ, situated as I have described it to be, should be one of the last convolute leaves of the axis which the theca terminates, bearing the same relation to the latter as the convolute bractea to the flower of Magnolia, or, to speak more precisely still, as the calyptriform bracteæ to the flower of Pileanthus? If the calyptra be anatomically examined, especially in such genera as Tortula and Dicranum, no difference in its tissue and that of the leaves will be observable; and that very common tendency to dehisce on one side only as the diameter of the theca increases, which characterises the dimidiate calyptra, may not unreasonably be understood to be the separation at the line where the margins of the supposed leaf united; in the mitriform calyptra this separation at a given line does not take place, and the consequence is an irregular laceration of its base. The analogy of the calyptra being of this nature, the next inference would naturally be, that the part it contains is analogous to a flower-bud. Upon this supposition, the external series of parts belonging to this supposed bud would be the operculum; the adhesion of this to the theca, which would answer to the apex of the axis, or to the tube of the calyx of flowering plants, would be analogous to that which obtains in Eucalyptus, or perhaps more exactly to that of Eschscholtzia; but it would remain to determine of how many parts, in a state of cohesion, it was made up. In the paragraph above quoted, it is stated to be one only; but I confess I have no better reason to offer for this than the absence of any trace of division upon its surface or in the substance of its tissue, and also perhaps the apparent identity of nature between it and the calyptra when both are young, in the Tortula and Dicranum genera already cited. With regard to the peristomium, I would beg attention to the following particulars : — The teeth, as they are called, occupy one or more whorls ; they are evidently not mere lacerations of a membrane, because they are in a constant and regular number in each genus, and that number is universally some multiple of 4, as the floral leaves of flowering plants are ordinarily of 3, 4, or 5 ; they have the power of contracting an adhesion with each other by their contiguous margins, as the floral leaves of flowering plants ; they alter their position from being inflexed with their points to the axis, to being recurved with their points turned outwards, — exactly what happens in flowering plants ; the teeth of the inner peristomium often alternate with those of the outer, thus conforming to the law of alternation prevalent in the floral leaves of flowering plants ; and, finally, if we compare the various states of the leaves of Buxbaumia aphylla with the teeth of Mosses, it is impossible not to be struck with the great similarity in the anatomical structure of the two.   These are the considerations which have led me to the conclusion, that the calyptra, the operculum, and the teeth of Mosses, are all modified leaves ; and hence that the theca is to be considered more analogous to a flower than to a seed-vessel. With regard to the membrane, or epiphragma, which occasionally closes up the orifice of the theca, it may be considered as formed by the absolute cohesion of the leaves of the peristomium, just as the operculum of Eudesmia is formed by the cohesion of the petals ; and this is confirmed, first, by Calymperes, in which the membrane ultimately separates into teeth, and by the fact that the horizontal membrane exists most perfectly in such genera as Polytrichum and Lyellia, in which there is no distinct peristomium.   It now remains to explain the internal structure of the theca consistently with the theory that has been advanced of the peristomium, operculum, and calyptra. I consider the theca to be merely the thickened apex of the axis, the sporules to be a partial dissolution of its cellular tissue, and the columella to be the unconverted centre.   That the end of the axis of plants frequently becomes much more incrassated than the theca of Mosses, requires no illustration for those who are acquainted with the spongy receptacle of Nelumbium, Rubus, and Fragraria, the dilated disk of Ochna, the curious genus Eschscholtzia, or Rosa, or Calycanthus, or, finally, the spadix of Arums.   That the tissue is frequently separated by nature for particular purposes, is proved by the production of pollen out of the cellular tissue of an anther, and by the general law of propagation that seems to prevail in flowerless plants, as Ferns, Lichens, Algæ, and Fungi ; the same phenomenon may be therefore expected in Mosses.   That the columella should be left in this dissolution of the tissue might be expected, from its being a continuation of the seta or axis of developement, the tissue of which is more compact, and of course less liable to separation, than the looser tissue that surrounds it ; this is analogous to the separation of the pollen from the connectivum of most plants, or from parts only of the anther of all those genera which, like Viscum, Ægiceras, or Rafflesia, have what are called cellular anthers ; and to the very common separation of the placenta, or a portion of it, from the dissepiments, as in Bignoniaceæ, Ericeæ, and many others.   That it is presumptuous in me, who lay no claim to reputation as a Cryptogamic botanist, to offer any opinion upon plants I have only occasionally studied, I am fully sensible ; but I hope for the indulgence of the skilful Cryptogamist, in consideration of this having been the first attempt to call his attention to the inquiry.

GEOGRAPHY.   Mosses are found in all parts of the world where the atmosphere is humid ; but they are far more common in temperate climates than in the tropics.   They are among the first vegetables that clothe the soil with verdure in newly-formed countries, and they are the last that disappear when the atmosphere ceases to be capable of nourishing vegetation. The first green crust upon the cinders of Ascension was minute Mosses, they form more than a quarter of the whole Flora of Melville Island, and the black and lifeless soil of New South Shetland is covered with specks of Mosses struggling for existence. How they find their way to such places, and under what laws they are created, are mysteries that human ingenuity has not yet succeeded in unveiling. About 800 species are known.

PROPERTIES. The slight astringency of Polytrichum and others caused them to be formerly employed in medicine, but they are now disused. In the economy of man they perform but an insignificant part; but in the economy of nature, how vast an end!

EXAMPLES. There is no settled arrangement of the genera, almost every writer having a method of his own. Much merit is due to several, especially to that of Greville and Arnott, published in the *Wernerian Transactions*, vols. 4. and 5.

Sphagnum, Hypnum, Bryum, Fontinalis, Gymnostomum, Dawsonia, Weissia, Phascum.

CCLXVIII. HEPATICÆ. THE LIVER-WORT TRIBE.

HEPATICÆ, *Juss. Gen.* 7. (1789); *Dec. Fl. Fr.* 2. 415. (1815); *Agardh Aph.* 104. (1822); *Greville Flora Edin.* xv. (1824); *Fée in Dict. Class.* 8. 131. (1825.)

DIAGNOSIS. Flowerless terrestrial plants, with their sporules contained in dehiscent thecæ, destitute of an operculum.

ANOMALIES. Riccia has indehiscent fruit immersed in the substance of the frond.

ESSENTIAL CHARACTER.— Plants growing on the earth or trees in damp places, composed entirely of cellular tissue, emitting roots from their under-side, and consisting of an axis or *stem*, which is either furnished with leaves, or leafless, and then bordered by a membranous expansion; these expansions sometimes unite at their margins, so as to form a broad lobed thallus. *Reproductive organs* of several kinds; either a 1- 2- or 4-valved theca, supported upon a membranous peduncle, covered when young by a leaf, through which it afterwards protrudes, and often containing spiral fibres, called Elateres, within which the sporules are intermixed; or a peltate stalked receptacle, bearing thecæ on its under surface; or sessile naked thecæ, either immersed or superficial. Besides these there are in Jungermannia " minute, spherical, membranous, reticulated bodies, supported upon short white peduncles," (*Grev.*); in Marchantia, " peltate receptacles, plane on the upper surface, and having oblong bodies imbedded in the disk;" and also " little open cups, sessile on the upper surface, and containing minute green bodies (gemmæ) which have the power of producing new plants, as well as the sporules;" and in Anthoceros, " small cup-shaped receptacles, containing minute, spherical, pedunculated, reticulated bodies."

AFFINITIES. The structure of the reproductive organs of this order is so exceedingly variable that no common character seems deducible from them; nor has it been found possible either to determine what analogy exists between the organs, or even to decide what their respective functions are. What are here called the thecæ are considered to be the cases of the sporules, properly so called, but the other bodies are of a more doubtful kind. Those who have sought for sexual organs in Cryptogamous plants have naturally taken the imbedded oblong bodies of Marchantia, and the pedunculated reticulated ones of Jungermannia, for anthers; but Dr. Hooker, in his beautiful *Monograph* of the latter genus, and also in his *British Flora* (p. 459.), is evidently unsatisfied as to their nature. Dr. Greville, in the *Flora Edinensis*, the most useful and original work upon British Cryptogamic plants that we yet possess, is clearly in a similar state of uncertainty; and Agardh admits nothing more in them than a resemblance to male organs, adopting the opinion that they are a particular form of gemmules. The bodies lying in the cup-shaped receptacles of Anthoceros have been said to be anthers, but upon no good evidence. In Jungermannia there is a third kind of reproductive matter, consisting of heaped clusters of little amorphous bodies, growing from the surface of the leaves, and called gemmæ.

The most remarkable point of structure in Hepaticæ is the spiral filament, as it is called, lying among the sporules within the theca. This consists of a single fibre, or of two, twisted spirally in different directions, so as to cross each other, and contained within a very delicate, transparent, perishable tube. They have a strong elastic force, and have been supposed to be destined to aid in the dispersion of the sporules, — a most inadequate end for so curious and unusual an apparatus. It is more probable that they are destined to fulfil, in the economy of these plants, some function of which we have no knowledge. Hepaticæ are intermediate between Mosses and Lichens, agreeing with the former in the presence of a distinct axis of growth, and frequently of leaves also, and in most cases in the sporules being contained in stalked thecæ, having a calyptra and a definite mode of dehiscence. Fée says they have no calyptra, which must have been an oversight. They differ from Mosses in the want of an operculum, by which Andræa, which forms the link between Hepaticæ and Mosses, is referred to the latter. Lichens are distinguished by their want of a distinct axis of growth, by their texture and colour, never assuming the rich lucid green of Hepaticæ, and by their sporules not being contained in distinct thecæ, but lying in membranous tubes or asci in the substance of the thallus. Riccia and Endocarpus form the connexion between them.

GEOGRAPHY. Natives of damp shady places in all climates; two were found in Melville Island. The only atmospheric condition to which they cannot submit is excessive dryness: thus, of the 237 species enumerated by Sprengel, 6 only are found in Africa, while 50 are cited from Java alone.

PROPERTIES. Nothing is known of them. Decandolle thinks it probable that the larger kinds will be found to resemble foliaceous Lichens in their qualities. A few are slightly fragrant.

EXAMPLES. Marchantia, Targionia, Sphærocarpus, Jungermannia.

CCLXIX. CHARACEÆ. THE CHARA TRIBE.

CHARACEÆ, *Rich. et Kunth in Humb. et Bonpl. N. G. Pl.* 1. 45. (1815); *A. Brongn. in Dict. Class.* 3. 474. (1823); *Grev. Fl. Edin.* xvii. (1824); *Dec. and Duby*, 533. (1828); *Hooker Brit. Fl.* 459. (1830.)

DIAGNOSIS. Submersed leafless water-plants, having slender verticillate branches and deciduous thecæ.

ANOMALIES.

ESSENTIAL CHARACTER.—Plants composed of an axis, consisting of parallel tubes, which are either transparent or encrusted with carbonate of lime, and of regular whorls of tubes, which may be either considered as leaves or branches. Organs of reproduction, round succulent *globules*, containing filaments and fluid; and axillary *nucules*, formed of a few short tubes, twisted spirally around a centre, which has the power of germinating.

AFFINITIES. The two genera of which this little order is composed are among the most obscure of the vegetable kingdom, in regard to the nature of their reproductive organs; and accordingly we find them, under the common name of Chara, placed by Linnæus among Cryptogamous plants near Lichens; then referred by the same author to Phænogamous plants, in Monœcia Monandria; retained by Jussieu and Decandolle among Naiades, by Mr. Brown at the end of Hydrocharideæ, and by Leman in Halorageæ; referred to Confervæ by Von Martius, Agardh, and Wallroth; and finally admitted as a distinct order, upon the proposition of Richard, by Kunth, Decandolle, Adolphe Brongniart, Greville, Hooker, and others. Such being the uncertainty about the place of these plants, it will be useful to give rather a detailed account of their structure, in which I avail myself chiefly of Ad. Brongniart's remarks in the place above referred to, and of Agardh's observations in the *Ann. des Sciences*, 4. 61. I have not seen Professor Nees v. Esenbeck's monograph of Characeæ in the *Transactions of the Ratisbon Society*, quoted by the latter author.

Characeæ are aquatic plants, found in stagnant fresh or salt water; always submersed, giving out a fetid odour, and having a dull greenish colour. Their stems are regularly branched, brittle, and surrounded here and there by whorls of smaller branches. In Nitella the stem consists of a single transparent tube with transverse partitions, and, as Agardh remarks, so like the tubes of some Algæ, as to offer a strong proof of the affinity of the orders. In Chara, properly so called, there is, in addition to this tube, many other external ones, much smaller, which only cease to cover the central tube towards the extremities. In the axillæ of the uppermost whorls of these branchlets the organs of reproduction take their origin; they are of two kinds, one called the nucule, the other the globule; the former has been supposed to be the pistillum, the latter the anther.

The nucule is described by Dr. Greville as being " sessile, oval, solitary, spirally striated, having a membranous covering, and the summit indistinctly cleft into 5 segments; the interior is filled with minute sporules." *Fl. Edin.* xvii. This is the general opinion entertained of its structure. But Ad. Brongniart describes it thus:—" Capsule unilocular, monospermous; pericarp composed of two envelopes; the outer membranous, transparent, very thin, terminated at the upper end by 5 spreading teeth; the inner hard, dry, opaque, formed of 5 narrow valves, twisted spirally." *Dict. Class.* l. c. He founds his opinion of the nucule containing but one germinating body upon the experiments of M. Vaucher, of Geneva, who ascertained that if ripe nucules of Chara, which have fallen naturally in the autumn, are kept through the winter in water, they will germinate about the end of April; at that time a little body protrudes from the upper end between the 5 valves, and gradually gives birth to one whorl of branches, which produce a second. Below these whorls the stem swells, and little tufts of roots are emitted. The nucule adheres for a long time to the base of the stem, even when the latter has itself begun to fructify. Hence it is reasonable to conclude that the nucule is really monospermous. M. Brongniart remarks, that it is true, when a fresh nucule of Chara is cut across, an infinite number of little white grains are squeezed out; but if these were really all reproductive particles, how would they ever find their way out of the nucule, which is indehiscent? he considers them rather of the nature of albumen. And he is the more confirmed in his opinion, because in Pilularia, the thecæ of which also contain many similar grains, but one plant is produced by each theca. Finally, Amici has described (*Ann. des Sc.* 2.) the nucule in another way. He admits it to be monospermous, but he considers the points of the 5 valves to be stigmata, and the valves themselves to be at once pericarp and style. It is not worth entering into any discussion upon the reasonableness of such a supposition, as it is not likely to find any advocates among botanists; but I may observe, that Amici's observations seem to shew that the 5 valves of the nucule, as they are called, are a verticillus of leaves, straight at first, and twisted afterwards; and that the nucule itself is, therefore, analogous to the bud of flowering plants.

The globule is described by Dr. Greville as " a minute round body, of a reddish colour, composed externally of a number of triangular (always?) scales, which separate and produce its dehiscence. The interior is filled with a mass of elastic transversely undulate filaments. The scales are composed of radiating hollow tubes, partly filled with minute coloured spherical granules, which freely escape from the tubes when injured." Vaucher describes them as " tubercles formed externally of a reticulated transparent membrane, containing, in the midst of a mucilaginous fluid, certain white articulated transparent filaments, and some other cylindrical bodies, closed at one end, and appearing to open at the other. These latter are filled with the red matter to which the tubercles owe their colour, and which disappears readily and long before the maturity of the nucule." The account of the globule by Agardh is at variance with both these. " Their surface," he remarks, " is hyaline, or colourless; under this membrane is observed a red and reticulated or cellular globe, which has not, however, always such an appearance; often, instead of this reticulated aspect, the globe is colourless, but marked by rosettes or stars, the rays of which are red or lanceolate. In the figures given by authors, one finds sometimes one of these forms, sometimes the other. I have myself found them both on the same species; and I am disposed to believe that the last state is the true kernel of the globule, concealed under the reticulated scale. (When the globule is very ripe, one may often succeed, by means of a slight degree of pressure, in separating it into several valves, as is very well shewn in Wallroth's figures, tab. 2. f. 3. and tab. 5. These valves are rayed, and no doubt answer to the stars, of which mention has been made.) The kernel contains some very singular filaments; they are simple (I once thought I saw them forked), curved and interlaced, transparent and colourless, with transverse striæ, parallel and closely packed, as in an Oscillatoria or Nostoc; but what is very remarkable, they are attached, several together, to a particular organ formed like a bell, which is itself also colourless, but filled with a red pigment. This bell, to the base of which on the outside they are fixed, differs a little in form in different species. It is slender and long in Chara vulgaris, thicker in C. firma, shorter in C. delicatula, and shorter still in C. collabens. I have not succeeded in determining the exact position of these bells in the kernel. I have often thought they were the same thing as the rays of the rosettes or stars upon the globule above mentioned; whence it would follow that they are placed near the surface, while the filaments have a direction towards the centre. The bells are not numerous; they often separate from the filaments, and readily part with their pigment, which renders it difficult to observe them, and has caused them to be overlooked." That these globules, whatever their nature may be, have no sort of analogy in structure with anthers, is clear from these descriptions, whichever may be eventually admitted. Wallroth, indeed, says he has sown them, and that they have germinated; but this observation requires to be verified.

It does not appear from the preceding descriptions that Chara has a marked affinity to any other plants. I incline to the opinion of those who consider it near Confervæ, chiefly on account of the organisation of the stems; for it does not seem that the reproductive organs of flowerless plants are of the same degree of importance in deciding affinities as the fructification of flowering plants. Its total want of vascular system renders it impossible to adopt the opinion of those who would place it near Ferns next to Marsileaceæ, and the regularity with which all the parts are formed round a common axis renders it equally impossible to refer it absolutely to the leafless section. I therefore place it on the limits of the latter, among Muscoideæ.

There are two other points deserving of attention in Characeæ: 1st, the calcareous incrustation of some species; and 2dly, the visible and rapid motion of the sap in the articulations of the stem.

Of the two genera, Nitella is transparent and free from all foreign matter; but Chara contains, on the outside of its central tube, a thick layer of calcareous matter, which renders it opaque. This incrustation appears, from the observations of Dr. Greville (*Fl. Edin.* 281), not to be a deposit upon the outside, and of an adventitious nature, but the result of some peculiar economy in the plant itself; and according to Dr. Brewster, it is analogous to the siliceous deposit in Equisetum, exhibiting similar phenomena.

Whatever is known of the motions of the fluids of vegetables has been necessarily a matter of inference, rather than the result of direct observation; for who could ever actually see the sap of plants move in the vessels destined to its conveyance? It is true that it was known to botanists that a certain Abbé Corti of Lucca, had, in 1774, published some remarkable observations upon the circulation of fluid in some aquatic plants, and that the accuracy of this statement had been confirmed by Dr. Treviranus so long ago as 1817; but the fact does not seem to have attracted general attention until the publication, by Amici, the celebrated professor at Modena, of a memoir in the 18th volume of the *Transactions of the Italian Society*, which was succeeded by another in the 19th. From all these observers it appears, that if the stems of any transparent species of Chara, or of any opaque one, the incrustation of which is removed, are examined with a good microscope, a distinct current will be seen to take place in every tube of which the plant is composed, setting from the base to the apex of the tubes, at the rate, in Chara vulgaris, of about two lines per minute (*v. Ann. des Sc.* 2. 51. line 9); and according to Treviranus this play is at any time destroyed by the application of a few drops of brandy, by pressure, or by any laceration of the tube. This is the nature of the singular phenomena which are to be seen in Characeæ, and which become the more interesting because they are not to be found in any other water-plants, with the exception of Naïas and Caulinia. Those who are anxious to become acquainted with the details of Amici's observations will find his first paper translated in the *Annales de Chimie*, 13. 384, and his second in the *Ann. des Sc.* 2. 41; that of Treviranus is to be found in the latter work, 10. 22. According to the last-named author, these facts lead to the conclusion that there is a primitive vitality in amorphous organic matter, which is antecedent to the formation of all organic beings, and is in its turn produced by them, to serve, according to circumstances, either for the support or enlargement of the individual, or for the production of a new organisation. This vitality is manifested in movements which may appear to take place without rule or object, but which are differently modified according to the differences of organic bodies; all which seems to shew that the vital principle is originally susceptible of a variety of modifications, without having occasion for the assistance of organs of various forms or structure.

Geography. The creation of plants of this order would appear to have been of a very recent date, compared with that of Ferns and Palms, or even Algæ, if we are to judge by their fossil remains, which are found for the first time in the lower fresh-water formation, along with numerous Dicotyledonous plants resembling those of our own times. In the recent Flora of the world they make their appearance every where in stagnant waters, in Europe, Asia, and Africa, in North and South America, in New Holland, and in either India. They are most common in temperate countries.

Properties. Unknown.

Examples. Nitella, Chara.

Tribe III. APHYLLÆ, or LEAFLESS FLOWERLESS PLANTS.

ACOTYLEDONEÆ, *Agardh Aph.* 72. (1821). — HOMONEMEA, *Fries Syst. Orb. Veg.* 33. (1825). — ACOTYLEDONES, Class I. *Ad. Brongn. in Dict. Class.* (1824). — CRYPTOGAMICÆ, 3d Circle, *T. F. L. Nees v. Esenbeck und Ebermaier Handb. der Med. Bot.* 1. 18. (1830.)

DIAGNOSIS. Flowerless leafless plants, destitute of vascular tissue, with no distinct axis of growth, the sporules simple and lying naked in the substance of the plant.

In this tribe we have arrived at the limits which separate the vegetable from the animal kingdom. We have not only passed beyond the dominion of the sexes, but we have no longer any trace, however ambiguous, of more than one form of reproductive matter. It is even uncertain whether this matter will reproduce its like, and whether it is not a mere representation of the vital principle of vegetation capable of being called into action either as a Fungus, an Alga, or a Lichen, according to the particular conditions of heat, light, moisture, and medium, in which it is placed; producing Fungi upon dead or putrid organic beings; Lichens upon living vegetables, earth, or stones; and Algæ where water is the medium in which it is developed. The nearest approach to animals is in the tribes of Algæ called Arthrodieæ and Chaodineæ, where it is perhaps impossible to decide whether some of the species are not actually animalcules.

It is not easy to settle the limits of the orders of this part of vegetation. Linnæus and Jussieu had but two divisions, viz., Algæ, including Lichens and Fungi; and they have been followed by some modern botanists, particularly Fries and Wahlenberg. Others have been satisfied with separating the Lichens from Algæ, which, indeed, was virtually done by most of those who acknowledged but two divisions, and with admitting three equally distinct groups. Some, on the contrary, have sought to multiply them, as Decandolle and others, by introducing a tribe called Hypoxyla; Dr. Greville by adopting the latter, Gastromyci, Byssoideæ, and Epiphytæ, and proposing a new group under the name of Chætophoroideæ; and finally, M. Adolphe Brongniart, who carries the number of groups in this division of Acotyledones as far as 12, viz. Lichens, Hypoxyla, Fungi, Lycopodiaceæ, Mucedineæ, Uredineæ, Fucaceæ, Ulvaceæ, Ceramiaceæ, Confervæ, Chaodineæ, and Arthrodieæ; part of which have originated with himself, and others with M. Bory de St. Vincent. I think, however, in the present state of our knowledge, it will be more prudent to admit only the three principal groups adopted by Agardh and Hooker; and even these are distinguishable by their general habit rather than by any very positive character of structure. Thus, Lichens are aerial plants, with distinct spaces upon their surface, in which their sporules are contained; Fungi differ from Lichens only in their fugacity and want of external receptacles of sporules; while Algæ are all aquatic.

The structure of leafless plants is among the most important subjects of contemplation for those who wish to become acquainted with the exact laws of vegetation. They represent the organised matter, of which all other plants are composed, both in its simplest state and when it begins to enter into a state of high composition. In short, it is here only that the physical properties of vegetable matter can be usefully studied.

LIST OF THE ORDERS.

270. Lichenes. | 271. Fungi. | 272. Algæ.

CCLXX. LICHENES. The Lichen Tribe.

Algæ, § 3. Lichenes, *Juss. Gen.* 6. (1789). — Lichenes, *Hoffm. Enumerat. Lichenum*, (1784); *Acharius Prodr. Lichen.* (1798); *Id. Methodus*, (1803); *Id. Lichenogr. Univers.* (1810); *Dec. Fl. Fr.* 2. 321. (1815); *Fries in Act. Holm.* (1821); *Agardh Aph.* 89. (1821); *Eschweiler Syst. Lich.* (1824); *Wallroth Naturgesch. der Flechten*, (1824); *Grev. Flora Edin.* xix. (1824); *Meyer über die Entwickelung, &c. der Flecht.* (1825); *Fée Méth. Lich.* (1825); *Fries Syst. Orb. Veg.* 224. (1825); *Martius in Bot. Zeitung*, 193. (1826); *Fée in Dict. Class.* 9. 360. (1826). — Hypoxla, in part, *Dec. Fl. Fr.* 2. 280. (1815); *Grev. Fl. Edin.* xx. (1824). — Graphideæ, *Chevallier Hist. des Graphidées.* (1824, &c.)

Diagnosis. Aerial, leafless, flowerless, perennial plants, with a distinct thallus, and external disk containing sporules.

Anomalies.

Essential Character. — *Perennial* plants, often spreading over the surface of the earth, or rocks or trees in dry places, in the form of a lobed and foliaceous, or hard and crustaceous, or leprous substance, called a thallus. This *thallus* is formed of a cortical and medullary layer, of which the former is simply cellular, the latter both cellular and filamentous; in the crustaceous species the cortical and medullary layer differ chiefly in texture, and in the former being coloured, the latter colourless; but in the fruticulose or foliaceous species, the medulla is distinctly floccose, in the latter occupying the lower half of the thallus, in the former enclosed all round by the cortical layer. *Reproductive* matter of two kinds; 1, *sporules* lying in membranous tubes (*theca*) immersed in nuclei of the medullary substance, which burst through the cortical layer, and colour and harden by exposure to the air in the form of little disks called shields; 2, the separated cellules of the medullary layer of the thallus.

Affinities. According to Fries, Lichens are types of Algæ born in the air, interrupted in their developement by the deficiency of water, and stimulated into forming a nucleus (or receptacle of sporules) by light. No Lichen is ever submersed; there is none of which the vegetation is not interrupted by the variable hygrometrical state of the atmosphere; and, finally, none that ever developes in mines, caverns, or places deprived of light. On this account, their shields are more rare in the fissures of mountains, or in shady groves, than in places fully exposed to light. In wet places, also, their shields are not produced; for so long as they are under the influence of water they are hardly distinguishable from Hydrophyceæ (forms of Algæ); as, for instance, Collema, &c. But these plants, when exposed to the sun, do perfect their shields, as is found by Nostoc Lichenoides, foliaceum, &c., which, when dry, are ascertained to be Collema limosum, flaccidum, &c., surcharged with water. By being acquainted with this rule, the same author says, he has succeeded in discovering many Swedish Lichens with shields, which have for many years been constantly found sterile; as Parmelia conoplea, lanuginosa, gelida, &c.; and he even asserts that he has succeeded artificially in inducing sterile Lichens to become fruitful, as Usnea jubata, and others. *Plant. Hom.* 224. Lichens consist, according to Eschweiler, of a medullary and a cortical layer of tissue, of which the former is imperfectly cellular or filamentous, and bursts through the latter in the form of shields (apothecia), which contain a nucleus, consisting of a flocculose-gelatinous substance, among which lie the cases of sporules. These cases (thecæ) are transparent membranous tubes, either simple or composed of several placed end to end, which either lie free in the nucleus, or are themselves contained in other membranous cases (asci). In the beginning Lichens are stated to be in all cases developed in humidity, and to be, in fact, at that time, mere Phyceæ or Confervæ; but as soon as the humidity diminishes, the under part dies, and an inert leprous crust is formed, which ultimately becomes the basis of the plant. Hence Lichens consist of two distinct sorts of tissue,—living cellules forming the vegetating part, and dead cellules the cohesion of which is lost; when separate, the former is Palmella botryoides, and the latter Lepraria. Of these two sorts of matter, the leprous is incapable of perpetuating the Lichen, while every part of the living stratum has been ascertained to become reproductive matter. See Fries, as above quoted, and Meyer *Ueber die Entwickelung, &c., der Flechten*. The investigations of the latter are exceedingly interesting. By sowing Lichens, he arrived at some curious conclusions, the chief of which are, that, like other imperfect plants, they may owe their origin either to an original elementary, or to a reproductive generation—the latter by the creation of parts capable of developement in conformity to the plant by which they are borne; that decomposed vegetable, and some inorganic, matter, are equally capable of assuming organisation under the influence of water and light; and that the pulverulent matter of Lichens is that which is subject to this kind of indefinite propagation, while the sporules lying in the shields are the only part that will really multiply the species. He further says, that he has ascertained, by means of experiments from seed, that supposed species and even some genera of Acharius, are all forms of the same; as, for instance, Lecanora cerina, Lecidea luteo-alba, and others of the common Parmelia parietina. As these remarks have not been, as far as I know, contradicted, they may now be considered established facts.

Agardh considers Lichens more nearly allied to Fungi than to Algæ: he remarks, that if Sphærias or Pezizas had a thallus, they would be Lichens, and that the same part is all that determines such genera as Calycium, Verrucaria, or Opegrapha to be Lichens, and not Fungi. He adds, that all the transitions from Algæ to the state of Lichens, which have been detected by modern inquirers, are mere degenerations into the form of the Lichen tribe, and by no means into Lichens themselves.

With regard to the arrangement of the genera of Lichens, that of Acharius has been adopted by lichenologists of this country and of most others; but, which is remarkable, not in Sweden; and it seems probable, from the investigations that have lately been instituted, that this celebrated system will, like the more general one of Linnæus, be wholly abandoned. In its room every writer upon Lichens has proposed a new one of his own; Meyer, Eschweiler, Wallroth, Agardh, Fries, Chevalier, Fée, have each brought forward methods of arrangement, of which it may be said, without disparagement to any of them, that it is impossible at present to say which will be eventually adopted.

The only point to which it is further necessary to advert, is the separation of the tribe called Hypoxyla from Lichens. In part, this is composed of Opegrapha and other Lichenoid, and of Sphæria and various Fungoid, genera: its character is to discharge a sporuliferous pulp from the nucleus. But it seems to be a prevalent opinion that this character is uncertain and unimportant, and consequently the supposed tribe will fall back in part into Lichens, and in part into Fungi, from which it sprung. Dr. Greville, however, adheres to the distinction.

GEOGRAPHY. Pulverulent Lichens are the first plants that clothe the bare rocks of newly-formed islands in the midst of the ocean, foliaceous

Lichens follow these, and then Mosses and Hepaticæ. *D'Urville Ann. Sc.* 6. 54. About 800 species are described by Acharius, the number of which is perhaps capable of some reduction; 200 are added by Fée, and great numbers are, no doubt, still undiscovered. They are found upon trees, rocks, stones, bricks, pales, and similar places; and the same species seem to be found in many different parts of the world : thus, the Lichens of North America differ little from those of Europe. Fée estimates the number actually known, either in herbaria or in books, at 2400.

PROPERTIES. Lichens have been remarked by Decandolle to possess two distinct classes of characters, the one rendering them fit for being employed as dyes after maceration in urine, the other making them nutritive and medicinally useful to man. M. Braconnot has ascertained that oxalate of lime, or oxalic acid, exists in great abundance in Lichens, particularly in those which are granular and crustaceous. The common Variolaria, which is found upon almost every old beech-tree, contains rather more than 29 per cent. *Ed. P. J.* 13.194. Lichens that grow on the summit of fir-trees have been found by Dr. John, of Berlin, to contain an uncommon proportion of oxide of iron, which may be viewed as illustrative of the formation of iron by the vegetable process. *Ibid.* 2. 394. Of those used in dyeing, the principal crustaceous kinds are, Lecanora perella, the Orseille de terre, or Perelle d'Auvergne of the French, Lecanora tartarea, or Cudbear, hæmatomma and atra, Variolaria lactea, Urceolaria scruposa and cinerea, Isidium Westringii, Lepraria chlorina; of the foliaceous species, Parmelia saxatilis, omphalodes, encausta, conspersa, and parietina, Sticta pulmonacea, Solorina crocea, and Gyrophora deusta and pustulata; but the most important are Roccella tinctoria and fusiformis, the dye of which is so largely used by manufacturers under the name of Orchall, or Archil, or Orseille des Canaries; there are other species capable of being employed in a similar manner, as Usnea plicata, Evernia prunastri, Alectoria jubata, Ramalina Scopulorum, and several Cenomyces. The nutritive properties of Lichens probably depend upon the presence of an amylaceous substance analogous to gelatine, which, according to Berzelius, exists in the form of pure starch or amylaceous fibre, to the amount of 80·8 per cent in Cetraria islandica. This plant, which is the Iceland Moss of the shops, is slightly bitter as well as mucilaginous, and is frequently used as tonic, demulcent, and nutrient; Cetraria nivalis, Sticta pulmonacea, and Alectoria usneoides, will all answer the same purpose. Tripe de Roche, on which the Canadian hunters are often forced to subsist, is the name of various species of Gyrophora; the Rein Deer Moss, which forms the winter food of that animal, is Cenomyce rangiferina. Parmelia parietina, Borrera furfuracea, Evernia prunastri, Cenomyce pyxidata and coccifera, are reputed astringents and febrifuges, and Peltidea aphthosa an anthelmintic; Sticta pulmonacea is used in Siberia for giving a bitter to beer; Evernia vulpina, called Ulfmossa by the Swedes, is believed by that people to be poisonous to wolves; but this requires confirmation. See *Decand. Essai Méd.* 318, and *Agardh Aph.* 94.

EXAMPLES. Parmelia, Sticta, Ramalina, Nephroma, Bæomyces.

CCLXXI. FUNGI. The Mushroom Tribe.

Fungi, *Juss. Gen.* 3. (1789); *Dec. Fl. Fr.* 2. 65. (1815); *Nees das System der Pilze und Schwämme*, (1817); *Fries Syst. Mycolog.* (1821); *Syst. Orb. Veg.* (1825); *Adolphe Brongn. in Dict. Class.* 5. 155. (1824); *Grev. Scott. Crypt. Fl.* 6. (1828); *Hooker British Flora,* 457. (1830).—Epiphytæ, *Link; Grev. Fl. Edin.* xxv. (1824).— Gasteromyci, *Grev. Fl. Edin.* xxiv. (1824).—Byssoideæ, *Grev. Fl. Edin.* xxv. (1824); *Fries Syst. Orb. Veg.* (1825); *Grev. Scott. Crypt. Fl.* 6. (1828).— Mycetes, *Spreng. Syst.* 4. 376. (1827).—Uredineæ, Mucedineæ, *and* Lycoperdaceæ, *Ad. Brongn. in Dict. Class. l. c.* (1824.)

Diagnosis. Aerial, leafless, flowerless plants, with no thallus or external sporuliferous disks.

Anomalies. Sphærias approach Lichens in their structure: they are known by their want of thallus.

Essential Character.— *Plants* consisting of a congeries of cellules, among which filaments are occasionally intermixed, increasing in size by addition to their inside, their outside undergoing no change after its first formation, chiefly growing upon decayed substances, frequently ephemeral, and variously coloured. *Sporules* lying either loose among the tissue, or enclosed in membranous cases called sporidia.

Affinities. These are only distinguished from Lichens by their more fugitive nature, their more succulent texture, their want of a thallus or expansion independent of the part that bears the reproductive matter, and by the latter being contained within their substance and not in hard distinct nuclei originating in the centre and breaking through a cortical layer. From Algæ there is no absolute character of division, except their never growing in water; in fact, it is, as has been before stated, rather the medium in which Fungi and Algæ are developed that distinguishes them, than any peculiarity in their own organisation: for instance, the aerial Byssaceæ, which are Fungi, are nearly the same in structure as the aquatic Hydronemateæ, which are Algæ. While there is so near an approximation of these families to each other, particularly in the simplest forms, it is important to remark that no spontaneous motion has been observed in Fungi, which, therefore, cannot be considered so closely allied to the animal kingdom as Algæ, notwithstanding the presence of azote in them, and the near resemblance of the substance by chemists called Fungin, to animal matter.

Fungi are almost universally found growing upon decayed animal or vegetable substances, and scarcely ever upon living bodies of either kingdom; in which respect they differ from Lichens, which very commonly grow upon the living bark of trees. They are, however, not confined to dead or putrid substances, as is shewn by their attacking various plants when in a state of perfect life and vigour. In their simplest form they are little articulated filaments, composed of simple cellules placed end to end; such is the mouldiness that is found upon various substances, the mildew of the Rose-bush, and, in short, all the tribes of Mucor and Mucedo; in some of these the joints disarticulate, and appear to be capable of reproduction; in others sporules collect in the terminal joints, and are finally dispersed by the rupture of the cellule that contained them. In a higher state of composition, Fungi are masses of cellular tissue of a determinate figure, the whole centre of which consists of sporules either lying naked among filaments, as in the Puff-balls, or contained in membranous tubes or sporidia, like the thecæ of Lichens, as in the Sphærias. In their most complete state they consist of two surfaces, one of which is even and imperforate, like the cortical layer in Lichens; the other separated into plates or cells, and called the hymenium, in which the sporules are deposited.

Upon this kind of difference of structure, Fungi have not only been divided into distinctly marked tribes, but it has been proposed to separate certain orders from them under the name of Byssaceæ, Gasteromyci, and Hypoxyla: the first comprehending the filamentous Fungi found in cellars, and similar plants; the second Lycoperdons and the like ; and the third species which approach Lichens in the formation of a distinct nucleus for the sporules, such as Sphæria. But it appears to me better to consider all these mere forms of one great vegetable group.

Some writers have questioned the propriety of considering Fungi as plants, and have proposed to establish them as an independent kingdom, equally distinct from animals and vegetables; others have entertained doubts of their being more than mere fortuitous developements of vegetable matter, called into action by special conditions of light, heat, earth, and air—doubts which have been caused by some remarkable circumstances connected with their developement, the most material of which are the following: they grow with a degree of rapidity unknown in other plants, acquiring the volume of many inches in the space of a night, and are frequently meteoric, that is, spring up after storms, or only in particular states of the atmosphere. It is possible to increase particular species with certainty, by an ascertained mixture of organic and inorganic matter exposed to well-known atmospheric conditions, as is proved by the process adopted by gardeners for obtaining Agaricus campestris; a process so certain, that no one ever saw any other kind of Agaricus produced in Mushroom-beds; this could not happen if the Mushrooms sprang from seeds or sporules floating in the air, as in that case many species would necessarily be mixed together; they are often produced constantly upon the same kind of matter, and upon nothing else, such as the species that are parasitic upon leaves : all which is considered strong evidence of the production of Fungi being accidental, and not analogous to that of perfect plants. Fries, however, whose opinion must have great weight in all questions relating to Fungi, argues against these notions in the following manner : " Their sporules are so infinite (in a single individual of Reticularia maxima I have counted above 10,000,000), so subtile (they are scarcely visible to the naked eye, and often resemble thin smoke), so light (raised, perhaps, by evaporation into the atmosphere), and are dispersed in so many ways (by the attraction of the sun, by insects, wind, elasticity, adhesion, &c.), that it is difficult to conceive a place from which they can be excluded." I give his words as nearly as possible, because they may be considered the sum of all that has to be urged against the doctrine of equivocal generation in Fungi ; but without admitting, by any means, so much force in his statement as is required to set the question at rest. In short, it is no answer to such arguments as those just adverted to. It seems to me that a preliminary examination is necessary into the existence of an exact analogy between all the plants called Fungi; a question which must be settled before any further inquiry can be properly entered upon. That a number of the fungus-like bodies found upon leaves are mere diseases of the cuticle, or of the subjacent tissue, is by no means an uncommon opinion ; that many more, such as the Byssaceæ in particular, are irregular and accidental expansions of vegetable tissue in the absence of light, is not improbable; and it is already certain that no inconsiderable number of the Fungi of botanists are actually either, as various Rhizomorphas, the deformed roots of flowering plants growing in cellars, clefts of rocks, and walls ; or mere stains upon the surface of leaves, as Venularia grammica; or the rudiments of other Fungi, as many of Persoon's Fibrillarias. Those who are anxious to inquire into these and other points, are referred to Fries' works generally, to the various writings of Nees von Esenbeck, and to the Scottish Cryptogamic Flora of Dr. Greville.

GEOGRAPHY. The Fungi by which most extra-tropical countries are inhabited are so numerous, that no one can safely form even a conjecture as to the number that actually exists. If they are ever fortuitous productions, the number must be indeterminable; if many are mere diseases and the remainder fixed species, then the knowledge of their nature must be reduced to a more settled state before any judgment upon their number can be formed. According to Fries, he discovered no fewer than 2000 species within the compass of a square furlong in Sweden; of Agaricus alone above 1000 species are described; and of the lower tribes the number must be infinite. Sprengel, however, does not enumerate in his *Systema Vegetabilium* more than between 2700 and 2800; but when we consider that his genus Agaricus does not go beyond number 646, although 1000 at least are described, it is not improbable that the rest of his enumeration is equally defective, and that the number of described Fungi perhaps amounts to between 4 and 5000. Of tropical species we know but little; their fugitive nature, the difficulty of preserving them, and perhaps the incuriousness of travellers, as well as their scarcity in the damp parts of equinoctial countries, have been the causes of the proportion in such climates between Fungi and other plants being unknown.

PROPERTIES. A large volume might be written upon the qualities and uses of Fungi, but in this place they can be only briefly adverted to in a very general way. They may be said to be important, either as food or as poison, or as parasites destructive to the plants upon which they grow. As food, the most valuable are the Agaricus campestris, or common Mushroom, the various species of Helvella or Morel, and Tuber or Truffle; but a considerable number of other kinds are used for food in various parts of the world, of which a useful account will be found in Decandolle's excellent *Essai sur les Propriétés Médicales des Plantes*, in Persoon's work *Sur les Champignons comestibles*, and in a paper by Dr. Greville in the 4th volume of the *Transactions of the Wernerian Society*.

It is necessary to exercise the utmost care in employing Fungi, the nature of which is not perfectly well ascertained, in consequence of the resemblance of poisonous and wholesome species, and the dreadful effects that have followed their incautious use. It is true that many kinds are named by Pallas as being commonly used by the Russians, which are plentiful in countries where they are not employed for food; but, in the first place, it is not perhaps quite certain that poisonous and wholesome species are not confounded under the same name; in the next place, climate may make a difference; and lastly, much depends upon the mode in which they are cooked. Upon this subject Delile observes, that it was ascertained by M. Paulet, in 1776, that salt and vinegar removed every deleterious principle from that most poisonous plant the Agaricus bulbosus; that it is the universal practice in Russia to salt the Fungi, and that this may be the cause of their harmlessness, just as the pickling and subsequent washing of the poisonous Agaric of the Olive renders it eatable in the Cevennes; but that nevertheless it is much wiser to run no risk with unknown Fungi, even taking such precautions; a remark to which he was led by the lamentable death of a French officer and his wife, in consequence of breakfasting off some poisonous Agarics, which were nevertheless eaten by other persons in the same house with impunity. It was probable that in that case a difference in the cooking was the cause of the difference in the effect of the Fungi; but it was a sufficient ground for distrusting all Fungi except the cultivated ones. So strongly did the late

Professor L. C. Richard feel the prudence of this, that, although no one was better acquainted with the distinctions of Fungi, he would never eat any except such as had been raised in gardens in mushroom beds. One of the most poisonous of our Fungi is the Amanita muscaria, so called from its power of killing flies when steeped in milk. Even this is eaten in Kamchatka, with no other than intoxicating effects, according to the following account by Dr. Langsdorff, as translated by Dr. Greville, from whom I borrow it.

" This variety of Amanita muscaria is used by the inhabitants of the north-eastern parts of Asia in the same manner as wine, brandy, arrack, opium, &c. is by other nations. These Fungi are found most plentifully about Wischna, Kamchatka, and Wilkowa Derecona, and are very abundant in some seasons, and scarce in others. They are collected in the hottest months, and hung up by a string in the air to dry: some dry of themselves on the ground, and are said to be far more narcotic than those artificially preserved. Small deep-coloured specimens, thickly covered with warts, are also said to be more powerful than those of a larger size and paler colour. The usual mode of taking the Fungus is, to roll it up like a bolus, and swallow it without chewing, which, the Kamchatkadales say, would disorder the stomach. It is sometimes eaten fresh in soups and sauces, and then loses much of its intoxicating property: when steeped in the juice of the berries of Vaccinium uliginosum, its effects are those of strong wine. One large, or two small Fungi, is a common dose to produce a pleasant intoxication for a whole day, particularly if water be drank after it, which augments the narcotic principle. The desired effect comes on from one to two hours after taking the Fungus. Giddiness and drunkenness result in the same manner as from wine or spirits; cheerful emotions of the mind are first produced; the countenance becomes flushed; involuntary words and actions follow, and sometimes at last an entire loss of consciousness. It renders some remarkably active, and proves highly stimulant to muscular exertion: by too large a dose, violent spasmodic effects are produced. So very exciting to the nervous system, in many individuals, is this Fungus, that the effects are often very ludicrous. If a person under its influence wishes to step over a straw or small stick, he takes a stride or a jump sufficient to clear the trunk of a tree; a talkative person cannot keep silence or secrets; and one fond of music is perpetually singing. The most singular effect of the Amanita is the influence it possesses over the urine. It is said that, from time immemorial, the inhabitants have known that the Fungus imparts an intoxicating quality to that secretion, which continues for a considerable time after taking it. For instance, a man·moderately intoxicated to day will, by the next morning, have slept himself sober, but (as is the custom), by taking a teacup of his urine he will be more powerfully intoxicated than he was the preceding day. It is, therefore, not uncommon for confirmed drunkards to preserve their urine as a precious liquor against a scarcity of the Fungus. This intoxicating property of the urine is capable of being propagated; for every one who partakes of it has his urine similarly affected. Thus, with a very few Amanitæ, a party of drunkards may keep up their debauch for a week. Dr. Langsdorf mentions, that by means of the second person taking the urine of the first, the third that of the second, and so on, the intoxication may be propagated through five individuals."

Of parasitical Fungi, the most important are those which are called dry rot, such as Polyporus destructor, Merulius lacrymans and vastator, &c., which are the pest of wooden constructions; next to these come the blight in corn, occasioned by Puccinia graminis; the smut and ergot, if they are really any thing more than the diseased and disorganised tissue of the plants affected; the rust, which is owing to the ravages of Æcidiums; and finally, in this class is to be included what we call mildew, minute simple articulated Mucors, Mucedos, and Byssi. The genus Rhizomorpha, which vegetates in dark mines far from the light of day, is remarkable for its phosphorescent properties. In the coal mines near Dresden the species are described as giving those places the air of an enchanted castle; the roofs, walls, and pillars, are entirely covered with them, their beautiful light almost dazzling the eye. The light is found to increase with the temperature of the mines. *Ed. P. J.* 14. 178. It is a most remarkable circumstance, and one which deserves particular inquiry, that the growth of the minute Fungi, which constitute what is called mouldiness, is effectually prevented by any kind of perfume. It is known that books will not become mouldy in the neighbourhood of Russia leather, nor any substance, if placed within the influence of some essential oil. *Ibid.* 8. 34. Boletus igniarius is used in India as a styptic, as well as for Amadou. *Ainslie,* 1. 5. The Boleti, when wounded, heal much in the same manner as the flesh of animals. *Edin. Philosoph. Journ.* 14. 369.

EXAMPLES. § Coniomycetes (Uredo, Æcidium, Mucor).
§ Gasteromycetes (Sclerotium, Physarum, Lycoperdon).
§ Pyrenomycetes (Hysterium, Sphæria).
§ Hymenomycetes (Agaricus, Boletus, Clavaria).
§ Byssaceæ (Racodium, Monilia, Erineum).

CCLXXII. ALGÆ. THE SEA-WEED TRIBE.

ALGÆ, *Juss. Gen.* 5. (1788); *Roth. Catalecta Botanica* (1797); *Dec. Fl. Fr.* 2. 2. (1815); *Agardh Synops. Alg.* (1817.); *Species Alg.* (1821–1828); *Syst. Alg.* (1824); *Greville Alg. Brit.* (1830). — PHYCEI, *Acharius* (1807 ?). — THALASSIOPHYTA, *Lamouroux Ann. Mus.* 20. (1812); *Gaillon in Dict. des Sc.* 53. 350. (1828). — HYDROPHYTA, *Lyngb. Tentam.* (1819). — ARTHRODIEÆ, *Bory in Dict. Class.* 1. 591. (1822). — HYDRONEMATEÆ, *Nees in Nov. Act. Nat. Cur.* 11. 509. (1823); *Ann. des Sc.* 13. 439. (1828). — CHAODINEÆ, CONFERVÆ, *and* CERAMIARIÆ, *Bory in Dict. Class.* 3. *and* 4. (1823). — CHÆTOPHOROIDEÆ, *Greville Fl. Edin.* 321. (1824). HYDROPHYCÆ, *Fries Syst. Orb. Veg.* 320. (1825.)

DIAGNOSIS. Aquatic leafless flowerless plants.
ANOMALIES.

ESSENTIAL CHARACTER. — Leafless flowerless plants, with no distinct axis of vegetation, growing in water, frequently having an animal motion, and consisting either of simple vesicles lying in mucus, or of articulated filaments, or of lobed fronds, formed of uniform cellular tissue. *Reproductive matter* either altogether wanting, or contained in the joints of the filaments, or deposited in thecæ of various form, size, and position, caused by dilatations of the substance of the frond. *Sporules,* with no proper integument, in germination elongating in two opposite directions.

AFFINITIES. Whatever ingenuity may be employed in determining the relative degree of dignity in the vegetable creation between Fungi, Lichens, and Algæ, it seems to me that the conclusion which is constantly arrived at is, that Algæ are absolutely distinguishable from the two others only by their living in water, and that, except for the influence which that medium exercises on them, they would be identical with Lichens on the one hand, and with Fungi on the other. The method under which the genera should be arranged, almost every observer having a method of his own, is a question still to settle; but in this place we have chiefly to consider the more remarkable facts connected with their organisation. Those who wish to make the order a special study will do well to take the excellent *Species Algarum* of Agardh for their guide, and to study the papers of Bory de St. Vincent, and Fries, for general ideas, and that most beautiful of all books, the *Algæ Britannicæ* of Dr. Greville, for the application of them to the Flora of this country.

Those who have ever examined the surface of stones constantly moistened by water, the glass of hothouses, the face of rocks in the sea, or of walls where the sun never shines, or the hard paths in damp parts of gardens after rain, cannot fail to have remarked a green mucous slime with which they are covered. This slime consists of Algæ in their simplest state of organisation, belonging to the genera Palmella, Nostoc, Red Snow, and the like, the Nostochinæ of Agardh, or Chætophoroideæ of Greville; they have been called Chaodineæ by Bory de St. Vincent, whose account of them is to the following effect: — The slime resembles a layer of albumen spread with a brush; it exfoliates in drying, and finally becomes visible by the manner in which it colours green or deep brown. One might call it a provisional creation waiting to be organised, and then assuming different forms, according to the nature of the corpuscles which penetrate it or develope among it. It may further be said to be the origin of two very distinct existences, the one certainly animal, the other purely vegetable. This matter lying among amorphous mucus consists in its simplest state, of solitary, spherical corpuscles, (such as are figured by Turpin in the *Mémoires du Muséum*, vol. 18. t. 5.; and as may be easily seen in the common green crust upon old pales, Palmella botryoides); these corpuscles are afterwards grouped, agglomerated, or chained together, so producing more complex states of organisation. Sometimes the mucus, which acts as the basis or matrix of the corpuscles, when it is found in water, which is the most favourable medium for its developement, elongates, thickens, and finally forms masses of some inches extent, which float and fix themselves to aquatic plants. These masses are at first like the spawn of fish, but they soon change colour and become green, in consequence of the formation of interior vegetable corpuscles. Often, however, they assume a milky or ferruginous appearance; and if in this state they are examined under the microscope, they will be found completely filled with the animalcules called Naviculariæ, Lunulineæ, and Stylariæ, assembled in such dense crowds as to be incapable of swimming. In this state the animalcules are inert. Are they developed here, or have they found their way to such a nidus, and have they hindered the developement of the green corpuscles? Is the mucus in which they lie the same to them as the albuminous substance in which the eggs of many aquatic animals are deposited? At present we have no means of answering these questions. According to M. Gaillon, many of these simple plants are certainly nothing but congeries or rows of the singular and minute animalculæ called Vibrio tripunctatus and bipunctatus by Muller, strung end to end. See Ferussac's *Bulletin, Feb.* 1824. He particularly applies this remark to Monema comoides.

Another form of Algæ, one which may be considered a higher degree of developement of the last, is that in which they assume a tubular state, containing pulverulent or corpuscular matter in the inside, and become what are called Confervæ, or, as M. Bory styles them, Arthrodieæ. These, which comprehend true Confervæ, Oscillatorias, and many Diatomeæ, are thus spoken of by the acute botanist last mentioned :—The general character of Arthrodieæ consists in filaments, generally simple, and formed of two tubes, of which one, which is exterior and transparent, offers no trace of organisation to the most powerful eye, so that it might be called a tube of glass, contains an inner articulated filament filled with colouring matter, often almost imperceptible, but at other times very intense green, purple, or yellowish ; these compound filaments present to the astonished eye the strangest and most different phenomena, all of which have the plainest characters of animal life, supposing that animal life is to be inferred from motions indicating a well-marked power of volition. The Arthrodia tribe usually inhabit either fresh or sea-water, and several are common to both. One of them, but a species referred to the tribe with some uncertainty, the Conferva ericetorum, grows on the ground, but in places that are very damp, and often inundated ; others among the Oscillating species cover the humid surface of rocks or earth, and the interstices in the pavement of cities ; some even grow in hot springs of a very high temperature. (Ulva thermalis lives in the hot springs of Gastein in a temperature of about 117° Fahrenheit. *Ed. P. J.* 4. 206.) The most remarkable are, 1st. The Fragillarias, to which Diatoma and Achnanthes belong ; these, when combined in the little ribandlike threads which are natural to them, have no apparent action ; but as soon as the separation of the joints takes place, a sort of sliding or starting motion may be seen between them. 2dly. The Oscillarias, some of which have an oscillatory movement, extremely active and perceptible ; and the Ulva labyrinthiformis and Anabaina, which, with all the appearance of a plant, has, according to Vauquelin and Chaptal, all the chemical characters of an animal. 3dly. The Conjugatæ, the filaments of which separate at one period, and unite again at another, and finally, by a mode of coupling completely animal, resolve themselves into a single and uniform being ; and, 4thly, the Zoocarpeæ, most extraordinary productions, in which the animal and vegetable nature follow each other in the same individual ; vegetables in the earlier period of their existence, but producing, in the room of sporules or buds, little microscopic animalcules, which become filamentous vegetables after a certain length of time. Dr. Greville, in his *Flora Edinensis*, adopted an opinion of Dr. Fleming and others, that many of the species referred to this group possess an animal structure ; such as Diatoma flocculosum, tenue, arcuatum, and obliquatum, and Fragillaria striatula and pectinalis ; and he believed Conferva stipitata, Biddulphiana and tæniæformis of Eng. Bot., together with the whole genus Echinella, to be equally dubious. But he altered this opinion after two or three years, if we are to judge from his *Cryptogamic Flora*, in which are beautiful figures of some of the very beings the animal nature of which is so much to be suspected. For example, Diatoma tenue, a little Confervoid plant with parallelogramic articulations, at first attached by their longest sides, and afterwards separating at their alternate extremities, so as to form a filiform tube. " The filaments," according to an interesting observation of the Rev. Mr. Berkley, " at a certain period seem to lose the squareness of their figure, to be attenuated at the extremities and dilated in the centre, to become cylindrical and opaque, and, in short, metamorphosed into a moniliform filament, with elliptical or oblong purple joints and colourless articulations." (Vol. vi. 354.) Agardh is of opinion that we have among these rudimentary Algæ not only a distinct passage to the animal, but even to the mineral kingdom : for he states that some of his Diatomeæ include vegetable **crystals** bounded by right lines, collected into a crystalliform body, and with no other difference from minerals than that the individuals have the power of again separating. *System*, xiii. The observations above quoted are those of naturalists of so high a reputation for accuracy, that they may safely be accepted as certain; but I do not know what to say of such as the following, by a German botanist of the name of Meyen, unless that they require to be verified by others, especially because those who have sought for the phenomena he mentions have not succeeded in finding them. This writer states that he has seen, very often, a spontaneous motion in Zygnema nitidum; and its filaments contract from the length of 10 inches to that of 4-6 lines; that the Oscillatorias move in a circle; that the globules contained in the filaments of Zygnema have a life partly vegetable, partly animal, and procreate similar globules, some of which become animals endowed with motion. See Agardh's *Species Algarum*, 2. 48., from which this account is extracted. Certain supposed Confervæ, called Bacillarias, are rejected from plants by M. Bory de St. Vincent, and placed in the lowest grade of the animal creation. See *Dict. Class.* 2. 128.

Other Algæ approach nearly to the structure of Lichens, lose entirely their animal properties, and become broad flat expansions, or finely divided vegetables, such as are seen in the ordinary state of Sea-weeds, Fuci, or marine Confervæ. Of the British species of these, and of their general nature, an excellent account has been given by Dr. Greville in his *Algæ Britannicæ*, from which the greater part of the following remarks is extracted. While the two first groups consist of microscopic objects inhabiting obscure places, shady paths, or half-immersed surfaces of stones and banks, the more complete Algæ comprehend species forming subaqueous forests of considerable extent in the vast ocean, emulating in their own gigantic dimensions the boundless element that enfolds them. Chorda filum, a species common in the North Sea, is frequently found of the length of 30 or 40 feet. In Scalpa Bay, in Orkney, according to Mr. Neill, this species forms meadows, through which a pinnace with difficulty forces its way. Lessonia fuscescens is described by M. Bory de St. Vincent as 25 or 30 feet in length, with a trunk often as thick as a man's thigh. But all these, and indeed every other vegetable production, is exceeded in size by the prodigious fronds of Macrocystis pyrifera. "This appears to be the sea-weed reported by navigators to be from 500 to 1500 feet in length: the leaves are long and narrow, and at the base of each is placed a vesicle filled with air, without which it would be impossible for the plant to support its enormous length in the water; the stem not being thicker than the finger, and the upper branches as slender as common packthread."

These remarks may be concluded by a reference to the following works, in which further information relating to the animal nature of certain Confervæ may be found: Nees von Esenbeck *Die Algen des Sussen Wassers* (1814); Treviranus in *Ann des Sc.* 10. 22. (1817); Gruithuisen in *Nov. Act. Acad. Leopold. Curios.* 10. 437.; Carus in *the same*, 11. 491. (1823); Gaillon in *Ann. Sc. Nat.* 1. 309. (1823); Desmazières in *the same*, 10. 42. (1825), and 14. 206, (1828); Unger in *the same*, 13. 431. (1828): all of which should be carefully consulted by those who wish to form any accurate judgment upon this most curious and interesting subject.

GEOGRAPHY. This has been treated upon carefully by Lamouroux in the *Annales des Sciences Naturelles*, vol. 7, and by Dr. Greville in the *Algæ Britannicæ*. Algæ are most important in the economy of nature for forming the commencement of soil by their deposit and decomposition. The basin of the ocean is said to be continually rising by the deposit of such plants, particularly of Conferva chthonoplastes, the closely aggregated slimy fibres of which form dense beds. *Ed. P. J.* 2. 392. The same circumstance occurs in lakes and ditches: the bottoms of some of the former, in this country, are no doubt increased by the curious production called Conf. ægagropila. To the peculiar distribution of Phænogamous plants into certain botanical regions, a fact familiar to all botanists, there is something analogous in the submersed Flora of the ocean. We find latitude, depth, currents, influencing the forms of Algæ in nearly the same way as latitude, elevation, and station, affect those plants which are more perfect; and as many of the latter are confined to small extent of country, so do several of the Algæ extend but to short distances in the sea. Thus Odonthalia dentata and Rhodomenia cristata are confined to the northern parts of Great Britain, while many others are peculiar to the southern parts; and, on the contrary, many are cosmopolites of an unbounded range, such as Codium and Ulvaceæ. The latter thrive best in the polar and temperate zones, Dictyoteæ increase as we approach the equator, Fuci particularly flourish between the parallels of 55° and 44°, and, according to Lamouroux, rarely approach the equator nearer than 36°. The articulated or imperfectly formed fresh-water Algæ are nearly confined to the temperate and northern parts of the world, being almost unknown or undescribed from within the tropics. The number of species is scarcely capable of being estimated.

PROPERTIES. For what wise purpose the Creator has filled the sea and the rivers with countless myriads of these plants, so that the Flora of the deep waters is as extensive as that of dry land, we can only conjecture; the uses to which they are applied by man are, doubtless, of but secondary consideration; and yet they are of no little importance in the manufactures and domestic economy of the human race. Dr. Greville describes them thus (*Algæ Britannicæ*, xix.): —

" Rhodomenia palmata, the dulse of the Scots, dillesk of the Irish, and saccharine Fucus of the Icelanders, is consumed in considerable quantities throughout the maritime countries of the north of Europe, and in the Grecian Archipelago; Iridæa edulis is still occasionally used, both in Scotland and the south-west of England. Porphyra laciniata and vulgaris is stewed, and brought to our tables as a luxury under the name of Laver; and even the Ulva latissima, or green Laver, is not slighted in the absence of the Porphyræ. Enteromorpha compressa, a common species on our shores, is regarded, according to Gaudichaud, as an esculent by the Sandwich Islanders. Laurentia pinnatifida, distinguished for its pungency, and the young stalks and fronds of Laminaria digitata (the former called Pepperdulse, the latter Tangle), were often eaten in Scotland; and even now, though rarely, the old cry, ' Buy dulse and tangle,' may be heard in the streets of Edinburgh. When stripped of the thin part, the beautiful Alaria esculenta forms a part of the simple fare of the poorer classes of Ireland, Scotland, Iceland, Denmark, and the Faroe Islands.

" To go further from home, we find the large Laminaria potatorum of Australia furnishing the aborigines with a proportion of their 'instruments, vessels, and food.' On the authority of Bory de St. Vincent, the Durvillea utilis and other Laminarieæ constitute an equally important resource to the poor on the west coast of South America. In Asia, several species of Gelidium are made use of to render more palatable the hot and biting condiments of the East. Some undetermined species of this genus also furnish the materials of which the edible swallows' nests are composed. It is remarked by Lamouroux, that three species of swallow construct edible nests, two of which build at a distance from the sea-coast, and use the sea-weed only as a cement for other matters. The nests of the third are consequently most esteemed, and sold for nearly their weight in gold. Gracillaria lichenoides is highly valued for food in Ceylon and other parts of the East, and bears a great resemblance to Gracillaria compressa, a species recently discovered on the British shores, and which seems to be little inferior to it; for my friend Mrs. Griffiths tried it as a pickle and preserve, and in both ways found it excellent.

"It is not to mankind alone that marine Algæ have furnished luxuries, or resources in times of scarcity. Several species are greedily sought after by cattle, especially in the north of Europe. Rhodomenia palmata is so great a favourite with sheep and goats, that Bishop Gunner named it Fucus ovinus. In some of the Scottish islands, horses, cattle, and sheep, feed chiefly upon Fucus vesiculosus during the winter months; and in Gothland it is commonly given to pigs. Fucus serratus also, and Chorda Filum, constitute a part of the fodder upon which the cattle are supported in Norway.

"In medicine we are not altogether unindebted to the Algæ. The Gigartina helminthocorton, or Corsican Moss, as it is frequently called, is a native of the Mediterranean, and held once a considerable reputation as a vermifuge. The most important medical use, however (omitting minor ones), derived from sea-weeds, is through the medium of Iodine, which may be obtained either from the plants themselves, or from kelp. French kelp, according to Sir Humphrey Davy, yields more Iodine than British; and, from some recent experiments made at the Cape of Good Hope by M. Ecklon, Laminaria buccinalis is found to contain more than any European Algæ. Iodine is known to be a powerful remedy in cases of goitre. The burnt sponge formerly administered in similar cases, probably owed its efficacy to the Iodine it contained; and it is also a very curious fact, that the stems of a sea-weed are sold in the shops, and chewed by the inhabitants of South America, wherever goitre is prevalent, for the same purpose. This remedy is termed by them Palo Coto (literally, goitre-stick); and, from the fragments placed in my hands by my friend Dr. Gillies, to whom I am indebted for this information, the plant certainly belongs to the order Laminarieæ, and is probably a species of Laminaria.

"Were the Algæ neither ' really serviceable either in supplying the wants or in administering to the comforts of mankind' in any other respect, their character would be redeemed by their usefulness in the arts; and it is highly probable that we shall find ourselves eventually infinitely more indebted to them. One species (and I regret to say that it is not a British one) is invaluable as a glue and varnish to the Chinese. This is the Gracilaria tenax, the Fucus tenax of Turner's *Historia Fucorum*. Though a small plant, the quantity annually imported at Canton from the provinces of Fokien and Tche-kiang is stated by Mr. Turner to be about 27,000 lbs. It is sold at Canton for 6d. or 8d. per pound, and is used for the purposes to which we apply glue and gum-arabic. The Chinese employ it chiefly in the manufacture of lanterns, to strengthen or varnish the paper, and sometimes to thicken or give a gloss to silks or gauze. In addition to the above account, the substance of which I have extracted from Mr. Turner's work, Mr. Neill remarks that it ' seems probable that this is the principal ingredient in the celebrated gummy matter called Chin-chon, or Hai-tsai, in China and Japan. Windows made merely of slips of Bamboo, crossed diagonally, have frequently their lozenge-shaped interstices wholly filled with the transparent gluten of the Hai-tsai.'

" On the southern and western coasts of Ireland, our own Chondrus crispus is converted into size, for the use of house-painters, &c.; and, if I be not erroneously informed, is also considered as a culinary article, and enters into the composition of blanc-mange, as well as other dishes. In the manufacture of kelp, however, for the use of the glass-maker and soap-boiler, it is that the Algæ take their place among the most useful vegetables. The species most valued for this purpose are, Fucus vesiculosus, nodosus, and serratus, Laminaria digitata and bulbosa, Himanthalia lorea, and Chorda Filum."

EXAMPLES. Protococcus, Chroolepus, Mesogloia, Batrachospermum, Conferva, Ulva, Fucus, Sargassum.

INDEX.

The names printed in Italics are only incidentally noticed; those in Roman letters form a principal subject at the page referred to.

Anthoxanthum odoratum, 303
Anthyllis cretica, 91
Antiaris, 95
Antidesma, 97
Antirhea, 205
Antirrhineæ, 228
Apeiba, 40
Apetalous plants, 2
Aphyllæ, 330
Aphyllanthes, 256
Apocyneæ, 210, 213
Apocyneæ, 202, 203, 206, 209
Aponogeton, 172
       distachyon, 290
Apostasia, 262
Apple, monstrous, 64, 84
Apple Tribe, 83
Apricot, 85
Aquifoliaceæ, 178
Aquilaria Agallochum, 77
      ovata, 77
Aquilarineæ, 77
Aquilarineæ, 75, 217
Aquilegia, 6, 8
Arabis, 18
    chinensis, 17
Arachis, 88, 89
Aralia, 4
Aralia Tribe, 4
    umbellifera, 4
Araliaceæ, 2, 5, 52, 208
Araliaceæ, 4
Araticu do Mato, 23
Araucaria, 249
      excelsa, 249
      Dombeyi, 250
Arayana, 237
Arbutus Unedo, 183
Archil, 333
Arctostaphylos, 182
      Uva Ursi, 182
      alpina, 381
Arctium Bardana, 200
Ardisia, 225
Areca Catechu, 282
Arenaria peploides, 157
Arethuseæ, 265
Argemone mexicana, 9
Arguziæ, 243
Arhizæ, 307
Aristolochia rotunda, 73
      longa, 73
      Clematitis, 73
      bracteata, 73
      indica, 73
      odoratissima, 73
      fragrantissima, 73
      serpentaria, 73
      serpentaria, 173
Aristolochiæ, 72
Aristolochia, 200
Armeria, 196
Arnica, 200

Arnotto Tribe, 152
Aroideæ, 286
Aroideæ, 2, 174, 175, 253, 285
Arracacha, 140
Arrack, 181, 282
Arrow-root Tribe, 267
Artabotrys odoratissima, 22
Artemisia chinensis, 199
      maderaspatana, 199
      indica, 199
      Dracunculus, 200
Arthrodieæ, 338
Artichoke, 200
      Jerusalem, 200
Artocarpeæ, 95
Artocarpeæ, 93, 99, 103, 182
Artocarpus incisa, 95
Arueira Shrub, 129
Arum Tribe, 286
Arum ovatum, 287
      esculentum, 287
      triphyllum, 287
      Colocasia, 287
      mucronatum, 287
      violaceum, 287
      maculatum, 287
      sagittifolium, 288
      cordatum, 288
      italicum, 288
Arundinaria, 304
Arundo arenaria, 303
Arvore de Paina, 36
Asarinæ, 72
Asarum canadense, 73
      europæum, 73
Ascarina, 173
Asclepiadeæ, 210
Asclepiadeæ, 55, 162
Asclepias decumbens, 213
      lactifera, 213
      aphylla, 213
      stipitacea, 213
      volubilis, 213
      tuberosa, 213
      curassavica, 213
Ash, 131
Ash, 224
Asimina triloba, 23
Asparagi, 273
Asparagin, 34
Asparagus, 274
Asparagus, 168
Asperifoliæ, 241
Asperula cynanchica, 203
      odorata, 203
Asphodeleæ, 273
Asphodel Tribe, 273
Asphodeleæ, 2, 55, 162, 256, 259, 271, 272, 277, 279
Aspicarpa, 119
Aspidium fragrans, 315
      Filix mas, 314
Assafœtida, 5

THE END.

LONDON:
I. MOYES, TOOK'S COURT, CHANCERY LANE.

ERRATA.

Page 18, line 23, *omit* Subularia.

47,   8, *for* Antholema, *read* Antholoma.

        26, *for* Laneritia, *read* Lancretia.

49,   9, *add after* irregular, and there are stipulæ.

59,   24, *from the bottom, omit* The Loosestrife Tribe.

72,   29, *for* Sarcocollim, *read* Sarcocollin.

90,   12, *from the bottom, for* Clove *read* Clover.

91,   32, *for* Guilandina Bonduccella, *read* the latter.

101,   5, *from the bottom, for* Juglans cathartica and cinerea are esteemed, *read* Juglans cathartica or cinerea is esteemed.

116,   16, *omit* or Brazil Nuts.

152,   5, *from the bottom, for* Sarracennieæ, *read* Sarracenieæ, *and make the same alteration throughout the work.*

153,   10, *from the bottom, for* (1814), *read* (1824).

172,   7, *from the bottom, take out* Saururus, Aponogeton, *which belong to the previous order.*

205,   13, *from the bottom, for* Weberea, *read* Webera.

210,   29, *for* Fragræa, *read* Fagræa.

228,   25, *for* Melampyraceæ, *read* Rhinanthaceæ.

240,   21, *from the bottom, omit* 200.

249,   6, *from the bottom, for* Kawie, *read* Kawrie.

255,   10, *from the bottom, for* CCXXIII. *read* CCXXXIII.

Printed in the United States
By Bookmasters